Lecture Notes on Data Engineering and Communications Technologies

Volume 63

Series Editor

Fatos Xhafa, Technical University of Catalonia, Barcelona, Spain

The aim of the book series is to present cutting edge engineering approaches to data technologies and communications. It will publish latest advances on the engineering task of building and deploying distributed, scalable and reliable data infrastructures and communication systems.

The series will have a prominent applied focus on data technologies and communications with aim to promote the bridging from fundamental research on data science and networking to data engineering and communications that lead to industry products, business knowledge and standardisation.

Indexed by SCOPUS, INSPEC, EI Compendex.

All books published in the series are submitted for consideration in Web of Science.

More information about this series at http://www.springer.com/series/15362

K. Ashoka Reddy · B. Rama Devi · Boby George ·
K. Srujan Raju
Editors

Data Engineering and Communication Technology

Proceedings of ICDECT 2020

 Springer

Editors
K. Ashoka Reddy
Kakatiya Institute of Technology
and Science
Warangal, Telangana, India

Boby George
Department of Electrical Engineering
Indian Institute of Technology Madras
Chennai, Tamil Nadu, India

B. Rama Devi
Department of Electronics
and Communication Engineering
Kakatiya Institute of Technology
and Science
Warangal, Telangana, India

K. Srujan Raju
CMR Technical Campus
Hyderabad, Telangana, India

ISSN 2367-4512   ISSN 2367-4520   (electronic)
Lecture Notes on Data Engineering and Communications Technologies
ISBN 978-981-16-0080-7   ISBN 978-981-16-0081-4   (eBook)
https://doi.org/10.1007/978-981-16-0081-4

This Springer imprint is published by the registered company Springer Nature Singapore Pte Ltd.
The registered company address is: 152 Beach Road, #21-01/04 Gateway East, Singapore 189721,
Singapore

Preface

Springer 4th International Conference on Data Engineering and Communication Technology (ICDECT-2020) was held at Kakatiya Institute of Technology and Science (KITSW), Warangal, Telangana, India, September, 25–26, 2020. Kakatiya Institute of Technology and Science, Warangal (KITSW), is a pioneer technical institute established under the umbrella of Ekasila Educational Society in 1980 with affiliation to Kakatiya University, Warangal, and it became autonomous institution in the year 2014. It has been illuminating lives through education over the last four decades. It has attracted academicians of proven competence onto its faculty, placed its products in reputed organizations all over the world and gained recognition among academic circles. The institute aims to prepare the students for meeting the challenges of the growing and changing needs of industry through delivering high-quality technical education blended with training and research. The college is approved by the All India Council for Technical Education (AICTE), accredited by NAAC 'A' Grade with a CGPA of 3.21, and all the UG engineering programs are accredited by the National Board of Accreditation (NBA), New Delhi.

ICDECT-2020: A major objective and feature of the conference is to provide a unique forum for the academic scientists, engineers, scholars, industry researchers and students to share recent developments and trends in engineering for research community. Experts from different parts of the globe deal in smart computing, real-time information systems, smart electronic systems using the Internet of Things (IoT) and artificial intelligence (AI). The papers presented in this conference were appended in the proceedings intended to share the valuable information to the researchers and authors on the current issues of trending technologies. The conference has six tracks, which are well balanced in content, manageable in terms of the number of contributions and create an adequate discussion space for trendy topics. On this occasion, two distinguished keynote speakers had delivered their outstanding research works in various fields of data engineering and communication technology. There were 69 oral presentations by participants which brought great opportunity to share their recent research works knowledge among each other graciously. We appreciate the efforts taken by peer reviewers for their critical reviews which helped to improve the quality of papers.

We are very grateful to the International/National Advisory Committee, Session Chairs, conference team, student volunteers and faculty from the Department of Electronics and Communication Engineering, who selflessly contributed to the success of this conference. Also, we are thankful to all the authors who submitted papers, because of which the conference became a grand success. It was the quality of their presentations and their passion to communicate with the other participants that really made this conference series a grand success.

Last but not the least, we are thankful to the management for supporting us in every step of our journey toward success. Their support was not only the strength but also an inspiration for the organizers.

Warangal, India   Dr. K. Ashoka Reddy
                                                         Conference Chair, ICDECT-2020

Organization

Data Engineering and Communication Technology
(ICDECT-2020)
September, 25–26, 2020

Estd-1980

KITSW

Organized by

Department of Electronics and Communication Engineering
Kakatiya Institute of Technology and Science, Warangal-15
(*An Autonomous Institute under Kakatiya University, Warangal*)
Yerragattugutta, Hasanparthy (M), Warangal – 506015, Telangana, India
Website: www.kitsw.ac.in

Chief Patron

Capt. V. Lakshmikantha Rao, Honorable M.P. (Rajyasabha), Secretary, KITSW

Patron

Sri. P. Narayana Reddy, Treasurer, KITSW

Conference Chair

Dr. K. Ashoka Reddy, Principal, KITSW

Program Chair

Dr. B. Rama Devi, Professor and HoD, ECE, KITSW

Honorary Chair

Dr. Suresh Chandra Satapathy, School of Computer Engineering, KIIT, Bhubaneswar

Conference Team

Conveners
Sri. E. Suresh, Associate Professor, Department of ECE, KITSW
Dr. V. Venkateshwar Reddy, Associate Professor, Department of ECE, KITSW

Co-conveners
Dr. M. Raju, Associate Professor, Department of ECE, KITSW
Dr. K. Sowjanya, Assistant Professor, Department of ECE, KITSW

Editorial Board
Dr. K. Ashoka Reddy, Principal, KITSW
Dr. B. Rama Devi, Professor and HoD, ECE, KITSW
Dr. Boby George, Professor, Department of EE, IIT Madras
Dr. K. Srujan Raju, Professor, Dean Student Welfare, CMR Technical Campus

International Advisory Committee

Dr. Mathini Sellathurai, Professor, School of Engineering and Physical Sciences, HWU, UK

Dr. Anitesh Barua, Professor, Department of Information Management, UT, USA

Dr. Kun-lin Hsieh, NTU, Taiwan

Dr. Ahamad J. Rusumdar, KIT, Germany

Dr. V. R. Chirumamilla, EUT, Netherland

Dr. Halis Altun, MU, Turkey

Dr. P. N. Suganthan, NTU, Singapore

Dr. Boka Kumsa, Wollega University, Ethiopia

National Advisory Committee

Dr. N. V. S. Narasimha Sarma, Director, IIIT Trichy

Dr. P. Lakshmi Narayana, Director, NERTU, Osmania University, Hyderabad

Dr. A. Govardhan, Professor and Rector, JNTUH, Hyderabad

Prof. T. Srinivasulu, Dean Faculty of Engineering and Technology and Principal, University College of Engineering and Technology for Women, KU-Warangal

Prof. P. Malla Reddy, Principal, KU College of Engineering and Technology, Warangal

Dr. E. Srinivasa Reddy, Dean (Faculty Affairs), Department of CSE, ANU

Dr. M. Sydulu, Professor, Department of EEE, NIT Warangal

Dr. T. Kishore Kumar, Professor, Department of ECE, NIT Warangal

Dr. L. Anjaneyulu, Professor, Department of ECE, NIT Warangal

Dr. G. Lakhmi Narayanan, Professor and Head, Department of ECE, NIT Tiruchirappalli

Dr. D. Sriram Kumar, Professor, Department of ECE, NIT Tiruchirappalli

Dr. P. Palany Swamy, Professor, Department of ECE, NIT Tiruchirappalli

Dr. B. Rajendra Naik, Professor and Head, Department of ECE, Osmania University

Dr. P. Rajesh Kumar, Professor, Department of ECE, Andhra University

Dr. G. Sashibhushan Rao, Professor and Head, Department of ECE, Andhra University

Dr. L. Pratap Reddy, Professor, Department of ECE, JNTU

Dr. G. Narsimha, Professor in CSE, JNTUH, Hyderabad.

Dr. M. Asharani, Professor, Department of ECE, JNTU

Dr. Ch. Srinivasa Rao Professor, Department of ECE, JNTUK, Vizianagaram

Dr. Dheeraj Sunhera, Professor, Department of ECE, JNTUH, Jagtial

Dr. T. Ramasri, Professor, Department of ECE, SV University, Tirupati

Dr. B. Ravi Kumar, Associate Professor, Department of ECE, IIT Hyderabad

Sri. K. V. Sridhar, Associate Professor, Department of ECE, NIT Warangal

Dr. D. Vakula, Associate Professor, ECE Department, NIT Warangal

Dr. V. Venkata Mani, Associate Professor, ECE Department, NIT Warangal

Dr. S. Anuradha, Associate Professor, Department of ECE, NIT Warangal

Dr. R. Pandeeswari, Associate Professor, Department of ECE, NIT Tiruchirappalli

Dr. B. Venu Gopal Reddy, Associate Professor, Department of EEE, NIT Goa

Dr. S. Aruna, Associate professor, Department of ECE, AUCE

Dr. Akkala Subba Rao, Assistant Professor, Department of ECE, MANIT, Bhopal.

Dr. K. Kishan Rao, Professor, Department of ECE, SNIST, Hyderabad

Dr. D. Jeya Mala, Director—MCA, Fatima College

Dr. K. Sivani, Professor, Department of EIE, Dean Academics, KITSW

Dr. T. Anil Kumar, Professor, Department of ECE, CMR IT

Dr. D. Krishna Reddy, Professor and Head, Department of ECE, CBIT, Hyderabad

Dr. Panyam Narahari Sastry, Professor, Department of ECE, CBIT

Humaira Nishat, Professor, Department of ECE, CVR College of Engineering

Dr. Ramesh Babu Vallabhaneni, Professor, Department of ECE, NRI Institute of Technology, Vijayawada

Dr. M. Sushanth Babu, Professor, Department of ECE, Matrusri Engineering College, Hyderabad

Dr. Bhavani Sankar, Professor and Head, Department of ECE, Anjalai Ammal–Mahalingam Engineering College

Dr. M. N. Tibdewal, Professor, Department of ECE, Shri Sant Gajanan Maharaj College of Engineering, Shegaon

Dr. B. Ramesh, Professor and Head, Department of ECE, KITS, Huzurabad, Karimnagar

Dr. Neelamadhab Padhy, GIET University, Department of CSE, Gunupur, Odisha

Dr. M. Sunil Kumar, Professor, Department of CSE, SVEC

Dr. S. Sumathi, Professor, Department of EEE, Mahendra Engineering College

Dr. B. R. Sanjeeva Reddy, Professor, Department of ECE, BVRIT, Hyderabad

Dr. P. Niranjan Reddy, Professor, Department of CSE, Dean R&D, KITSW

Dr. G. Raghotham Reddy, Professor, Department of ECE, Dean Students Affairs, KITSW

Dr. K. Venumadhav, Professor and Head, Department of EIE, KITSW

Dr. V. Shankar, Professor and Head, Department of CSE, KITSW

Dr. C. Venkatesh, Professor and Head, Department of EEE, KITSW

Dr. P. Kamakshi, Professor and Head, Department of IT, KITSW

Dr. V.Chandra Shekar Rao, Associate Professor, Department of CSE, KITSW

Website and Publication Committee

Chairman
Dr. S. Narasimha Reddy, Associate Professor, Department of CSE, KITSW
Members
Mr. M. Kishore, Assistant Professor, Department of IT, KITSW
Mr. Syed Zaheeruddin, Assistant Professor, Department of ECE, KITSW

Mr. Md. Abdul Muqueem, Assistant Professor, Department of ECE, KITSW
Mr. P. Sreenivasa Rao, Programmer, Department of CSE, KITSW

Technical Review Committee

Chairman
Dr. K. Srujan Raju, Professor and Head, Department of CSE and IT, CMR Technical Campus

Members
Dr. D. Vakula, Associate Professor, ECE Department, NITW
Dr. Jitendra V. Themburne, Assistant Professor, Department of CSE, IIIT Nagpur
Dr. K. Bikshalu, Assistant Professor, Department of ECE, KU Campus, Kothagudem
Dr. C. Venkatesh, Professor and HoD, Department of EEE, KITSW
Dr. V. Shankar, Professor and HoD, Department of CSE, KITSW
Dr. G. Raghotham Reddy, Professor in ECE, Dean SAC, KITSW
Dr. V. Rajagopal, Professor, Department of EEE, KITSW
Dr. V. Chandra Shekar Rao, Associate Professor, Department of CSE, KITSW
Dr. S. Narasimha Reddy, Associate Professor, Department of CSE, KITSW
Dr. T. Senthil Murugan, Associate Professor, Department of IT, KITSW
Dr. G. Rajender Naik, Associate Professor, Department of EEE, KITSW
Dr. G. Sudheer Kumar, Associate Professor, Department of EEE, KITSW
Dr. B. Vijay Kumar, Associate Professor, Department of EEE, KITSW
Dr. Akkala Subba Rao, Assistant Professor, Department of ECE, MANIT, Bhopal
Dr. K. Kishan Rao, Professor, Department of ECE, SIST, Hyderabad
Dr. Tummala Rangababu, Professor and Head, Department of ECE, RVR and JC CE
Dr. T. Sai Kumar, Associate Professor, Department of ECE, CMR Technical Campus
Dr. N. V. Koteswara Rao, Professor, Department of ECE, CBIT
Dr. J. Tarun Kumar, Professor and Head, ECE Department, SR Engineering College
Dr. P. Prasad Rao, Principal, Vaagdevi Engineering. College
Dr. E. Srinivasa Rao, Professor and Head, Department of ECE, Vasavi Engineering College
Dr. Shaik Jakeer Hussain, Professor, Head, Vignan University
Dr. Ram Desh Mukh, Professor and Head, Department of EEE, SR Engineering College
Dr. B. Ramesh, Professor and Head, Department of ECE, KITS, Huzurabad
Dr. Ravindra Babu Kallam, Professor and Head, Department of CSE, KITS, Huzurabad
Dr. Sunny Dial, Professor, School of Electronics Engineering, VIT, AP
Dr. M. Sushanth Babu, Professor, Department of ECE, Matrusri Engineering College
Dr. M. Raju, Associate Professor, Department of ECE, KITSW
Dr. M. Chandrasekar, Assistant Professor, Department of ECE, KITSW
Dr. K. Sowjanya, Assistant Professor, Department of ECE, KITSW

Dr. B. Dhanalaxmi, Assistant Professor, Department of ECE, KITSW

Dr. P. Krishna, Professor, Department of ECE, VEC

Dr. E. Hari Krishna, Assistant Professor, Department of ECE, KUCE, Kothagudem

Dr. S. Umamaheshwar, Associate Professor, Department of ECE, SR Engineering. College

Mr. K. Ramudu, Assistant Professor, Department of ECE, KITSW

Mr. B. Narsimha, Assistant Professor, Department of ECE, KITSW

Mr. V. Raju, Assistant Professor, Department of ECE, KITSW

Mr. P. Chiranjeevi, Assistant Professor, Department of ECE, KITSW

Mr. R. Srikanth, Assistant Professor, Department of ECE, KITSW

Dr. M. Sunil Kumar, Assistant Professor, Department of ECE, SVEC

Dr. D. Jeya Mala, Director-MCA, Fathima College

Humaira Nishat, Professor, Department of ECE, CVR College of Engineering

Dr. Panyam Narahari Sastry, Professor, Department of ECE, CBIT

Dr. Ramesh Babu Vallabhaneni, Professor, Department of ECE, NRI Institute of Technology, Vijayawada

Dr. D. Krishna Reddy, Professor and Head of the Department, Department of ECE, CBIT, Hyderabad

Dr. Bhavani Sankar, Professor and Head, Department of ECE, Anjalai Ammal–Mahalingam Engineering College

Dr. M. N. Tibdewal, Professor (Ph.D. IIT Kharagpur), Department of ECE, Shri Sant Gajanan Maharaj College of Engineering, Shegaon

Dr. S. Aruna, Associate professor, Department of ECE, AUCE

Dr. S. Sumathi, Professor, Department of EEE, Mahendra Engineering College

Dr. Varun Revuri, Associate Professor, Department of CSE, Vaagdevi Engineering College, Warangal

Dr. Prakash Pareek, Assistant Professor, Department of ECE, Vaagdevi Engineering College, Warangal

Dr. Korlapati Keerti Kumar, Associate Professor, Department of ECE, Vaageswari College of Engineering, Karimnagar

Dr. Guntha Karthik, Associate Professor, Department of ECE , Stanley College of Engineering and Technology for Women Engineering College

Dr. K. C. T. Swamy, Associate professor, Department of ECE, GPCET

Mrs. Rasmita panigrahi, GIET University, Gunupur, Odisha

Mr. Sabyasachi Pramanik, Assistant Professor, Haldia Institute of Technology, Haldia

Dr. B. Rajitha, Associate professor, Department of CSE, MNNIT, Allahabad

Dr. Ravi Boda, Assistant Professor, Department of ECE, KL University, Hyderabad

Dr. K. Srinivas, Associate Professor, Department of CSE, Vasavi College of Engineering, Hyderabad

Dr. K. Ravi Kumar, Assistant Professor, Department of ECE, SCCE, Karimnagar

Dr. Santhosh Kumar Allemki, Associate Professor and Head, Department of ECE, SCIT, Karimnagar

Dr. K. Mahender, Associate Professor and Head, Department of ECE, SRITW, Warangal

Dr. R. Mohandas, Professor, Department of ECE, BITS, Narsampet, Warangal

Dr. Giriraj Kumar Prajapati, Professor, Department of ECE, SCIT, Karimnagar
Dr. Vallem Sharmila, Professor, Department of ECE, VBIT, Hyderabad
Mr. Shyamsunder Merugu, Assistant Professor, Department of ECE, SRITW, Warangal
Dr. Kuncham Sreenivasa Rao, Professor, Department of CSE, VBIT, Hyderabad
Dr. G. Sreeram, Professor, Department of CSE, VBIT, Hyderabad
Dr. Mahesh Kumar Porwal, Professor, Department of EEE, SCCE, Karimnagar
Dr. R. Saravanan, Professor, Department of EEE, BITS, Narsampet, Warangal

Organizing Committees

I. **Program Committee**
    Dr. G. Raghotham Reddy, Professor
    Smt. S. P. Girija, Associate Professor
    Smt. A. Vijaya, Associate Professor
    Dr. M. Chandrasekhar, Assistant Professor
    Mr. Ch. Pavan Kumar, Assistant Professor
II. **Technical Committee: (25th afternoon)**

    Track 1: Mr. V. Raju, Assistant Professor (Main coordinator)
                   Mr. V. Shobhan Reddy, Assistant Professor
    Track 2: Mr. Syed Zaheeruddin, Assistant Professor (Main coordinator)
                   Mr. P. Yugendar, Assistant Professor

    Technical Committee: (26th morning)

    Track 3: Mr. B. Narsimha, Assistant Professor (Main coordinator)
                   Mr. K. Kranthi Kumar, Assistant Professor
    Track 4: Mr. A. Srinivas, Assistant Professor (Main coordinator)
                   Mr. Ch. Pavan Kumar, Assistant Professor

    Technical Committee: (26th afternoon)

    Track 5: Mr. A. Pavan, Assistant Professor (Main coordinator)
                   Mr. D.Santhosh, Assistant Professor
    Track 6: Mr. R.Srikanth, Assistant Professor (Main coordinator)
                   Mr. D.Srinivasa Rao, Assistant Professor

III. **Live Streaming Coordination Committee: (FB live and YouTube live)**
    Dr. K. Ramudu, Assistant Professor
    Mr. D. Venu, Assistant Professor
    Mr. P. Chianjeevi, Assistant Professor
IV. **Souvenir Committee**
    Dr. M. Raju, Associate Professor
    Mr. G. Kranthi Kumar, Assistant Professor

V	Springer Book Preparation Committee

V   **Springer Book Preparation Committee**
Dr. V. Venkateshwar Reddy, Associate Professor
Dr. M. Chandrashekar, Assistant Professor
Smt. Shailaja (operator)

VI.   **Certificates Committee: Certificate Designing and Printing and Distribution**
Dr. K. Ramudu, Assistant Professor
Mr. S. Pradeep Kumar, Assistant Professor
Dr. B. Dhanalaxmi, Assistant Professor
Mr. Md Muqheem, Assistant Professor

VII.   **Press and Publicity Committee**
Mr. B. Komuraiah, Assistant Professor
Mr. V. Shoban Reddy, Assistant Professor
Dr. D. Prabhakara Chary, PRO

VIII.   **Banner, Mementos, Badges, Bouquets Committee**
Mr. A. Srinivas
Mr. J. Sheshagiri Babu
Smt. E. Sushmitha
Mr. P. Yugander

IX.   **Hospitality Committee (Tea Snacks and Lunch)**
Dr. G. Raghotham Reddy
Dr. K. Ramudu
Mr. A. Srinivas

X.   **Validictory**
Mr. P. Chiranjeevi
Mr. S. Pradeep Kumar
Dr. B. Dhanalaxmi

Contents

About the Editors

Dr. K. Ashoka Reddy Komalla is acting as Principal, Kakatiya Institute of Technology and Science, Warangal (KITSW). He studied Bachelor of Engineering in Electronics and Instrumentation Engineering at KITSW and received B.Tech. degree in 1992 from Kakatiya University, Warangal, Telangana. He received M.Tech. degree in 1994 from Jawaharlal Nehru Technological University, Kakinada (JNTUK), Andhra Pradesh. He did research on Pulse Oximeters and received Ph.D. in Electrical Engineering in 2008 from Indian Institute of Technology Madras (IITM), Chennai, India. He received Innovative Project Award in 2008 from Indian National Academy of Engineering (INAE) for his Ph.D. work. His teaching and research interests include signal processing and instrumentation. He has authored over 15 research papers in refereed journals and 50 papers in conferences proceedings. He received Rs. 10 lakh funding under RPS from AICTE, New Delhi, during 2013–2016; received 7.5 lakhs from AICTE in 2017. He is a reviewer for *IEEE Transactions on Measurements & Instrumentation, IEEE transactions on Biomedical Engineering*, and *IEEE Sensors Journal*. He is also a member of IEEE, a life member of ISTE, a member of IETE and a member of CSI.

Dr. B. Rama Devi received the Ph.D. from JNTUH College of Engineering, Hyderabad, Telangana, India, in April 2016. She completed M.Tech.—Digital Communication Engineering from Kakatiya University, Warangal in 2007.

She joined the faculty of Electronics and Communication Engineering, KITSW in 2007. Currently, she is working as Professor and Head, Department of ECE. She published more than 40 papers in various journals and conferences, and filed four patents. She published three books and acted as Session Chair for various international conferences. Her areas of interest include wireless communication, wireless networks, Signal processing for communications, medical body area networks, and Smart grid. She is an active reviewer for *IEEE Transactions on Vehicular Technology* (TVT), Elsevier, *Wireless Personal Communications* and Springer journals.

Dr. Boby George received the M.Tech. and Ph.D. degrees in Electrical Engineering from the Indian Institute of Technology (IIT) Madras, Chennai, India, in 2003 and 2007, respectively. He was a Postdoctoral Fellow with the Institute of Electrical

Measurement and Measurement Signal Processing, Technical University of Graz, Graz, Austria from 2007 to 2010.

He joined the faculty of the Department of Electrical Engineering, IIT Madras in 2010. Currently, he is working as a Professor there. His areas of interest include magnetic and electric field-based sensing approaches, sensor interface circuits/signal conditioning circuits, sensors and instrumentation for automotive and industrial applications. He has co-authored more than 70 IEEE transactions/journals. He is an Associate Editor for *IEEE Sensors Journal, IEEE Transactions on Industrial Electronics*, and *IEEE Transactions on Instrumentation and Measurement*.

Dr. K. Srujan Raju is currently working as Dean Student Welfare and Heading Department of Computer Science and Engineering and Information Technology at CMR Technical Campus. He obtained his Doctorate in Computer Science in the area of Network Security. He has more than 20 years of experience in academics and research. His research interest areas include computer networks, information security, data mining, cognitive radio networks and image processing and other programming languages. Dr. Raju is presently working on 2 projects funded by Government of India under CSRI & NSTMIS, has also filed 7 patents and 1 copyright at Indian Patent Office, edited more than 14 books published by Springer Book Proceedings of AISC series, LAIS series and other which are indexed by Scopus, also authored books in C Programming & Data Structure, Exploring to Internet, Hacking Secrets, contributed chapters in various books and published more than 30 papers in reputed peer-reviewed journals and international conferences. Dr. Raju was invited as Session Chair, Keynote Speaker, a Technical Program Committee member, Track Manager and a reviewer for many national and international conferences also appointed as subject Expert by CEPTAM DRDO—Delhi & CDAC. He has undergone specific training conducted by Wipro Mission 10X & NITTTR, Chennai, which helped his involvement with students and is very conducive for solving their day-to-day problems. He has guided various student clubs for activities ranging from photography to Hackathon. He mentored more than 100 students for incubating cutting-edge solutions. He has organized many conferences, FDPs, workshops and symposiums. He has established the Centre of Excellence in IoT, Data Analytics. Dr. Raju is a member of various professional bodies, received Significant Contributor Award and Active Young Member Award from Computer Society of India and also served as a Management Committee member, State Student Coordinator & Secretary of CSI—Hyderabad Chapter.

Innovative Light Management System for Outdoor Applications

B. Rama Devi, K. Ashoka Reddy, Amreen Tabassum,
and Md. Abdul Muqueem

Abstract Outdoor luminaries' management is one of the prominent areas for smart cities that improve the living standards of the citizens by reducing the emission of CO_2, energy consumption, and cost. In this paper, a motion detection (MD)-based outdoor luminary intensity control system using sensors is proposed. This prototype designed controls luminary intensity during nighttime using different intensity controls, and especially at midnight time, it operates in motion detection mode using sensors and saves a lot of power wastage. It supports motion detection of high-speed vehicles on highways up to 120 kmph or more. The prototype is designed by using simple combinational circuits, integrated with renewable energy sources, and consumes less power.

Keywords Energy · LED · Luminary · Light intensity · Motion detection · Ultrasonic sensor

1 Introduction

The outdoor luminaries require 19% of bulk energy and 3/4th demand of the bulk load [1, 2]. In a European Country, electricity cost per 1 lakh population is around 1–3 M\$. In India, 35 million street lights consume 3,400 MW power/year, which is

B. Rama Devi (✉) · K. Ashoka Reddy · A. Tabassum · Md. A. Muqueem
Department of Electronics and Communication Engineering, Kakatiya Institute of Technology
and Science, Warangal, Telangana, India
e-mail: ramadevikitsw@gmail.com

K. Ashoka Reddy
e-mail: kareddy.iitm@gmail.com

A. Tabassum
e-mail: amreentabassum1996@gmail.com

Md. A. Muqueem
e-mail: mq.ece@kitsw.ac.in

18% of the total energy consumption of the world [3] and emits 1.6 billion tons of CO_2 per annum.

Most of the countries adopt smart street lighting (SSL) to reduce costs, CO_2 emission, and glare, which improves visibility. It also improves energy efficiency, luminary lifetime, quality of life, citizen's comfort, and security.

In this paper, a novel MD-based light intensity control (MDLIC) is proposed, and the main objectives are listed below.

- To choose proper LED luminary,
- To build a prototype of MDLIC circuit,
- Develop a prototype of a central control mechanism,
- Testing and verification of overall functioning.

The paper is organized as follows. The rationale of the study and background information is given in Sect. 2. Section 3 discusses the proposed work. In Sect. 4, the experiment with the developed prototype is verified. The conclusions are consolidated in Sect. 5.

2 Rationale of the Study and Background Information

The first LED smart lighting control [4] in the world was implemented in Estonia street with half maintenance cost and saves 65–85% energy. In Budapest (Hungary Capital), outdoor luminaries used LED. LEDs' advantages are economical and save power.

In India, outdoor luminaries are replaced with LEDs of one thousand four hundred megawatt demand, which conserve nine thousand million kWh per year which saves eight fifty million dollars. Energy efficiency services limited (EESL) replaces street lighting in smart cities with LED (e.g., Nashik Municipal Corporation).

The luminary intensity control with pointer control can be done with motion/speed/direction sensors, and open central management software with MD was illustrated in [3]. Adaptive LED (ALED) streetlights of turning fixtures can be used to save a maximum of 50% energy using MD. Energy supply to LED luminaries with solar panels [5] is economical in long run. Energy saving can be done by using dimming with motion sensor and up to 70% of energy can be saved using LED arrays with dimming [6].

Few smart street light control techniques using Zigbee [7], wireless sensors [8, 9], pulse width modulation [PWM] [10], Global System for Mobile Communication [GSM] [11], adaptive control [12], and Internet of things [IOT] [13] can be used for various outdoor applications. Smart luminary intensity control using Ethernet was given in [14]. The comparison of various luminaries is given in Table 1.

Table 1 Comparison of various luminaries

	LED	ALED	LED + Network control	ALED + Network control	Post top ALED + Network control	Proposed system
Street lighting	Good	Better (100–20% intensity control)	Better	Best	Best	Best
Energy saving	72%	88%	80%	91%	93%	75% + 60–85% at midnight
Area and post top lighting	Non-compliant	Good	Better	Best	Best	Best
Cost in corers	$370	$420	$595	$645	$525	1/5th of least cost

3 Innovative Light Management System for Outdoor Applications

The motion sensor-based street light cannot tackle high-speed vehicles on highways. The aim of the Innovative Light Management System for Outdoor Applications (ILMSOA) is to develop a prototype module of the MDLIC module with a central control system to support up to 120 kmph speed. The lights have special LED array structures with 0–100% intensity control and a specially designed control unit with MD sensors using a zonal control mechanism.

The proposed ILMSOA promotes indigenous technology with simple hardware modules to integrate entire city street lights. The novelty of the proposed work is consolidated in Table 2.

Key features of the ILMSOA are:

- Light intensity (LI) control with MD of vehicles up to 120 kmph spccd.
- It controls LI based on MD.
- LI control using MD can save energy consumption more than 85% using control mode at midnight.
- It uses simple digital hardware circuits and cost effective for implementation.
- ILMSOA eliminates complex software control mechanisms.

The basic block diagram of the MDLIC circuit with a zone timer (ZT) is given in Fig. 1. MDLIC circuit considers four successive luminaries sensors S_{i-2}, S_{i-1}, S_i, and S_{i+1}. Here, 'S_i' indicates the sensor output on 'ith' luminary, 'S_{i-1}' means sensor output from luminary before 'ith' luminary, 'S_{i+1}' indicates the sensor output of successive luminary to 'i', and 'S_{i-2}' indicates sensor output from luminary before '$i - 1$.' When motion is detected, the sensor generates output high (logic 1).

Table 2 Novelty of MDLIC

	Proposed ILMSOA	Sensor MD	Network-based
Cost	Less (45L @5 km*)	High	Very high
Operation principle	MD (using sensors)-based intensity control	-Same-	Internet-based
Energy saved	70% by intensity control + 30–40% additional at midnight	70%	70–90%
Modes of operation	Multi-modes: Timer, MD, zonal, hybrid with priority-based implementation	MD & timer	Single, based on input data
Reliability	Very good	Good	Good
Speed	120 kmph	Not support high speed	Okay
Architecture	Uses simple logic circuits; eliminates complicated network architectures; software and protocols	Little complicate	Complex architecture and software protocols
Complexity	Moderate	High	Extremely high
Renewable sources	Yes (solar panel)	Yes	Yes
Lighting applications	Rural, urban, public place, gated community, hilly, and tourism areas	-Same but in different ways-	-Same-
Skill required	Nontechnical person can handle it	Need trained persons	Need expert
Economic advantage	Returns investment in 2–3 years and improves GDP of the country	Investment returns in 5 years	Investment returns in 6 years
Cost	Less (45L @5 km*)	High	Very high

The sensor output from each sensor is converted into an AC signal using a monostable multivibrator and transmitted to the successive luminaries. These signals are decoded back at the luminary and applied to the MDLIC circuit.

The three ZT inputs T_{NM1}, T_{NM2}, and T_{CM} are generated by the ZT or zone mode generator (ZMG) based on the time zones. From these control input signals and sensor inputs, the MDLIC produces three output control signals D2D, DIM 1, DIM 2 which controls the LI of the luminary using MD or on external control input signals.

Based on the vehicle or passenger position, the sensors on successive luminaries generate sensor control inputs. These sensor inputs control the LI to 25, 50, and 100% in controlled mode (CM). The simplification of the MDLIC circuit using K-maps is given in Fig. 2.

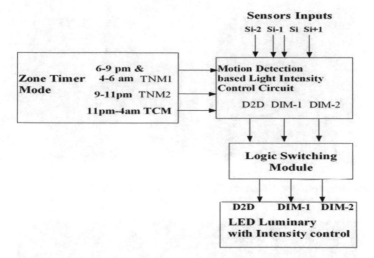

Fig. 1 MDLIC circuit

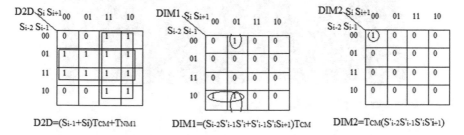

Fig. 2 MDLIC circuit simplification using K-maps

The LI can be controlled directly by external timing control input signals (T_{NM1}, T_{NM2}, and T_{CM}) without using MD also.

4 Results

In a laboratory test, a solar panel is integrated with OSRAM LED luminary as shown in Fig. 3. The ultrasonic sensor HCSR04 with the Arduino board is used for MD. The MDLIC with sensors in an open field setup for MD with outputs 1001 is given in Fig. 4 (indicated by LEDs).

Similarly, an MDLIC circuit with T_{NM1} T_{NM2} T_{CM} S_{i-2} S_{i-1} S_i S_{i+1} inputs 001 0100 produces D2D DIM 1 DIM 2 output 100 (100% LI) as shown in Fig. 5. The zone mode LI control system is implemented successfully and its overall functioning is verified. The overall functioning of the MDLIC is consolidated in Table 3.

Fig. 3 Laboratory test setup-OSRAM LED luminary with solar panel

Fig. 4 Laboratory setup—'$S_{i-2}, S_{i-1}, S_i, S_{i+1} = 1001$'

Fig. 5 MDLIC with T_{NM1}
T_{NM2} T_{CM} S_{i-2} S_{i-1} S_i S_{i+1}
İnput 001 0100 and D2D
DIM 1 DIM 2 output 100

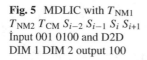

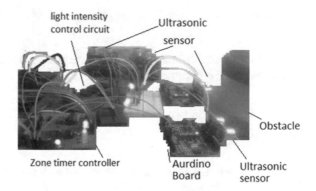

5 Conclusion

In this work, a prototype module of Innovative Light Management System for Outdoor Applications is developed. A specially designed MDLIC module at an LED array is used to detect the motion of the vehicle using ultrasonic sensors. The proposed

Table 3 ZMG circuit

ZT outputs T_{NM1} T_{NM2} T_{CM}	Sensor inputs S_{i-2} S_{i-1} S_i S_{i+1}	LI	LED luminary inputs circuit D2D DIM 1 DIM 2
000	XXXX	0%	000
100	XXXX	100%	100
010	XXXX	50%	010
001	0000	25%	001
001	0001	50%	010
001	0010	100%	100
001	0011	100%	100
001	0100	100%	100
001	0101	100%	100
001	0110	100%	100
001	0111	100%	100
001	1000	50	010
001	1001	50	010
001	1010	100	100
001	1011	100	100
001	1100	100	100
001	1101	100	100
001	1110	100	100
001	1111	100	100

X don't care combination

MDLIC module supports the motion detection of high-speed vehicles on highways. The developed prototype system can save 85% energy during the midnight time using control mode operation.

The proposed technique optimizes energy consumption, cost-effective, eco-friendly, and abolishes the inconvenience of the citizens. This work is highly recommended for outdoor luminary intensity control in smart cities, a potential area to invent and implement world-class technology. This work improves the sustainable growth of the nation in rural and urban development.

References

1. Aleksandra AF, Tatyana AB, Dmitry AS (2017) Outdoor lighting system upgrading based on smart grid concept. Energy Procedia 111:678–688
2. Carlos V, Jesús G, Rubén N, Alberto R (2017) Photovoltaic lighting system with intelligent control based on ZigBee and Arduino. J Renew Energy Res 7:1
3. Reinhard M, Andreas R (2011) An energy efficient pedestrian aware smart street lighting system. J Pervas Comput Commun 7(2):147–216

4. Dheeraj S (2015) A sensor-less and energy efficient street light control system. J Adv Res Electr Electron Instrum Eng 4(3):1805–1812
5. Pilar E, Ignacio A, Asier P, Aitor C, Ignacio JGZ, Asier M, Leire A, José JA, Francisco F, Jesús V (2013) An easy to deploy street light control system based on wireless communication and LED technology. J Sens 2013:6493–6523
6. Gul S, Heekwon Y, Arbab WA, Chankil L (2016) Energy-efficient intelligent street lighting system using traffic-adaptive control. J IEEE Sens 16(13):1–1
7. Zeeshan K, Ishtiaq A, Chankil L (2014) Smart and energy efficient LED street light control system using zigbee network. In: IEEE 12th international conference on frontiers of information technology. IEEE Press, Islamabad, Pakistan, pp 361–365
8. Felipe LDL, Leandro TM, Felipe CF (2014) Energy efficiency and system control of street lighting using wireless sensors network and actuators. J Key Eng Mater 605:267–270
9. Yusaku F, Noriaki Y, Akihiro T (2013) Smart street light system with energy saving function based on the sensor network. In: e-Energy'13: fourth international conference on future energy systems. USA ACM, California, pp 271–272
10. Sharath PGS, Rudresh SM, Kallendrachari K, Kiran Kumar M, Vani HV (2015) Design and implementation of automatic street light control using sensors and solar panel. J Eng Res Appl 5(6) (Part–1):97–100
11. Chaitanya A, Ashutosh N, Paridhi H, Rahul K (2013) GSM based autonomous street illumination system for efficient power management. J Eng Trends Technol 4(1):54–60
12. Soledad E, Jesús C, Maria CM, Stefano C (2014) Estimating energy savings in smart street lighting by using an adaptive control system. Int J Distrib Sens Netw 1–17
13. Claudio R, Manuel G, Antonio D, Fabrizio D (2016) AURORA: an energy efficient public lighting IoT system for smart cities. J ACM SIGMETRICS Perform Eval Rev 44(2):76–81
14. Alexandru L, Valentin P, Ilie F, Daniel S (2012) The design and implementation of an energy efficient street lighting monitoring and control system. Przegląd Elektrotechniczny (Electr Rev) 312–316

Authentic User-Based Luminary Intensity Control System

K. Ashoka Reddy, B. Rama Devi, E. Susmitha, and D. Santhosh Kumar

Abstract In this work, authentic user-based outdoor lighting control technique is proposed. This technique controls outdoor light intensity based on different time zones and as well as by authentic user. The designed prototype will run in two modes: (i) autorun self-test mode and (ii) normal working mode. The normal working mode is used for infield light intensity control and supports two different modes: (i) timer mode and (ii) external authentic user control mode. The working condition is examined, and the results observed are satisfactory for the timer and authentic user control in normal mode and autorun self-test mode.It promotes the modern technologies in the world and improves the living standards of rural and urban citizens.

Keywords Energy efficiency · Intensity control · LED array · Luminary · Solar panel

1 Introduction

Street lighting is a ubiquitous utility. The heavy financial and environmental burden using outdoor lighting can be reduced by minimizing energy consumption. In modern cities and highways, the lights are replaced with low energy consumption street lights like LEDs. It will reduce energy consumption, still inefficient as most of the lights are "ON" during the midnight time when there was no traffic too. Most of the

K. Ashoka Reddy · B. Rama Devi (✉) · E. Susmitha · D. Santhosh Kumar
Department of Electronics and Communication Engineering, Kakatiya Institute of Technology
and Science, Warangal, Telangana, India
e-mail: ramadevikitsw@gmail.com

K. Ashoka Reddy
e-mail: kareddy.iitm@gmail.com

E. Susmitha
e-mail: esu.ece@kitsw.ac.in

D. Santhosh Kumar
e-mail: ds.ece@kitsw.ac.in

© The Author(s), under exclusive license to Springer Nature Singapore Pte Ltd. 2021   9
K. A. Reddy et al. (eds.), *Data Engineering and Communication Technology*,
Lecture Notes on Data Engineering and Communications Technologies 63,
https://doi.org/10.1007/978-981-16-0081-4_2

street lights are "ON" and half of the energy wasted during late-night hours. Light dimming during non-traffic hours avoids energy wastage and supports the safety of citizens. Light dimming can be useful in highways. On highways, traffic density-based adaptive outdoor light intensity control using wireless networks or software protocols can be adapted to support real-time traffic requirements. The cost for such implementation is too high and requires wireless network connectivity on highways.

The main aim of this proposed work is to develop a luminary intensity (LI) control system for highways. The light intensity of the luminary is controlled based on the time zones or by using external authentic user control. It uses a specially designed intensity control module and controls the intensity of the street lights. The street lights have special LED array structures and support various intensity control modes. The primary objectives of the proposed work are:

- To select proper LED luminary with intensity control provision and integrating luminary with solar panel,
- To develop the LI control module,
- Designing autorun self-test mode and normal working mode control,
- To verify the LI control with the designed control module.

The proposed work is organized as follows. Outdoor lighting current status in the literature is explained in Sect. 2. The system model and implementation of the LI control module are given in Sect. 3. Section 4 explains the verification results. Conclusions are drawn in Sect. 5.

2 Outdoor Lighting Current Status in Literature

For outdoor lighting, fluorescent, incandescent, or high-intensity discharge (HID), LEDs, organic LEDs (OLEDs), and polymer LEDs (PLEDs) are used. The comparison of various outdoor lights was given in [1]. Linear intensity control of sodium luminary at airport visual landing was proposed in [2]. Dimming of magnetic ballast-driven high-intensity discharge (HID) lamps using a central control [3] and lighting control by ballast using DC supply for outdoor lighting system was given in [4]. Among various luminaries, LED array lights are energy-efficient, economic, and have a long lifetime [5]. Hence, most of the outdoor lights are now replaced with LEDs which save energy 65–85% and reduce the economic burden to half. The dimming of the light by means of LED panels saves up to 70% energy [6]. In addition, intelligent planning with smart readers according to traffic density saves 50% more energy during traffic hours. Light dimming is another good option. Timer-based outdoor LED LI control was given in [7]. Light intensity control using sensors [8–10], Internet of Things (IoT) [11], temperature-humidity sensor with Arduino board for IoT [12], illumination and humidity using RISC processor [13], GSM [14] and Wi-Fi [15, 16] were discussed. Various smart, adaptable, and automatic outdoor lighting control systems were described in [17–19].

3 System Model and Implementation

This work focuses on the design and implementation of a LI control module, and the proposed module is shown in Fig. 1. The proposed LI control module can have eight major blocks. They are (i) DC power supply, (ii) positive/negative output logic selector, (iii) real-time clock, (iv) mobile transmitter and receiver with connecting cable, (v) DTMF decoder, (vi) autorun self-test/normal mode selector, (vii) micro-controller, (viii) LED driver module with 5 V relays (or logic switching module), and (ix) LED luminary array with intensity control inputs.

DC power supply has the AC supply input, step-down transformer, bridge rectifier, filter, and 7805 voltage regulator with DC power supply (P/S) output ports V_{CC} and GND. The controlled outputs from timer T_1, T_2, T_3, and DTMF decoder C_1, C_2, C_3, C_4 can be observed either in positive logic (PL) or active low (output is active low for "1") using a positive or negative logic selector. It provides flexibility by generating the control outputs either in positive or negative logic and supports for LI control if its intensity control inputs available in either of the logics.

Different time zones are generated by using the DS1307 real-time clock (RTC).The RTC crystal oscillator frequency is 32 kHz. The 24 h of the day are majorly divided to generate four time zones. They are: (i) T_1: 6–9 pm and 4–6 am, (ii) T_2: 9–11 pm, (iii) T_3: 11–4 pm, and (iv) other: 6 am–6 pm. In PL mode, timer outputs T_1, T_2, and T_3 produce logic high output during the prescribed time intervals (only one of the output is active high in T_1, T_2, T_3), and from 6 am to 6 pm, all outputs are zero.

The authentic user control can be done by a GSM module or mobile using a DTMF decoder. In this method,the desired light intensity control by sending a properly control input via a mobile key button. If we use a mobile receiver, charging the mobile receiver and battery backup is a problem. Instead of mobile receiver, a GSM module SIM800L is used. The received control input from the GSM module or mobile receiver is given to the DTMF decoder. The HT9170 DTMF decoder decodes this control input and generates four control outputs C_1, C_2, C_3, and C_4. These control

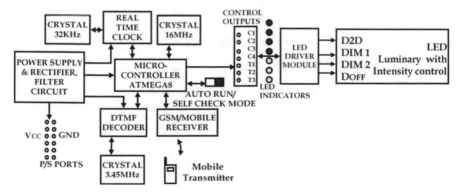

Fig. 1 Block diagram of the proposed LI control module

outputs are used to control the intensity of the luminary as per the authentic user input. The HT9170 DTMF decoder crystal oscillator operates at 3.45 MHz. The control outputs from timer T_1, T_2, T_3, and DTMF decoder C_1, C_2, C_3, and C_4 can be observed by the LED indicators at the control output port as shown in Fig. 1. These control outputs are applied to the LED driver module with 5 V relays, which acts as a logic switching module to control inputs DIM1, DIM2, and D2D of luminary LED array.

Autorun self-test/normal mode selector is used to operate the module in two modes. They are (i) autorun self-test mode and (ii) normal mode. The autorun self-test mode will generate the control outputs C_1, C_2, C_3, C_4, and T_1, T_2, T_3 one after the other with a prescribed delay of 2 s. This autorun self-test mode is helpful to check and monitor the LI control with the proposed module in the laboratory and clear its working condition as "TESTED & WORKING" for infield application. In normal mode, the proposed module generates the control outputs as per timer or external authentic user via mobile.

The RTC, DTMF decoder, autorun self-test, or normal mode selector and positive or negative logic selector inputs are connected to the programable ATMEGA8 micro-controller. Its crystal operates at 16 MHz. Light intensity is controlled depending on mode selection, authentic user control input, or timer input.

4 Results

The LED panel with intensity adjustment with 100%, 50% and 25% brightness is used in this experiment. For primary examination in the laboratory, the luminary is connected to the solar panel with a 15 V battery. Switch 1 (SW1) is connected to the solar panel and switch 2 (SW2) is connected to the battery. The normal working operation of the luminary (ON or OFF) with these control switches is given in Table 1. Testing LI control inputs by connecting jumper are shown in Table 2. After primary examination, connect the designed module to the luminary and verify its operation in different modes.

Table 1 Luminary working condition checking using solar panel and battery

SW1	SW2	Luminary (L) operation	Comment
ON	X	OFF	Its day time, L is OFF
OFF	ON	ON	Its evening/nighttime, L is ON via battery
OFF	OFF	OFF	Both switches are OFF, luminary is OFF

Note X don't care

Table 2 Testing LI control inputs by connecting jumper

DIM1	DIM2	DL1	DL2	DL3	D2D	LI %
O	O	O	O	O	O	100
S	O	O	O	O	O	50
O	S	O	O	O	O	25

S jumper connected, *O* opened

4.1 Autorun Self-Test Mode

Autorun self-test mode is used to check the working condition of the luminary when integrated with the control module at the laboratory. It is used to check and clear the products, control module/luminary at the laboratory level, and clear them by mentioning "TESTED & WORKING" for infield application.

Let us first verify the autorun self-test mode operation. Keep autorun self-test or normal mode selector switch in autorun self-test mode. In autorun self-test mode, the control outputs from C_1, C_2, C_3, C_4, T_1, T_2, and T_3 are high one by one after each 2 s successively in PL mode (reverse in negative logic mode).

The luminary used to test in the laboratory works with positive logic, no change if we apply negative logic outputs. Hence, choose a PL mode, connect the control outputs C_1, C_2, C_3 to the luminary control inputs D2D, DIM1, DIM2 respectively, and repeat the same by connecting T_1, T_2, T_3 to D2D, DIM1, DIM2. The luminary intensity control in autorun self-test mode is observed and given in Table 3 and shown in Fig. 2.

Table 3 Luminary intensity control in autorun self-test mode

Selector	Mode	T_1 T_2 T_3	C_1 C_2 C_3 C_4	LI (%)
PL	A	000	1000	100
		000	0100	50
		000	0010	25
		000	0001	0
		100	0000	100
		010	0000	50
		001	0000	25
		000	0000	0

PL positive logic, *A* autorun self-test mode, *N* normal mode

Fig. 2 Control outputs—T_1 T_2 T_3 C_1 C_2 C_3 C_4 observed by LED indicators (outputs are in negative logic 111 1101 in autorun self-test mode test case)

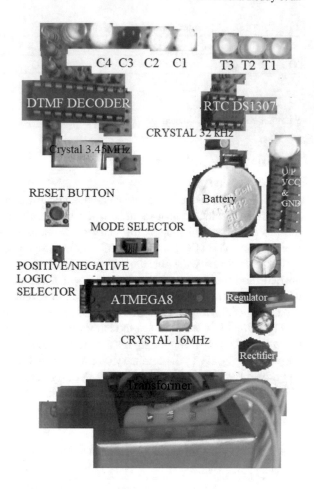

4.2 Testing Normal Mode Operation

Once the test is cleared, the module operates in normal mode and ready for in infield application. For infield application, keep autorun self-test or normal mode selector switch in normal mode. Select PL using a jumper. The normal mode has (i) timer mode and (ii) external authentic user control mode. Luminary intensity control in timer mode is given in Table 4. External authentic user control mode gives the flexibility to control the luminary intensity by the external authentic user. In external authentic user control mode, connect the mobile receiver to the DTMF decoder and verify the authentic user access control via transmitter mobile. C_1, C_2, C_3, and C_4 are generated based on the mobile key 8, 4, 2, 1 pressed by the authentic user. Connect control outputs C_1, C_2, C_3 to luminary intensity control inputs.

For example, when 8 is pressed the code word "C_1 C_2 C_3 C_4—1000" is generated. When key "4" pressed, 0100 code word generated is equivalent to binary code. C_4

Table 4 Luminary intensity control in in normal mode

Selector	Mode	Control mode	MB	T_1 T_2 T_3	C_1 C_2 C_3 C_4	LI (%)
PL	N	E	8	000	1000	100
			4	000	0100	50
			2	000	0010	25
			1	000	0001	0
		T	–	100	0000	100
			–	010	0000	50
			–	001	0000	25
			–	000	0000	0

PL positive logic, *N* normal mode, *MB* mobile button, *T* timer mode, *A* authentic user control

is used to switch OFF the luminary. If the user pressed other buttons by mistake, the DTMF decoder will not respond to it and the control output remains unchanged. The test results are given in Table 4.

5 Conclusion

The basic goal of the proposed work is to design an energy-efficient, cost-effective central outdoor lighting control using simple logic circuits which are suitable for rural and urban roads without using the Internet. It provides the flexibility of the luminary intensity control using authentic user controlling. When the module is in autorun self-test mode, it generates time zone output signals and four control mode output signals by enabling each output successively with a certain delay. It is used to test the working condition of each light intensity control in the laboratory. The normal working mode, timer, or external authentic user control mode is used for real-time light intensity control.

The future scope of this work is to develop a central outdoor street lighting control system suitable for various conditions in the rural and urban environment using motion detection.

References

1. Boddu RD (2017) Effective road and street light load demand control in a smart city. J Appl Adv Sci Res (Special Issue)
2. Latin M, Weston RF, Lambert GK (1963) Linear sodium lighting with intensity control applied to airport visual landing aids. IET J Mag 110(6):1037–1043
3. Wei Y, Hui SYR, Henry SHC (2009) Energy saving of large-scale high-intensity-discharge lamp lighting networks using a central reactive power control system. IEEE Trans Industr Electron 56(8):3069–3078

4. Vanesa R, Andrea P, Rafael D, Gabriel P (2013) Light control for electronic ballast powered by a DC power supply. In: Brazilian power electronics conference, pp 1166–1170
5. Alexandru L, Valentin P, Ilie F, Daniel S (2012) The design and implementation of an energy efficient street lighting monitoring and control system. In: IEEE 2012 international conference and exposition on electrical and power engineering (EPE), pp 312–316
6. Gul S, Heekwon Y, Arbab WA, Chankil L (2016) Energy-efficient intelligent street lighting system using traffic-adaptive control. IEEE Sens J 16(13):1–1
7. Mukul J, Rajashri M, Shruti S, Abhinav G, Sonawane DN (2013) Time based intensity control for energy optimization used for street lighting. In: Texas instruments India educators' conference, pp 211–215
8. Xudan S, Wenjun L, Lingling S, Siliang G (2010) A new streetlight monitoring system based on wireless sensor networks. In: IEEE 2010 2nd international conference on information science and engineering, pp 6394–6397
9. Felipe LDL, Leandro TM, Felipe CF (2014) Energy efficiency and system control of street lighting using wireless sensors network and actuators. Key Eng Mater 605:267–270
10. Yusaku F, Noriaki Y, Akihiro T (2013) Smart street light system with energy saving function based on the sensor network. In: Fourth international conference on future energy systems (e-Energy'13), pp 271–272
11. Claudio R, Manuel G, Antonio D, Fabrizio D (2016) AURORA: an energy efficient public lighting iot system for smart cities. ACM SIGMETRICS Perform Eval Rev 44(2):76–81
12. Dheena PPF, Greema SR, Gopika D, Jinny SV (2017) IOT based smart street light management system. In: 2017 IEEE international conference on circuits and systems (ICCS), pp 368–371
13. Sung-Il H, Chi-Gook I, Hyun-Sik R, Ji-Chul P, Dal-Hwan Y, Chi-Ho L (2011) A development of LED-IT-sensor integration streetlight management system on Ad-hoc. In: IEEE region 10 conference TENCON, pp 1331–1335
14. Chaitanya A, Ashutosh N, Paridhi H, Rahul K (2013) GSM based autonomous street illumination system for efficient power management. J Eng Trends Technol 4(1):54–60
15. Rifki M, Yudha HA (2017) Development of street lights controller using WIFI mesh network. In: IEEE 2017 international conference on smart cities, automation and intelligent computing systems (ICON-SONICS), pp 105–109
16. Babak F, Alexis K, Ali D, Robert SB, Morgan K (2011) Charge it, in smart grid: reinventing the electric power system. IEEE Power Energy Mag
17. Tejaswini A, Shere VB (2016) Modern LED street lighting system with intensity control based on vehicle movements and atmospheric conditions using WSN. J Innov Res Adv Eng 3(5):10–15
18. Abinaya R, Varsha V, Kaluvan H (2017) An intelligent street light system based on piezoelectric sensor networks. In: IEEE 4th international conference on electronics and communication systems (ICECS), pp 138–142
19. Pilar E, Ignacio A, Asier P, Aitor C, Ignacio JGZ, Asier M, Leire A, José JA, Francisco F, Jesús V (2013) An easy to deploy street light control system based on wireless communication and LED technology. J Sens 13(5):6493–6523

A Simple Method to Detect Partial Shading in PV Systems

U. Malavya, R. Chander, and K. Sumalatha

Abstract The photovoltaic systems are suffering with the diminishing output power due to the partial shading conditions. The conventional maximum power point tracking (MPPT) methods fail to track global maximum power point. Hence, a new method is developed to detect the partial shading conditions. So that the system can decide to call either conventional MPPT tracker or global MPPT tracker based on the partial shading condition status. The proposed method has developed based on the voltage at the first peak in P–V characteristics curve to decide the partial shading status. The theoretical performance has been validated with MATLAB/Simulink simulations under various partial shade conditions to show the effectiveness of the proposed method.

Keywords Partial shade detection · Maximum power tracking · Photovoltaic systems · And non-uniform insolation

1 Introduction

Due to the drastic usage of conventional sources, there is raising an enormous impact on environment. Burning of these fossil fuels leads to an emission of harmful gases like CO_2 into the atmosphere causing greenhouse effect [1]. So, there is a need to generate power through renewable energy sources. The generation of power through renewable energy sources is by mixing the technology with the hydro, wind and the solar energy which has been combined in the power system [2]. In view of the solar power generation, the utilization of solar energy has become popular in

U. Malavya (✉)
Department of EE, RGUKT, IIIT, Basar, Telangana, India
e-mail: udugulamalavya@gmail.com

R. Chander
Department of EEE, CITS, Warangal, Telangana, India

K. Sumalatha
Department of EEE, UCE(KU), Warangal, Telangana, India

© The Author(s), under exclusive license to Springer Nature Singapore Pte Ltd. 2021   17
K. A. Reddy et al. (eds.), *Data Engineering and Communication Technology*,
Lecture Notes on Data Engineering and Communications Technologies 63,
https://doi.org/10.1007/978-981-16-0081-4_3

building solar wall system and stand-alone system [3]. Due to the variation in light intensity and temperature from solar irradiations, tracking of maximum power is adopted by various methods. There are many algorithms such as microcontroller-based maximum power point tracking (MPPT) [4] and fuzzy logic [5]. Under some environmental conditions, the maximum power point may track wrongly such as variation of insolation, change in temperature and partial shading. So, researchers developed global maximum power point tracking methods to track maximum power under non-uniform irradiation condition under non-uniform irradiation, the PV string creates multiple peaks in the P–V characteristics [6] because of by-pass diodes. Under such conditions, the MPPT techniques like the P&O instead of tracking utmost power the technique are tracking local maxima. So, global MPP needs to be tracked.

In the present years, different methods are proposed for tracking the GMPP during uneven irradiation [7]. Different Methods include in the area of search algorithm such as Fibonacci search [8] and the metaheuristic algorithms like PSO [9]. More methods like PSO and P&O and differential evolution of PSO are proposed. A global algorithm called Fireworks algorithm is also one of the method for tracking MPP [10], by Tan and Zhu. The explosion of fireworks in the sky is modelled in this algorithm [11].

After detailed study, the following problems are identified to track maximum power under uneven irradiations conditions.

(i)   During multiple peaks, the conventional maximum power tracker is not tracking GMPP.
(ii) This global maximum power point tracker is taking more time due to more space to search under normal operating conditions. Under uncertain conditions, tracking the utmost power is a problem.

This paper has prepared as follows: Sect. 2 explains the equivalent circuit of PV model, and the simple method to detect partial shading is explained in Sect. 3. The results in Sect. 4 and the end of paper give conclusions and references.

2   System Configuration

2.1   Modelling of a PV Cell

The ideal PV cell is a grouping of current source, and a diode is in parallel with series and shunt resistance as given in Fig. 1. The PV cell characteristic cells are obtained by subtracting the diode nonlinear current from solar cell current.

$$I_{ph} = I_{ph\ STC} \frac{G}{G_{STC}} (1 + \alpha_I (T - T_{STC})) \tag{1}$$

where α_I is the temperature coefficient of the current, and it is defined in Standard Test Conditions (STC) as follows.

Fig. 1 Equivalent circuit of
a solar PV cell

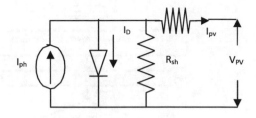

$$\alpha_I = \frac{dI}{dt}_{\text{ at STC}} \tag{2}$$

As a consequence, the I–V characteristics take from the following equation

$$I = I_{\text{ph}} - I_{\text{sat}}\left(e^{\frac{v}{\eta V_t}} - 1\right) \tag{3}$$

3 Proposed Method

The proposed method has been explained with algorithm and flowchart as follows.

3.1 Modelling of a PV Cell

Step 1   Read PV array as 'm' rows and 'n' columns.
Step 2   Call conventional MPPT tracker.
Step 3   Store the value of the first peak power and corresponding voltage in P_1 and V_1, respectively.
Step 4   Compare V_1 with V_{mpp} value.
Step 5   Check whether V_1 is greater than V_{mpp}.
Step 6   If Yes, partial shading has occurred and multiple peaks are present in PV system call GMPPT tracker for global peaks.
Step 7   If No, partial shading has not occurred and single peak present in PV system.
Step 8   End.

3.2 Flowchart

The proposed method is explained with the flowchart as given in Fig. 2.

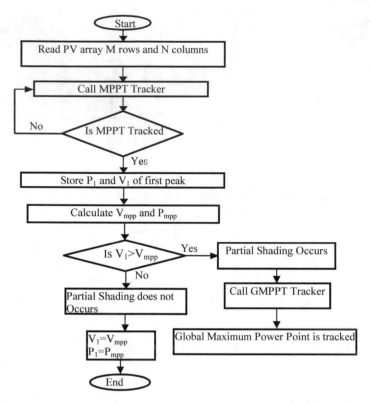

Fig. 2 Flowchart for the proposed GMPPT strategy

4 Partial Shading Detection in PV Array

Table 1 gives specifications of PV cell under Standard Test Condition.

The string characteristics procure by summing the voltage of every module, where the currents remain the same. The transition of radiation is explained in four ways.

1. Changeover from uniform to uniform insolation.
2. Changeover from uniform to non-uniform insolation.
3. Changeover from non-uniform to uniform.

Table 1 PV module specifications at 1000 W/m², 25 °C

Parameter	Value
PV power	80 W
Open-circuit voltage	21.6 V
Short-circuit current	4.7 A
Nominal voltage	18 V
Nominal current	4.4 V

Fig. 3 I–V curves of strings when transition from uniform to uniform insolation

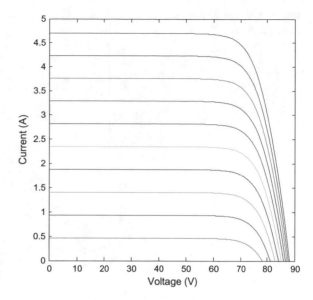

4.   Changeover from non-uniform to non-uniform insolation.

4.1 Change Over from Constant to Constant Insolation

The insolation is 1000 W/m² for all modules and is decreased in steps to 100 W/m². From the I–V curves shown in Fig. 3, we observe that the voltage decreases and current decreases with decrease inirradiation. In the P–V curves shown in Fig. 4, we observe that the voltage decreases and power decreases with decrease in irradiation. The transition is that the maximum power point shifts left side from the standard irradiation.

If the curve shifts to left side, then we will tell that there is a transition which is uniform to uniform transition. With increase in radiation, the voltage increases and power also increases and has single peak as there are no different irradiations.

4.2 Simulated Waveform Unifrom to Unifrom Isolation

Figure 5 explains that when there is decrease in irradiation, the current and power decreased with slight decrease in voltage. Figure 6 explains that when there is increase in irradiation, the current and power increase with slight increase in voltage.

Fig. 4 P–V curves of string
under constant irradiation

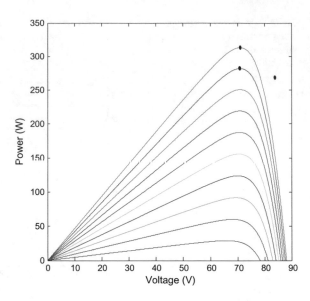

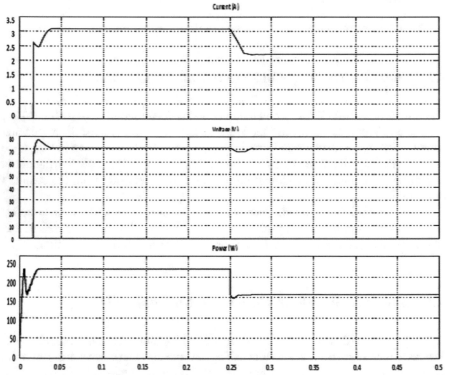

Fig. 5 Simulated results for transition from uniform to uniform irradiation with decreased
irradiation

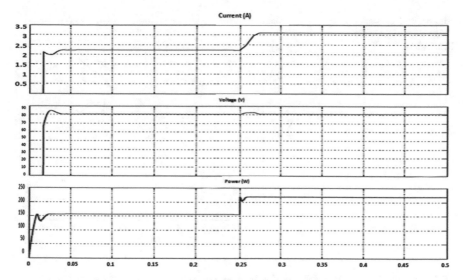

Fig. 6 Simulated results for transition from constant to constant irradiation with increased irradiation

4.3 Change Over from Constant to Uncertain Iradiation

Assume that a module receives a different insolation. The insolation is 1000 W/m^2 to each module and 500 W/m^2 to one of the module.

The module with less irradiation is shaded module. From the I–V curves shown in Fig. 7, we observe that the voltage increases and current decreases with decrease in irradiation. In the P–V curves given in Fig. 8, we observe that the voltage increases and power decreases with decrease in irradiation.

Fig. 7 I–V characteristics of a PV string under non-uniform irradiation

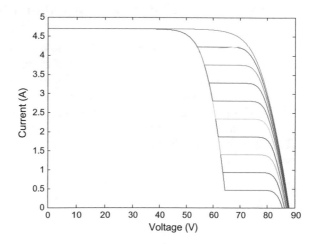

Fig. 8 P–V curves of a
string under uncertain
irradiation

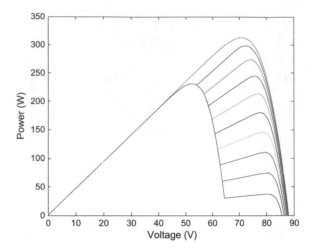

4.4 Simulated Waveform Uniform to Nonuniform Insolation

Figure 9 explains that for the first peak when there is decrease in irradiation, the
current and voltage of a string decreased, but the voltage of a shaded string decreased
more compared to unshaded string, so the power also decreased. Depends on number
of different irradiations in a row decides number of peaks. Figure 10 explains that for
the first peak, see when there is decrease in irradiation, the current slightly decreased
and voltage of a string decreased, but the voltage of a shaded string decreased more
compared to unshaded string, so the power also decreased.

4.5 I–V Curves When Two Different Irradiations Have Equal
Power Under Uniform to Non-uniform Irradiation

Figure 11 explains I–V curves where three modules receive same irradiation and the
left over one module receives a different irradiation; so due to partial shading, one
receives less irradiation where there is a possibility to have two peaks of equal power.
Totally, we considered here four modules; out of which different modules receives
irradiation as mentioned above by which instead of unique peak, we have two peaks
due to two different irradiations. The initial curve starts with short-circuit current of
4.7 A which is the standard test current.

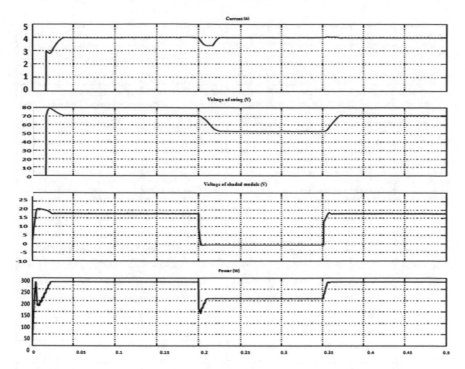

Fig. 9 Simulated results for transition from uniform to uniform irradiation with decreased irradiation for the first peak

5 Results and Discussion

The circuit for the proposed strategy consists of a PV string, boost converter and load. The whole converter is given to GMPPT controller. The schematic circuit for the GMPPT controller is shown in Fig. 12. Here, GMPPT is to track the global maximum power point when there is a difference in the irradiation occurs; those are described in Table 2 for different values of irradiation. In the below figure, we considered a PV string of four modules having a boost converter with duty ratio control. Table 2 shows the curve number and the irradiation levels whether they are uniform or any partial shade occurs on the panel. From the test PV curves, we will observe that the PV module designed has nine cases.

Starting from $C1$–$C5$ will have only one peak as it is same insolation to all panels. The curve $C6$ has four panels with three panels of same irradiation and one with different irradiation so it has first peak. Figure 13 shows simulated PV curves which need to create eight subsystems of PV module where initial case is tested with the first main system, and for the remaining eight cases, we need to have eight system; for every subsystem, we have to give different irradiations as input which we will see nine curves of different irradiations. So, how the curve is shifting for uniform irradiation, and non-uniform irradiation has been explained.

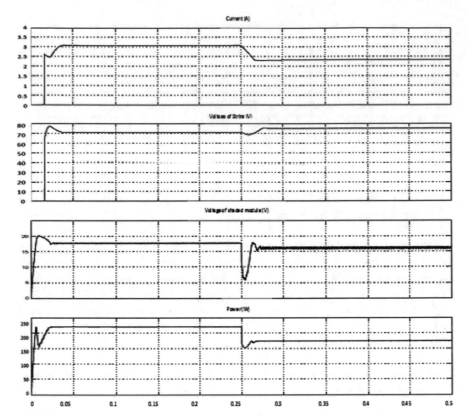

Fig. 10 Simulated results for transition from constant to constant irradiation with decreased irradiation for the second peak

Fig. 11 I–V curves for two peaks having equal power

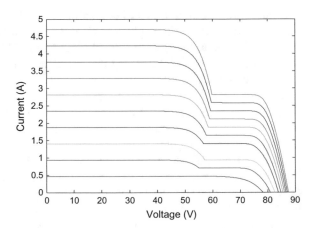

Fig. 12 Circuit for the proposed GMPPT controller

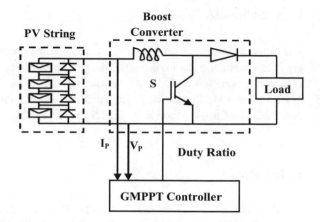

Table 2 Test different irradiation for PV strings

Curve	$G1$ (W/m^2)	$G2$ (W/m^2)	$G3$ (W/m^2)	$G4$ (W/m^2)	GMPP
$C1$	950	950	950	950	Unique
$C2$	750	750	750	750	Unique
$C3$	550	550	550	550	Unique
$C4$	350	350	350	350	Unique
$C5$	150	150	150	150	Unique
$C6$	1000	1000	1000	900	First peak
$C7$	1000	1000	800	700	Second Peak
$C8$	1000	950	350	250	Third peak
$C9$	110	110	110	110	First peak

Fig.13 Test P–V curves used for validating the proposed method

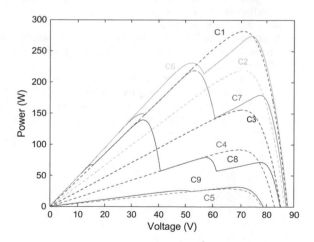

6 Conclusion

The proposed method has successfully detected the partial shading. The analysis had given to understand the logic behind the proposed method with P–V characteristics successfully verified with PV system and boost converter. It had reduced the search space by knowing the partial shade condition status. The theoretical and practical performance of this method is verified by MATLAB/Simulink simulations.

References

1. Mohsenian-Rad AH, Leon-Garcia A (2010) Optimal residential load control with price prediction in real time electricity pricing environments. IEEE Trans Smart Grid 1(2):120–133
2. Rani BI, Ilango GS, Nagamani C (2013) Enhanced power generation from PV array under partial shading conditions by shade dispersion using Su Do Ku configuration. IEEE Trans Sustain Energy 4(3):594–660
3. Srinivasa Rao P, Ilango GS, Nagamani C (2014) Maximum power from PV Arrays using a fixed configuration under different shading conditions. IEEE J Photovolt 4(2):679–686
4. Rakesh N, Madhava Ram TV, Ajith K, Rajendra Naik G, Nagarjun Reddy P (2015) A new technique to enhance output power from solar PV array under different partial shaded conditions. In: IEEE conference on electron devices and solid state circuits (EDSSC), Singapore, pp 345–348
5. Bidram A, Davoudi A, Balog S (2012) Control and circuit techniques to mitigate partial shading effects in photovoltaic arrays. IEEE J Photovolt 2(4):532–546
6. Saheb-Koussa D, Haddadi M, Belhamel M (2009) Economic and technical study of hybrid system for rural electrification in Algeria. Energy 86:1024–1030
7. Veerasamy B, Thelkar AR, Ramu G, Takeshita T (2016) Efficient MPPT control for fast irradiation changes and partial shading conditions on PV systems. In: IEEE international conference on renewable energy research and applications, pp 358–363
8. Miyatake M (2001) Maximum power point tracking control employing fibonacci search algorithm for photovoltaic power generation system. In: International conference on power electronics, Austria, pp 622–625
9. Ishaque K, Salam Z (2013) A deterministic particle swarm optimization maximum power point tracker for photovoltaic system under partial shading conditions. IEEE Trans Ind Electron 60(8):3195–3206
10. Femia N, Petrone G, Spagnuolo G, Vitelli M (2005) Optimization of perturb and observe maximum power point tracking method. IEEE Trans Power Electron 20:963–973
11. Manickam C, Raman GP, Raman GR, Ilango GS, Nagamani C (2016) Fireworks enriched P&O algorithm for GMPPT and detection of partial shading in PV systems. IEEE Trans Power Electron 32(6):4432–4443

Capsule Networks for Classifying Conflicting Double-Handed Classical Dance Gestures

S. Shailesh and M. V. Judy

Abstract A challenging application of computer vision is the classification of gestures in classical dance. Classical dance is one of the most accepted forms of arts in India. Due to the complexity of various gestures used to convey unique meaning to the audiance, computational analysis of the dance gestures is a complex task. Hence, it requires efficient machine learning techniques to solve this problem. In this paper, we introduce a classification model for identifying a subclass of hastas called Samyuta hastas (double-handed gestures which preserve hierarchy) where the mudras or hastas are unique hand gestures that imply some actions and ideologies. A classification technique that preserves the object location within an image is necessary for this problem. Hence, we use capsule networks, which maintain equivariance in an image and can analyze the hierarchy and order of objects within the image. A dataset comprising of around 2400 images distributed evenly among six classes of Asamyutha hastas is used for the work. The classification was performed by the deep learning pipelines using typical convolutional neural networks (CNN), transfer learning, and the proposed capsule network (hasta CapsNet). The results show that since CNN loses the location information of an object due to its pooling layers that extract only the vital features, it tends to wrongly classify the double-handed gestures, whereas the capsule networks have achieved better accuracy on the dataset

Keywords Classical dance · Hastas · Mudras · Hand gestures · Capsule networks · Dynamic routing · Classification

S. Shailesh (✉) · M. V. Judy
Department of Computer Applications, Cochin University of Science and Technology, Kalamserry, Kochi, Kerala, India
e-mail: shaileshsivan@gmail.com

M. V. Judy
e-mail: judy.nair@gmail.com

1   Introduction

Recent efforts to digitize art have eased the difficulties of archiving in maintaining the history of art forms. Indian classical dance or Shastriya Nritya is a familiar art form. Lucid interpretations of its theories and practices can be seen in the works like Natya Shastra, Abhinaya Darpana, and Thandavalakshana. The digitization of these works opens up technical, scientific chances as the scope is potential. The range of computational analysis in this archived data, especially in the classical dance domain, is enormous. In classical dance, hastas or mudras are hand gestures or symbols that are a primary mode of communication. The necessary intention of hastas is to convey some ideas, thought, or to feel to the viewers. They are two kinds of hastas in dance. They are single-handed gestures called Asamyuta hastas and two-handed gestures called Samyuta hastas. There are 28 Asamyuta hastas. Samyuta hastas are also termed as combined hand gestures and different from Asamyuta hastas; these gestures are realized using the palm of both hands. There are 24 Samyuta hastas. While considering the visual perception of Asamyuta hastas, some of them are visually similar to a slight difference, and this makes the classification task more difficult. Also, in a real-time scenario, the orientation of these gestures may change by rotation, a translation like transformation and lead to conflicts. This paper focuses on creating a classification model for such conflicting Asamyuta hastas. The dataset considered for this work consists of around 2400 images, which are evenly distributed to six Asamyuta hasta classes(Anjali, Kapotham, Swastika, Garuda, Paasha, and Keelikam). Figure 1 represents these six Asamyuta hastas. As we know, the state-of-art method for image classification is Convolutional Neural Network (CNN), it has improved the classification of image data through efficient image recognition. Besides, the new challenges have also emerged along with the continuous development of CNN techniques for representation of image content. CNN always perform classification using the extracted feature maps from each region of an image, so it fails to address the logical connection between these regions. In the case of Asamyuta hastas, we can find different hastas with similar sub-regions like Swasthika and Garuda. In the proposed work, a new method is used to overcome this problem, which works with the concept of capsule networks.

| Anjali | Kapotham | Swasthikam | Garuda | Paasha | Keelikam |

Fig. 1   Six Samyuta hastas

2 Literature Review

The existing methods in mudra classification and related techniques are analyzed by conducting a literature review. The gist of papers is as given under Hahriharan [1] who suggested a two-level transformation-invariant classification model for identifying hand gestures of a Bharathanatyam dancer. They use orientation filters at first level, and gradients form shilloute in the second level as features. Mozarkar [2] used pattern recognition and image processing approaches to recognize the mudra sequences, thereby interpreting a few static Bharatnatyam mudras. They highlighted the object from the background using the hypercomplex representation of the image and extracted the salient features of the static double-handed mudra image. For classification, k-nearest neighbor algorithm is used. Mampi [3] introduced a simple two-level classification approach for single-handed gestures. In the first level, they grouped twenty-nine class hand gestures into three categories based on their structural similarity. In addition, the proposed method extracts medial axis transformation (MAT) from the images to identify the groups. In the second level, hastas are identified from the database inside the group; Kumar [4] introduced a technique in which classical dance mudras are identified with histogram of oriented (HOG) features and SVM as classifier. And these gestures are finally converted to text messages. Anami [5] presents a three-level process for recognizing 24 double-handed mudras. The primary task in the proposed method is to preprocess the image. Then, the cell features are extracted, and in the last stage, the given image is mapped into any of the 24 classes of mudras using a newly developed classifier. In another paper [6], Anami proposed a three-step process for the identification of single-handed mudra images. The initial stage is the same as the previous one, and in the second stage, features such as Hu-moments and eigenvalues were extracted. The third stage is for classification of mudras using artificial neural networks. Further, Anami [7] illustrates a three-level classification pipeline for pin pointing the double-hand mudra images of Bharatanatyam. This method also considered the features mentioned in the previous paper, which are contours, cell features Hu Moments, and then, three classifiers, namely rule-based, artificial neural network, and k-nearest neighbor, are used for classification.

3 Capsule Networks

Convolutional neural networks [8, 9] have improved the methodologies for image classification and related tasks to a great extend. But still there exist some root issues that it cannot preserve the logical connection between the sub-regions in an image, due to mistakes in pooling operations. The capsule networks are first introduced by Geoffrey E Hinton, Nicholas Frosst, and Sara Sabour. Also, they proposed a 'dynamic routing algorithm' [10–12] that can be used to intelligent routing between

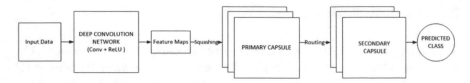

Fig. 2 General CapsNet pipeline

capsules. In Fig. 2, a typical structure of a capsule network is illustrated. Important components in the capsule networks are explained in the following section.

3.1 Capsules

A capsule is a collection of neurons which perform collective input computations internally. Then, the results are wrapped into small vectors that can precisely represent the properties of the same object as it is. Capsules are composed of two principal units. One of them is the probability of an object being present, which is locally steady. The next is the bunch of equivariant instantiating variables, sometimes called pose. It is necessary to have these two elements because by understanding their sections, they help to understand the whole thing.

3.2 Squashing Function

Capsule vector lengths are normalized using the squashing function. Same like ReLu and sigmoid squashing function is a function of nonlinearity.The method squashes capsules, which are basically activation vectors, maps the output vector to 0 if the vector is short and to 1 otherwise. The squashing function is set out in Eq. (1).

$$v_j = \frac{\left\|s_j\right\|^2}{1 + \left\|s_j\right\|^2} \frac{s_j}{\left\|s_j\right\|} \tag{1}$$

3.3 Dynamic Routing Algorithm

To solve which capsule is enabled for incoming data, a routing algorithm is used to. The capsule at the bottom level transfers its input to a higher level, which is in line with its data. Through this agreement, weight matrices are modified between the two capsular rates.

3.4 Decoder

The decoder processes the class capsule outcome, and from it, the object is reconstructed.The decoder works after the final capsule layer. This helps to boost the capsule network's encoding capability, and thus, the object can be represented in a smaller domain effectively.

$$LR = \frac{1}{w \times h} \sum_{i \in \text{pixelgrid}} (X_i + R_i)^2 \tag{2}$$

Equation (2) gives a single object's loss, where X represents the input image and R represents the reconstructed image. Through this agreement, weight matrices are modified between the two capsular rates.

3.5 Loss Function

To improve the capsule network preparation, Sabour et al. [6] applied margin losses. The per capsule margin loss is described as given [6]:

$$L_k = T_k \max(0, m^+ - \|v_k\|)2 + \lambda(1 - T_k) \max(0, \|v_k\| - m^-)^2 \tag{3}$$

where k represents the class capsule index, $T_k = 1$ for true class. As in [6], we set $m^- = 0.1$, $m^+ = 0.9$ and $\lambda = 0.5$. Therefore, for such C number of classes, the overall loss function for a batch of N size is defined as:

$$\text{loss} = \frac{1}{N} \sum_{i=0}^{N} \sum_{i=0}^{N} L_k^i + \gamma L_R^i \tag{4}$$

Reconstruction loss is used as a regularization parameter.

4 Proposed Hasta Capsule Network

Our proposed capsule network architecture for hasta classification model is shown in Fig. 3. The input to the network was color images of size 32×32, so the input is a vector with dimension $32 \times 32 \times 3$, where the last dimension represents the RGB channels. Then, the input vector is feed into 2D convolution networks for extracting the feature maps. In the given architecture, there are two convolution networks both having kernel size 9×9 and strides 1 and 2, respectively. The intensities in the input vector are used by these convolutional layers for activating and extracting the local

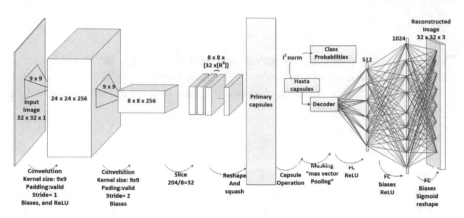

Fig. 3 Hatsa CapsNet architecture

features into a feature map, and it is used as the input for the next layer. The second layer consists of primary capsule layers. And it is basically a convolutional capsule layer containing 32 channels of convolutional 8D capsules (i.e., a 9 × 9 kernel and a stride set to 2 is used with each primary capsule containing eight convolutional units). The outputs of all 256 × 81 Conv1 units are visible to each primary capsule. Each capsule applies eight 9 × 9 × 256 convolutional kernels (of stride 2) to the 24 × 24 × 256 input and thus produces 8 × 8 × 8 output tensor. 32 such capsules create the output volume with the shape of 8 × 8 × 8 × 32. HastaCaps layer has six-digit capsules, one dedicated for each hasta. An 8 × 8 × 8 × 32 tensor is taken as the input for each capsule. That is, there are 8 × 8 × 32 8-dimensional vectors, which are 2048 input vectors totally. Each input vector gets its own 8 × 16 weight matrix, which is used to map 8-dimensional input space to the 16-dimensional capsule output space. This output is channeled to two layers; one is the class prediction layers, which can be used to obtain the output class mapping. The output at this layer is several N 6 × 16 × 1 vectors; each represents the predicted class of the N input images. This can be converted to class probabilities using L2 normalization, and the result will be N × 6 vectors. The second channel leads to a decoder layer, which is used for regularization as well as loss calculation. The output from the HastaCapsules will be given to decoder layers; after passing through three fully connected layers in the decoder network, it will reconstruct the image having the same dimension of the input image. By using this reconstructed image, the loss can be calculated as mentioned in Sect. 3.5. In dynamic routing, the higher level capsule receives input from lower level capsule output that promises the matching of dimensions, and this happens between the primary capsule and the hasta capsule layers.

5   Experimental Results

Before the implementation of these hasta capsule networks, the same dataset was feed to a convolutional neural network model and also in the transfer learning model.The implementation of these hasta capsule networks [13] was done with Pytorch using Python. First, we just tried to train the model in CPU, but it failed due to the demanding computational power and then got switched to a more efficient computing environment (64 GB RAM,Nvidia GeForce GTX 1080Ti GPU,Intel Core i7-7700K CPU). It took around 40 min to train 2400 images with our setup.

The result after the training is quite significant. The proposed capsule network model achieved 100% accuracy, while CNN and transfer learning models have 96% and 98%, respectively. The detailed results and comparison of these models are illustrated using confusion matrices in Fig. 4 and classification reports in Table 1. From the confusion matrix of CapsNet, it is clear that the miss classifications is zero. That is, this model possesses an accuracy of 100%. The other evaluation metrics measured are precision, F1 score, recall, accuracy and support for the three models and were illustrated in Table 1 HYPERLINK "SPS:extnid::1" . While building a machine learning model, its training loss and accuracy attainment depend on the batch size and number of epochs. The following graphs in Fig. 5 represent the change in training loss and the accuracy at each epoch for a batch of size 16. From Fig. 5, it is observed that for the first few epochs, there is a gradual decrease in training loss, and afterward, the changes are quite negligible and end up with a training loss of 0.6528. In the case of accuracy also, for the first few epochs, the change is very drastic, but within a short span, the model attained an accuracy of 100%

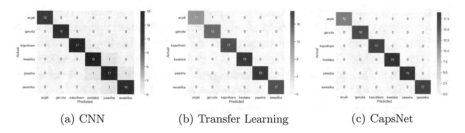

(a) CNN   (b) Transfer Learning   (c) CapsNet

Fig. 4 Confusion matrices

Table 1 Evaluation metrics of three models

Models	Convolutional Neural Networks				Transfer Learning				Capsule Network			
Classes	Precision	Recall	F1-score	Support	Precision	Recall	F1-score	Support	Precision	Recall	F1-score	Support
Anjali	1.00	1.00	1.00	12	0.92	0.92	0.92	12	1.00	1.00	1.00	12
Garuda	1.00	1.00	1.00	16	0.94	0.94	0.94	16	1.00	1.00	1.00	16
Kapotham	1.00	1.00	1.00	17	1.00	1.00	1.00	17	1.00	1.00	1.00	17
Keelaka	0.95	0.95	0.95	19	1.00	1.00	1.00	19	1.00	1.00	1.00	19
Paasha	0.89	0.89	0.89	19	1.00	1.00	1.00	19	1.00	1.00	1.00	19
Swastika	0.94	0.94	0.94	17	1.00	1.00	1.00	17	1.00	1.00	1.00	17
Accuracy			0.96	100			0.98	100			1.00	100
Macro avg	0.96	0.96	0.96	100	0.98	0.98	0.98	100	1.00	1.00	1.00	100
Weight avg	0.96	0.96	0.96	100	0.98	0.98	0.98	100	1.00	1.00	1.00	100

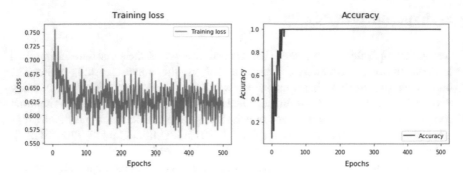

Fig. 5 Training loss and accuracy

6 Conclusion and Future Scope

In this approach, the performance of the capsule network was compared with convolutional neural networks and transfer learning models in the classification of Samyuta hastas (double-handed gestures) particular category of hastas. The proposed capsule network considered knowledge about the pose and its associated parameters to identify the constructs; hence it was able to overcome some drawbacks of CNNs. Though the capsule networks are based on activity vectors that consume more memory, it can be used in situations where the training dataset can be lesser capacity. The activation procedure is an association on the hierarchy of parts; hence, with a small number of hyper-parameters, it could efficiently classify Samyuta hastas. As in every machine learning task, the availability of the dataset was an issue, but we have created a dataset of 2400 images with six classes captured from four dancers. The capsule networks proposed in this work outperformed the state-of-the-art methods like CNN and transfer learning models for hasta classification by achieving a classification accuracy of 100%. The capsule network is still in the research and development phase, and it is not completely accustomed to real-time scenarios. But it is not too far that capsule networks will achieve a milestone in image classification and emerge as a promising technique with refined architectures.

References

1. Hariharan D, Acharya T, Mitra S (2011) Recognizing hand gestures of a dancer. In: Lecture notes in computer science (including subseries lecture notes in artificial intelligence and lecture notes in bioinformatics) 6744 LNCS, pp 186–192
2. Mozarkar S, Warnekar D (2013) Recognizing Bharatnatyam Mudra using principles of gesture recognition. Int J Comput Sci Netw 02(4):46–53
3. Devi M, Saharia S (2016) A two-level classification scheme for single-hand gestures of Sattriya Dance. In: International conference on accessibility to digital world, ICADW 2016—proceedings, pp 193–196

4. Kumar KV, Kishore PV (2018) Indian classical dance Mudra classification using HOG features and SVM classifier. Smart Innov Syst Technol 77(5):659–668
5. Anami BS, Bhandage VA (2018) A vertical-horizontal-intersections feature based method for identification of Bharatanatyam double hand Mudra images. Multim Tools Appl 77(23):31021–31040
6. Anami BS, Bhandage VA (2019) A comparative study of suitability of certain features in classification of Bharatanatyam Mudra images using artificial neural network. Neural Process Lett 50(1):741–769
7. Anami BS, Bhandage VA (2019) Suitability study of certain features and classifiers for Bharatanatyam double-hand Mudra images. Int J Arts Technol 11(4):393–412
8. Pinto RF, Borges CD, Almeida AM, Paula IC (2019) Static hand gesture recognition based on convolutional neural networks. J Electr Comput Eng
9. Pyo J, Ji S, You S, Kuc T (2016) Depth-based hand gesture recognition using convolutional neural networks. In: 2016 13th international conference on ubiquitous robots and ambient intelligence, URAI 2016, pp 225–227
10. Gol'din P, Gladilina E (2015) Dynamic routing between capsules Sara. Aquat Biol 23(2):159–166
11. Hinton G, Sabour S, Frosst N (2018) Matrix capsules with EM routing. In: 6th international conference on learning representations, ICLR 2018—conference track proceedings
12. Sahu SK, Kumar P, Singh AP (2018) Dynamic routing using inter capsule routing protocol between capsules. In: Proceedings—2018 Uksim-AMSS 20th international conference on modelling and simulation, Uksim 2018, pp 1–5
13. Pechyonkin M (2017) Understanding Hinton's capsule networks. Part III: dynamic routing between capsules

Wormhole Attack Detection in Wireless Sensor Network Using SVM and Delay Per-hop Indication

Asmita Singh, Atul Kumar Sah, Arun Singh, Bhavnesh Jaint, and S. Indu

Abstract In this age of rapid technological development, as the popularity of WSNs is increasing, wormhole attacks are also becoming more popular as a way to steal data or disrupt functionality of the system. However, small may the chance be, the aftermath of such attacks can be catastrophic in cases where WSNs are deployed in a sensitive area. In this paper, a technique which uses SVM classification along with the delay per-hop indication, to detect the nodes which are under the wormhole attack, is proposed. The use of SVM seems to be most convenient and more accurate technique for detecting failures in wireless sensor networks. The use of delay per-hop indication does not make it important for the mobile nodes to be equipped with special hardware; thus, it provides higher power efficiency. The support vector machine classification technique combined with delay per-hop indication is used to ensure a good accuracy and a minimal prediction time.

Keywords Wormhole attack · Support vector machine · Delay per-hop indication

1 Introduction

WSN in simple terms is a network of really small and low power consuming nodes called the sensor nodes which communicate within themselves through a wireless

A. Singh · A. K. Sah · A. Singh · B. Jaint (✉) · S. Indu
Delhi Technological University, New Delhi 110042, India
e-mail: bhavneshmk@gmail.com

A. Singh
e-mail: asmitasingh53@gmail.com

A. K. Sah
e-mail: atulkesha@gmail.com

A. Singh
e-mail: 0arun0singh0@gmail.com

S. Indu
e-mail: s.indu@dce.ac.in

© The Author(s), under exclusive license to Springer Nature Singapore Pte Ltd. 2021   39
K. A. Reddy et al. (eds.), *Data Engineering and Communication Technology*,
Lecture Notes on Data Engineering and Communications Technologies 63,
https://doi.org/10.1007/978-981-16-0081-4_5

medium, which records data (some physical quantity). Wireless sensor networks are becoming popular betokens by visually examining a more immensely massive area as they enhance the data analyzing capabilities of a system while being simple to establish and operate at an economically viable rate. Due to this, immensely massive wireless sensor networks can be deployed for a particular task. Such an exhilarating technology does come with its drawbacks.

In many situations, the technology can be misused by exploiting its drawbacks which can lead to catastrophic failures. In a WSN, it is most important that the data being recorded and transmitted is true. Security is a major concern as in case of an assailment on WSNs, the data will be compromised or even altered, which can lead to misinformation and various other quandaries that will follow. These intentional attacks are launched by rivals, adversaries, terrorist organizations, etc., to harm the other party. These attacks work in many ways, some blocking the data, obviating the system from working, others victual in erroneous information to discombobulate the enemy, others glom the data being recorded in the network. Wormhole attack is one such attack that can cause momentous harm to the sensor network and most importantly to the message to be transmitted across the network. Wormhole attacks are deployed against WSNs to launch a denial of service (DoS) attack and make the network slow by congesting the data flow channels, reading traffic flow and dropping packets. Authors in [1] describe the two different types of wormhole attacks, i.e., hidden and expose. In simple terms, these can be explained as: Under the hidden wormhole attack, the legitimate or normal nodes are not aware of the fact that the damaged nodes are forwarding the data packets and are participating in data transmission, whereas in exposed wormhole attack these normal nodes are aware of the fact that these damaged nodes are forwarding data packets but they do not know that these nodes are malicious. Two more malicious nodes are used to create a tunnel in the network through which packets can be transferred. These malicious nodes behave as normal nodes but disrupt communication in one way or another. These malicious nodes can transfer packets among themselves multiple times to create a delay, drop messages, advertise false routes, create an illusion of shorter paths, etc. These attacks are hard to detect as they do not disclose their identity. Cryptographic methods are also not as effective as the accomplishment of the attack is free of the cryptographic method utilized (Fig. 1).

In this paper, we propose a technique which combines the support vector machine classification and the delay per-hop indication to detect these pernicious wormhole attacks in the network and classify the nodes as damaged (under attack) and normal. SVM is a popular classification technique which can be best utilized with the right kernel function. Its ability to scale the given data points to a relatively higher dimensional area is exceptional and that is why it is being used here. It can give the best possible accuracy even when the data is unstructured or semi-structured such as text. Delay per-hop technique is one of the techniques which can be used for the detection of wormhole attacks. It is famous for the fact that it requires the minimum maintenance of the nodes and is not too expensive. The only drawback it has is that it does not work well when the number of nodes is high. That is why SVM comes into

Fig. 1 Wormhole attack in a
WSN

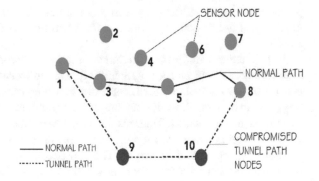

play along with this technique so that better accuracy can be obtained even when the
number of nodes is high.

2 Security Measures Against the Wormhole Attack on WSNs

2.1 Related Work

All operations in a WSN, primarily implemented in a hostile environment such as a
war zone, are autonomous and prone to many security breaches. Due to the sensitive
nature of such technology, they are always at some risk of infiltration. To date, many
researches are conducted on different malicious node detection methods in a wireless
sensor network.

A hybrid technique WRHT is used in [2] to calculate wormhole presence probability, which utilizes a double wormhole detection mechanism of calculating probability factor time delay and packet loss probability of the conventional way. In
[3], a wireless network coding scheme called algebraic watchdog provides a global
self-checking network, giving the sender an active role in checking the downstream
nodes and detecting malicious behavior probabilistically. In [4], the authors study
network coding systems and their vulnerabilities in different situations like pollution attacks and wormhole attacks, which undermine these systems' performance.
They have quantized the impact of the wormhole effect and propose an algorithm
called DAWM (distributed detection algorithm against wormhole in wireless network
coding systems). Tsitsiroudi et al. [5] proposes a visual-assisted tool called EyeSim,
a human interactive visual-based anomaly detection system for exposing security
threats in IP-enabled WSNs. Authors show through simulation results that EyeSim
is capable of detecting multiple wormhole attacks accurately in real time. Biswas et al.
[6] addresses the security concerns in mobile ad hoc networks (MANETs) and the
threat of wormhole attacks. The authors propose modifications in the AODV routing

protocol to detect and remove wormhole attacks in MANETs. WADP, an identi-fication and prevention algorithm for the wormhole, is actualized in this modified AODV.

In [7], authors study the difficulties in detecting wormhole attacks that do not require any protocol or cryptographic information to launch an attack while using private channels, making them invisible to other sensor nodes. This paper has proposed two-phase detection techniques for the detection of wormhole attacks in dynamic sensor networks. Their results indicate that the proposed approach provides a good detection accuracy. Hayajneh et al. [8] proposes a simple protocol to detect wormhole attacks called "deworm". This protocol does not require any unique equip-ment, severe location information, or synchronization. Deworm uses discrepancies in routing information between neighbors to detect wormholes. Deworm can detect wormholes with high detection rate, low false-positive rate, and low overhead while being simple localized and capable of detecting various kinds of wormholes such as physical layer wormholes. In [9], the authors also demonstrates and observes that nodes attacked by a similar wormhole are either 1-hop neighbors or 2-hop neighbors. Three nodes are non-1-hop neighbors with a high probability of the convergence of two neighbors. This algorithm gives significant detection probability, low system overhead, and low false cautions and low miss detection, which appeared through simulation results. In the portrayed work, the most crucial factor determining the effectiveness of the model proposed comes out to be the accuracy with which it detects the attack and the prediction time, and the number of nodes the wireless sensor network has.

The proposed work keeps all these factors in mind and proposes the best possible solution to increase accuracy and decrease the prediction time.

2.2 Working Model

In this section, we illustrate the working model, as this approach to determine the nodes under wormhole attack in a WSNs makes efficient use of support vector machines. We use the delay information as well as the hop count of the disjoint paths, and then delay/hop data of the nodes in the proposed network serves as the indicator for detection of wormhole attack. Under the attack, propagation across the false neighbors results in a considerably high delay as compared to that of the normal propagation path [1]. SVM is used as a tool to ensure excellent accuracy in very little time to determine the nodes under attack.

The term hop count in a network can be elucidated as the number of intermediate nodes through which a given piece of information must travel between the source and destination. Wormhole attack detection technique, presented in this paper, the delay per-hop information of each sensor node is being observed, it is also assumed that the distance between the two disjoint nodes is equal so, hop count is equal from the sender node to any of the intermediate receiving node.

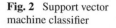

Fig. 2 Support vector
machine classifier

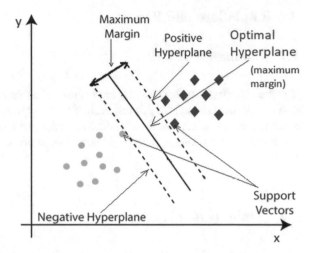

SVM determines a decision boundary that maximizes the margin between the different classes. If the data is linearly separable (as in Fig. 2) for a particular classification problem, then the hyperplane with the least training error is found. At that point, the decision boundary is adjusted in such a manner that the partition between the boundary and the data points nearest toward the hyperplane is minimized.

In this way, the maximum margin classification is used for a finer classification of the data. The support vectors are the parameters that are used for defining the hyperplanes. The algorithm works as shown in Fig. 3.

In the proposed model, the radial basis kernel function (RBF) is used to classify the nonlinear data and also provide excellent prediction time and better accuracy.

The dataset used in the proposed model consists of the information about the sensor nodes and their delay per hops at different instants of time. Whenever the delay per hop of the particular node at a particular instance is greater than the threshold value, it is considered to be under the wormhole attack. SVM is used to classify these nodes, the desired values have been obtained by means of trial and error method, and the SVM toolbox available in MATLAB is being used to carry out the research work.

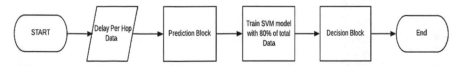

Fig. 3 Process flow of predicting the malicious nodes

3   Simulations and Results

3.1   Simulation Results

The multi-hop dataset that was prepared consisted of various parameters of the nodes, and one of them is used for this research. The delay per-hop indication of each node after a particular interval of time is being recorded, and whenever a node under attack is involved, the TAG value at that particular instant becomes '1'. Otherwise it remains '0'.

3.2   Training and Testing

The total data has been sliced into training data of 80% and testing data of 20%. During each execution, random 80% of data is used as training data and the rest of the data is used for prediction.

Figure 4 represents the time taken by algorithm for prediction of the node under attack for one of the simulations for five iterations.

Table 1 summarizes the results of different sets of nodes, i.e., its accuracy and prediction time.

Figure 5 shows the results of the simulation. It displays the comparison between the predicted label and true label, which is an error in prediction. In predicted and

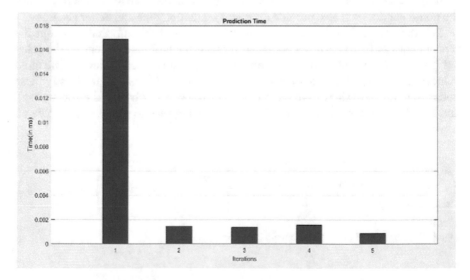

Fig. 4 Prediction time of SVM

Table 1 Prediction time for different numbers of nodes

Nodes	Average accuracy (%)	Prediction time (ms)
50	98.0	0.0049
100	98.0	0.0038
150	98.6027	0.0036
200	98.5	0.0053
250	98.8	0.0045
300	98.3333	0.0058
350	99.2471	0.0051
400	99.5	0.0052
450	99.7778	0.0051
500	99.6	0.0052

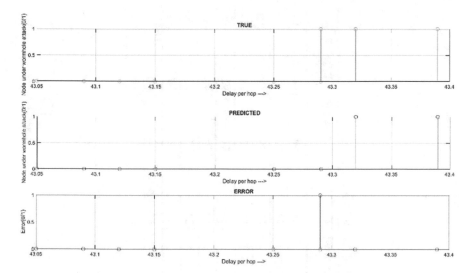

Fig. 5 Error graph

true graph, TAG '0' represents "true reading" whereas TAG '1' represents "malicious reading", while in error graph '0' stands for "no error" and '1' stands for "error".

To evaluate our proposed technique, we compared it with previously proposed wormhole detection techniques in WSNs. This comparison is essentially based on the accuracy.

Table 2 shows a comparison of our technique to others in terms of accuracy. The last column in Table 2 shows the accuracy that our technique has achieved is better than various other techniques used to date. Even when the number of nodes has increased, the model gives excellent accuracy. This implies that a support vector machine is an excellent tool that can be used in WSNs to achieve remarkable accuracy.

Table 2 Comparing accuracies of various methods of detecting the wormhole attack

Nodes	Chui and Lui [1]	Hayajneh et al. [3]	Wang et al. [4]	Kim et al. [5]	Ji et al. [6]	Tsitsiro-udi et al. [7]	Biwas et al. [8]	Patel and Agarwal [9]	WRHT [2]	SVM
50	74	80	82	85	90	92	91	94	95	98
100	75	77	81	83	87	91	93	95	97	98
150	76	79	72	85	89	92	95	94	98	98.62
200	77	82	84	87	92	94	97	98	99	98.5
250	76	81	82	84	91	92	94	94	98	98.8
300	74	82	80	82	92	90	92	93	96	99.3
350	77	79	79	84	88	89	94	96	99	99.2
400	75	79	80	82	89	90	92	95	97	99.5
450	74	82	79	83	92	89	93	93	95	99.77
500	76	77	84	87	87	94	97	95	98	99.6

4 Conclusion

In WSNs, a wormhole attack is one of the significant security threats, which results in disrupting many of the routing protocol and data theft/loss.

This paper has proposed a technique to detect a wormhole attack using the machine learning approach based on the support vector machine (SVM). Our results show that it does a great job of detecting the attack with reasonable accuracy. Also, this algorithm shows good performance within a few milliseconds.

In the future, we would like to proceed with our work on the security parts of the WSNs to make their utilization increasingly suitable in practical situations.

References

1. Chiu HS, Lui KS (2006) DelPHI: wormhole detection mechanism for ad hoc wireless networks. In: 1st international symposium on wireless pervasive computing. IEEE Press, Phuket, Thailand, pp 6
2. Singh R, Singh J, Singh R (2016) WRHT: a hybrid technique for detection of wormhole attack in wireless sensor networks. Mob Inf Syst 2016:1–13
3. Kim M, Médard M, Barros J (2011) Algebraic watchdog: mitigating misbehavior in wireless network coding. IEEE J Select Areas Commun 29(10):1916–1925
4. Ji S, Chen T, Zhong S (2015) Wormhole attack detection algorithms in wireless network coding systems. IEEE Trans Mob Comput 14(3):660–674
5. Tsitsiroudi N, Sarigiannidis P, Karapistoli E, Economides AA (2016) EyeSim: a mobile application for visual-assisted wormhole attack detection in IoT-enabled WSNs. In: 9th IFIP wireless and mobile networking conference (WMNC), Colmar, pp 103–109
6. Biswas J, Gupta A, Singh D (2014) WADP: a wormhole attack detection and prevention technique in MANET using modified AODV routing protocol. In: 9th international conference on industrial and information systems (ICIIS), Gwalior, pp 1–6
7. Patel MM, Aggarwal A (2016) Two phase wormhole detection approach for dynamic wireless sensor networks. In: International conference on wireless communications, signal processing and networking (WiSPNET), Chennai, pp 2109–2112
8. Hayajneh T, Krishnamurthy P, Tipper D (2009) DeWorm: a simple protocol to detect wormhole attacks in wireless ad hoc networks. In: Third international conference on network and system security. IEEE Press, Australia, pp 73–80
9. Wang Y, Zhang Z, Wu J (2010) A distributed approach for hidden wormhole detection with neighborhood information. In: IEEE fifth international conference on networking, architecture, and storage, Macau, pp 63–72

Performance Enhancement of Dielectric Pocket-Based Dual-Gate FinFET

Vikram Singh, Pradeep Kumar, and Aditya Mudgal

Abstract Over the past few years, MOSFET has played a tremendous role in IC technology. In this paper, we have proposed a novel device dielectric pocket-enhanced FinFET which we have is based on TFET whose characteristics are found to be better than conventional devices. Various performance parameters were compared like I_{ON}, I_{OFF}, I_{ON}/I_{OFF} ratio, subthreshold swing, transconductance and results were compared which gave us the best device out of the four we created which is DGDP. We found subthreshold swing was 0.0159547 mV/decade which is the best out of the four devices we created. This is simulation-based work performed on ATLAS device simulator.

Keywords Dual-gate dielectric pocket (DGDP) · Threshold voltage (VT) · Short channel effects (SCE) · Dual-gate without dielectric pocket (DGWDP) · Single-gate dielectric pocket (SGDP) · Single-gate without dielectric pocket (SGWDP) · Transconductance (gm)

1 Introduction

MOSFET seems promising but with Moore's law staying true, as the MOSFET is made smaller and smaller, the channel shortening leads to certain drawbacks called short channel effects. These effects occur when the depletion layer width and the channel length due to shortening become so close that they are comparable [1].

TFET is one of the modifications that enable on the subthreshold swing of less than 60 mV/decade based on multiple performance factors of traditional MOSFETs, and

V. Singh (✉) · P. Kumar · A. Mudgal
Department of ECE, ASET, Amity University, Noida, Uttar Pradesh, India
e-mail: vikram.versusv1@gmail.com

P. Kumar
e-mail: pkumar4@amity.edu

A. Mudgal
e-mail: amudgal@amity.edu

© The Author(s), under exclusive license to Springer Nature Singapore Pte Ltd. 2021   49
K. A. Reddy et al. (eds.), *Data Engineering and Communication Technology*,
Lecture Notes on Data Engineering and Communications Technologies 63,
https://doi.org/10.1007/978-981-16-0081-4_6

its construction is also similar to that of MOSFET [2]. Characteristics of TFET are its ultra-low power and ultra-low voltage, resistance to short channel effects, leakage current reduction, speed because of tunneling effect, and the ability to operate at sub-thresholds and super-threshold voltages. After considering the above factors, MOSFET can be substituted by tunnel effect field transistors (TFET) targeting energy-efficient, low-power and high-speed applications in the field of ICs [3]. Also, the basic conversion mechanism is different, making the device a good candidate for low-power electronics.

The two major advantages of TFET are as follows:

1. It does now require high power. This characteristic makes it very desirable.
2. It has a sub-threshold swing less than 60 mV/dec, which is way better in comparison with MOSFET.

Tunneling in TFETs is called band-to-band tunneling. For it to be BBT, the electron in the valence band of semiconductor tunnels is from the band gap (BG) to the conductor band without the help of a trap [2]. The BG acts as a barrier (potential) for crossing cell tunnels. In tunneling, the electron travels directly from the valence band to the conduction band without emitting or absorbing photons [3]. This is the most important point to consider because it means no energy is lost or has to be provided since no photons are absorbed or emitted. A tunneling cell conveys a change in velocity by absorbing or releasing the phonon during the tunneling process (indirect).

Fin field-effect transistor: A fin field-effect transistor (FinFET) is a type of MOSFET. It is a three-dimensional implementation of the conventional planar FET in which the gate surrounds the conducting channel (MOSFET) or tunnel (TFET) which is uplifted in a way that the gate wraps the channel or tunnel on three sides. This gives a better command on the junction [4, 5]. This implementation leads to better control over the channel as it raises and enwraps the channel with the gate providing better control over the channel which otherwise was suffering due to short channel effects which makes it hard to deplete the channel; in other words switch off the transistor [5, 6]. Using a dielectric pocket or dielectric enhancement in the FinFET provides with better characteristics as the charge-holding properties of the dielectric lead to better switching off capabilities of the three-dimensional FET device [7].

We worked on 3D implementation (FinFET) of a tunnel field-effect transistor, because besides overcoming the SCE faced in MOSFET, it has the following advantages:

1. Lower sub-threshold swing
2. Low-power consumption
3. Better characteristics.

The major roadblock in the semiconductor industry has always been power dissipation. In comparison with MOSFET, FinFET offers lower leakage current which leads to reduced static power leakage but its dynamic power leakage is high due to its higher parasitic capacitances since it is a direct function of the capacitance of gate which is higher due to the same reason of parasitic capacitances [8]. This is controlled with introducing a dielectric pocket which helps in tackling this problem and also helps in improving the device transconductance, the threshold voltage and I_{on}–I_{off} ratio. We use SiO_2 as the dielectric since it has better availability, and device parameters do not need to be altered since it has a K value of 3.9 with band gap of 9 eV as compared to very strong K values of 25 and 80 in case of HfO_2 and TiO_2. HfO_2 does have its own issues though. It is not as steady as SiO_2 and can frame unfortunate portable imperfections and experience stage changes all the more effectively. TiO_2 is also not very widely available and hard to fabricate from Ti which is a very costly process and with TiO_2 device parameters need to be drastically changed to compensate for its high dielectric constant. It is also hard to fabricate TiO_2 of thickness selected in case of SiO_2.

1.1 TFET with DP

Tunnel FET designs have been introduced to a dielectric pocket (DP) at junction between channel and source of conventional PIN configuration tunnel FET. The design proposed, i.e., tunnel FET with a DP (having dielectric pocket at junction between the source and the channel) has the potential of delivering comparatively suppressed sub-threshold swing, higher I_{ON}, lower V_{th} (threshold voltage), high I_{on}/I_{off}, higher transconductance, with a relatively lower value of parasitic capacitance. It also benefits in the suppuration of parasitic capacitance values and also helps in improvement of V_T, gm, on state current (I_{on}), I_{ON}/I_{OFF} ratio and sub-threshold swing. The benefit of proposed configuration tunnel FET with DP over the conventional structure which lacks a pocket is the tunneling rate of electron at the barrier between channel and source which has been carefully observed. It should be noted that BBT is high in case of tunnel FET with inclusion of dielectric pocket as compared to the conventional design. The cause for this improvement is the result of high field and electron concentration near the junction region between source and channel which gives off better on current values. Therefore, the inclusion of a dielectric pocket highly aids the performance of a TFET.

The paper is organized into four sections:

1. Introduction
2. Proposed design
3. Results
4. Conclusion.

First, we introduce the device in Sect. 1 and then we move on to talk about the proposed design in Sect. 2. Section 3 talks about the results obtained through the device in Sect. 2. Then we move on to talk about the conclusion in Sect. 4. After that references are shared.

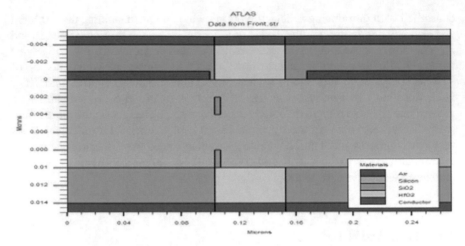

Fig. 1 Front side cross section of DGDP device

2 Proposed Design

We propose a dual-gate FinFET based on a TFET as it overcomes most of the drawbacks discussed so far. The structure of FinFET was made which results in higher output current per input voltage, also the switching speed is higher and the power consumption is lower due to lower equivalent input capacitance and channel quantization effects also due to channel quantization effects; there is better contrast in I_{ON}/I_{OFF} and also reduces short channel effects due to more effective physical separation of the source and drain regions. In comparison with MOSFET and TFET, FinFET offers lower leakage current which leads to reduced leakage of static power but its leakage of dynamic power is high due to its higher parasitic capacitances since it is a direct function of the capacitance of gate which is higher due to the same reason of parasitic capacitances. Dual-gate dielectric pocket: Figs. 1, 2, 3, 4 and 5 presented below are the basic structure of a DGDP-TFET which had been obtained through ATLAS device simulator. It is clearly visible that it is dual gate, and hence, it is called a dual-gate device. Hafnium dioxide has been used in the oxide layer under the gates, and silicon dioxide has been used under the drain and source electrodes.

3 Results

The energy band diagram of DGDP device is shown below, and it showed that DGDP device consumes less energy (Figs. 6 and 7).

The final device is the dual-gate dielectric pocket TFET-based FinFET. Below are the comparison between the transfer characteristics of four devices. It can be observed from the graph that the threshold voltage of DGDP is achieved before the

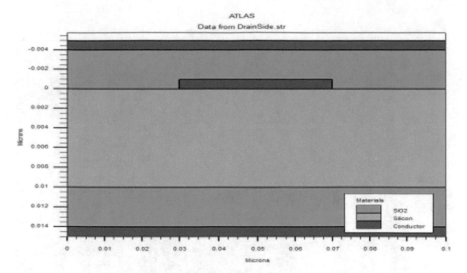

Fig. 2 Drain side cross section of DGDP device

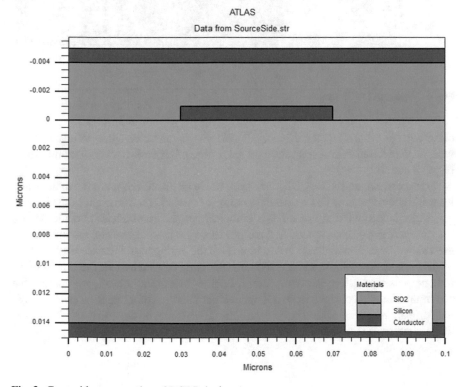

Fig. 3 Cross side cross section of DGDP device

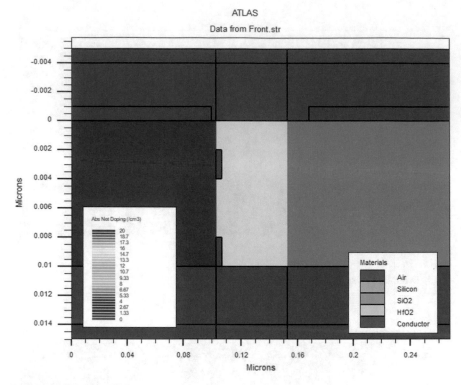

Fig. 4 Doping in DGDP device

other devices. Therefore, it can be concluded that for any given gate voltage, DGDP will result in a higher drain current than that of the other three devices in comparison (Figs. 8 and 9).

Transconductance is one of the measures to check the efficiency of the device, and its value shows us how fast a transistor can turn on and off. The higher the value of gm, faster is the switching speed. It is evident from the graph above that DGDP has the best performance compared to the other three devices. Therefore, the increase in the value of transconductance verily increases the efficiency of the device.

The graph given above depicts the concentration of electrons in the tunnel of all four devices, and observing the graph, it is clear that at any given gate voltage the concentration of electrons in a DGDP is much higher than other three devices. Below we compare different parameters of the four devices and observe that DGDP shows the best parameter values compared to the other three devices (Table 1).

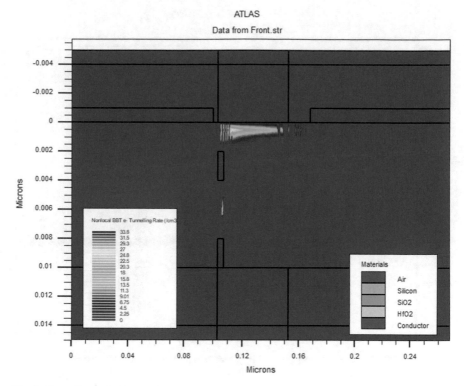

Fig. 5 Tunneling in DGDP device

4 Conclusion

The structure of dielectric enhanced dual-gate FinFET based on TFET (DGDP) was made, it results in higher output current per input voltage, also the switching speed is higher, and the power consumption is lower, and also due to channel quantization effects there is better contrast in I_{ON}/I_{OFF} as can be seen in table in Sect. 3. The voltage–current graph showed a huge improvement, and other parameters are also presented in order to support the same which can be observed in the graphs in Sect. 3. Dielectric pocket inclusion into TFET-based FinFET provides with a solution to solve various shortcomings of a conventional TFET design. This also helps in making the device ready not only for analog applications but also for the digital applications as it is highly efficient in terms of energy. The DP solution to conventional design brings with itself great potential in terms of high I_{ON}/I_{OFF}, low I_{OFF}, high transconductance (gm), high I_{ON}, lower I_{AMB}, lower SS and low threshold voltage in contrast to the conventional design.

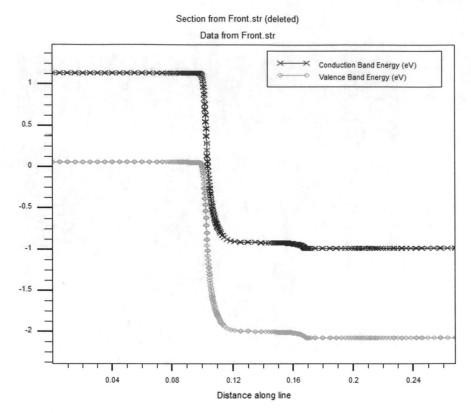

Fig. 6 Energy band diagram of DGDP device

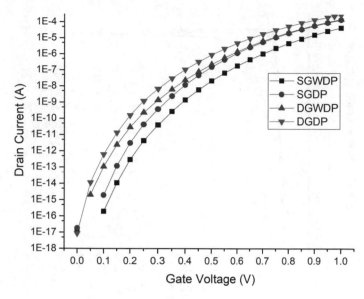

Fig. 7 Comparison of the transfer characteristics of four devices

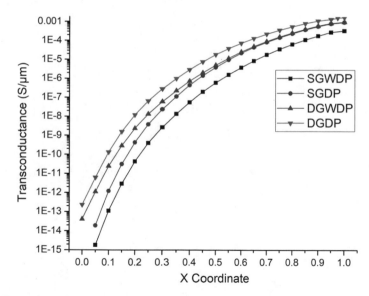

Fig. 8 Comparison of the transconductance of four devices

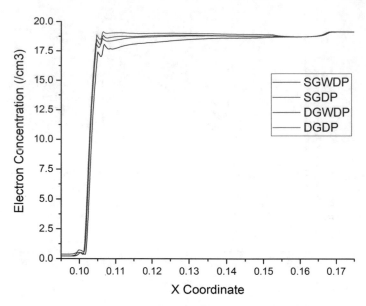

Fig. 9 Comparison of the electron concentration of four devices

Table 1 Comparison of parameters of devices

Parameters	SGWDP	SGDP	DGWDP	DGDP
Threshold voltage	0.570883	0.482088	0.461596	**0.398037**
Subthreshold swing	0.0285025 V/decade	0.0185301 V/decade	0.0257151 V/decade	**0.0159547 V/decade**
I_{OFF}	1.40535e−017	1.83483e−017	−2.31115e−017	**8.39457e−018**
I_{ON}	4.34569e−005	0.000136612	0.000145863	**0.000275882**
I_{ON}/I_{OFF}	3.09225e+012	7.44549e+012	6.31127e+012	**3.28643e+013**
Avss	0.0579446	0.049514	0.0479024	**0.0395035**

References

1. Wang X, Peterson J, Majhi P, Gardner MI, Kwong DL (2005) Impacts of gate electrode materials on threshold voltage (V/Sub th/) instability in nMOS HfO/Sub 2/Gate stacks under DC and AC stressing. IEEE Electron Device Lett 26(8):553–556
2. Mohata D, Rajamohanan B, Mayer T, Hudait M, Fastenau J, Lubyshev D, Liu AW, Datta S (2012) Barrier-engineered arsenide-antimonide heterojunction tunnel FETs with enhanced drive current. IEEE Electron Device Lett 33(11):1568–1570
3. Narang R, Saxena M, Gupta M (2014) Switching performance analyses of gGate material and gate dielectric engineered TFET architectures and impact of interface oxide charges. In: 2014 2nd international conference on devices, circuits and systems (ICDCS). IEEE, pp 1–6
4. Kang CY, Sohn C, Baek RH, Hobbs C, Kirsch P, Jammy R (2013) Effects of layout and process parameters on device/circuit performance and variability for 10 nm node FinFET technology. In: 2013 symposium on VLSI technology. IEEE, pp T90–T91
5. Colinge JP (2008) FinFETs and other multigate transistors. Springer, New York
6. Guillorn M, Chang J, Bryant A, Fuller N, Dokumaci O, Wang X, Newbury J, Babich K, Ott J, Haran B, Yu R (2008) FinFET performance advantage at 22 nm: an AC perspective. In: 2008 symposium on VLSI technology. IEEE, pp 12–13
7. Das R, Baishya S (2017) Dual stacked gate dielectric source/oxide overlap Si/Ge FinFETs: proposal and analysis. In: 2017 devices for integrated circuit (DevIC). IEEE, pp 66–70
8. Gill A, Madhu C, Kaur P (2015) Investigation of short channel effects in bulk MOSFET and SOI FinFET at 20 nm node technology. In: 2015 annual IEEE india conference (INDICON). IEEE, pp 1–4
9. Mookerjea S, Mohata D, Mayer T, Narayanan V, Datta S (2010) Temperature-dependent I–V characteristics of a vertical $In_{0.53}Ga_{0.47}As$ tunnel FET. IEEE Electron Device Lett 31(6):564–566

Cooperative Spectrum Sensing for PU Detection in Cognitive Radio Using SVM

Surendra Solanki, Vasudev Dehalwar, and Jaytrilok Choudhary

Abstract In the world of cognitive radio, spectrum sensing is among the challenging and essential tasks. For dynamic allocation of spectrum, the primary task is the detection of the primary user's (PU) presence over the spectrum. Primary users can be detected through a single node (CR user) but there are various limitations in this method like multipath fading, hidden terminal problem and shadowing. For overcoming these limitations, cooperative spectrum sensing was proposed. In this paper, a cooperative spectrum sensing (CSS) based on the support vector machine (SVM) is being proposed. Firstly, we have generated a data set through energy detection method containing energy vector of a signal as a feature. Random over sampler is further used in order to balance the classes. After balancing, we have used various machine learning algorithms and made comparison between them to classify if a primary user is present over the spectrum. The comparison clearly shows that the best accuracy is given by the proposed SVM algorithm for classifying the presence of primary user. Simulation results also show the better capability, robustness and superior efficiency of SVM method as compared to other algorithms over the detector which we have used in this paper.

Keywords CR · SVM · PU · SU · Spectrum sensing · Energy detection

1 Introduction

With the evolution of communication technologies, the demand for wireless spectrum has increased. Due to large-scale development of the 5G network, rapid growth of

S. Solanki (✉) · V. Dehalwar · J. Choudhary
Department of CSE, Maulana Azad National Institute of Technology, Bhopal, India
e-mail: Surendra1.iet@gmail.com

V. Dehalwar
e-mail: vasudevd@manit.ac.in

J. Choudhary
e-mail: jaytrilok@manit.ac.in

© The Author(s), under exclusive license to Springer Nature Singapore Pte Ltd. 2021
K. A. Reddy et al. (eds.), *Data Engineering and Communication Technology*,
Lecture Notes on Data Engineering and Communications Technologies 63,
https://doi.org/10.1007/978-981-16-0081-4_7

Internet of things, cyber-physical communication and numerous emerging technologies, requirement and demand of wireless spectrum has become urgent [1, 2]. It has been observed through studies in 2003 by the Federal Communications Commission (FCC) that there exists great underutilization of spectrum [3]. Because of static allocation schemes, there is spectrum underutilization and spectrum scarcity problem. Therefore, some dynamic allocation management scheme is required which can identify underutilized spectrum; hence, cognitive radio has been suggested as an opportunistic spectrum usage technology that can mitigate the spectrum underutilization [4].

Cognitive radio (CR) appeared as a promising technology allows secondary users (SUs) to access the licensed band without having an impact on the primary user transmission. Major function of the cognitive radio is spectrum sensing, which is a very important function for gaining real-time awareness of spectrum utilization along with the presence of licensed users [5, 6].

In recent years, to address the spectrum sensing task, the attraction for artificial intelligence (AI) and machine learning (ML) application to the CR network has increased significantly. In [7], the authors proposed ANN-based spectrum sensing method, which use energy of signal and likelihood ratio test statistic as a feature to train a model, and this method enhanced the performance compared to energy detection method. Tang et al. [8] also proposed ANN-based spectrum sensing method which used energy and cyclostationary features of signal as features for training an ANN, model improving the detection performance in low SNR condition. In [9], authors proposed a deep CSS method based on CNN which operates without regard for the type of sensing decision at individual SUs. It can be either hard combining or soft combining, and through the simulation it is proved that proposed method achieves higher sensing accuracy compared to conventional approaches.

Although, the neural network-based methods work well for cooperative spectrum sensing, it suffers the inherent problem of overfitting. The models are basically black boxes along with being computationally expensive and time consuming. Support vector machines, on the other hand, perform quite well for classification and are memory efficient. Due to this, here we intend to focus on SVMs for the task of CSS. Authors in [10] proposed a novel algorithm based on machine learning techniques for pattern classification. supervised (SVM, weighted KNN) and unsupervised (K-means clustering, GMM) are implemented for CSS. Energy vector received at the CR device is considered as feature vector and then fed into the classifier for categorization of channel available class and channel not available. Training time, classification delay and ROC performance have been used for classification algorithm comparison. In [11], the authors proposed SVM and clustering-based method in which a probability vector with low dimensionality is used as a feature vector.

The motivation behind consideration of cooperative spectrum sensing is correct identification of underutilized spectrum and improvement of the detection performance of PU. This paper makes the major contributions of preparation of dataset using energy detection method and proposing SVM-based detector for prediction. Comparison of these classifiers based on the performance measures is done using confusion matrix, accuracy, area under curve of receiver operating characteristics (ROC-AUC) and receiver operating characteristics (ROC).

The remainder of the paper has been organized as follows. Section 2 discusses system design and local sensing techniques. Proposed model is discussed in Sect. 3, which includes data preparation, data pre-processing and classification model. Section 4 discusses the results, and the conclusion is represented in Sect. 5.

2 System Model and Assumption

A cognitive radio network (CRN) is considered, consisting of a primary user and N secondary users indexed by $n = 1, \ldots, N$ which are randomly distributed in a given area. For CSS, each secondary user evaluates energy level and does the reporting to the central node (fusion centre) which is responsible for information combining and making global decision about the presence of PU. This paper presents a scheme for centralized CSS in which all the secondary users transmit their decision to fusion centre (FC), and energy detection technique is utilized as spectrum sensing at each SU node.

The energy detector is a blind detector that is independent of the signals' features. This is a simple method to quickly determine the presence of signal present channel, except, for weak signal case. For estimation of the energy level, SU performs energy detection. Energy detector's operation is based on the binary hypothesis testing problem which is performed by every SU for deciding whether the PU is present or absent. Statistically, from the binary hypothesis test, the inferences include H_0, which represent PU absence and H_1, representing PU presence. The ith SU's received signal at n sample index is written as follows:

$$x_i(n) = \begin{cases} w_i(n) H_0 \\ h_i(n)s(n) + w_i(n) H_1 \end{cases}$$

where $w_i(n)$ is AWGN, $s(n)$ is PU signal, and $h_i(n)$ is the channel gain between PU and SUs.

3 SVM-Based Cooperative Detection

The paper proposed a cooperative spectrum sensing based on SVM classifiers. Figure 1 denotes the block diagram of the SVM-based cooperative detection method. The proposed model is described further in the following sections divided into three subsections. First section describes spectrum sensing simulation and data set generation while the second part is data set pre-processing. After the pre-processing, in last section, PU signal is classified by SVM classifier.

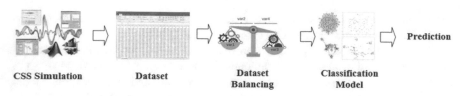

CSS Simulation   Dataset   Dataset   Classification
                                             Balancing   Model   Prediction

Fig. 1 Block diagram representing the proposed method

3.1 CSS Simulation and Data Generation

For dataset generation, a cognitive radio network is developed, with the help of
MATLAB, which consists of a single PU and N SUs. Monte Carlo simulation is
used, simulating spectrum sensing techniques and generating 10,000 signals, each
having 2000 samples. Each SU user gets this PU signal, and with the help of energy
detection method energy of signal is calculated. Now, comparing this energy to
threshold, the signal is classified as PU signal or noise only signal. After the decision
is taken over signal, binary information is sent to FC. Now, the FC is responsible
for taking global decision by using AND rule. This global decision corresponding
to each signal is saved, thus preparing the data set.

3.2 Balancing of Data Set

The data set that is prepared consists of the class imbalance problem, i.e., the classes
are not present in an equal and balanced manner. As such, for pre-processing and
balancing, 'random oversampling' method is used. Random oversampling balances
the data by oversampling the minority class randomly, which implies that using the
majority class the instances corresponding to the minority class are increased by
replicating them to a specific rate. This method does not lead to any information loss
which is an added advantage.

3.3 Classification of Signal

Once the dataset is pre-processed, the proposed SVM classifier along with various
classification algorithms for comparison such as KNN, random forest, decision tree,
Naive Bayes and logistic regression is used to classify the PU signal from noise
only signal over the spectrum. Grid search is used for parameter tuning of various
classification algorithms. Figure 2 represents the classification model adopted in our
proposed mechanism.

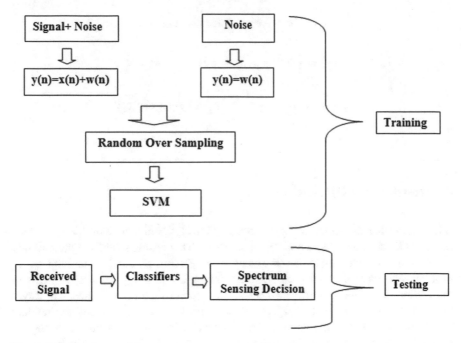

Fig. 2 Classification model

SVM Classifier. Support vector machine, unlike other machine learning models, is not dependent on data dimensionality [12]. It consists of special components, kernel and decision boundary. The decision boundaries are constructed across the data for the separating hyperplane to be formed with the support vectors for maximizing the distance between the classes during the task of classification. The data may not be always separable linearly due to which the kernel trick was introduced. Kernels perform the transformation of the input vectors to a feature space, i.e., it takes the data from a lower dimension to higher dimensions. Kernels can be of different types, the radial basis function (RBF), the linear kernel or the polynomial kernel. Kernel trick is utilized by the support vector machines for finding optimum boundary. This optimum boundary is called the decision boundary as it has the maximum distance from input instances of the different classes. Hence, effective classification of the data is done according to this boundary. For an input, say x, to the SVM belonging to either class (+1) or (−1), the mapping into a feature space is done firstly. For training vectors, $i = 1, \ldots, n$, belonging to two classes:

$$\hat{Y}(x) = W^{\mathrm{T}}\phi x + b \begin{cases} \geq 1 \text{ if } Y = 1 \\ \leq 1 \text{ for } Y = -1 \end{cases}$$

where (+1) and (−1) on the RHS can be denoted by α and $-\alpha$.

$$Y_i\left(W^{\mathrm{T}}\phi x + b\right) > 1 \text{ for } i = 1, \dots, n$$

$$\hat{Y}(x) = \left(W^{\mathrm{T}}\phi x + b\right) = c \text{ for } -1 < c < 1$$

$$\arg\max_{\alpha} \sum_{j} \alpha_j - \frac{1}{2}\sum_{j,k} \alpha_j, \alpha_k y_j y_k\left(x_j \cdot x_k\right)$$

4 Results and Discussion

Initially, the data set which was generated had the class imbalance problem in which the minority class is '1' and majority class is '0'. This problem leads to the majority class prediction resulting in poor accuracy. In order to eliminate this class imbalance problem, we used random over sampler. Tables 1 and 2 represent the results of the various experiments performed where LR represents logistic regression, Bayes denotes Bayes classifier, DT represents decision tree, and RF represents random forest. Table 1 shows the accuracy of various classifiers with random oversampling as well as without random oversampling. As shown in Fig. 5, it can be clearly seen that the very high increase in accuracy of random forest classifier and SVM

Table 1 Performance comparison of various machine learning algorithm

	LR	Bayes	DT	KNN	RF	SVM
Without	65.44	53.83	62.96	61.34	74.28	75.00
With	65.52	64.31	81.26	62.65	94.60	95.23

Table 2 Performance comparison of various machine learning algorithm

Performance metrics/classifiers	LR	Bayes	DT	KNN	RF	SVM
Sensitivity/true positive rate (TPR)	70.00	63.27	93.53	75.22	90.19	90.55
Specificity/True negative rate (TNR)	60.96	65.37	68.77	49.86	99.78	100
Precision/positive predictive value (PPV)	64.61	65.04	75.30	60.43	99.76	100
Accuracy	65.52	64.31	81.26	62.65	94.94	95.23
F1 score	67.20	64.14	83.43	67.02	94.73	95.04
AUC	65.48	64.43	81.15	62.54	94.99	95.27
Training time (s)	07.78	66.31	43.62	0.79	04.14	831.20
Prediction speed (s)	7.5436	4.5291	9.9456	0.0384	3.4554	0.0368

Fig. 3 Confusion matrix of SVM

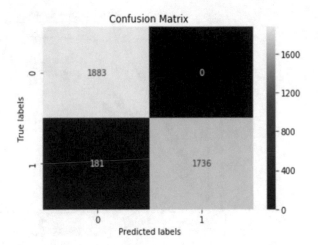

classifier is after random oversampling. The average increase in the accuracy is 12.3% considering all the six classifiers which we have compared.

For comparison of the results of various machine learning algorithms to detect the presence of the primary user, we have used various parameters which are accuracy, sensitivity, specificity, precision, false positive rate, F-1 score and AUC. After scrutinizing the results, we have concluded that the proposed SVM model gives best accuracy among all other machine learning model as shown in Table 2. SVM gives the best accuracy because the data set which we have used is having two classes (0/1), i.e., either a primary user is present or absent over the spectrum. SVM gives better results as we can easily tune the parameters of the algorithm. We have also compared the computational cost of various machine learning algorithms as shown in Table 2.

The confusion matrix for SVM model with random oversampling classification is shown in Fig. 3. The total number of instances which is used in test data set is 3800. The correct classification is done for 3619 test data set instances while the incorrect ones are only 181 instances of test data set. Figure 4 is the ROC curve of various machine learning algorithms which we have compared with each other. The ROC curve for SVM is slightly better as compared with the curve for random forest, showing the better classification of two classes. Hence, in order to classify whether a primary user is present or not present, SVM classifier gives the best result as compared to other classifiers. Although the computational time of SVC is greater than the other classifier, but the prediction speed is very less. Hence, our proposed model can be operated in real-time work (Fig. 5).

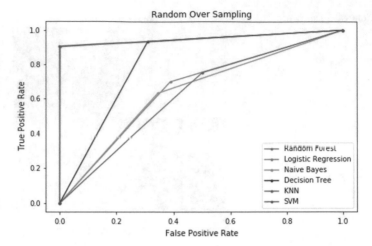

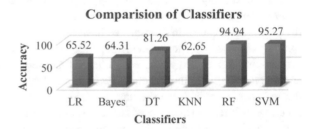

Fig. 4 ROC curve of different classifiers

Fig. 5 Comparison of different classifiers with SVM

5 Conclusions

The detection ability of the PU signal in an environment with low SNR is very important for cognitive radio. Traditionally, various techniques of spectrum sensing were already proposed with limitations which include computation cost and primary user detection in low SNR conditions. In this research, a novel SVM classifier-based PU detection method is proposed. The results prove that the proposed classifier can be operated in real time with better accuracy as compared to other methods.

Reference

1. Gupta A, Jha RK (2015) A survey of 5G network: architecture and emerging technologies. In: IEEE Access, pp 1206–1232
2. Al-Fuqaha A, Guizani M, Mohammadi M, Aledhari M, Ayyash M (2015) Internet of things: a survey on enabling technologies, protocols, and applications. In: IEEE communications surveys and tutorials, pp 2347–2376

3. Cabric D, Mishra S, Brodersen R (2004) Implementation issues in spectrum sensing for cognitive radio. In: Conference record—Asilomar conference on signals, systems and computers
4. Marcus MJ (2012) Spectrum policy for radio spectrum access. In: Proceedings of the IEEE, pp 1685–1691
5. Mitola J, Maguire GQ (1999) Cognitive radio: making software radios more personal. In: IEEE personal communications, pp 13–18
6. Haykin S (2005) Cognitive radio: brain-empowered wireless communications. IEEE J Select Areas Commun 201–220
7. Vyas MR, Patel DK, Lopez-Benitez M (2017) Artificial neural network based hybrid spectrum sensing scheme for cognitive radio. In: 28th annual international symposium on personal, indoor, and mobile radio communications (PIMRC). IEEE, Montreal, pp 1–7
8. Tang Y, Zhang Q, Lin W (2010) Artificial neural network based spectrum sensing method for cognitive radio. In: 6th international conference on wireless communications networking and mobile computing (WiCOM). Chengdu, pp 1–4
9. Lee W, Kim M, Cho D (2019) Deep cooperative sensing: cooperative spectrum sensing based on convolutional neural networks. IEEE Trans Vehic Technol 3005–3009
10. Thilina KM, Choi KW, Saquib N, Hossain E (2013) Machine learning techniques for cooperative spectrum sensing in cognitive radio networks. IEEE J Select Areas Commun 2209–2221
11. Zhu Y, Lu P, Wang D, Fattouche M (2016) Machine learning techniques with probability vector for cooperative spectrum sensing in cognitive radio networks. In: IEEE wireless communications and networking conference. Doha, pp 1–6
12. Goodfellow L, Bengio Y, Courville A (2015) Deep learning. MIT Press

Number Plate Detection Using Morphology and Geometrical Properties

Vangala Sneha and Narasimha Reddy Soora

Abstract With the endless increase in number of vehicles, it is becoming extremely tedious to detect License Plate (LP) for security, traffic management, parking control, etc. To solve these issues, numerous LP detection techniques have been proposed in recent times, and most of them have failed to succeed when there is a complex background or in illumination and they need some enhancements. So, we have proposed a new LP detection technique to detect the multiple LPs of vehicle regardless of background of number plate which is from different angles. This designed system may improve for those who are violating the stop line, disobey traffic rules, and raise serious threat to pedestrians which are major concerns. The proposed technique was tested using 200 images with numerous LP variations and achieved good results. Apart from our own data set, we have used publicly available Medialab benchmark LP data set having more than 740 images and achieved good results.

Keywords Number plate detection · Geometrical region shapes · Bernsen local image thresholding · Mathematical morphology operations

1 Introduction

Vehicles play a crucial role in modern transportation systems as there is an epidemic increase in vehicle usage. Due to this, traffic management can turn into a huge issue to control and leads to an increase in various crimes like violating traffic rules and regulations and might lead to accidents, theft of vehicles, hit and ran accidents, etc. Manual monitoring of vehicles is hefty and error-prone because of weak visual perception and unreliable human memory. Therefore, adopting a number plate detecting system

V. Sneha (✉) · N. R. Soora
Department of Computer Science and Engineering, Kakatiya Institute of Technology and Science, Warangal, Telangana, India
e-mail: snehavangala789@gmail.com

N. R. Soora
e-mail: snreddy75@gmail.com

is necessary. LP detection system [1] is helpful in many sectors like vehicle thefts, toll collection, traffic monitoring, parking lot areas, etc.

The image which has a complex background with an uneven illumination is considered as noise. When these kinds of images are captured, the noise will affect the LP extraction success rate. So, the work in this paper is to detect the LP by removing noise with the help of Bernsen algorithm. The proposed system works as follows: Cameras captured images will be stored temporarily for further processing. The image will be processed using preprocessing techniques such as Bernsen algorithm and morphological image processing (IP) techniques to remove the noise and extracts the probable LPs with the help of geometrical properties (GPs) of LPs. Image preprocessing techniques such as Bernsen algorithm and morphological IP techniques are used as filtering techniques to remove unwanted things and to improve the images for further processing. By adapting various preprocessing techniques for LP detection, LP problem can be justified by various researches which are helpful to attain the ultimate goal of detecting LP from an image and achieve acquired results.

Image preprocessing techniques enhance the image quality. It is helpful for humans by improving the visual effect of the image using preprocessing techniques. For IP, the input may be a low-quality image and output will be an effective and quality improved image.

2   Literature Review

In this section, we describe some of the well-known LP detection and recognition methods from the literature and the features used to detect and extract the LP from the input images.

Work in the paper [2] is regarding the detection and recognition of the number plate. This paper used various edge detection algorithms to detect the edges and applied various normalization techniques to remove unwanted edges. Here recognition of the detected LP is also developed. In the papers [1, 2], object recognition was performed by using template matching. Work in the paper [3] is about the detection of the LP with the help of distinct morphology operations and with set operations like dilation and erosion. Work in the paper [4] explains how IP techniques are crucial for analysis, extraction of LP, and noise removal. Through Hough transform technique, Hough lines are opted from the grayscale image in paper [5], and this is helpful for determining the edges.

The techniques in the papers [6–8] make utilization of morphological operations along with other technique for LP localization, however, the papers [3] and [4] specifically used for morphological processing. In the paper [9], an LP detection system was developed and helpful for finding the vehicles crossing the red line. A work on Sobel edge detection is used in this work. Authors Sarbjit Kaur and Sukhvir Kaur presented a paper [10] about iterative bilateral filtering for noise removal, an adaptive histogram, and morphological operations for enhancements and used Sobel edge detection to find edges of the probable LPs from an input.

Optical character recognition (OCR) technique is to analyze the image of the vehicle LP for its character recognition. Work in the paper [11, 12] is about OCR with the usage of IP techniques. The authors in the paper [12, 13] proposed novel feature extraction techniques for OCR and reported good results using publicly available benchmark data sets. Authors in the paper [14] applied various filtering techniques to remove the noise and further applied the Sobel edge detection algorithm but this may not solve the problem and further used dilation and erosion to improve the results.

Work in the paper [15] is highlighted about how morphological IP techniques and Sobel operators are implemented to detect and extract the LP regions. Through Sobel operator, vertical edges were detected. Work on paper [16] explains how to extract the LP with the help of adaptive thresholding algorithm to convert to binary and applied intensity unwanted-line elimination algorithm (ULEA) and vertical edge detection for desired LP detection. Work in the paper [17] is about LP detection using various geometrical-based clustering techniques which are scale, color, and rotation independent. It works for the image taken in any weather conditions and is able to detect one or more than one LP from the given single input image.

We observed that most of the papers used edge detection, morphological IP operations, and color information to detect more than one LP and learned so many intestine topics.

3 Proposed System

In the literature, there were so many LP detecting systems, and many of them used edge detection systems, different morphological operations, clustering techniques, genetic algorithms, IoT, etc., to improve performance. Based on the existing systems pitfalls, we have proposed this design which takes less time for finding more than one LP from an input image using morphological IP techniques and GPs of LPs. The flowchart of the proposed method is shown in Fig. 1.

We have used Bernsen local image thresholding technique to convert grayscale image to a binary image for removing noise, shadows, and illuminations. Morphological IP techniques like dilation and tophat are applied to the image to filtering noise and for enhancement of the input image. To the filtered image, we have applied geometrical technique to find rectangular region to find probable LPs. Through this approach, we can detect multiple LPs of different vehicles present in an input image.

Steps:

1. Read the image and convert the input RGB image to a gray scale. Usually, a grayscale image has an 8-bit color format, whereas the color image has a 24-bit representation with 8-bits for red, 8-bits for blue, and 8-bits for green information.

Fig. 1 Flowchart for LP
detection from an input
image

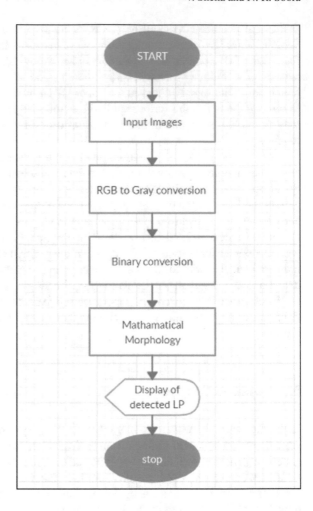

2. Grayscale image must be converted to binary image using Bernsen's local image thresholding to remove uneven illuminations or shadows for further processing which reduces computational complexity.

3. Next, apply morphological IP operations (dilation, erosion, and tophat) for eliminating contrast features and detect required components from an image by filling or removing small holes through preserving the size and shape of the objects in an image.

4. Next, find the GPs of the components of the binary image and find the components which are rectangle in nature and crop the rectangular components which are the probable LPs.

Fig. 2 Example of input image

Fig. 3 Grayscale image

3.1 RGB to Gray Conversion

Convert the color map image shown in Fig. 2 to a grayscale image shown in Fig. 3 using the rgb2gray() function for eliminating the hue and saturation information. A grayscale image is an image with the combination of shades from white to black present in it. The gray color has a similar intensity for all RGB components in RGB space. So, instead of specifying three intensity values for each pixel, it is required to specify a single intensity value for a pixel in the color image.

To store a color map image, we need 8 * 3 = 24 bits (8 for each component RGB) but when we convert to grayscale image, only 8-bit is required to store a single picture of the image. Grayscale images are effortless to work with any morphological operations or image segmentation. Convert the grayscale image shown in Fig. 3 to binary image shown in Fig. 4 using Bernsen local image thresholding for removal of shadow and illuminations. This method uses a user-provided contrast threshold.

3.2 Morphological Image Processing Techniques

Mathematical morphology (MM) or morphological IP techniques are used to analyze the input image and to find the shapes of various geometrical structures like rectangle,

Fig. 4 Binary image

square, and so on and also used as a filter to remove noise present in the image. MM operations were originally developed for binary images and were later extended to grayscale images. When morphological operations are applied on the image, each pixel value is adjusted based on its neighborhood pixels. Based on the size and shape of the input image, we can modify the image and can retrieve the required components. Morphological IP techniques deal with tools for extracting required values, information from the image that is useful in the description of shape and representation.

By erosion, dilation, tophat, and morphological IP techniques, we get filtered image so that we can extract the required components or shapes from the input image as shown in Fig. 5. Erosion, dilation, and tophat are shown in Eqs. (1), (2), and (3), respectively.

$$A(-)B = \left\{ z/(B)_z \bigcap A^c = \phi \right\} \tag{1}$$

$$A \oplus B = \left\{ z/\left(\hat{B}\right)_z \bigcap A \neq \phi \right\} \tag{2}$$

$$T(A) = A - (A(-)B) \oplus B \tag{3}$$

Fig. 5 Resultant image after morphological IP techniques (erosion, dilation, and tophat) applied

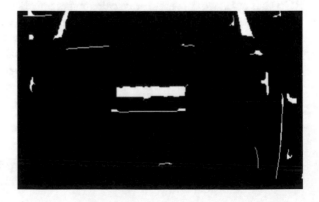

where in Eqs. (1), (2), and (3), A is the input image and B is the structuring element (SE). Erosion is denoted by $(-)$ the symbol as shown in Eq. (1), and dilation is denoted by $(+)$ symbols as shown in the Eq. (2). Equation (1) shows the erosion of the input image A by the SE B results in the set of all points z contained in A. Equation (2) shows the dilation of the input image A by the SE B results in reflecting B about the origin and shifting this reflection by z. Equation (3) shows tophat of input image A by SE B.

Next step is to find the expected LP/LPs from the filtered image as shown in Fig. 5 by applying region properties and extracting the components which are rectangle in nature. This will result in the component with probable LP/LPs. Next find whether it contains few more components in the region extracted. The components with few objects present in it are our expected LP/LPs region for the given input image. The method proposed in this paper failed to detect the LP or LPs if the LP is not rectangle in nature or LPs are distorted.

4 Experimental Results

In this paper, we have developed a novel LP detection system with the help of morphological IP techniques and GPs of the LPs present in the image. The proposed system is tested using proprietary data set with more than 200 images with different LP variations and found around 97% recognition accuracy. We have used Eq. (4) to find the recognition accuracy of the proposed system where M is the total number of plates successfully for detection and N is the total number of plates present in the input images. In order to test the efficiency of the proposed LP detection methods, we have used publicly available Medialab LP benchmark data set [18] and achieved 92% recognition accuracy. The recognition accuracy using Medialab License Plate benchmark data set is less as compared to the paper [17], but the processing time for the proposed system for finding LP/LPs is very fast as compared to [17] as shown in Table 1.

$$\text{accuracy} = M/N \qquad (4)$$

This system may fail to detect single or multiple LPs from the input image when there is the LP region not properly visible, or LP regions contain lot of illumination. In these situations, the system may fail to extract LP successfully. Figure 6 shows an example output of the LP where the LP is not visible properly in the input image. The experiments on the data sets were conducted on an Intel Core i5 using MATLAB R2019b.

The proposed system detects multiple LPs in a single image. The proposed system took on an average 9 s to detect the LP and took on an average 14 s to detect two LPs present in an image, whereas the paper [17] took almost more than 10 s to detect the LP and took more than 20 s to detect two LPs present in the image. The proposed system is not matched with the performance result of the paper [17] using publicly

Table 1 Proposed LP detection method performance comparison with competitive method from the literature

Reference	License plate data set	No. of images in license plate data set	Vehicle types present in the LP data set	Accuracy (%)	Time taken for LP detection
[17]	Medialab LP benchmark data set	741	cars, vans, and trucks	97.3	10 s for images with one LP. 20 s for images with more than one LP
[17]	Proprietary LP data set	159	trucks, cars, and motorcycles	98.74	Not mentioned
Proposed method	Medialab LP benchmark data set	741	vans, trucks, and cars	92	9 s for images with one LP 14 s for images with more than one LP
Proposed method	Proprietary LP data set	200	cars and motorcycles	97	–

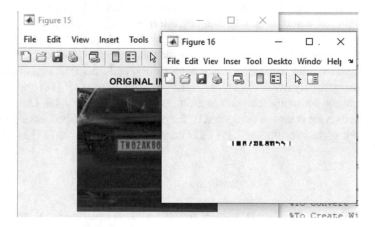

Fig. 6 Example output if the LP is not visible properly in the input image

available Medialab benchmark LP data set, but it took less time for LP detection as compared to paper [17]. Medialab LP benchmark data set contains maximum of three LPs in an image. Figure 7 shows few of the results from the proprietary data set.

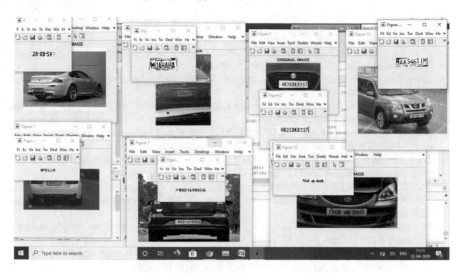

Fig. 7 Few of the results from the proprietary data set

5 Conclusion

This paper presents the result of an algorithm developed for a successful working of LP detection. This paper uses Bernsen binary conversion algorithm to convert the gray scale to binary image and applied dilation, erosion, and tophat morphological operations to filter out unwanted components from the binary image. On the resultant image, we have extracted the GPs to get the rectangular regions to find the probable LP/LPs. The proposed system's performance was tested using proprietary data sets and publicly available Medialab LP benchmark data sets with different LP variations and achieved encouraging results.

References

1. Puranic A, Deepak KT, Umadevi V (2016) Vehicle number plate recognition system: a literature review and implementation using template matching. Int J Comput Appl 134(1):12–16
2. Saleem N, Muazzam H, Tahir HM, Farooq U (2016) Automatic license plate recognition using extracted features. In: 2016 4th international symposium on computational and business intelligence (ISCBI). IEEE, pp 221–225
3. Mohanan N, Ahmad A, Al-Busaidi SS, Khiriji L, Abdulghani A, Al Nadabi M (2018) Use of mathematical morphology in vehicle plate detection. Oriental J Comput Sci Technol 11(4):195–200
4. Islam R, Sharif KF, Biswas S (2015) Automatic vehicle number plate recognition using structured elements. In: 2015 IEEE conference on systems, process and control (ICSPC). IEEE, pp 44–48
5. Prabhakar P, Anupama P, Resmi SR (2014) Automatic vehicle number plate detection and recognition. In: 2014 international conference on control, instrumentation, communication

and computational technologies (ICCICCT). IEEE, pp 185–190
6. Ktata S, Benzarti F (2012) License plate detection using mathematical morphology. In: 2012 6th international conference on sciences of electronics, technologies of information and telecommunications (SETIT). IEEE, pp 735–739
7. Hung KM, Hsieh CT (2010) A real-time mobile vehicle license plate detection and recognition. J Sci Eng 13(4):433–442
8. Krishna PS (2015) Automatic number plate recognition by using MATLAB. Int J Innov Res Electron Commun (IJIREC) 2:1–7
9 Saha S, Basu S, Nasipuri M, Basu DK (2009) License plate localization from vehicle images: an edge based multi-stage approach. Int J Recent Trends Eng 1(1):284
10. Kaur S, Kaur S (2014) An efficient approach for number plate extraction from vehicles image under image processing. Int J Comput Sci Inf Technol 5(3):2954–2959
11. Sulaiman N, Jalan SNHM, Mustafa M, Hawari K (2013) Development of automatic vehicle plate detection system. In: 2013 IEEE 3rd international conference on system engineering and technology. IEEE, pp 130–135
12. Soora NR, Deshpande PS (2015) Robust feature extraction technique for license plate characters recognition. IETE J Res 61(1):72–79
13. Soora NR, Deshpande PS (2017) Novel geometrical shape feature extraction techniques for multilingual character recognition. IETE Tech Rev 34(6):612–621
14. Owens R Mathematical morphology. Computer Vision Lectures. https://homepages.inf.ed.ac.uk/rbf/CVonline/LOCAL_COPIES/OWENS/LECT3/node3.html
15. Lalimi MA, Ghofrani S, McLernon D (2013) A vehicle license plate detection method using region and edge based methods. Comput Electr Eng 39(3):834–845
16. Al-Ghaili AM, Mashohor S, Ramli AR, Ismail A (2012) Vertical-edge-based car-license-plate detection method. IEEE Trans Veh Technol 62(1):26–38
17. Soora NR, Deshpande PS (2016) Color, Scale, and rotation independent multiple license plates detection in videos and still images. In: Mathematical problems in engineering
18. Anagnostopoulos IE, Psoroulas ID, Loumos V, Kayafas E, Anagnostopoulos C-NE Medialab LPR database. Multimedia Technology Laboratory, National Technical University of Athens, https://www.medialab.ntua.gr/research/LPRdatabase.html

Fast and Robust Segmentation of Images Using FRFCM Clustering with Active Contour Model

Dupati Ravali and E. Suresh

Abstract The segmentation of images is the basis on which it is immediately implemented by the computer, and within the comprehension of images, it is still in the correct position. For image segmentation, there are distinct approaches. The hottest tool used for image segmentation is grouping. Since the fuzzy c-means clustering (FCM) algorithm is noise-sensitive, the normal feedback of local spatial information is to a function which is objective. The implementation of local spatial knowledge, however, leads to high complexity in computing. If you suggest an FCM algorithm that is better than morphological reconstruction and my brothers' filter (FRFCM) that is dramatically faster and more stable than FCM, to solve this problem. Second, with the implementation of the morphological reconstruction operation, local spatial imaging information is introduced into FRFCM to take immunity from noise and retain image detail. Second, based on the distances between the walls in the room premises and the grouping centers, the alteration of the membrane information whether it is replaced by a local membership filter which depends only on the membership division's space neighbors. Using the active contour model (ACM) for detailed segmentation and robustness to optimize the FRFCM clustering outcomes. With active contour model efficiency, which is superior to the current FCM algorithm, the FRFCM is proposed

Keywords Image segmentation · Fuzzy c-means clustering (FCM) · Local spatial information · Morphological reconstruction (MR) · Active contour model.

D. Ravali (✉) · E. Suresh
Department of Electronics and Communication Engineering, Kakatiya Institute of Technology and Science, Warangal, Telangana 506515, India
e-mail: ravalidupati@gmail.com

E. Suresh
e-mail: esuresh_77@yahoo.com

© The Author(s), under exclusive license to Springer Nature Singapore Pte Ltd. 2021   81
K. A. Reddy et al. (eds.), *Data Engineering and Communication Technology*,
Lecture Notes on Data Engineering and Communications Technologies 63,
https://doi.org/10.1007/978-981-16-0081-4_9

1   Introdudction

In order to analyze an image by separating it into non-overlapping areas, image segmentation is an important image processing technique. And with uniform and homogeneous features such as strength, color, tone or texture, etc., these regions are distinct. Several different segmentation strategies are proposed. Grouping, regional expansion, active contour model, average displacement, graph split, etc. are the technologies. Of these technologies, one of the most common approaches used for image segmentation is clustering. Clustering may be a system in which objects or models are grouped in such a way that the samples from an identical group are more like each other than samples from separate groups. In general, classification strategies are often grouped into hierarchical, theoretical diagrams, broken down into a function of density, and simplified to an objective function. In this paper, we will specialize in grouping methods assisted by image segmentation by minimizing an objective function [1–4].

Since traditional classification is difficult, image segmentation results in bad outcomes. Bezdek also suggested fuzzy c-means classification, which endorsed fuzzy's pure mathematics (FCM). The FCM algorithm is the hottest method used for image segmentation among the fuzzy classification methods. The grouping algorithm performs better than usual traditional grouping because it has a greater uncertainty threshold and more knowledge about the first picture persists. Fuzzy c-means. For simple textured and background images, FCM is effective, and it does not segment corrupted noise images because it only considers the awareness at the gray level without taking spatial information into account. The primary common concept is to enter local spatial knowledge into an analytical function to beat the drawback of FCM. Various algorithms that implement local spatial knowledge are recommended. None of these, however, are useful because the incorporation of local spatial information contributes to high computational complexity and low noise immunity. For image segmentation, we propose a considerably quick and robust algorithm. For different images with a coffee computing cost, the proposed algorithm will perform good segmentation results. In our proposed algorithm, before operations, we use two Segmentation is a morphological reconstruction and filter for membership [5–8].

2   Literature Review on FCM Clustering

Fuzzy function-based grouping techniques deal with groupings whose boundaries cannot be precisely defined. Through the fuzzy grouping, you can find out whether the data objects belong wholly or partially to the clusters based on their membership in their different clusters. Of the fuzzy grouping methods, fuzzy c-means (FCM) is the best known algorithm as it has the advantage of obscuring information about clusters. Fuzzy c-mean clustering was first reported by Joe Dunn for a distinct case

($m = 2$). The common case (for each m larger than 1) was developed by Jim Bezdek [9–12].

In FCM, a dataset is grouped into k groups, where each data object can be related to each group with some degree of membership in that group. Membership of a data object in a cluster can range from 0 to 1. The sum of the members for each data point must be units.

FCM cracks to divider set gathering of facts into assembly of C fuzzy clusters through esteem to certain assumed measures. Before using FCM algorithm, the following parameters must be specified

i.   Number of clusters 'C'
ii.   Fuzziness exponent 'm' (Fig. 1).

FCM is well organized for images with meek qualities and backgrounds; you cannot segment complex textured and background images or noise corrupted images

Fig. 1 Flowchart of FCM

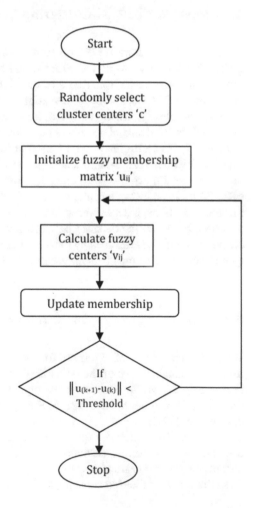

because you only consider gray-level substantial deprived of seeing spatial information. To overcome the difficulties of FCM, we introduce local spatial information into an objective function. Various algorithms have been proposed that incorporate local spatial information. But none of these is efficient because this outline of confined spatial information indications to great computational complication and they also have low noise immunity [13–16].

In this paper, we develop the FCM algorithm in two methods: One is to arrive local spatial information using a new method with low computational complexity, and the other is to adapt the membership of the pixels without trusting on the deviousness of the distance between the pixels inside the space local neighbors and combination centers. The proposed image segmentation algorithm drive stand executed efficiently with little computational cost.

3   Proposed FRFCM Clustering Method

Based on the literature review on fuzzy c-means clustering (FCM) algorithm, it is sensitive to noise; local spatial information is often introduced to an objective function to improve the robustness of the FCM algorithm for image segmentation. However, the introduction of local spatial information often leads to a high computational complexity, arising out of an iterative calculation of the distance between pixels within local spatial neighbors and clustering centers. To address this issue, an improved FCM algorithm based on morphological reconstruction and membership filtering (FRFCM) that is significantly faster and more robust than FCM is proposed in this paper. Firstly, the local spatial information of images is incorporated into FRFCM by introducing morphological reconstruction operation to guarantee noise immunity and image detail preservation. Secondly, the modification of membership partition, based on the distance between pixels within local spatial neighbors and clustering centers, is replaced by local membership filtering that depends only on the spatial neighbors of membership partition [17–20].

3.1   Morphological Reconstruction

We introduced the FCM algorithm for morphological reconstruction (MR) to optimize the characteristics of the data distribution before applying the grouping. MR is able to preserve the contour of the object and eliminate noise without knowing the type of noise beforehand, useful for optimizing the characteristics of the data distribution [21–24].

Here are two simple morphological reconstruction operations, morphological dilation and erosion reconstruction. Depending on the composition of the erosion and expansion reconstructions, it is possible to obtain some reconstruction operators with a higher filtering capacity, such as the opening and closing morphological reconstructions defined in the Eq. (1)

$$R^C(f) = R^\varepsilon_{R^\delta_f(\varepsilon(f))}\big(\delta\big(R^\delta_f(\varepsilon(f))\big)\big) \tag{1}$$

3.2 Membership Filtering

A membership matrix is defined based on Eqs. (3and1) and is given in Eq. (2).

$$\max\big\{U^x - U^{x+1}\big\} < \xi \tag{2}$$

In this process, consider median filtered-based membership partition matrix is defined in the following Eq. (3).

Finally, the updated membership partition matrix and cluster centers are defined in the following Eqs (3) and (4), respectively.

$$U_{\text{new}} = \text{med}\{u_{ik}\} \tag{3}$$

$$v_{\text{updated}} = \frac{\sum_{i=1}^{q} U_{\text{new}}\gamma_l x_k}{\sum_{i=1}^{q} U_{\text{new}}\gamma_l} \tag{4}$$

4 Implementation to Active Contour Model (ACM)

The results of the FRFCM pool are defined to overcome the disadvantages of the previous phase; it is necessary to use an active contour model to refine the segmentation results in the FRFCM pool. In this phase, FRFCM grouping results that implement online region-based active model outline (ORACM) eliminate these two disadvantages as they run too slowly and require parameters that have a significant impact on the results and are adjusted depending on the input images. Using a new set of binary level formulas and a new regularization operation such as morphological opening and closing. Without changing the precision of segmentation, ORACM requires no parameters and less time than traditional ACMs. Experiments on synthetic and real images show that the computational cost of ORACM with morphological operations is on average 3.75 times less than traditional ACM. The mathematical formula for the active contour evolves without parameters are defined as [25–30].

$$\frac{\partial \phi}{\partial \chi} = H(spf(I_{\text{FRFCM}}(x))) \cdot \phi(\chi) \tag{5}$$

$$\mathrm{spf}(I_{\mathrm{FRFCM}}(x, y)) = \frac{I_{\mathrm{FRFCM}}(x, y) - \frac{c1+c2}{2}}{\max\left| I_{\mathrm{FRFCM}}(x, y) - \frac{c1+c2}{2} \right|} \tag{6}$$

$$c1(\phi) = \frac{\int_l (\chi) \cdot H(\phi)\mathrm{d}\chi}{\int_H (\phi)\mathrm{d}\chi} \tag{7}$$

$$c2(\phi) = \frac{\int_l (\chi) \cdot (1 - H(\phi))\mathrm{d}\chi}{\int_H (1 - H(\phi))\mathrm{d}\chi} \tag{8}$$

5 Simulation Results and Discussions

The performance of the proposed method fast and robust FCM (FRFCM)-based active contour model is analyzed the parameters such as CPU time, pixels covered and dice similarity coefficient (DSC) over the existing FCM clustering algorithm. The simulation results of the existing as well as proposed methods tested on synthetic and medical which is taken from MRI brainweb science dataset and online available synthetic dataset, respectively. The simulations carried out and implemented with MATLAB 2019b. The proposed algorithm is superior, accurate, and faster segmentation results over the existing conventional FCM segmentation results. The simulation results carried out both existing FCM and proposed FRFCM with active contour model in the following figures.

Simulation results on color Image (Image 1):
  See Figs. 2 and 3.

Simulation results on synthetic Image (Image 2):
  See Figs. 4 and 5.

Simulation results on MRI brain image (Image 3):
  See Figs. 6, 7 and Table 1.

6 Conclusion

The most general approach used for image segmentation is clustering, the fuzzy c-means clustering (FCM) algorithm is the most commonly used segmentation algorithm in image processing, the key downside of the FCM algorithm is noise-sensitive, and an objective function is also applied to local spatial knowledge. However, the introduction of local spatial data results in high computational complexity. An improved FCM algorithm based on morphological reconstruction and membership filtering (FRFCM), which is substantially faster and more stable than FCM, is suggested in this paper to solve this problem. Firstly, by incorporating morphology,

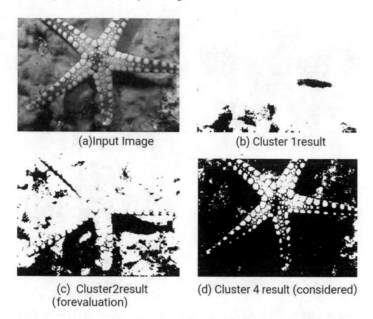

(a)Input Image

(b) Cluster 1result

(c) Cluster2result
(forevaluation)

(d) Cluster 4 result (considered)

Fig. 2 Simulation results on color image (octopus image) using conventional FCM algorithm

(a) Input Image

(b) Segmentation Results using FRFCM

(c) Clustered Image of(b)

(d) Segmentation results using Level Set method

Fig. 3 Simulation results on color image (octopus image) using FRFCM based active contour model

the local spatial picture data is introduced into FRFCM. Secondly, the membership partition adjustment, based on the distance between pixels within local spatial neighbors and clustering centers, is replaced by local membership filtering that relies only on the membership partition's spatial neighbors. Using the active contour model

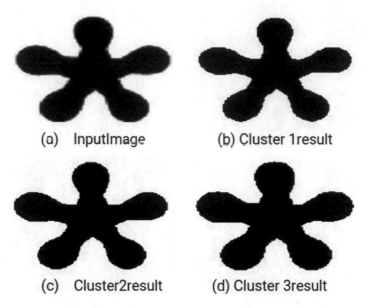

Fig. 4 Simulation results on synthetic image (Image 2) using conventional FCM algorithm

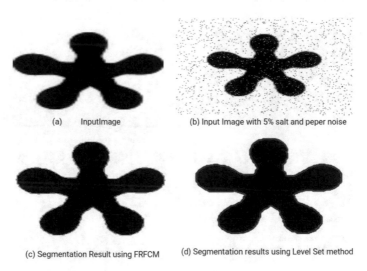

Fig. 5 Simulation results on synthetic image (image2) using FRFCM-based active contour model

(ACM) to optimize the effects of the FRFCM clustering to obtain precise segmentation and power. The proposed FRFCM is superior to the current FCM algorithm, with active contour model efficiency.

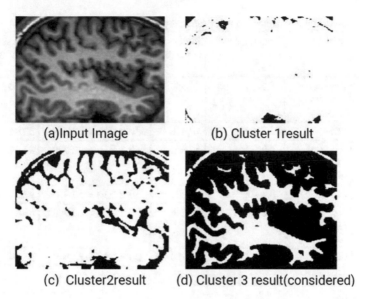

(a)Input Image   (b) Cluster 1result

(c) Cluster2result   (d) Cluster 3 result(considered)

Fig. 6 Simulation results on MRI brain image (Image 3) using conventional FCM algorithm

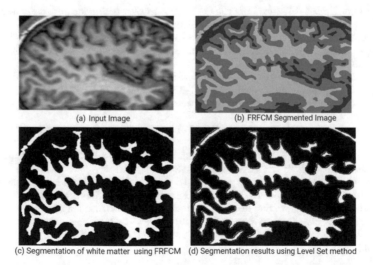

(a) Input Image   (b) FRFCM Segmented Image

(c) Segmentation of white matter using FRFCM   (d) Segmentation results using Level Set method

Fig. 7 Simulation results on MRI brain image (Image3) using FRFCM-based active contour model

Table 1 Comparision of existing and proposed aproach

Image	Method	Running time (s)	Pixels covered	Ground truth pixels	Dice similarity index (DSC) (%)
Image 1	Proposed FRFCM + ACM	4.7437	46,357	230 × 230 = 52,900	93.40
	FCM	11.7757	10,440	230 × 230 = 52,900	32.96
Image 2	Proposed FRFCM + ACM	0.9074	50,063	230 × 230 = 52,900	97.24
	FCM	3.3043	50,031	230 × 230 = 52,900	97.21
Image 3	Proposed FRFCM + ACM	4.8861	41,686	230 × 230 = 52,900	88.14
	FCM	11.2504	19,789	230 × 230 = 52,900	54.44

References

1. Li P, Chen Z, Yang LT, Zhao L, Zhang Q (2017) A privacy-preserving high-order neuro-fuzzy c-means algorithm with cloud computing. Neurocomputing 256:82–89
2. Gupta S, Arbeláez P, Girshick R, Malik J (2015) Indoor scene understanding with RGB-D images: Bottom-up segmentation, object detection and semantic segmentation. Int J Comput Vis 112(2):133–149
3. He L et al (2008) A comparative study of deformable contour methods on medical image segmentation. Image Vis Comput 26(2):141–163
4. Gibou F, Fedkiw R (2002) A fast hybrid k-means level set algorithm for segmentation. Stanford University, Stanford, CA, USA
5. Tuia D, Muñoz-Marí J, Camps-Valls G (2012) Remote sensing image segmentation by active queries. Pattern Recognit 45(6):2180–2192
6. Yuan J, Wang D, Li R (2014) 'Remote sensing image segmentation by combining spectral and texture features.' IEEE Trans Geosci Remote Sens 52(1):16–24
7. Mylonas SK, Stavrakoudis DG, Theocharis JB (2013) 'GeneSIS: a GA-based fuzzy segmentation algorithm for remote sensing images. Knowl Based Syst 54(4):86–102
8. Pal M, Maxwell AE, Warner TA (2013) Kernel-based extreme learning machine for remote-sensing image classification. Remote Sens Lett 4(9):853–862
9. Gan G, Ma C, Wu J (2007) Data clustering theory, algorithms, and applications. Soc Industr Appl Mathe
10. Li F, Ng MK (2010) Kernel density estimation based multiphase fuzzy region competition method for texture image segmentation. Commun Comput Phys 8(3):623–641
11. Zhu SC, Yuille AL (1996) Region competition: unifying snakes, region growing, energy/bayes/mdl for multi-band image segmentation. IEEE Trans Pattern Anal Mach Intell 18:884–900
12. Zhang M, Jiang W, Zhou X, Xue Y, Chen S (2017) A hybrid biogeography-based optimization and fuzz yC-means algorithm for image segmentation. Soft Comput 1:1–14
13. Gade M, Scholz J, von Viebahn C (2000) On the detectability of marine oil pollution in European marginal waters by means of ERS SAR imagery. Proc. IGARSS 2000, vol 6, pp. 2510–2512
14. Lei T, Jia X, Zhang Y, He L, Meng H, Nandi AK (2018) Significantly fast and robust fuzzy C-Means clustering algorithm based on morphological reconstruction and membership filtering. IEEE Trans Fuzzy Syst 25(5):3027–3041

15. Chuang KS, Hzeng HL, Chen S, Wu J, Chen TJ (2006) Fuzzy c-means clustering with spatial information for image segmentation. Comput Med Imag Graph 30(1):9–15
16. Srinivas KRA, Rangababu T (2016) Segmentation of satellite and medical imagery using homomorphic filtering based level set model. Indian J Sci Technol 9(S1). https://doi.org/10.17485/ijst/2016/v9iS1/107818
17. Ramudu K, Babu TR (2015) Layers and dark sand dunes segmentation of MARS satellite imagery using level set model. IETE J 56(2):59–67. ISSN: 0974-7338
18. Reddy GR, Kama R, Srikanth R, Rao R (2016) Level set segmentation of oil spill images using non-separable wavelet transform. Int J Comput Sci Inf Secur (IJCSIS) 14(7):682–693. ISSN 1947-5500
19. Chan T, Vese L (2001) Active contours without edges. IEEE Trans Image Process 10(2):266–277
20. Pinheiro M, Alves JL (2015) A new level-set-based protocol for accurate bone segmentation from CT imaging. IEEE Access 3:1894–1906
21. Ronford R (1994) Region based strategies for active counter models. Int J Comput Vis 3(2):229–251
22. Fatih Talu M (2013) ORCAM: online region-based active contour model. Exp Syst Appl 40:6233–6240. www.elsevier.com/locate/eswa
23. Gongt H, Li Y, Liu G, Wu W, Chen G (2016) A level set method for retina image vessel segmentation based on the local cluster value via bias correction. In Proceedings of IEEE international congress on image signal processing, pp 413–417
24. Xin X, Wang L, Pan C, Liu S (2015) Adaptive regularization level set evolution for medical image segmentation and bias field correction. In: Proceedings of IEEE international conference on image processing, pp 1006–1010
25. Abdelsamea MM, Gnecco G, Gaber MM (2015) An efficient self-organizing active contour model for image segmentation. Neurocomputing 149:820–835
26. Zhang K, Zhang L, Song H, Zhou W (2010) Active contours with selective local or global segmentation: a new formulation and level set method. Image Vis Comput 28:668–676
27. Ramudu K, Kalyani C, Babu TR, Reddy GR (2019) Segmentation of soft tissues and tumours from biomedical images using optimized K-means clustering via level set formulation. Int J Intell Syst Appl (IJISA) 11(9):18–28
28. Ramudu K, Babu TR (2019) Level set evolution of biomedical MRI and CT scan images using optimized fuzzy region clustering. In: Computer methods in biomechanics and biomedical engineering: imaging and visualization, vol 7, issue 1, pp 99–110. Taylor and Francis group. ISSN: 2168-1163 (Print) 2168-1171 (Online)
29. Ramudu K, Babu TR (2018) Segmentation of tissues from MRI bio-medical images using kernel fuzzy PSO clustering based level set approach. Current Med Imag Rev Int J 14(3):389–400. ISSN (print): 1573-4056 and ISSN (online): 1875-6603
30. Kalyani C, Kama R, Reddy GR (2018) A review on optimized K-means and FCM clustering techniques for biomedical image segmentation using level set formulation. Biomed Res (India) Int J Med Sci 29(20):3660–3668. ISSN: 0970-938X

A Microstrip Patch Antenna Design with Two Ground Slots for 5G Mobile Applications

B. Komuraiah, V. Raju, L. Meghana Florence, and M. S. Anuradha

Abstract In this paper, we proposed a microstrip patch antenna with two rectangular ground slots for 5G mobile applications. This antenna has dimensions 2.69 mm × 4.55 mm, and less space occupation is suitable to 5G mobile applications in the current world. The two ground slots are used for better impedance matching, since the better impedance matching gives better results (return loss, VSWR, gain, and radiation). The proposed design has an improved return loss of −23.17 dB and gain 6.92 dB. We have taken FR4 substrate and created a two rectangular ground slots at an operating frequency of 38 GHz in this paper.

Keywords Microstrip patch · 5G · Mobile · Return loss · Ground slots

1 Introduction

From the past decade, there is a rapid development in mobile communication technology, the developed generations were from 0G, 1G, 2G, 3G, and now as users requirements are increasing, 4G is unable to fulfill the customers, as the challenges are increasing each day, a new generation 5G (5th Generation) was introduced which has high data rate to fulfill the requirements. The range of the 5G spectrum is the

B. Komuraiah (✉) · V. Raju · L. Meghana Florence
Department of ECE, KITS, Warangal, Telangana, India
e-mail: kvvkomraiah@gmail.com

V. Raju
e-mail: rajureddyv@gmail.com

L. Meghana Florence
e-mail: megha.erza@gmail.com

M. S. Anuradha
Department of ECE, AU College of Engineering for Women, Visakhapatnam, Andhra Pradesh, India
e-mail: radhamsa@gmail.com

© The Author(s), under exclusive license to Springer Nature Singapore Pte Ltd. 2021   93
K. A. Reddy et al. (eds.), *Data Engineering and Communication Technology*,
Lecture Notes on Data Engineering and Communications Technologies 63,
https://doi.org/10.1007/978-981-16-0081-4_10

range of radiofrequency in the sub-6 GHz, and millimeter-wave frequency range is
24.25 GHz and above.

There are two sets of frequency bands for 5G:

* Frequency range 1 (FR1) which is called as sub-6 GHz range having a frequency
  range from 450 MHz to 6 GHz.;
* And frequency range 2 (FR2) which is called mmWave (millimeter-wave) having
  a frequency range from 24.25 to 52.6 GHz.

For the 5G networks, the peak data rates are as 20 GB/s downlink and 10 GB/s
uplink.

There are different antenna designs for 5G mobile communication using different
feeding techniques and different antenna technologies [1, 2]. Shows a simple patch
antenna with feeding line technique having good results. A small microstrip patch
antenna with a compact antenna shows a minimum fabrication and accuracy is shown
in [3–5]. But these could not fully reach the requirements, so to solve these problems millimeter waves are used to increase the efficiency for 5G technology. In [6],
millimeter waves are used in an antenna design for better channel capacity to achieve
wider band width and higher efficiency compared to other antenna designs.

A single antenna design of any shape can have different operating frequencies,
each operating frequency can have different applications. Instead of using different
antennas for each application, a single antenna can be used for different applications
such as WLAN, WSN, Wi-Fi, Wi-Max [7–9]. A dual band can be achieved by etching
a U-shaped slot on the patch which is able to transmit either of the two different
frequency range. It is also flexible and has non-overlapping channels.

In this paper, a rectangular patch antenna operating at high frequency with two
slots etched on the ground plane is presented. In Sect. 2, the designing procedure of
the proposed antenna is explained. In Sect. 3, the simulation results of the design are
discussed. And at the end, conclusion is given in Sect. 4.

2 Antenna Structure and Design

The design of the proposed antenna structure has an inset feeding technique with two
slots etched on the ground plane. The substrate of the design is selected according
to the requirements for the proposed design. In this design, FR4 substrate is used
having relative permittivity as 4.4 and dielectric loss tangent as 0.02. The thickness
of the substrate is 1.6 mm, and the dimensions of the substrate are 6 mm × 6 mm. A
rectangular patch is placed on the substrate with dimensions 2.69 mm × 4.55 mm,
these are calculated by using the Eqs. (1) and (4), and the operating frequency f_0 is
taken as 38 GHz. This microstrip patch antenna is fed with a microstrip line having
50 ohms input impedance, and the inset feeding length (Y_0) is 0.78 mm given by
Eq. (6).

2.1 Calculation of the Proposed Antenna Design

The proposed antenna is designed by the following equations:

- Calculating the width of the patch (W), the width of the patch effects the radiation efficiency

$$W = \frac{c}{2f_0}\sqrt{\frac{2}{\varepsilon_r + 1}} \tag{1}$$

where c is free space velocity of light, f_0 is resonant frequency and ε_r is the dielectric constant.

By calculating, the width of the antenna is 4.55 mm.

- Calculation of effective dielectric constant

$$\varepsilon_{reff} = \frac{(\varepsilon_r + 1)}{2} + \frac{(\varepsilon_r - 1)}{2}\left[1 + 12\frac{h}{w}\right]^{-1} \tag{2}$$

where h is the height of the substrate or thickness of the substrate, after calculating ε_{reff} is given as 3.44.

- Calculation of the extension length ΔL

$$\frac{\Delta L}{h} = 0.412\frac{(\varepsilon_{reff} + 0.3)\left(\frac{w}{h} + 0.264\right)}{(\varepsilon_{reff} - 0.3)\left(\frac{w}{h} + 0.8\right)} \tag{3}$$

- Calculation of the actual length (L) of the patch antenna is given by

$$L = \frac{\lambda_0}{2f_0\sqrt{\varepsilon_{reff}}} - 2\Delta L \tag{4}$$

where the effective length of the patch is given by

$$L_{eff} = \frac{\lambda_0}{2f_0\sqrt{\varepsilon_{reff}}} \tag{5}$$

- Calculation of the inset length of the feedline (Y_0) is given by

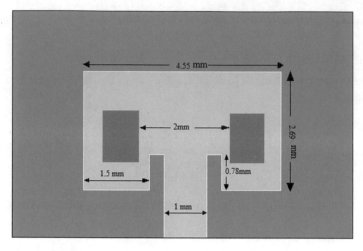

Fig. 1 Dimensions of the proposed antenna

$$Y_0 = \frac{L}{\pi} \cos^{-1}\left(\sqrt{Z_{\text{in}}/R_{\text{in}}}\right) \tag{6}$$

where Z_{in} is the resonant input impedance and R_{in} is the resonant input resistance.

The Y_0 is calculated as 0.78 mm.

By using the above calculations, the patch antenna is designed in the HFSS software.

2.2 Ground Slots

Figure 1 shows the dimensions of the proposed patch antenna with two slots etched on either side of the ground or the patch antenna, length and width of the slot are taken as 2 mm and 1 mm, respectively. The rectangular slots are separated by a distance of 2 mm.

The two ground slots behave as a load so, by adding slots to the antenna, it brings the input impedance closer to the characteristic impedance (Fig. 2).

3 Simulation Results and Discussion

3.1 Return Loss of the Proposed Antenna

Figures 3 and 4 show the return loss of the patch antenna without and with slots, respectively. As the reflected signal increases, the signal delivered decreases, so as

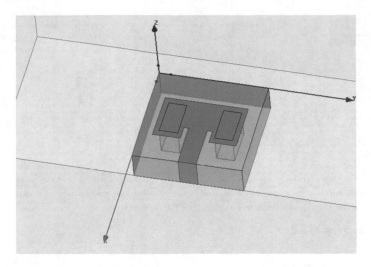

Fig. 2 Proposed design using HFSS

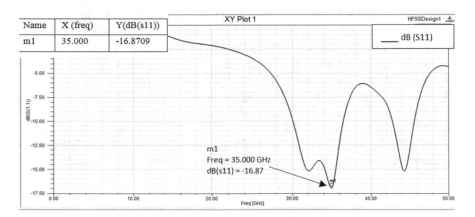

Fig. 3 Return loss without slots

to transmit the data without any loss impedance should be equal, if there is good impedance matching the reflected signal which is the return loss will be less. As the return loss decreases, the signal transmitted to the load increases. Here, by adding slots as they can be used as a load, the input impedance is brought nearer to the characteristic impedance; the impedance is matched, and hence, we get a better return loss (S_{11}). By this we can see that the antenna with slots has better return loss which is -23.17 dB compared to the antenna without slot (-16.8 dB), since without slots, the impedance is not matched properly.

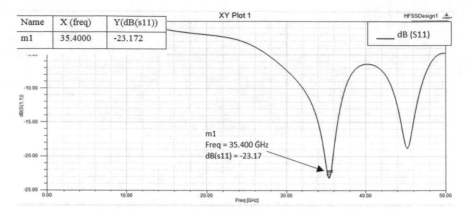

Name	X (freq)	Y(dB(s11))
m1	35.4000	-23.172

Fig. 4 Return loss with slots

3.2 Antenna Radiation of the Proposed Antenna

Radiation is due to the fringing fields around the antenna of a microstrip antenna. Since the patch antenna is taken as the open-circuited transmission line, the reflection coefficient is 1 and the current at the end and the center of half of the patch are zero and maximum, respectively. Due to this, the voltage and current are out of phase; this leads to maximum voltage at the end of the patch and minimum voltage at the beginning of the patch by which fringing of the fields takes place around the edges of the patch. In Figs. 5 and 6, the radiation patterns of the patch without and with the slots are shown, respectively. There will be an effect on the radiation pattern due to the slots, and they can rearrange the current distribution of the antenna. The antenna designed has a gain of 6.92 dB with slots and 7.8 dB without slots. By adjusting the

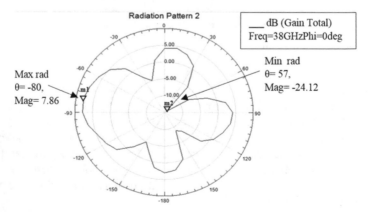

Fig. 5 Radiation pattern without slots

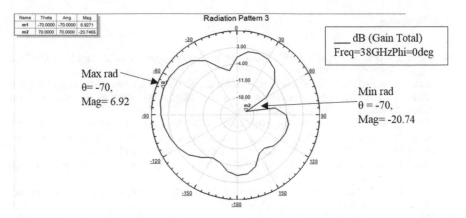

Fig. 6 Radiation pattern with slots

Table 1 Comparison between references and proposed antenna

References	Dimensions (mm)	Frequency (GHz)	Return loss (dB)	Gain (dB)
[3]	10.2 × 7	10.15	−18.27	4.46
[5]	2.69 × 4.55	38	−21.16	7.2
[7]	3.6 × 4.6	24.25,38	−25.20, −13.62	7.23, 3.69
[9]	16 × 16	28,38	−	5.42, 6.25
Proposed antenna with slots	2.69 × 4.55	38	−16.8	7.86
Proposed antenna w/o slots	2.69 × 4.55	38	−23.17	6.92

slots of the antenna according to the current distribution, we can improve the gain of the antenna (Table 1).

4 Conclusion

In this paper, a reduced small-sized antenna designed and simulated for 5G wireless communication is successfully analyzed. The proposed antenna is reduced in size compared to the previous antenna having resonant frequency 38 GHz with slots which have a return loss of −23.17 dB and a gain of 6.92 dB, and this concludes that the reflection coefficient is good with the slots compared to the antenna without slots, and the received signal has less reflection of the transmitted signal for the designed antenna.

Acknowledgements This work was supported by Dept. of ECE, Kakatiya Institute of Technology and Science, Warangal (KITSW).

References

1. Gupta A, Gupta A, Gupta S (2013) 5G: the future mobile wireless technology by 2020. Int J Eng Res Technol (IJERT) 2(9)
2. Jandi Y, Gharnati F, Said AO (2017) Design of a compact dual bands patch antenna for 5G applications. In: 2017 international conference on wireless technologies, embedded and intelligent system (WITS). IEEE 19–20 Apr 2017
3. Verma S, Mahajant L, Kumar R, Singh HS, Kumar N (2016) A small microstrip patch antenna for future 5G application. In: IEEE 5th international conference on reliability, Infocom technologies and optimization (ICRITO) (trends and future directions), 7–9 Sept 2016, pp 460–463
4. Sam CM, Mokayef M (2016) A wide band slotted microstrip patch antenna for future 5G. EPH-Int J Sci Eng 2(7):19–23. ISSN: 2454-2016
5. Annalakshmi E, Prabakaran D (2017) A patch array antenna for 5G mobile phone applications. Asian J Appl Sci Technol (AJAST) 1(3):48–51
6. Mohammed Jajere A (2017) Millimeter wave patch antenna design antenna for future 5G applications. Int J Eng Res Technol (IJERT) 6(02):289–291. https://www.ijert.org. ISSN: 2278–0181
7. Outerelo DA, Alejos AV, Sanchez M, Isasa MV (2015) Microstrip antenna for 5G broadband communications: overview of design issues. IEEE Commun Mag 2443–2444
8. Mohan GP, Patil S (2017) Design and development of fractal microstrip patch antenna for 5G communication. Adv Wirel Mob Commun 10(1):13–25. ISSN 0973–6972
9. Rafique U, Khalil H, Rehman SU (2017) Dual-band micristrip patch antenna array for 5G mobile communications. In: 2017 progress in electromagnetic research symposium-fall (PIERSFALL), 19–22 Nov 2017, pp 55–59

An Intelligent ANFIS Controller based PV Custom Device to enhance Power Quality

Y. Manjusree and A. V. V. Sudhakar

Abstract The trend setting indicated paper in intelligent adaptive-neuro fuzzy interface system (ANFIS) controller-based photovoltaic (PV) system device to solve power quality (PQ) issues are presented. Optimal power electronic converters (PEC) are employed to get controlled balanced load voltage/current. The presence of nonlinear loads and sudden addition and removal of sensitive loads on distribution will always create harmonics in the network. It will cause to produce dip in voltage (sag) or rise in voltage (swell) means severe oscillation in load voltage. These PQ issues are suppressed to a large extent by employing solar custom devices with optimal cost over FACTS controllers, which are available in the literature. To get maximum output power from PV system, fuzzy MPPT technique is adopted. The best soft switching of series voltage source converter (SeVSC) is done by employing the proposed method. The results obtained are compared with other well-known conventional control techniques such as PI controller, fuzzy-based proportional integral (PI) controller, robust ECKF (RECKF) and proves that the proposed control strategy works very well.

Keywords ANFIS controller · PQ issues · PV system · PI controller

1 Introduction

The most effective challenge of control engineer in the field of power system (PS) is to suppress the power quality issues. It is much significant to provide harmonic-free power supply to the loads irrespective of its nature (linear/nonlinear load). The presence of nonlinear loads always originates disturbances and uncertainties in the PS network [1, 2]. If these nonlinear loads are available for the longer time period in

Y. Manjusree (✉)
Department of EEE, Kakatiya Institute of Technology and Science, Warangal, India
e-mail: manju547sree@gmail.com

A. V. V. Sudhakar
Department of EEE, S R Engineering College, Warangal, Telangana, India
e-mail: sudheavv@gmail.com

© The Author(s), under exclusive license to Springer Nature Singapore Pte Ltd. 2021   101
K. A. Reddy et al. (eds.), *Data Engineering and Communication Technology*,
Lecture Notes on Data Engineering and Communications Technologies 63,
https://doi.org/10.1007/978-981-16-0081-4_11

the PS network, it is very difficult to provide smooth power supply to the loads. When the power feeds to these nonlinear loads (converters, laptop/mobile chargers, etc.), it creates harmonics in to the power system network and fault conditions. Sudden insertion or removal of load also create disturbance in the power system network to the severe extent. If duration of these instable issues is continuous for long time that may lead to damage of the loads.

These PQ issues are investigated past in the literature, using series voltage controllers [3], shunt voltage controllers [4], hybrid active power filters [5] and integrated series-shunt controllers [6]. Generally, these techniques are reactive power compensation approaches which injects constant voltage or current at CPI. These conventional FACTS controllers supply variable voltages to the CPI and balance the load voltage by employing different control strategies. These FACTS controllers are more cost-effective methodologies to solve PQ problems. They need backup sources like battery, super-ultra capacitor, fly wheel energy or superconducting magnetic energy storage (SMES). This problem can be overcome by adopting custom-based PV systems which act as input source to inject voltage in the PS network.

The controller-integrated PV custom devices are working as benchmark to enhance the PQ of the network under different fault tolerant conditions. It is available with PI controllers [7], fuzzy controllers [8], robust extended complex Kalman filter (RECKF) [9]. Compared to these control algorithms, the proposed ANFIS system works well to improve PQ. PS network with FACTS devices to enhance power quality is described in [10–13]. From the obtained simulation results concluded that the proposed ANFIS controllers work well.

2   System Configuration

The configuration of intelligent controller-based PV system with PS integration is shown in Fig. 1.

Different topologies are available to get maximum/lower power from PV such as ripple correlation technique (RCT), perturb and observe (P&O) and incremental conductance (IC). These performances sense the amplitude of the load current and make changes in input current quickly. The main drawback of these MPPT techniques is its transient response of variation in current is sluggish. This can be overcome by fuzzy MPPT intelligent control algorithm [8]. To maintain a constant voltage across capacitor (Cdc) and maximum utility, DC-DC converters are operated with fuzzy controller. This fuzzy-based DC converter provides less oscillatory voltage to the series VSCs which improves dynamic performance and overall efficiency of the PS network. The structure of fuzzy-based direct current converter is shown in Fig. 2.

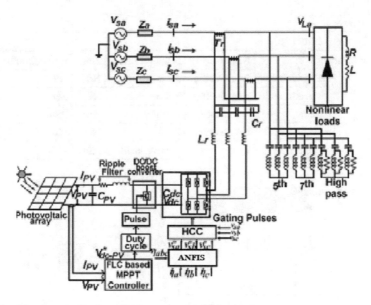

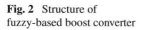

Fig. 1 Intelligent controller-based PV system with PS integration

Fig. 2 Structure of
fuzzy-based boost converter

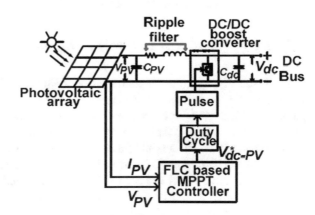

3 Control Scheme

The overall system configuration depends on series HAPF. It will operate with various control strategies that are illustrated below. A power system is usually characterized by complex nonlinear large-scale systems. However, it is possible to linearize the system about its operating point to solve PQ issues.

Table 1 Fuzzy rule set

	Error (E)					
Change in error (ΔE)		NegativeBig	NegativeSmall	Zero	PostiveSmall	PB
	NegativeBig	NegativeBig	NegativeBig	NS	NSm	Zero
	NegativeSma	NegativeBig	NSm	NS	Zero	PSm
	Zero	NegativeSm	NSm	Zero	PSm	PSm
	PostiveSmall	NegativeSm	Zero	PSm	PSm	PBig
	PostiveBig	Zero	PostiveSmall	P3m	PBig	PBig

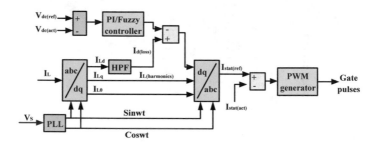

Fig. 3 Control strategy of fuzzy PI-based VSC

3.1 Fuzzy-Based VSI Controller Design

Fuzzy-based VSI controller implementation is similar to the fuzzy MPPT algorithm. In this, error is treated as set of fuzzy rules. These fuzzy sets provide PI control parameters by selecting rules, shape of membership and de-fuzzification. The fuzzy rules set is mentioned in Table 1. The fuzzy logic rationale contrasts with both idea and substance from conventional frameworks such on the point of negative big (NB), negative small (NS), zero (Z), positive big (PB) and positive small (PS).

The actual value of voltage across (Vdcact) CPI point is contrast with reference DC voltage (Vdc) that error is optimized with fuzzification then error is rectified send to the system after de-fuzzification. The control strategy of operate VSC with fuzzy PI controller is shown in Fig. 3.

3.2 RECKF-Based VSI Controller Design

It is an advanced soft computational algorithm to be operated with VSC. This is extended version of ECKF, in this injection of voltage at CPI is derived from its nonlinear state equations. In these approaches, error is estimated or pre-defined algorithm. The structure of RECKF is mentioned in Fig. 4.

Fig. 4 Structure of RECKF

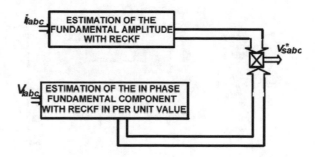

3.3 ANFIS-Based VSI Controller Design

The architecture of integration of ANN and fuzzy-based PV system to operate VSC is shown in Fig. 5. These methods provide some heuristics and thumb rules for tuning of PI controllers for soft switching of VSC. Many of these approaches involve simplification of higher order transfer functions into lower order approximations. As with any intelligent-based approach, guarantee that the tuning provided by these approaches will result in satisfactory performance for all systems. To design this ANN human brain network, five layers are considered. In this, layer1 has 25 neurons, layer2 to 4 have 16 and fifth layer has 2 neurons, respectively, and shown in Fig. 6. It trained approximately 300 iterations, intention assemble waits for the output. Later, the completion of iterations output response is fed to fuzzy to optimize the error. This ANN with fuzzy will give optimized PI parameters gave better outputs contrast with other traditional methods.

Fig. 5 Architecture of ANFIS-based VSC

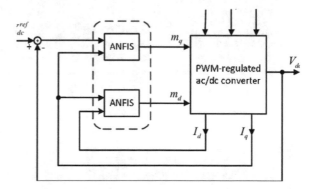

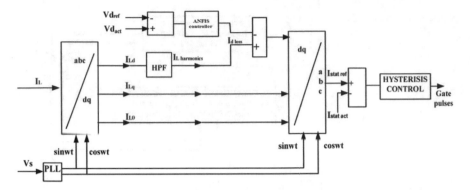

Fig. 6 Configuration of ANFIS

4 Simulation Discussion

4.1 *Without Any Control Scheme*

The PS network without any controllers is analysed in MATLAB. The obtained output simulation output waveform is shown in Fig. 7. Here, the voltage sag is occurred from $t = 0.5$ s to $t = 0.7$ s and voltage swell occurred in between 1.5 and 1.7 s.

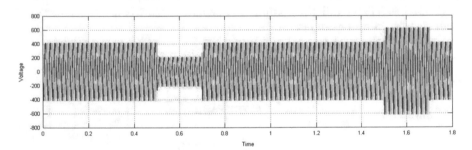

Fig. 7 Simulation results of load voltage

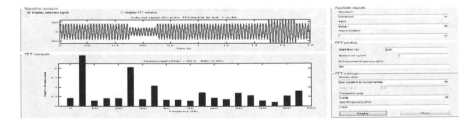

Fig. 8 THD plot of load voltage without employing any control strategy

Figure 8 exposes the total harmonic distortion (THD) plot of load voltage under sag and swell, and it is around 6.67%. This THD content output voltage at load side is not acceptable and it will lead to damage the loads.

4.2 With FLC-PI Controller

In this approach, PI controller operated with fuzzy control algorithm improves the transient/dynamic switching action of VSC and enhances PQ. The obtained simulation results, i.e. load voltage, load current, grid current and DC link voltage are mentioned from Figs. 9, 10, 11 and 12, respectively.

Figure 13 shows the THD plot of load voltage with unbalance nonlinear loads is around 1.46%.

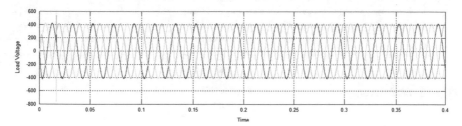

Fig. 9 Load voltage

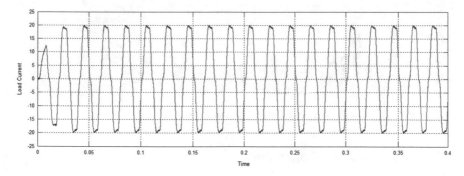

Fig. 10 Load current

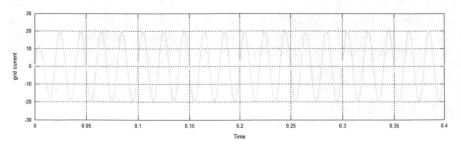

Fig. 11 Grid current

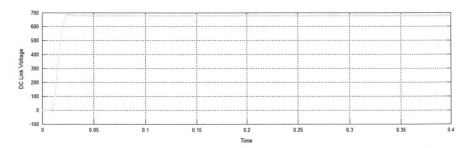

Fig. 12 DC link voltage

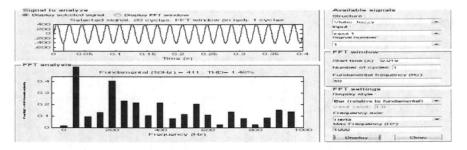

Fig. 13 THD plot of load voltage with fuzzy PI control strategy

4.3 With RECKNF Controller

Mostly, similar results of fuzzy-based PI controller simulation output results are obtained and improved load voltage and load current profile much extent. The output load voltage simulation is shown in Fig. 14.

From Fig. 15, it is clear that the load voltage is almost balanced. The THD plot of RECKF-based PV system for PS network is shown in Fig. 17. Obtained THD is around 0.96%.

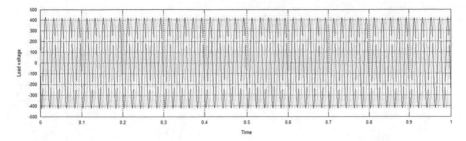

Fig. 14 Load voltage

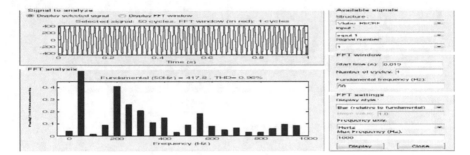

Fig. 15 THD plot of load voltage with RECKF control strategy

4.4 Proposed Simulation Results

The proposed ANFIS-based PV system integration of PS network is to enhanced PQ. Compared to other discussed approaches, the best output results are in obtained with the proposed method. The obtained simulation results of load voltage, load current and grid current DC link voltage are shown in Figs. 16, 17, 18 and 19.

By adopting the proposed control strategy with help of reduced rating of DC link capacitive voltage, the PQ has been improved. The THD value of load voltage is almost near zero (=0.33%) that shown in Fig. 20.

The performance of THD comparison of load voltage is described in Table 2.

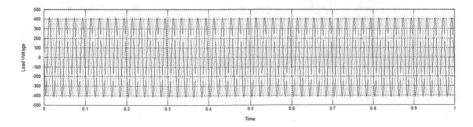

Fig. 16 Load voltage

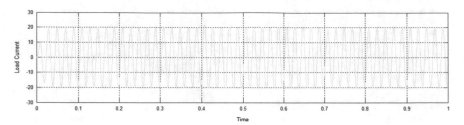

Fig. 17 Load current

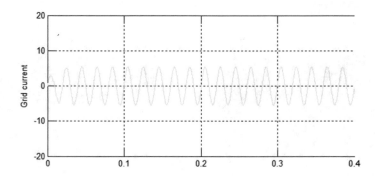

Fig. 18 Grid current

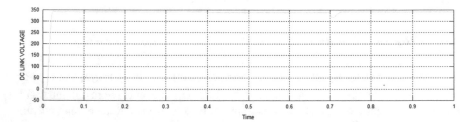

Fig. 19 DC link voltage

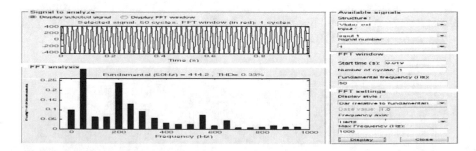

Fig. 20 THD plot of load voltage with ANFIS control strategy

Table 2 Comparative study of THD with various control schemes

Method	%THD
Without controller	6.67
PI controller	2.93
Fuzzy PI controller	1.46
RECKF controller	0.96
ANFIS controller	0.33

5 Conclusion

In this article, an ANFIS controller-based PV custom device to improve PQ is discussed. Series VSC converter is effectively operated with the proposed controller and minimized voltage and voltage swells much effectively. The obtained simulation results are proven harmonic-free output load voltage, and load current is achieved with integration of ANN control algorithm with fuzzy controllers. Almost neglected THD value (mentioned in Table 2) is obtained with the proposed method compared to the other fuzzy PI controllers and RECKF control approaches.

References

1. Omar R, Rahim NA (2012) Voltage unbalanced compensation using dynamic voltage restorer based on super capacitor. Int J Electr Power Energy Syst 43:573–581
2. Fitzer C, Arulampalam A, Barnes M, Zurowski R (2002) Mitigation of saturation in dynamic voltage restorer connection transformers. IEEE Trans Power Electron 17:1058–1066
3. Hashim HF, Omar R, Rasheed M (2016) Design and analysis of a three phase series active power filter (SAPF) based on hysteresis controller. In: 4th IET clean energy and technology conference (CEAT 2016), Kuala Lumpur, pp 1–5
4. Kumar P (2011) Simulation of custom power electronic device DSTATCOM—a case study. In: International conference on power electronics (IICPE), pp 1–4
5. Luo Z, Su M, Sun Y, Zhang W, Lin Z (2016) Analysis and control of a reduced switch hybrid active power filter. IET Power Electron 9(7):1416–1425
6. Gyugyi L (1992) A unified power flow control concept for flexible AC transmission system. IEE Proc C 139(4):323–331
7. Faisal M, Alam MS, Arafat MIM, Rahman MM, Mostafa SMG (2014) PI controller and Park's transformation based control of dynamic voltage restorer for voltage sag minimization. In: 9th International forum on strategic technology (IFOST), Cox's Bazar, pp 276–279
8. Raveendra N, Madhusudhan V, Laxmi AJ (2017) PI and fuzzy controlled D-STATCOM based on power quality theory. In: International conference on energy, communication, data analytics and soft computing (ICECDS), Chennai, pp 357–362
9. Ray PK, Das SR, Mohanty A (2019) Fuzzy-controller-designed-PV-based custom power device for power quality enhancement. IEEE Trans Energy Convers 34(1):405–414
10. Kumar BS, Praveena S, Kumar KR (2018) Power quality improvement using custom power devices (AVC, DVR, APC). In: 2018 International conference on current trends towards converging technologies (ICCTCT), pp 1–5, Coimbatore
11. Moghbel M, Masoum MAS, Fereidouni A, Deilami S (2018) Optimal sizing, siting and operation of custom power devices with STATCOM and APLC functions for real-time reactive power and network voltage quality control of smart grid. IEEE Trans Smart Grid 9(6):5564–5575

12. Parastvand H, Bass O, Masoum MA, Chapman A, Lachowicz S (2020) Cyber-security constrained placement of FACTS devices in power networks from a novel topological perspective. IEEE Access 8:108201–108215
13. Choudhury SR, Das A, Anand S, Tungare S, Sonawane Y (2019) Adaptive shunt filtering control of UPQC for increased nonlinear loads. IET Power Electron 12(2):330–336

Privacy by Design Approach for Vehicular Tripdata Using *k*-Anonymity Perturbation

Nanna Babu Palla, B. Kameswara Rao, Kaladi Govinda Raju, and A. Vinaya Babu

Abstract Vehicular communication in intelligent transport system offers data dissemination among vehicles in rapid transmission of road incident log to trusted entities. The adversary attacks having background knowledge are often a side effect due to re-identity and linkage attacks by innocuous public data sharing provisions. The proposed work spotlight on attacks with background knowledge who attempts to extract individual's data using high end data extraction algorithms by linking with the vehicular trip database. Enhancing location privacy and individual privacy is achievable with *k*-anonymity perturbation scheme applied on vehicular database, which shows trivial for data leakage attacks, and the proposed algorithm significantly reduces the vehicle uniqueness to zero in achieving the sanitization process of vehicular database to avoid pre-knowledge attacks by intruder of ITS. This approach shows resilience in implementing the privacy preservation with lightweight process implementation and non-polynomial time complexity for DENIM messages. The data distortion caused due to perturbation is analyzed and reported in this work.

Keywords Anonymity · Adversary · Attacks · Background knowledge · Intelligent transport system · Re-identification · Sanitization · Vehicular database · Uniqueness

N. B. Palla (✉) · K. G. Raju
Department of CSE, Aditya Engineering College, Surampalem, India
e-mail: nanibabup@rediffmail.com

K. G. Raju
e-mail: govindarajukynm@gmail.com

B. K. Rao
Department of CSE, Aditya Institue of Technology and Management, Tekkali, India
e-mail: kamesh3410@gmail.com

A. Vinaya Babu
Stanley College of Engg. & Technology, Hyderabad, India
e-mail: nanibabup@rediffmail.com

© The Author(s), under exclusive license to Springer Nature Singapore Pte Ltd. 2021   113
K. A. Reddy et al. (eds.), *Data Engineering and Communication Technology*,
Lecture Notes on Data Engineering and Communications Technologies 63,
https://doi.org/10.1007/978-981-16-0081-4_12

1 Introduction

The published WHO statistics on road accidents in town and national highways is an eye-opener to government and allied authorities for procuring intelligent vehicular communication(VC) and surveillance system infrastructure development and establishment on roads. The IT-based infrastructure will curb the menace by immediate alarming, and incident attending onto the road mishap will greatly save the lives when road hazards occur. The road crashes causing the loss of 1.36 million a year according to recent statistics of [1] and loss of 3% GDP, often occurring in low- and middle-income countries.The integration of roadside unit (RSU), OBU and GPS enables vehicle-to-vehicle communication (V2VC) and offers a cooperative intelligent transport system (C-ITS) that will greatly preserve and revive the risked lives by integrating the vehicular incident data to application server followed by emergency response force (ERF) for life sustenance. The deployment of safety-intended VANETs may seem vulnerable if anonymity is not compliance according to PKI standards in wireless communication. PKI offers elegant framework for security concerns covering V2VC, vehicle 2 infrastructure(V2I) and vehicle 2 PKI (V2PKI) for effective C-TS as shown in Fig. 1. The sensitive data which includes vehicle number(VN), location details (longitude,latitude) may jeopardize the privacy and leads to privacy infringement at few contexts if this data is available to untrusted party (UTP) and adversary. The tracking of vehicle by intruder is achievable with macro-, micro- and fine-grained approaches.

In C-ITS, the vehicles are capable of dedicated short-range communication(DSRC) facility with IEEE 802.11p standard with multi-hop message forwarding among mobility vehicles [2]. Albeit establishing the communication link between vehicle arrived to communication channel and roadside infrastructure is tentative, and the validity is for short interval. The C-ITS abundantly serves ample services include road safety and response force alert system, commerce, advertisements and toll operations. Figure 2 shows allied services extended with C-ITS.

A. Black Zone: It is the area on road with high probability of accident occurrence identified based on archived data with 90% probability of road mishap. If any road crash occurred, it is to be communicated to vehicles on road (VANET) to caution

Fig.1 Types of vehicular communication in VANET in C-ITS

Fig. 2 Allied services of intelligent services in VC

other vehicles as well passing an incident rescue operation immediately to ERF. The RSU and GPS Internet communication to the surveillance processing server (SPS) and incident categorization at application server at speculated regular intervals on real-time basis are highly significant. It should deploy lightweight data structure for rapid processing and incident classification using data mining techniques [3]. The clustering of beacon messages generated by vehicular networks into proactive and reactive data of road incidents [3] is to perform efficiently on lightweight data (LWD) arrived. The structure of LWD is mentioned in Table 1 with limited attributes for rapid data transfer to ERF and SPS. SPS manipulates the data, avoids/processes noisy data and applies data mining techniques such as clustering. The vehicle identification is granted using public key infrastructure (PKI) using enrollment authority (EA) and authorization authority (AA). The proposed work shows a road map for inclusion of anonymity layer (AL) at vehicular database that mitigate the re-identity attacks and protecting the privacy for sensitive data of drivers based on location and vehicle number from adversary.

B. Enrollment Authority (EA): The VC with C-ITS ensures a vehicle can be trusted.

C. Authorization Authority (AA): Grants the authorization tickets (ATs) for vehicles for certain services on arrival.

D. Anonymity Layer (AL): The motive of this work is to enforce pseudonymity for vehicular data using lightweight data structure perturbative approach to comply with privacy norms and guidelines. The anonymity can be enforced after processing the incident details and rescue operation performed based on V-2-V communication payload. The proposed work ensures implementing pseudonymity of sensitive data such as {loc-code, vehicle#}. The vehicle will move from the location after sending the beacon message to RSU and GPS-based ITS comprising the incident occurred.

Table 1 LWD attributes and their description for vehicular dataset

Attribute	Description
Vn	Vehicular number
LOC,B	Each location have associated code
Z, d	Zone id specifies the type of road structure
Sum*	start time of trip data
Ec«	Finished time of trip data
direction	Longitude and latitude values
Attribute	Description
VN	Vehicular number
Loccode	Each location have associated code
Zid	Zone id specifies the type of road structure
Stime	Start time of trip data
Etime	Finished time of trip data
Direction	Longitude and latitude values

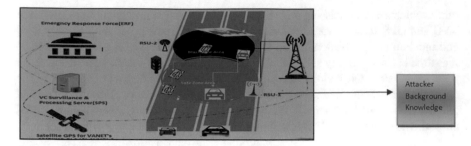

Fig. 3 An abstract view of C-ITS—vehicular communication for road infrastructure and attack scenario

The proposed model for controlled data disclosure in V2VC and V2I is implemented using k-anonymity model for VC-ICTS addressing privacy concerns. The aim of this work is to treat re-identity attacks by adversary due to data linkage as shown in Fig. 3.

2 Types of Message Standardization Service Models(MSSM)

The Car2Car-Communication Consortium (C2C-CC) [4] addressed and modeled the message standards in two categories, and these two models were standardized by European Telecommunication Standards (ETS) [5].

A. Cooperative Awareness Message(CAM): This is intended to broadcast messages in neighbor stations about vehicle dynamic location and speed at diverse requency from 1 to 10 Hz using IEEE 802.11p protocol standard. Special emergence vehicles such as ambulance, fire and police can use optional containers according to Brigetti et al. [6]. Security and privacy vulnerabilities are addressed in [7]. The privacy breach in cooperative vehicular communication is specified in [8, 9].

B. Decentralized Incident Notification Message (DENIM): This category of messages is transmitted when road mishap or hazard is detected by vehicles. The DENIM message comprises IDI, IDTime, location code ,direction, zone ID in the form of broadcast message forwarded at frequency of 1–10 HZ.

3 Proposed Work and Methodology

A. Anonymity for CAM Model

The anonymity of CAM messages in ITS by service providers disseminating available local services is promptly available among I-2-V model. Enforcing anonymity at CAM might be trivial as the communication between V2RSU and V2I is tentative.

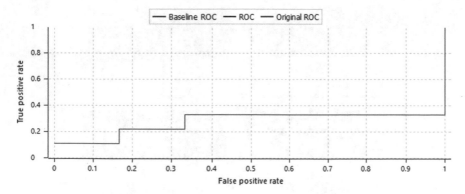

Fig. 4 Graph representing the ROC characteristics

B. Anonymity for DENIM Model

The thrust of the proposed work is to implement anonymity and sensitive data disclosure control by deploying the proposed perturbation anonymity algorithm for the messages broadcasted by V2X to comply with privacy protection for the vehicle database (V_D). The ROC characteristics were studied, and the vehicular details were suppressed from pristine data to offer pseudonymity as shown in Fig. 4. The AP and TS vehicles were considered for ROC generation with baseline AUC = 27.777% for AP* and AUC = 22.222 for TS* vehicles.

C. Algorithm

The adversary will re-attempt to re-identify the individual's sensitive data by linking the vehicular data V_D with public data P_D shared by transport department with the help of partial identifiers (P_i). The adversary will have pre-knowledge about vehicle and might successful at individual data extraction. To refrain from such background knowledge attacks, the vehicle database V_D is anonymized using the proposed perturbation algorithm rendering PV_D which is a sanitized data free from adversary attacks. The proposed work and process flow are shown in Fig. 5, intended to reduce the re-identity attack reducing the uniqueness of vehicles by offering anonymity on the sensitive attributes broadcasted by incident-identified vehicle (IIV) to TP conveying the road mishap dissemination among the peer vehicles for incident awareness. The k-anonymity applied and the ROC characteristics were evaluated and the distinction of quasi-identifiers for attribute data elicited followed by re-identity attacks. The Loc_{code} identifying data and V_N are fixed to be quasi-variable.

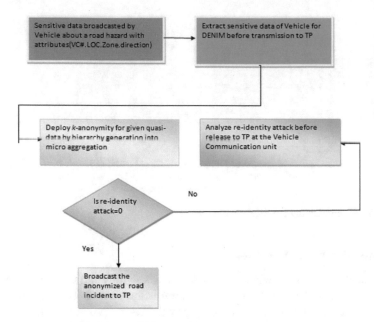

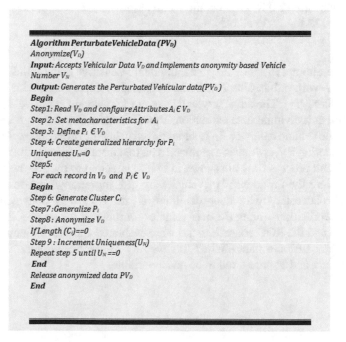

Fig. 5 Process flow diagram for VC-ITS anonymity implementation

Algorithm PerturbateVehicleData (PV$_D$)
Anonymize(V$_D$)
Input: *Accepts Vehicular Data V$_D$ and implements anonymity based Vehicle*
Number V$_N$
Output: *Generates the Perturbated Vehicular data(PV$_D$)*
Begin
Step1: Read V$_D$ and configure Attributes A$_i$ Є V$_D$
Step 2: Set metacharacteristics for A$_i$
Step 3: Define P$_i$ Є V$_D$
Step 4: Create generalized hierarchy for P$_i$
Uniqueness U$_N$=0
Step5:
For each record in V$_D$ and P$_i$ Є V$_D$
Begin
Step 6: Generate Cluster C$_i$
Step7 :Generalize P$_i$
Step8 : Anonymize V$_D$
IfLength (C$_i$)==0
Step 9 : Increment Uniqueness(U$_N$)
Repeat step 5 until U$_N$ ==0
End
Release anonymized data PV$_D$
End

	date		LocCode		Zone		vehicle#	incident-category
#d1		*		SA1		AP**		Road-block
#d7		*		SA1		AP**		Accident-towing
#d12		*		SA1		AP**		Accident-towing
#d2		*		SA2		AP**		Accident-Road
#d3		*		SA2		AP**		Accident-Ambulance
#d6		*		SA2		AP**		Accident-ambulance
#d8		*		SA2		AP**		Roadblock-treefall
#d11		*		SA2		AP**		Accident-ambulance
#d15		*		SA2		AP**		Accident-ambulance
#d4		*		SA1		TS**		Road-block
#d9		*		SA1		IS**		Road-fire
#d13		*		SA1		TS**		traffic-jam
#d5		*		SA3		TS**		Find-gasoline
#d10		*		SA3		TS**		Find-restaurant
#d14		*		SA3		TS**		Find-godown

Fig. 6 Output data after anonymization of trained data

D. k-anonymity Implementation

The 2-anonymity model is applied on the given training data, and the equivalence classes were constructed to distribute the vehicular data to more generalized hypernymity. The ROC characteristics were studied, and the vehicular details were suppressed from pristine data to offer pseudonymity as shown in Fig. 6.

The AP and TS vehicles were considered for ROC generation with baseline AUC = 27.777% for AP* and AUC = 22.222 for TS* vehicles. The output data after k-anonymity with masked Loc_{code} and V_n masked partially indicating the state transport codes having 60 and 40% maximal class, minimal class size having 2 –equivalence classes is shown in Fig. 6.

4 Results Discussion

The experiments were conducted on Intel PIV 64-bit processor having 2.49 GHz clock speed. The vehicular data V_D (Loc_{code}, zone, V_N, Inc_{cat}) was provided as input database to the proposed algorithm considering V_N as partial identifier (P_i). The k-anonymity is based on privacy approach adapted to perturbate vehicular data with sample size = 15 using the concept hierarchy generalization. The conducted experiments yield faster results in producing the equivalence clusters in two iterations, and the sample uniqueness (S_U) is reduced to zero by that the re-identity attacks by adversary are trivial, which is shown in Fig. 7. The risks associated with attacker are evaluated using Eq. (1).

A. Attacker risk(A_r): *Assume,for a given vehicular data V_D, and perturbated vehicular data (PV_D) then, the number of perturbated records is $P_r \in PV_D$, having n records, and then, the risk caused due to re-identity (Ar) is estimated,*

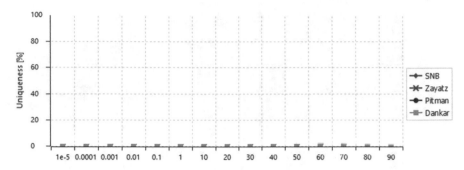

Fig. 7 Sample uniqueness for perturbated vehicular trip data (PV$_D$)

	Attacker risk parameter	Risk value (%)
Table 2 Evaluated risk values for vehicular data after perturbation	Records affected by minimal risk	60
	Lowest risk	11.11
	Highest pre-knowledge risk	16.66
	Records affected by maximal risk	40

$$A_r = \frac{P_r}{n} \tag{1}$$

The background knowledge attack by adversary about vehicular pre-knowledge is evaluated, and the perturbated vehicular data shows reduced vulnerability from 100 to 16.66%, which is shown in Table 2.

The equivalence class obtained with high performance shows robustness toward adversary background knowledge attack abreast minimal data distortion to original data V_D. This approach uses minimal data distortion to actual vehicle trip data, and the cluster arrangement for perturbated data is visualized in Fig. 8. The data utility aspects were evaluated obtaining minimal distortion, and leveraged entropy is shown in Fig. 9.

5 Conclusion

Implementing vehicular communication in intelligent road transport System significantly offers an efficient and elegant mechanism for storing and processing the vehicular data by trusted parties. Sharing the vehicular data leads may jeopardize privacy and leads to privacy breaches and corresponding attacks. The proposed algorithm shows a beneficial methodology in sharing the vehicle relevant data for DENIM messages among cooperative vehicles in PKI with reduced re-identity attack by adversary having pre-knowledge about the vehicle. This algorithm shows a direction in yielding immune to attacks using perturbation approach due to efficient

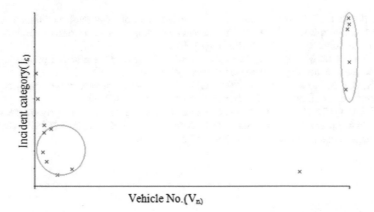

Fig. 8 Visualization of cluster for vehicular trip data

Fig. 9 Visualization of data quality metrics

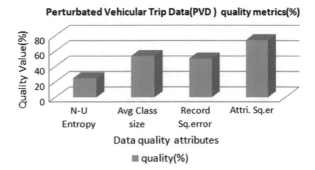

sanitization. The anonymity implemented with minimal distortion balancing data utility factor for vehicular repository services in the future research and demographic analysis. This work can be further extendable for location privacy in pervasive computational paradigms using *l*-diversity model.

References

1. Road traffic injuries. https://www.who.int/news-room/fact-sheets/detail/road-traffic-injuries
2. IEEE 1609 Working Group. IEEE Standard for Wireless Access in Vehicular Environments-Security Services for Applications and Management Messages (2016)
3. Haas JJ, Hu YC, Laberteaux KP (2009) Design and analysis of a lightweight certificate revocation mechanism for VANET. In: Proceedings of the sixth ACM international workshop on VehiculAr InterNETworking, pp 89–98
4. Car2Car Communication Consortium. https://www.car-car/org
5. ETSI EN 102 637–2. Intelligent Transport Systems (ITS); Vehicular communications; basic set of applications; Part 2: Specification of cooperative awareness basic service

6. Lonc B, Cincilla P (2016) Cooperative its security framework: Standards and implementations progress in europe. In: 2016 IEEE 17th international symposium on a world of wireless, mobile and multimedia networks (WoWMoM), pp 1–6. IEEE
7. Hoh B, Gruteser M, Xiong H, Alrabady A (2006) Enhancing security and privacy in traffic-monitoring systems. IEEE Pervasive Comput 5(4):38–46
8. Lin X, Sun X, Ho PH, Shen X (2007) GSIS: a secure and privacy-preserving protocol for vehicular communications. IEEE Trans Veh Technol 56(6):3442–3456
9. Lu R, Lin X, Zhu H, Ho PH, Shen X (2008) ECPP: efficient conditional privacy preservation protocol for secure vehicular communications. In: IEEE INFOCOM 2008-The 27th conference on computer communications, pp 1229–1237. IEEE Press

Self-powered Implantable Device Using Thermo Electric Generator with DC-DC Converter for Elimination of Bradycardia

M. Shwetha and S. Lakshmi

Abstract DC–DC converters with TEG for self-powering implantable devices is presented in this paper. A healthy heart rate is a key factor for the longevity of human beings. Bradycardia is a slow heartbeat that falls out as one of the major issues. An implantable device that deals with this disease are pacemaker. Battery oriented implantable pacemaker sustains until the life of the battery. Replacing implants due to limitation of battery life demands surgical operations leads to a painful process on patients and the area consumed by these batteries tend to 60–70%. The larger size of the battery is another issue due to which implant size also increases. In this research work to eliminate the battery, a self-powered device in which harvest energy within the body is introduced. Thermoelectric generators (TEG) replace the need for batteries in the pacemaker. The thermal energy of the human body is used as the source for TEG which gets converted to electrical form. Environmental condition issues come into picture which affects TEG operation. Thus, a new model of a DC–DC boost converter with TEG is designed for achieving self-powering for the pacemaker to overcome battery oriented problems. The proposed work is designed and validated using OrCad/PSPICE tool.

Keywords Self-powering pacemaker · Boost converter · Thermo electric generators

1 Introduction

A self-powered biomedical device is a remarkable subject intended for serious research by the researcher for several years. Life-supporting medical devices have changed from massive and persistent equipment to portable implantable devices,

M. Shwetha (✉) · S. Lakshmi
Department of Electronics and Communication, Sathyabama Institute of Science and Technology, Chennai, India
e-mail: shwetha.m1691@gmail.com

S. Lakshmi
e-mail: slakmy@yahoo.co.in

© The Author(s), under exclusive license to Springer Nature Singapore Pte Ltd. 2021   123
K. A. Reddy et al. (eds.), *Data Engineering and Communication Technology*,
Lecture Notes on Data Engineering and Communications Technologies 63,
https://doi.org/10.1007/978-981-16-0081-4_13

liberating patients from the unpleasantness of permanent hospitalization. Mobility inevitably requires moveable energy solutions, a role presently filled by energy storage technologies such as batteries. Evolution in energy investigation does not chase Moore's law; a massive amount of energy-harvesting approaches may ultimately permit to keep it up.

The self-powered biomedical devices are implantable and wearable electronics [1]. The regular resting heart rate is 60–100 BPM for adults. The pacemaker can be fitted to treat the problem of bradycardia, refers to the slow heart functionality implanted just below the collarbone. Slow heart rate can cause problems in the body and brain, which may not get enough blood flow and a variety of symptoms [2]. The pacemaker can be designed with the battery, computer memory, electronic circuits, and it provides the electrical signals [3, 4]. To attain a long period, performance of pacemaker devices must be powered with the help of the peripheral source. In the biomedical implant system, providing the required energy to the device over a long time is troublesome [5, 6].To boost the life span of pacemakers and replace the battery problems, different sources are developed by the researcher such as RF sources [7].

Thermoelectric generators are used to harvest the waste heat generated from the human body which is about 50 mV. This small voltage has to be boosted up to 2.5 V which is the binding operating voltage of the pacemakers. Alhawari et al. [8] projected a capable thermal energy-harvesting IC (EHIC) that supports a battery-less μ Watt SoC. Siouane et al. [9] presented a thermoelectric generator under constant heat flow conditions. Thermoelectric generators can be used under a stable temperature gradient. From the utilization of TEG, the obtained output voltage is very low; so, it is interrupted power supply to the pacemaker. It is hardly affected the human. So, the DC–DC converters [10] are used for enhancing the low voltage from the TEG with a boost converter.

2 Objectives

An imbalanced pulse rate is a common health issue with infant and mature group. It can be artificially healed by implanting an everlasting pacemaker within the body of the patient. Energy source or battery is a significant factor to drive the pacemaker; the source should last long and maintenance-free. For this reason, development of a better energy system using sustainable energy for permanent pacemakers is a major concern. The following objectives will be fulfilled with the proposed methodology.

- Lithium iodine battery is the source of an existing pacemaker. The life span of the battery is limited to about 10–12 years [11]. Replacement of battery demands for surgical operations with an average period of 10 years. The utilization of self-powered implants solves this condition.
- Usage of TEG as a source to supply a cardiac pacemaker.

- The influence of environmental factors may create an unstable generation of power (i.e., human body cooling and heat) [12].

3   Proposed Methodology

In this research work, to extract maximum power from TEG under the environmental conditions, the DC–DC boost converter is designed and developed. To achieve this, the following are a few steps to be carried out. At the first, pacemakers with the battery as sources are to be tainted to substitute thermoelectric generators in the place of Lithium battery.

Human body temperature is used as a key source of energy for TEG which is to be included within the existing pacemaker. TEG consists of a hot source, cold source, and several thermal power units linked at the center. The hot and cold sources are used to the thermoelectric unit and the particular ends of the loop current generation, as illustrated in Fig. 1. To attain high-quality thermoelectric conversion efficiency, the semiconductor with high ZT values is chosen. A single thermoelectric device is generally joined by copper conducting plates with columns of n/p type thermoelectric pairs made from high purity alumina ceramics. To link the components of the thermoelectric generator, the unique solder material is preferred.

3.1   Modeling of Thermoelectric Generator Mathematically is Given as Follows:

The energy conservation and charge continuity ruler equations are mentioned below,

$$\rho C_P \frac{\partial t}{\partial T} + \nabla \cdot \vec{X} = Q \tag{1}$$

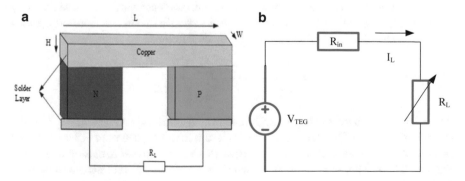

Fig. 1 **a** Structure of thermoelectric unit and **b** equivalent circuit of TEG

And,

$$\nabla \cdot \left(\vec{j} + \frac{\partial \vec{d}}{\partial T} \right) = 0 \tag{2}$$

ρ: Density, X: Heat flux, Q: Volumetric heat, CP: Definite heat at consonant pressure, j: the flow of charge intensity. The heat flux and current intensity is computed based on equations given below

$$\vec{X} = P \cdot \vec{j} - S \cdot \nabla t \tag{3}$$

$$\vec{j} = \mu \cdot \left(\overrightarrow{IE} - \mu \cdot \nabla t \right) \tag{4}$$

P: Peltier coefficient. To figure out dielectric flux density, dielectric equations are

$$\vec{d} = \varepsilon \cdot \overrightarrow{IE} \tag{5}$$

$\overrightarrow{IE}$ is the electric field strength, and ε is the dielectric coefficient. Electric field strength is irrational; the EMF remains unchanged for the time which can be obtained as the gradient of electric energy φ.

$$\overrightarrow{IE} = -\nabla \varphi \tag{6}$$

Steady-state derives the thermoelectric coupling formulations given below

$$\nabla \cdot \left(P \cdot \vec{j} \right) - \nabla \cdot (S \cdot \nabla t) = \vec{j} \cdot \overrightarrow{IE} \tag{7}$$

$$\nabla \cdot (\delta \cdot \mu \cdot \nabla t) + \nabla \cdot (\delta \cdot \nabla \varphi) = 0 \tag{8}$$

The efficiency and maximum power of the thermoelectric unit can be realized by connecting a voltage load at the hot and cold side. Based on the thermodynamics theory, the calculation of efficiency can be evaluated by

$$\eta = \frac{\Delta t \cdot r_l}{t_1(r_1 + r_2) - \frac{\Delta t \cdot r}{2} + \frac{(r + r_l)^2}{2r}} \tag{9}$$

$\Delta t = t_1 - t_2$ where t_1 and t_2 are hot side and cold side temperature, r is the internal resistance of the semiconductor, r_l is the load resistance. In order to reduce complexity, Thomson effect can be omitted that refers independent temperature μ. The conversion efficiency is obtained when $\frac{r_l}{r} = \sqrt{(1 + zt)}$ and has its maximum value as

$$\eta = \Delta t \cdot \frac{\sqrt{(1 + zt)} - 1}{t_1\sqrt{(1 + zt)} + t_2} \qquad (10)$$

where $t = t_1 + t_2$. In the condition, $r_l = r$ the maximum output power is denoted as follows,

$$P^{\max} = \mu^2 \frac{(t_1 + t_2)^2}{4R} \qquad (11)$$

To establish a stable power out of TEG, boost converters are taken into the picture, and this converter is placed along with the TEG in the pacemaker. Thus, modeling the boost converter is necessary. DC–DC boost converter enhances low voltage and acts as a supplementary element source for the thermoelectric generators. The architecture of the boost converter-based TEG for a pacemaker is given in Fig. 2.

In this planned process, the boost converter is the interface with TEG and load to regulate the voltage levels and to follow the peak power of the TEG. The circuit diagram of the boost converter is illustrated in Fig. 3. At the final stage output power generated by the boost, the converter is given to the designed pacemaker circuit which thereby done with analysis and validated based on the efficiency (Figs. 4 and 5).

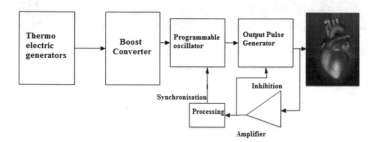

Fig. 2 Illustration of proposed design

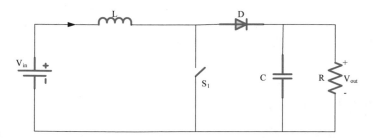

Fig. 3 Circuit diagram of the boost converter

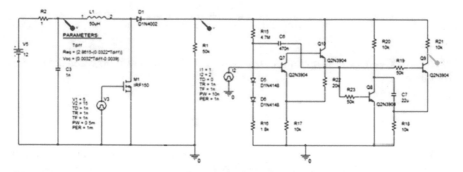

Fig. 4 Circuit design of proposed TEG with DC-DC boost converter

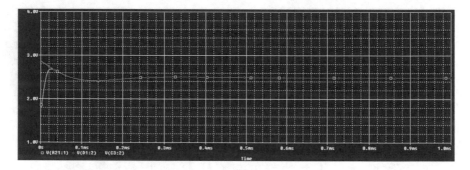

Fig. 5 Simulation result of proposed design

4 Conclusion

The proposed self-powered pacemaker has greater reliability, noiseless operation. Due to the elimination of its battery requirement, the size of this self-powering device will enhance the remarkable requirement of a bigger area in the device as well as tedious surgical operations whenever battery life elapses. Hence, the cost for operations gets erased. TEG which is a sustainable energy source eliminates the need for external sources. DC boost converter with TEG boost up low input voltage and makes perfect regulator for lower temperatures.

References

1. Awolusi I, Marks E, Hallowell M (2018) Wearable technology for personalized construction safety monitoring and trending: review of applicable devices. Autom Construct 85:96–106
2. Liu C, Jiang C, Song J, Chau KT (2018) An effective sandwiched wireless power transfer system for charging implantable cardiac pacemaker. IEEE Trans Industr Electron 66(5):4108–4117
3. Tondato F, Bazzell J, Schwartz L, Mc Donald BW, Fisher R, Anderson SS, Galindo A, Dueck AC, Scott LR (2017) Safety and interaction of patients with implantable cardiac defibrillators

driving a hybrid vehicle. Int J Cardiol 227:318–324
4. Kumar A, Kumar M, Komaragiri R (2018) Design of a biorthogonal wavelet transform based R-peak detection and data compression scheme for implantable cardiac pacemaker systems. J Med Syst 42(6):102
5. Jackson N, Olszewski OZ, O'Murchu C, Mathewson A (2017) Shock-induced aluminum nitride based mems energy harvester to power a leadless pacemaker. Sens Actuat A 264:212–218
6. Kumar A, Berwal D, Kumar Y (2018) Design of high-performance ecg detector for implantable cardiac pacemaker systems using biorthogonal wavelet transform. Circuits Syst Signal Process 37(9):3995–4014
7. Maury P, Rollin A, Monteil B, Mondoly P, Capellino S (2017) High density mapping of inappropriate sinus tachycardia further looks into potential mechanisms. Indian Pacing Electrophysiol J 17(4):116–119
8. Alhawari M, Kilani D, Mohammad B, Saleh H, Ismail M (2016) An efficient thermal energy harvesting and power management for μWatt wearable BioChips. In: 2016 IEEE international symposium on circuits and systems (ISCAS), pp 2258–2261. IEEE
9. Siouane S, Jovanović S, Poure P (2016) Equivalent electrical circuit of thermoelectric generators under constant heat flow. In: 2016 IEEE 16th international conference on environment and electrical engineering (EEEIC), pp 1–6. IEEE
10. . Xu Q, Zhang F, Xu L, Leung P, Yang C, Li H (2017) The applications and prospect of fuel cells in medical field: a review. Renew Sustain Energy Rev 67:574–580
11. Bouzelata Y, Kurt E, Uzun Y, Chenni R (2018) Mitigation of high harmonicity and design of a battery charger for a new piezoelectric wind energy harvester. Sens Actuat A 273:72–83
12. Watson TC, Vincent JN, Lee H (2019) Effect of DC-DC voltage step-up converter impedance on thermoelectric energy harvester system design strategy. Appl Energy 239:898–907

High Performance AXI4 Interface Protocol for Multi-Core Memory Controller on SoC

Ahmed Noami, B. Pradeep Kumar, and P. Chandrasekhar

Abstract The main issue for SoC design in this era is not only which IP cores SoC integrate, but also these IP cores or blocks how they will be interconnected to get high performance data exchange. AXI4 interface protocol is one of the widely SoC interfaces used in several recent designs. In this paper, we designed a memory controller for multi-core processors using the AXI4 interface protocol to improve the performance of the SoC design of the existing work. The multi-core processors represented by JTAG-to-AXI Master IP core and the memory controller represented by AXI BRAM Controller IP core which manages the write and read transactions of multi-core processors while accessing the main memory. Regardless of the type of main memory, Block RAM IP core represents the main memory. Each core processor can access only individual Block RAM which represents the address space of the core processor using the AXI4 interface protocol. The design is successfully implemented on FPGA ZedBoard (xc7z020clg484-l) with a maximum speed frequency of 100 MHz.

Keywords SoC · AMBA AXI4 protocol · Write transaction · Read transaction

1 Introduction

The SoC provides high speeds and good power efficiency, thus, these features are widely used in a variety of embedded systems such as smartphones and laptops. For efficient data transaction between the blocks on SoC, the interface reacts with

A. Noami (✉) · B. Pradeep Kumar · P. Chandrasekhar
Department of Electronics and Communication Engineering, College of Engineering, Osmania University, Hyderabad, India
e-mail: mrahmedyahya883@gmail.com

B. Pradeep Kumar
e-mail: pradeep.boge@gmail.com

P. Chandrasekhar
e-mail: sekharpaidimarry@gmail.com

© The Author(s), under exclusive license to Springer Nature Singapore Pte Ltd. 2021   131
K. A. Reddy et al. (eds.), *Data Engineering and Communication Technology*,
Lecture Notes on Data Engineering and Communications Technologies 63,
https://doi.org/10.1007/978-981-16-0081-4_14

specific standards are used. ARM has introduced the AMBA standard of the interface, Advanced eXtensible Interface (AXI) interface has introduced as the third-generation. High performance and high clock frequency system design is the main target of this third-generation interface. AMBA AXI protocol standard specifies a description of the signal such as addressing options for data transfers, response signaling, and channel handshake. It also specifies the process and properties of the transactions such as the size of data in each transfer in bytes, burst mode type and protection type, and the number of data transactions in a burst. AMBA AXI has separate data channels and address control. It is also supported for unaligned data transactions using byte strobes and the data transactions that are out of order [1]. The previous versions of this protocol are AXI and AXI3. The new version now of this AMBA AXI interface is AXI4. The AXI4 interface is widely used in several systems to increase system performance and decrease the latency of SoC design. AXI4 interface protocol helps in defining the five channels of transactions namely: write and read data channels, write and read address channels, and write response channel [2–5]. All the data transferred through these channels describes the nature of data transactions whether these transactions are read type or write type. The data transaction takes place between one/many masters and one/many slaves through the AXI interconnect IP core. The AXI protocol version 4 is available now in three types such as AXI4-Lite, AXI4, and AXI4 Stream [6].

2 Literature Survey

Nowadays, the interface between the different IP cores on the same chip became the big challenge while all the researchers trying to design the proper interface protocol to interconnect all the IP cores and get the high performance of data exchange. The AXI4-Lite interface protocols designed as a master interface to exchange data with a customized memory [7]. The memory controller for a single-core processor using the AXI4-Lite interface is designed [8]. A united memory controller for multi-core processors using the AXI4-Lite interface is designed [9]. The AXI4-Lite interface that used in all the previous works supports only 1 beat data transfer and fixed burst length (only 1 burst) with a transfer size 32 bits.

The main contribution of this work is using the AXI4 interface protocol to improve the system performance of the memory controller for multi-core processors on SoC. In this paper, we proposed the AXI4 interface protocol for the same memory controller for multi-core processors of the previous work [9]. The proposed interface in this paper can support up to 256 data transfers for INCR burst type with data transfer size 32 bits width.

3   Proposed Model

The proposed design model of the multi-core memory controller SoC is shown in Fig. 1. A different intellectual property (IP) existing in the design which represents all components of the memory controller for multi-core processors. The first IP is JTAG-to-AXI [10]. It is an IP core using to initiate the write and read transactions as multi-core processors using the AXI4 interface protocol mode. The second IP is AXI Block RAM (BRAM) controller [11]. It is a united memory controller that receives requests from multi-core processors and manages the requests for accessing the off-chip memory (which is represented by the BRAM in this work). This memory controller can also support different interface protocols such as AXI3, AXI4-Lite, and AXI4. We used four AXI BRAM Controllers with AXI4 interface protocol to manage the requests from four core processors. Each core processor can manage by an independent AXI BRAM Controller with the AXI4 interface protocol. The third IP core is block memory generator [12]. It is an IP core that creates the block memory (BRAM) which represents a portion of the off-chip memory that only one processor can access it. In other words, we used four BRAM in the design which represents four portions of the off-chip memory and each processor can access only one portion. The fourth IP is AXI interconnect [13]. It is an IP core that allows connects one or more of AXI master (core processor/s) and one or more AXI slave (AXI BRAM Controller/s), which can be different in data width, clock domain, and AXI protocol (AXI3, AXI4, or AXI4-Lite). We used this IP core to connect one master (JTAG_to_AXI Master) that can be representing the multi-core processors and multiple slaves (AXI BRAM Controller) using the AXI4 interface protocol to improve the system performance of the design. The last IP core is processor system reset [14]. We used this IP core to reset all the different IP cores available in our design.

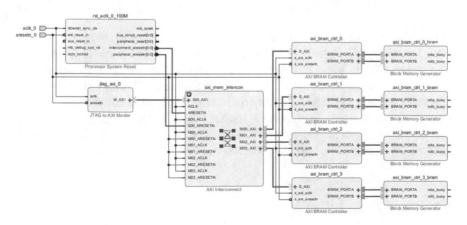

Fig. 1 Proposed model

The four core processors in the proposed design can write and read using the AXI4 interface protocol in different scenarios such as different burst lengths, but with the same data transfer size and burst type INCR as shown in the real-time implementation results section. Independent channels are for write and read transactions with different signals for each channel. The independent write and read channels of the AXI4 interface protocol feature make the sequential transaction easier. The entire write and read transaction signals scenario of the proposed design is discussed in the following two sub-sections.

3.1 Write Transaction

All the signals of the write transaction for multi-core processors using the AXI4 interface protocol with INCR type are discussed in this section. In the beginning, when the clock (aclk_0) and reset (areset_0) signals of the proposed design are high, the multi-core processors (JTAG_to_AXI Master) can start the write transaction when the signals of write address channel AWVALID and AWREADY are high, which represents that the write address from the master is valid and the AXI BRAM Controller (Slave) ready to receive the write address from the master, respectively. The signal AWBURST from master to salve indicates that the write transaction will be in the burst mode and represented by binary value $2'b01$ (INCR). The first data byte transaction will be for the first address of the burst transaction $ADDR_1$, otherwise, the data byte transaction will be for the $ADDR_{NEXT}$. To find the $ADDR_{NEXT}$, it should first to find the aligned address. The start address, number of bytes, and burst length of the transaction are the parameters that can calculate the aligned address. Finally, we can find the $ADDR_{NEXT}$ using the following formula:

$$ADDR_{NEXT} = Aligned_{ADDR} + (N-1) \times B \tag{1}$$

where N is the number of data bytes per transaction counted. When N is equal to the burst length (AWLEN + 1), WLAST is asserted. The variable B represents the number of data bytes transferred ($2^{\wedge}AWSIZE$). The write data transaction is transferred when the write channel signals WVALID and WREADY are high, which indicates that the data transmitted from master to slave is valid and the salve is ready to receive the data transmitted from the master. Then, WDATA represents the actual data transferred from the master to the slave. The last signals BVALID, BREADY, and BRESP represent the write response channel of the write transaction.

3.2 Read Transaction

All the signals of the read transaction for multi-core processors using the AXI4 interface protocol with INCR type are works on the same scenarios of All AXI4 interface signals for write transactions. We can compare the data write and data read for each transaction by using the "report_property" of the Tcl Console Command as shown in the real-time implementation results.

4 Real-Time Implementation Results

Debugging the memory controller for multi-core processors is done on FPGA ZYNQ-7 ZC702 Evaluation Board (xc7z020clg484-1). In our proposed model, there are four core processors that can write and read to/from the main memory (BRAM) through the independent AXI BRAM Controller for each core processor. The write and read transactions to/from the main memory in sequential mode for all core processors using the AXI4 interface protocol to improve the system on chip performance. Each core processor can write and read to/from the main memory in different scenarios. The write and read transactions of each core processor can initiate in different burst length with the same data transfer size and burst type "INCR". The JTAG-to-AXI IP core initiates the real-time write and reads transactions using the AXI4 interface protocol at the debugging design stage on FPGA by using the Tcl Console Command of the Vivado tool.

The Tcl Console Command of write transaction using AXI4 interface written in the form such: "create_hw_axi_txn write_txn [get_hw_axis hw_axi_1] -address xxxxxxxx -data {zzzzzzzz} -burst length -type write". This command indicates the start address of write transaction, data write transaction, burst length of the write transaction, and then the type of transaction. For the read transaction using AXI4 interface protocol, the command written such: "create_hw_axi_txn read_txn [get_hw_axis hw_axi_1] -address xxxxxxxx -burst length -type read". This command indicates the start address of the read transaction, burst length of a read transaction, and then the type of transaction only. These Tcl Console Commands for both write and read transactions using the AXI4 interface protocol are only supports the INCR type of the burst mode. After writing Tcl Console Commands for write and read transactions for cach core processor, there are additional Tcl Console Commands to run the write and read transactions on FPGA hardware. These commands are: run_hw_axi [get_hw_axi_txns write_txn] to run write transactions on FPGA and run_hw_axi [get_hw_axi_txns read_txn] to run read transactions on FPGA. Finally, we used the following Tcl Console Commands to get the final report about write and read transactions. These commands are: "report_property [get_hw_axi_txnswrite_txn]" and "report_property [get_hw_axi_txns read_txn]" for write and read transactions, respectively.

In this FPGA implementation results section, we have implemented our proposed model in different scenarios for each core processor as shown in screenshots of Figs. 2, 3, 4, and 5 of the FPGA implementation write and read transaction results using the AXI4 interface protocol. Our model has proposed four core processors and each core processor can only access a specific memory address space represented by Block RAM using the AXI4 interface protocol in different modes of write and read

Fig. 2 Write and read transactions of the first core processor using the AXI4 interface

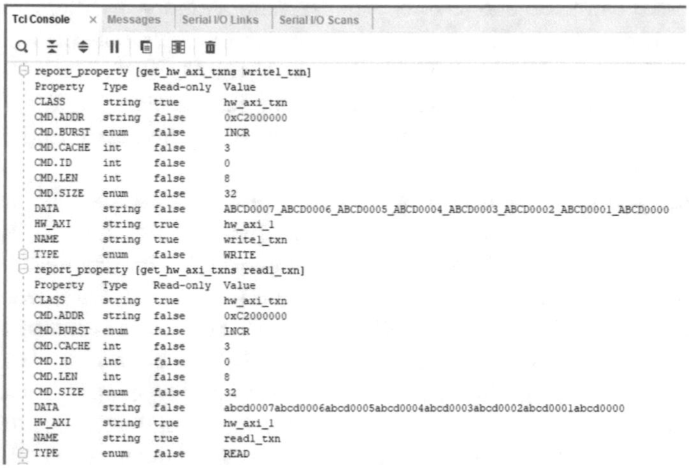

Fig. 3 Write and read transactions of the second core processor using the AXI4 interface

transactions. The start addresses according to our design are C0000000, C2000000, C4000000, and C6000000 of the first, second, third, and fourth core processors, respectively.

The first and second core processors can access the main memory by memory address space with starting address 0xC0000000 and 0xC2000000, respectively. The main specifications of the AXI4 interface protocol of the first and second core processors to access the main memory are: burst lengths are 2 and 4, respectively, data transfer size of both cores is 32 bits width, and the burst type of both core processors is INCR type. The final report of the write and read transactions of the first and second core processors is shown in Figs. 2 and 3, respectively.

```
Tcl Console × Messages Serial I/O Links Serial I/O Scans
report_property [get_hw_axi_txns write2_txn]
Property   Type    Read-only Value
CLASS      string  true      hw_axi_txn
CMD.ADDR   string  false     0xC4000000
CMD.BURST  enum    false     INCR
CMD.CACHE  int     false     3
CMD.ID     int     false     0
CMD.LEN    int     false     4
CMD.SIZE   enum    false     32
DATA       string  false     ABCD0003_ABCD0002_ABCD0001_ABCD0000
HW_AXI     string  true      hw_axi_1
NAME       string  true      write2_txn
TYPE       enum    false     WRITE
report_property [get_hw_axi_txns read2_txn]
Property   Type    Read-only Value
CLASS      string  true      hw_axi_txn
CMD.ADDR   string  false     0xC4000000
CMD.BURST  enum    false     INCR
CMD.CACHE  int     false     3
CMD.ID     int     false     0
CMD.LEN    int     false     4
CMD.SIZE   enum    false     32
DATA       string  false     abcd0003abcd0002abcd0001abcd0000
HW_AXI     string  true      hw_axi_1
NAME       string  true      read2_txn
TYPE       enum    false     READ
```

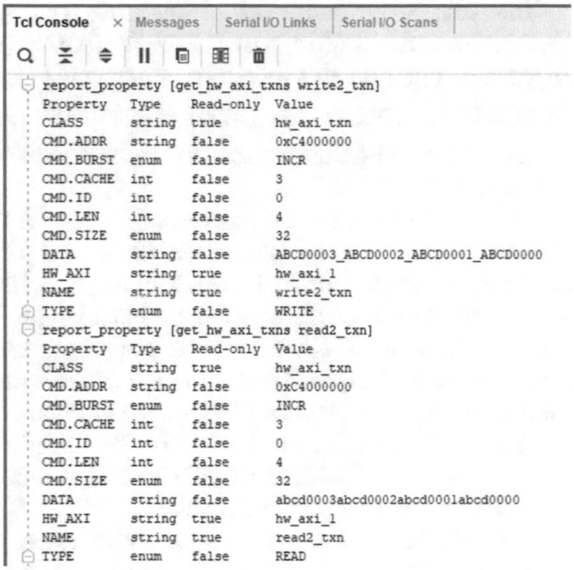

Fig. 4 Write and read transactions of the third core processor using the AXI4 interface

```
Tcl Console × Messages Serial I/O Links Serial I/O Scans
report_property [get_hw_axi_txns write3_txn]
Property   Type    Read-only Value
CLASS      string  true      hw_axi_txn
CMD.ADDR   string  false     0xC6000000
CMD.BURST  enum    false     INCR
CMD.CACHE  int     false     3
CMD.ID     int     false     0
CMD.LEN    int     false     2
CMD.SIZE   enum    false     32
DATA       string  false     ABCD0001_ABCD0000
HW_AXI     string  true      hw_axi_1
NAME       string  true      write3_txn
TYPE       enum    false     WRITE
report_property [get_hw_axi_txns read3_txn]
Property   Type    Read-only Value
CLASS      string  true      hw_axi_txn
CMD.ADDR   string  false     0xC6000000
CMD.BURST  enum    false     INCR
CMD.CACHE  int     false     3
CMD.ID     int     false     0
CMD.LEN    int     false     2
CMD.SIZE   enum    false     32
DATA       string  false     abcd0001abcd0000
HW_AXI     string  true      hw_axi_1
NAME       string  true      read3_txn
TYPE       enum    false     READ
```

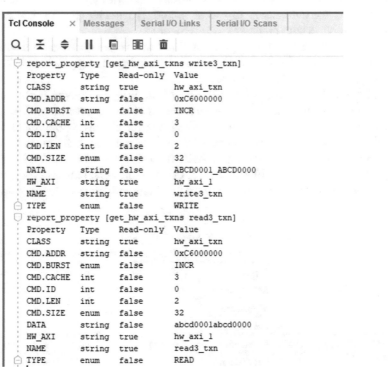

Fig. 5 Write and read transactions of the fourth core processor using the AXI4 interface

The data write and read of the first and second core processors are: {ABCD0015_ABCD0014_ABCD0013_ABCD0012_ABCD0011_A BCD0010_A BCD0009_ABCD0008_ABCD0007_ABCD0006_ABCD0005_ABCD0004_A BCD0003_ABCD0002_ABCD0001_ABCD0000}, and {ABCD0007_ABCD000 6_ABCD0005_ABCD0004_ABCD0003_ABCD0002_ABCD0001_ABCD0000}, respectively.

The third and fourth core processors can access the main memory by memory address space with starting address 0xC4000000 and 0xC6000000, respectively. The main specifications of the AXI4 interface protocol of the third and fourth core processors to access the main memory are: the burst lengths are 8 and 16, respectively, the data transfer size of both cores is 32 bits width, and the burst type of both core processors is INCR type. The final report of the write and read transactions of the third and fourth core processors are shown in Figs. 4 and 5, respectively.

The data write and read of the third and fourth core processors are: {ABCD0003_ABCD0002_ABCD0001_ABCD0000}, and {ABCD0001_ABCD0000}, respectively.

The real-time implementation results of write and read transactions the four core processors using the AXI4 interface protocol shown in the above screenshots are shown only the transactions of {16, 8, 4, and 2} burst lengths for the first, second, third, and fourth core processors, respectively.

The remaining real-time implementation results of another different burst lengths for the four core processors of the proposed model are directly written in Table 1 and compared with the previous work [9].

The FPGA logic utilization summary of the proposed model is shown in Table 2. We observed that the utilization percentages of the proposed model are increased than the previous work [9]. The main reasons for increasing the logics utilization in the proposed model are to increase the burst length of data transfer size of each core processor transaction using the AXI4 interface protocol as shown in screenshots results and remaining results in Table 1. Increase the burst length of the data transfer size transaction for each core processor needs more lookup tables, registers, slices,

Table 1 AXI4 and AXI4-Lite interface protocol comparison

Type of interface	Core processors	Data transfer size	Burst length	Burst type
AXI4-Lite interface protocol [9]	First core	32 bits width	1 transfer	INCR
	Second core			
	Third core			
	Fourth core			
AXI4 interface protocol [proposed]	First core	32 bits width	256 transfers	INCR
	Second core		128 transfers	
	Third core		64 transfers	
	Fourth core		32 transfers	

Table 2 FPGA utilization summary

Logic utilization	Available	Used	Utilization percentage (%)
Slice LUTs	53,200	12,759	23.98
Slice registers	106,400	21,652	20.35
Slice	13,300	6719	50.52
LUT as memory	17,400	2933	16.86
Block RAM tile	140	134	95.71
Bounded IOB	200	2	1

lookup tables as memory, and block RAMs to accommodate all these data transactions. Unlike the AXI4-Lite interface protocol that transfers only one burst length of data transaction.

Nowadays, all the heterogeneous multi-core processors in SoC look forward to improving the entire system performance using the proper SoC interface protocol such as AXI4 full memory map.

5 Conclusion

In this work, we have implemented the AXI4 interface protocol for a multi-core memory controller on FPGA ZedBoard (xc7z020clg484-l) with a maximum speed frequency of 100 MHz. Each core processor write and read to/from the main memory (which is represented by the Block RAM in this work) using the AXI4 interface protocol in different scenarios of burst length with the same data transfer size 32 bits width and burst type "INCR". The first core processor can transfer the data in 32 bits data transfer size with 16/256 different bursts length, respectively. The second core processor can transfer the data in 32bits data transfer size with 8/128 different bursts length, respectively. The third core processor can transfer the data in 32 bits width with 4/64 diffcrent bursts length, respectively. The fourth core processor can transfer the data in 32 bits data transfer size with 2/32 different bursts length, respectively. The proposed model improves the system performance of the entire multi-core memory controller on SoC.

Acknowledgements This work has been supported by the Indian Council for Cultural Relations (ICCR), New Delhi and the TEQIP-III official, University College of Engineering, Osmania University, Hyderabad.

Reference

1. Tidala N (2018) High performance network on chip using AXI4 protocol interface on an FPGA. In: 2018 second international conference on electronics, communication and aerospace technology (ICECA), pp. 1647–1651. IEEE
2. Anitha HT, Nataraj KR (2012) Implementation of multi-slave interface for AXIBus. In: Fourth international conference on advances in recent technologies in communication and computing (ARTCom2012), pp 297–301. IET
3. Krishnaih K, Ravinder Y (2016) Design of memory controller with AXI bus interface. Int J Eng Sci Generic Res (IJESAR)
4. Gindin R, Cidon I, Keidar I (2007) NOC-based FPGA: architecture and routing. In: First International Symposium on Networks-on-Chip (NOCS'07), pp 253–264. IEEE
5. Xiao FM, Li DS, Du GM, Song YK, Zhang DL, Gao ML (2009) Design of AXI bus based MPSoC on FPGA. In: 2009 3rd international conference on anti-counterfeiting, security, and identification in communication, pp 560–564. IEEE
6. AXI Reference Guide. UG1037 (v4.0) (2017). https://www.amba.com
7. Sainath Chaithanya A, Sulthana S, Yamuna B, Haritha C (2020) Design of AMBA AXI4-lite for effective read/write transactions with a customized memory. Int J Emerg Technol 11:396–402
8. Using the AXI4 VIP as a Master to Read and Write to an AXI4-Lite Slave Interface-confluence. https://www.xilinx-wiki.atlassian.net
9. Noami A, Kumar BP, Chandrasekhar P (2020) Design and implementation of a united multi-core memory controller using AXI4-lite interface protocol. Int J Emerg Technol 11:468–475
10. JTAG to AXI MasterIP Product Guide. https://www.xilinx.com
11. AXI BRAM Controller IP Product Guide. https://www.xilinx.com
12. AXI Block Memory Generator IP Product Guide. https://www.xilinx.com
13. AXI Interconnect IP Product Guide. https://www.xilinx.com
14. AXIProcessor System Reset Module IP Product Guide. https://www.xilinx.com

Compensation of Delay in Real-Time HILS for Aerospace Systems

L. A. V. Sanhith Rao, P. Sreehari Rao, I. M. Chhabra, and L. Sobhan Kumar

Abstract The sub-systems of aerospace vehicle can be developed and tested using a platform known as hardware-in-the-loop simulation (HILS). All flight subsystems of vehicle are introduced in HILS test bed for their performance evaluation in real-time scenario. The HILS results are suffering with unwanted diverging oscillations, which limit the HILS test bed effectiveness in evaluating the actual performance of embedded systems of aerospace vehicle. It is found that the oscillations are due to time delay introduced in HILS test bed. This paper describes the analysis of the delay in HILS results and available delay compensation methods. The suitable delay compensation method was implemented to achieve actual HILS results without unwanted delay effect.

Keywords HILS · 6DOF · Embedded system · Delay · Seeker · FMS · Model · Analysis

1 Introduction

The avionic subsystems of aerospace vehicle must be evaluated in respect of their functionality which is very critical just before the integration with each other. The performance of hardware and software is to be validated before going to the actual fields tests. HILS is a place where this validation can be done in real-time closed-loop simulated environment which is nearer to the realistic field scenario. HILS is basic to reduce the number of field tests.

Initially, the aerospace vehicle 6DOF model was developed in a PC and validated thoroughly by inputting proper FLIP (Flight Input Profile) data. The output of 6DOF model should match with all digital simulation reference data given by systems group

L. A. V. Sanhith Rao (✉) · I. M. Chhabra · L. Sobhan Kumar
RCI, DRDO, Hyderabad, India
e-mail: sanhith139@rediffmail.com

P. Sreehari Rao
NIT Warangal, Warangal, India

© The Author(s), under exclusive license to Springer Nature Singapore Pte Ltd. 2021   141
K. A. Reddy et al. (eds.), *Data Engineering and Communication Technology*,
Lecture Notes on Data Engineering and Communications Technologies 63,
https://doi.org/10.1007/978-981-16-0081-4_15

(designer group). The hardware subsystems like onboard computer (OBC), actuator, and sensor package are interfaced with 6DOF model PC through some I/O cards.

The criticality in this configuration is the timing aspect. The exact timing should be achieved between the hardware subsystem and plant model to perform the HILS run.

The plant model has to interact with hardware subsystems as per their defined timing cycles. Lot of programming skill is required to achieve this timing cycles. Moreover, the realistic hardware dynamics matters during this real-time HILS run.

During augmentation of sensor package in HILS test bed, the important surprise is diverging oscillations. This is not due to sensor package dynamics, but it is due to the delay offered by flight motion simulator (FMS) of HILS test bed on which the sensor package is mounted.

The thorough analysis of all these issues was presented. The main critical problem of 'delay' was highlighted in this paper. The delay compensation methods available in the literature were studied thoroughly. A suitable compensation method was implemented to overcome this diverging oscillations. The HILS results with this compensation are efficient and effective in evaluating the subsystems of aerospace vehicle perfectly.

Section 2 describes the different HILS configuration and analysis of HILS results.

Section 3 describes the problems observed during HILS runs. Section 4 describes the different delay compensation methods. Section 5 describes the implementation of suitable delay compensation method in aerospace HILS. Section 6 gives the conclusions.

Detailed literature survey was carried out regarding hardware-in-the-loop simulation [1–29].

2 Analysis of HILS Configurations, Results, and Issues

The main aim of this section is to describe the different configurations of HILS test bed and to understand the issues when actual flight hardware introduced into real-time HILS.

2.1 OBC-Sensor-In-The-Loop Simulation

This is to evaluate the performance of sensor system along with OBC in dynamic condition and to check the effect of its lag and noise on flight performance. The three rate gyros of sensor system get excited in the HILS test bed. The sensor package mounting on flight motion simulator (FMS) along with HILS test setup is shown in Fig. 1. The comparison of input rates to FMS and sensed rates of sensor system are shown in Fig. 2.

Fig. 1
OBC-sensor-in-the-loop
simulation

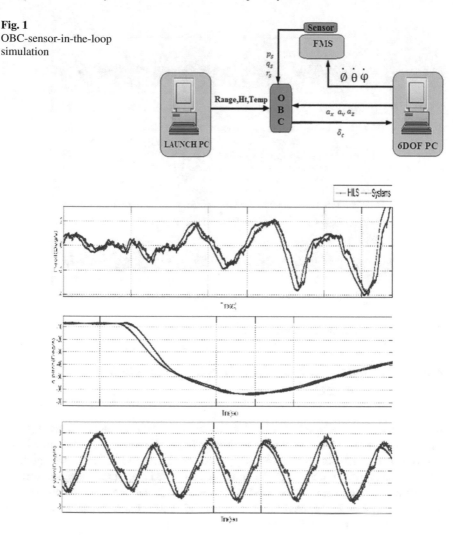

Fig. 2 Input rates versus sensed rates

From the results, it can be observed that there is nearly 15 ms time delay between the inputs and outputs. This delay is the contribution of the flight motion simulator (FMS). During HILS run, these delayed sensed rates are input to the OBC and will be used by control and guidance algorithm. Based on these sensed rates and demanded rates in the algorithm, OBC generates four deflection commands which are again inputs to the model actuator in 6DOF plant PC. The HILS run results are shown in Fig. 3.

144

L. A. V. Sanhith Rao et al.

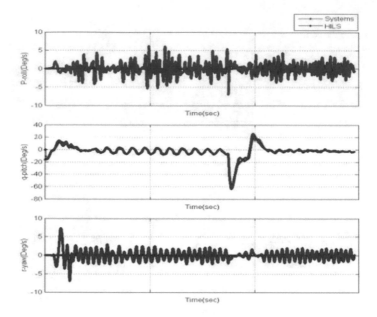

Fig. 3 OBC-sensor-in-the-loop simulation results

2.2 OBC-Sensor-Actuator-In-The-Loop Simulation (Complete HILS)

In this configuration, the model actuator and model sensor in 6DOF PC are replaced by real hardwares. This is known as complete HILS since all the models like OBC, actuator, and sensor in were replaced by real hardware OBC, hardware actuation system, and hardware sensor system and interconnected like flight configuration. All real-time communication is established among OBC, actuator and sensor system. The sensor system is mounted on FMS and OBC, and actuator is placed on the test bench. The configuration diagram is shown in Fig. 4.

The HILS runs were conducted and results shown in Fig. 5.

These results are very critical for analysis of problematic issues in HILS setup. The amplitude and frequency of the parameter roll rate are increased, i.e., ±15 °/s @ 9 Hz when compared with other configurations results, and these oscillations are higher than all digital simulation, i.e., systems results. Moreover, a 60 HZ frequency component also entered into the HILS results. The current drawn by the hardware actuator is increased like anything (max 7Amps) and the deflection feedbacks generated are very noisy with inherent frequency of 10 Hz.

These HILS results are not at all acceptable and the complex embedded systems like OBC, and actuator cannot be developed and evaluated under this testing environment. The objective of HILS cannot be achieved with this test bed limitation.

Fig. 4 OBC-Sensor-
Actuator-In-the-Loop
simulation

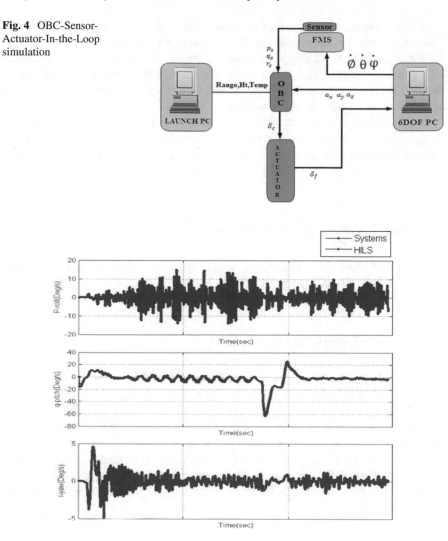

Fig. 5 OBC-Sensor-Actuator-In-the-Loop simulation results

3 Problem Discussion of HILS Results

As shown in the HILS results, the oscillations in the parameters observed high in complete HILS configuration (Fig. 5) only where all the flight hardware are introduced into HILS test bed. When sensor system and OBC in loop run are conducted, the oscillations observed are not prominent even though there is a time delay of 15 ms observed between input and output rates of sensor package. This time delay is due to the lag offered by FMS.

But when sensor package and actuation system both are introduced in HILS test bed, the oscillations problem in HILS runs is predominant.

The critical HILS parameters like roll rate are turned into oscillatory and putting high demand on control and guidance algorithm to stabilize the roll rate of aerospace vehicle. This is due to insufficient phase and gain margins of control and guidance algorithm to overcome the closed-loop delay when all subsystems are in loop.

Hence, the solution for the oscillations problem is compensation of 15 ms time delay offered by FMS. The thorough literature survey was conducted to understand different time delay compensation methods. The brief description about some important methods is given in following section.

4   Delay Compensation Methods

Different delay compensation methods were employed by different people [13–29]. Some important methods are described like below.

4.1   Inverse Compensation Method

This method was introduced by Chen and Ricles [19] and based on simple modeling and to minimize the effect of delay in the system. In this method, the idealization of system response can be considered as linear under the commanded signal. For the system to achieve the command signal from the numerical model, δt is the sampling time and td is the time interval. *td* is equal to $\alpha \, \delta t$. α is the delay constant, which is greater than 1.0. The system time delay is equal to $(\alpha - 1)\delta t$. The compensated signal comprises of command signal of current and previous time steps. In the time domain, it can be understood that this method is an extrapolation using the previous command signals.

4.2   Polynomial Extrapolation Method

Delay compensation is by forward prediction concept. Command signal is a sine wave, and the signal affected by time delay t is the resultant signal. Between the desired command, and the response an error is occurred. The command signal will be shifted forward by K time steps of Δt, such that $K \Delta t$, should be predicted to compensate for this time delay effect. The resulting output is almost nearer to the estimated one by inputting the predicted signal to the system. Polynomial with a proper order is used to predict the steps ahead of present time. It is helpful for obtaining an accurate estimation of future steps approximately. Time delay and sample time were used to estimate the command, and this was introduced by Horiuchi et al. [19].

4.3 Smith Predictor Strategy

This method proposes time delay compensation by comparing transfer functions of the closed-loop systems with time delay and without time delay [28]. If you introduce a Smith predictor in a delayed system transfer function, then the resultant system will be same as ideal system without any time delay. It is assured that the system will be stable with smith predictor. This method has a disadvantage of requiring an accurate model of physical system transfer function. The advantage of this method is it will not distort the original closed-loop system.

4.4 Static and Dynamic Compensation

The phase lead technique is used in this approach to compensate the delay in the force measurement system. The response error-based force compensation technique is used to compensate the dynamic delay introduced in the system by using motion simulators [29]. The proposed approach can effectively compensate the simulation divergence and guarantee the reproduction fidelity. This is confirmed by experiments and simulation results.

5   Implementation of Delay Compensation

The different delay compensation methods were explored thoroughly from the available literature. The inverse compensation method proposed by Chen and Ricles [19] is chosen for delay compensation for aerospace HILS. The gist of inverse compensation method is given like below.

The compensated signal
$$= \alpha * \text{present delayed signal} - (\alpha - 1) * \text{previous delayed signal};$$

where α = delay constant > 1.0;
The value of α chosen is 16 based on following calculation
Delay $= (\alpha - 1)\delta t$; δt = sampling time;
Since the delay is 15 ms & sampling time is 1 ms,
$\alpha = 16$ as per calculation;

Hence, compensated signal $= 16 * \text{present delayed signal} - (16 - 1)$
$$* \text{previous delayed signal} \tag{1}$$

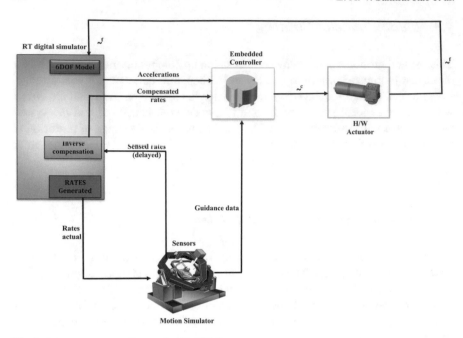

Fig. 6 Inverse compensation method in HILS

$$i.e., \text{Compensated signal} = 16 * \text{present delayed signal} - 15$$
$$* \text{previous delayed signal} \tag{2}$$

Figure 6 shows the block diagram for implementation of inverse compensation method in HILS test bed.

The sensed body rates which are already delayed due to FMS lag are fed to the inverse compensation model residing in real-time simulation PC. Equation (2) was implemented programmatically in the inverse compensation model using above-mentioned equation and the compensated body rates along with model accelerations will go as inputs to embedded controller OBC.

The outputs of OBC, i.e., deflection commands, are used to control the aerospace vehicle flight trajectory. The HILS run will be continued for the complete run time like this, and the results were plotted to analyze the compensation effect.

The comparison of uncompensated HILS results with compensated HILS results is shown in Fig. 7.

The roll rate observed with delay compensation is reduced from ±15 to ±5 °/s which is very less in value and acceptable by the control and guidance algorithm phase and gain margins to maintain the stability of flight vehicle. The total time delay of 15 ms is compensated. The oscillations observed in yaw rate and pitch rate also controlled with this compensation.

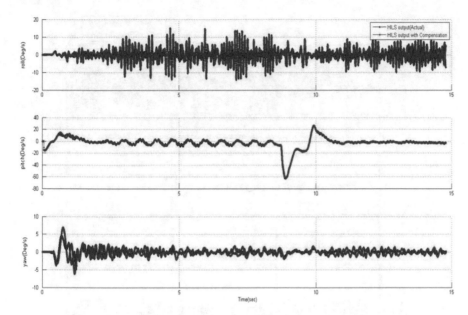

Fig. 7 Compensated versus Uncompensated HILS results

The compensated HILS results were compared with the OBC-in-loop results where the delay effect is not there as shown in Fig. 8. It can be observed that there is a close match between both of them. From this experiment, it is understood that the inverse compensation scheme worked effectively for Aerospace HILS.

The comparison of the HILS results with actual delay and with inverse compensation method is shown in Table 1

From Table 1, it can be concluded that the HILS results with inverse compensation are containing less diverging oscillations when compared to HILS results with delay, and they are more close to the real-time OBC-in-the-loop simulation results where the delay issue is not there.

6   Conclusions

The main cause for unwanted oscillations observed in HILS results was understood in depth. It is concluded that the different systems in HILS test bed profoundly contributing for this delay. Mainly, the delayed response of FMS for input excitation is creating considerable oscillations in HILS runs. This phenomenon is explained with supporting HILS results also.

The inverse compensation method was implemented during HILS runs, and the HILS results are summarized. The compensated results are closely matching with digital simulation results where the delay is not there. With this implementation of

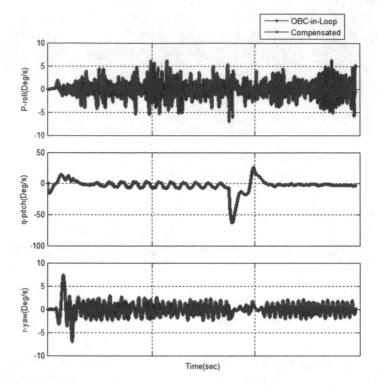

Fig. 8 Compensated HILS results versus OBC in loop HILS results

Table 1 Comparison of HILS results

Typical HILS parameter of aerospace system	With actual delay	With delay compensation
Roll rate	±25 °/s	±5 °/s
Yaw rate	±8 °/s	±2 °/s

delay compensation, the efficiency of HILS test bed is increased and the near flight test scenario was simulated.

Acknowledgements The authors acknowledge and express gratitude to Sri BHVS Narayana Murthy, Director RCI for his encouragement and support to write this paper.

References

1. LeSueur KG, Jovanov E (2009) Hardware-in-the-loop testing of wireless sensor networks. ITEA J 30:333–338

2. Matar M, Saeedifard M, Etemadi A, Iravani R (2011) FPGA-based hardware-in-the-loop simulator for multilevel converter systems
3. Chaudhuri SK, Venkatachalam G, Prabhakar M (1997) Hardware-in-loop simulation for missile guidance & control systems
4. Cosic K, Kopriva I, Kostic T, Slamic M, Volarevic M (1998) Design and Implementation of a hardware-in-the loop simulator for a semi-automatic guided missile system
5. Underwood RC (2007) An open framework for highly concurrent hardware-in-the-loop simulation
6. Matar M, Karimi H, Etemadi A, Iravani R (2012) A high performance real-time simulator for controllers hardware-in-the-loop testing
7. Lizarraga MI, Dobrokhodovy V, Elkaimz GH, Curryx R, Kaminer I (2009) Simulink based hardware-in-the-loop simulator for rapid prototyping of UAV control algorithms. In: Unmanned...Unlimited Conference, Seattle, Washington
8. Duman E (2014) FPGA based hardware-in-the-loop (HIL) simulation of induction machine model. Zhang H, Zhang W, Wu Y, Wang J (2014) Background modeling in infrared guidance hardware-in-loop simulation system. In: Guidance, navigation and control conference (CGNCC), pp 553–557. IEEE Chinese
9. Shen N, Su Z, Wang X, Li Y (2009) Robust controller design and hardware-in-loop simulation for a helicopter. In: 2009 4th IEEE conference on industrial electronics and applications, pp 3187–3191. IEEE
10. Qi Z, Pan F, Hu S (2011) Hardware-in-loop simulation analysis of spacecraft attitude control system. In: 2011 2nd International conference on artificial intelligence, management science and electronic commerce (AIMSEC), pp 1160–1163. IEEE
11. Krenn R, Schaefer B (1999) Limitations of hardware-in-the-loop simulations of space robotics dynamics using industrial robots. European Space Agency-Publications-ESA SP, vol 440, pp 681–686
12. Horiuchi T, Inoue M, Konno T, Namita Y (1999) Real-time hybrid experimental system with actuator delay compensation and its application to a piping system with energy absorber. Earthquake Eng Struct Dynam 28(10):1121–1141
13. Wallace MI, Wagg DJ, Neild SA (2005) An adaptive polynomial based forward prediction algorithm for multi-actuator real-time dynamic substructuring. Proc R Soc A Math Phys Eng Sci 461(2064):3807–3826
14. Chen C, Ricles JM (2009) Improving the inverse compensation method for real-time hybrid simulation through a dual compensation scheme. Earthquake Eng Struct Dyn 38(10):1237–1255
15. Horiuchi T, Konno T (2001) A new method for compensating actuator delay in real–time hybrid experiments. Philos Trans R Soc London. Series A: Math Phys Eng Sci 359(1786):1893–1909
16. Chen C, Ricles JM (2009) Analysis of actuator delay compensation methods for real-time testing. Eng Struct 31(11):2643–2655
17. Ahmadizadeh M, Mosqueda G, Reinhorn AM (2008) Compensation of actuator delay and dynamics for real-time hybrid structural simulation. Earthquake Eng Struct Dyn 37(1):21–42
18. Agrawal AK, Yang JN (2000) Compensation of time-delay for control of civil engineering structures. Earthquake Eng Struct Dyn 29(1):37–62
19. Montazeri-Gh M, Nasiri M, Rajabi M, Jamshidfard M (2012) Actuator-based hardware-in-the-loop testing of a jet engine fuel control unit in flight conditions. Simul Model Pract Theory 21(1):65–77
20. Osaki K, Konno A, Uchiyama M (2010) Delay time compensation for a hybrid simulator. Adv Robot 24(8–9):1081–1098
21. Zebenay M, Boge T, Krenn R, Choukroun D (2015) Analytical and experimental stability investigation of a hardware-in-the-loop satellite docking simulator. Proc Inst Mech Eng Part G J Aeros Eng 229(4):666–681
22. Abiko S, Satake Y, Jiang X, Tsujita T, Uchiyama M (2014) Delay time compensation based on coefficient of restitution for collision hybrid motion simulator. Adv Robot 28(17):1177–1188

23. Ananthakrishnan S, Teders R, Alder K (1996) Role of estimation in real-time contact dynamics enhancement of space station engineering facility. IEEE Robot Autom Mag 3(3):20–28
24. Chang T, Cong D, Ye Z, Han J (2007) Time problems in HIL simulation for on-orbit docking and compensation. In: 2007 2nd IEEE conference on industrial electronics and applications, pp 841–846. IEEE
25. Cao R, Gao F, Zhang Y, Pan D (2015) A key point dimensional design method of a 6-DOF parallel manipulator for a given workspace. Mech Mach Theory 85:1–13
26. Zhang J, Gao F, Yu H, Zhao X (2012) Use of an orthogonal parallel robot with redundant actuation as an earthquake simulator and its experiments. Proc Inst Mechan Eng Part C J Mech Eng Sci 226(1):251–272
27. Hashemi SR, Montazeri M, Nasiri M (2014) The compensation of actuator delay for hardware-in-the-loop simulation of a jet engine fuel control unit. Simulation 90(6):745–755
28. Qi C, Gao F, Zhao X, Ren A, Qian W (2016) A delay compensation approach for hardware-in-the-loop simulation of space collision. In: 2016 35th Chinese control conference (CCC), pp 6211–6216. IEEE
29. Force-based Delay Compensation for Hardware-in-the-loop Simulation Divergence of 6-DOF Space Contact (2017)

Malware Detection System Using Ensemble Learning: Tested Using Synthetic Data

Raghav Kaushik and Mayank Dave

Abstract Malwares refer to the malicious programs that are used to exploit the target system's vulnerabilities, such as a bug or a legitimate software. Malware infiltration can have disastrous consequences on any corporation which includes stealing confidential data, damaging network devices, and crippling of network systems. So, there is a need to filter the malware out from the network and this is achieved with the help of intrusion detection systems; the malware detection model sits at the core of those systems. This paper aims to design a malware detection model using machine learning techniques and training the model on synthetically generated data. In this paper, we first harness or generate synthetic dataset using a tool named CICFlowMeter. CICFlowMeter is a network traffic flow generator tool that captures the network traffic to produce a featured dataset of the network. We first capture the data using the tool, and then, this data would be used to produce synthetic dataset of the network. After that we use various machine learning techniques to build our malware detection model, which is trained and tested using synthetically produced dataset.

Keywords Malware detection · Ensemble learning · K-nearest neighbor · Decision tree · Synthetic data · Intrusion detection system

1 Introduction

Viruses, trojans, worms, and other malicious programs can enter our network and can gain access to our system. After gaining the access, they can steal our confidential data, delete our important files, or can even damage our system devices. So, there is a need to detect the malware as soon as they enter your network. This is done

R. Kaushik (✉) · M. Dave
Department of Computer Engineering, National Institute of Technology, Kurukshetra, India
e-mail: raghav_11710531@nitkkr.ac.in

M. Dave
e-mail: mdave@nitkkr.ac.in

with the help of a network security device named intrusion detection systems (IDS). These devices constantly scan the network traffic flow for malware detection. They use malware detection models at the root level for the detection of malware. So there is a need to design an efficient malware detection model, that could identify the malware on the network accurately and efficiently which we aspire to achieve out of this paper. Attacking a network is a punishable under law, and hence, its detection to most accurate precision is very necessary. It would help the cyber crime agencies to identify the real attackers so that they could be punished under law. Each network is different and has its own vulnerabilities which should be considered while designing the malware detection model. To capture these vulnerabilities, we are going to generate synthetic dataset with the help of the original data this data will be used to train our malware detection model, and in this way, vulnerabilities of the network are captured by the model. There are various machine learning models such as linear regression classification, decision tree classification, and random forest classification, which help in identifying if a particular packet over the network is malware infected or not. The work presented in this paper provides us a way to generate synthetic data from real data captured from the network; this synthetic data would be used to train our model. Then various machine learning techniques are used to design our model, and then the performance of the model is tested against the real labeled dataset. The paper is structured into five sections. Section 2 is for background and past researches done in the field. Section 3 discusses on a technique to create synthetic data for the network and various learning techniques to test this data. Section 4 analyzes the results of the work and touches upon the further work that can be done to extend the work. The last section concludes the proposed work.

2   Literature Work

Technology has become a vital component in today's life. Every sector depends on technology whether its education, defense, health, or business. However, these advancements in the technology also exposed us to threats by unleashing opportunities to various attackers. In the past few years, malware has become the primary security threat for every user over the internet. Malware is a program that intervene into the network with bad intentions and when it executes it reproduces, infects a network poses threat to all the resources available on that network. Malware can damage your network in large number of ways, and they could steal confidential data from the network, delete its important resources, or make those resources unavailable to the intended users. Malware can enter your network from different sources, and it could be from a file downloaded from the Internet, from some external storage devices or any other source. We can classify malwares in various categories like virus, spyware, backdoors on the basis of their propagation technique [1–3]. All this leads us to a question that how can we protect ourselves from malware, how can we prevent it from entering our network, and so on. The answer to all those questions is by designing a robust malware detection system which will help us to filter the malware out of our network which is the aim of this paper.

Earlier the work of malware detection is done with the help the signature matching techniques, in which the signatures of various malwares are recorded and when similar signature is found somewhere else the systems detect them. Snort [4] is an intrusion detection system which uses signature matching technique to identify attacks. Previous studies have shown various advantages and disadvantages of Snort [5, 6]. Snort is perfect for systems in which the attacks are clearly defined. But with growing era of technology the malwares have become smarter. They mutate their signature timely, to prevent them from being caught, making signature matching techniques incapable to detect them. Modern malwares use various ports and protocols which make their detection even more difficult [7]. However, there are traditional malware detection methods like deep packet inspection (DPI) and port-based methods which use port numbers and other packet information for malware detection, but they do not work well for modern malwares. They cannot be used in large networks, because in case of large networks there are large number of packets flowing over the network in real time and inspection of a large network load takes significant amount of time. So we need efficient methods for malware detection in systems where large traffic flow takes place in real time. In this paper, we have proposed a method which uses fast machine learning algorithms for malware detection which are most accurate methods when size of data is large.

In the modern era, the malware detection is done with the help of intrusion detection systems [8]. These systems first study the system to understand its "normal" behavior, and then abnormal behaviors are detected when there is any deviation from "normal" behavior [9]. These devices are capable to identify attacks done by unrevealed malwares. As the signatures of these malwares are not known to us they will produce abnormal behaviors. The major disadvantage of these systems is high false positives rate, and this is because these anomaly-based systems first identify the "normal" behavior of the network and with increasing complexity of the modern networks, and it has become extremely hard to accurately recognize the normal behavior of the network. So there needs to design the intrusion detection system that would give positive only for malware infected applications and not the legitimate applications. Actually, it is a very difficult task to completely remove the false positives from malware detection systems, but their number can be reduced to a greater extent with the help of machine learning techniques, that uses more powerful and efficient algorithms to detect the anomalies, which are used in our work.

In recent years [10], however, several machine learning (ML) techniques like logistic regression, K-nearest neighbor, decision tree classification have been used in the field of malware detection with the hope of improving detection rates and adaptability. These techniques use the latest knowledge of attacks for malware detection. But still as per the recent reports, over 0.35 million new malwares and potentially undesired applications are discovered around the globe [11] which confirms the need of designing an efficient and more robust system for the malware detection.

3 Proposed System

This paper proposes design of a robust malware detection system using various machine learning techniques. Since machine learning systems require data for training so the first milestone for designing a malware detection system is generating a good dataset. To generate our training dataset, we have first captured the real data of our college LAN network and then we used it to make synthetic dataset. Now, here the capturing of our network data is important since we are designing our machine learning model (core of malware detection system) for malware detection on local area network of our college. The machine learning models learn from the training data and predict results on the basis of the relationships they learn from the training data, and we know every network has its own characteristics or relationships and data captured from the network stores those characteristics, so it is important to feed the machine learning model with correct characteristics so that they could make good predictions in future. For capturing the data, we have used a tool named CICFlowMeter [12]. CICFlowMeter is a tool which is used to capture large number of features of live data packets flowing over the network. The reason behind using this tool is that it captures the live network packets and shows 84 different features of that packet and more the information we have about the packet better and more the information could be used by our malware detection model to learn the relationships or characteristics of our network. The most common attack that is possible over the LAN network is DDOS attack. We are basically designing malware detection system to identify this attack. Then we studied all the 84 features of the dataset and picked those features that are relevant for determining that whether a particular packet over the network is responsible for the DDOS attack on network or not. Some relevant features are destination port, total forward packets, total length of forward packets, average packet size, flow bytes/sec, backward IAT standard, acknowledgement flag count, etc. Then this dataset with relevant features can be used to synthesize a dataset which could be used as training data for our malware detection system. For this, firstly, we studied the data packets and then using some python libraries (random etc.) to generate a synthetic dataset. Now this dataset is unlabeled and for training of our machine learning model, we first need to label it and labeling is the most important task as on this the performance of our model depends. So we read about the relationships, every feature has with each other and how they change when a data packet contribute to DDOS attack and then labeled this dataset. Then we used this dataset as a training data for our machine learning model. The dataset is prepared, and there is a need to design a machine learning model for malware detection model. For this, we analyzed how the currently existing malware detection models are working. We found that all the existing malware detection models are using an algorithm at its core to classify whether a given packet is infected or not. We found out that the already existing models are using logistic regression classification, decision tree classification, support vector machine classification at the core to identify the packets for malware detection. Some terms you need to know before moving further:

Table 1 Structure of confusion matrix

	BENIGN (0)	MALWARE (1)
BENIGN (0)		
MALWARE (1)		

TRAINING DATASET: Dataset which would be used by us to train our model. TEST DATASET: Dataset which would be used by us to test our model.

CONFUSION MATRIX: It is a table whose order is of kind $n * n$ (n is the count of possible labels for our data). It plays an important role in determining the performance of our model. In our case, there are two labels BENIGN (for non-infected packets) and MALWARE (for packets that will contribute to DDOS attack over the network) (Table 1).

Each cell of the table contains a numerical value which tells the count of cells that our model predicted with the label on the top and the actual label is the label on the left so more is the number where the left and top labels matches the better our models. The machine learning algorithms that proved to be most efficient in malware detection are K-nearest neighbor classification [13] and decision tree classification [14]. K-nearest neighbor classification is a technique in machine learning in which class of a new case is predicted on the basis of its similarity measure with available cases. Decision tree classification model studies the training data relationships in depth. The end result of the algorithm is a decision tree consisting of decision nodes and class nodes, where decision nodes are non-leaf nodes that store conditions used for classification and class nodes are the leaf nodes that stores values of predicted classes. Then we use this decision tree to predict the class for the test data.

In the proposed system, we have used ensemble learning techniques to design our machine learning model for intrusion detection. Ensemble learning is an approach in machine learning in which the system is designed using multiple machine learning algorithms. Ensemble learning approach is better than traditional machine learning techniques as it builds a set of hypothesis using training data and later combine them to form a new hypothesis; unlike traditional techniques in which whole training data is used to form a single hypothesis [14]. Ensemble learning approach is used in the systems with large training data size. In the large data-sized systems, the data is partitioned into a number of subsets and each subset is used to train a separate classifier, then an appropriate combination rule is used to combine different classifiers. On the other hand, in the small data-sized systems, different classifiers are building using basic learners and then all the classifiers are trained using whole system's data. Later these classifiers are combined using appropriate combination rules. Irrespective of the size of the system, main objective of ensemble learning is to enhance the performance of the predictive model by using a combination of classifiers. Previous researches and studies have shown that generally the use of ensemble learning increases the accuracy and robustness of the systems [15]. In our work, we first made the machine learning model for malware detection using K-nearest neighbor classification and decision tree classification individually and then we designed an ensemble machine learning

Table 2 Confusion matrix for K-nearest neighbor classification model

	0	1
0	5771	11
1	206	12

Table 3 Confusion matrix for decision classification model

	0	1
0	5554	228
1	204	14

model for malware detection using both of these algorithms (K-nearest neighbor classification and decision tree classification). All the three models are trained and tested against synthetically generated dataset.

A. *K-Nearest Neighbor Classification Model*

When we made model using K-nearest neighbor classification algorithm, choosing 70% of our dataset as training dataset and 30% as test dataset. We got the below confusion matrix (Table 2).

So KNN classification makes 5783 correct predictions and 215 incorrect predictions. So accuracy is about 96.18%; however, it depends on the training and test data size.

B. *Decision Tree Classification Model*

When we made a model using decision tree classification algorithm, with training dataset and test dataset size 70 and 30%, respectively, we got the following results (Table 3).

So decision tree classification makes 5568 correct predictions and 432 incorrect predictions. So accuracy is about 92.8%.

C. *Ensemble Model (using both K-Nearest Neighbor Classification and Decision Tree Classification)*

The prior two are the models which are used in the present systems used for intrusion malware detection. So to move one step forward, our ensemble learning model has used both of these algorithms. To implement ensemble learning, we use VotingClassifier class of sklearn.ensemble class of python. This helps in ensembling these two models together for our system. It takes into consideration values predicted by all models individually for each input and goes with value predicted by most of the models. This indeed helps in improving the accuracy of model. Keeping the training data and test data size as 70, 30%, we got the following results (Table 4).

So ensemble model makes 5817 correct predictions and 183 incorrect predictions. So accuracy is about 97% (approx).

Table 4 Confusion matrix for ensemble model

	0	1
0	5770	10
1	173	47

4 Result Analysis and Further Work

To analyse the performance of the ensemble model, we need to compare its perfor
mance with the two component models, i.e., K-nearest neighbor classification model
and decision tree classification model. We are going to use the accuracy, precision,
and recall measure of the model to compare its performance. For better visualization,
we are also plotting the results for each model. However, we have more than two
attributes for each packet (more than 2 dimensions) so to plot the results in 2D we
used a dimensionality reduction technique named [16] component analysis (PCA).
It takes into account the effect of all the variables on the label and makes two new
attributes PC-1, PC-2. For each point in the figure, the model predicted a label either
Benign or Malware, so the light red region in the background represents the data
points for which our model have predicted Benign label and light green region (not
visible in the plot) represents the data points for which it predicted the Malware label.
The orange dots are the packets in the dataset with actual label Benign and the green
dots are for the packets in the dataset which have Malware label.

1.   K-Nearest Neighbor Model

True positive	False positive	True negative	False negative
12	206	5771	11

$$\text{Precision} = (\text{no. of true positives})/(\text{ no. of true positives} + \text{no. of false positives})$$
$$= 0.055 \text{ (approx)}$$

$$\text{Accuracy} = (\text{no. of true positives} + \text{no. of true negatives})$$
$$/(\text{total number of predictions made})$$
$$= 96.18\%$$

$$\text{Recall} = (\text{no. of true positives })/(\text{no. of true positives} + \text{no. of false negatives})$$
$$= 0.52 \text{ (approx)}$$

Now the accuracy is quite high but the low precision and recall values indicate
that we need a better model (Figs. 1 and 2).

Fig. 1 Visualization of training set results for *K*-nearest neighbor classification model

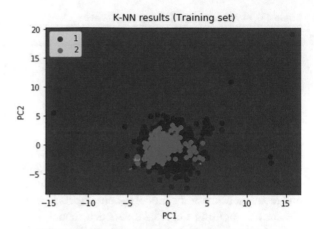

Fig. 2 Visualization of training set results for *K*-nearest neighbor classification model

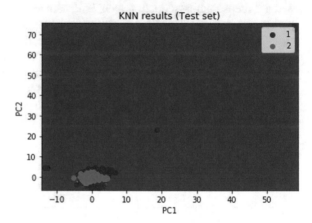

2.   Decision Tree Classification Model

True positive	False positive	True negative	False negative
14	204	5554	228

Precision = (no. of true positives)/(no. of true positives + no. of false positives)
= 0.064(approx)

Accuracy = (no. of true positives + no. of true negatives)
/(total number of predictions made)
= 92.8 %

Recall $=$ (no of true positives)$/$(no of true positives $+$ no of false negatives)

$\qquad = 0.57$ (approx)

Here, we observe a slight increase in the recall and precision values but the value of accuracy is significantly decreased which indicates the use of a better machine learning model (Figs. 3 and 4).

3. *Ensemble Model*

True positive	False positive	True negative	False negative
47	173	5770	10

Precision $=$ (no. of true positives)$/$(no. of true positives $+$ no. of false positives)

Fig. 3 Visualization of training set results for decision tree classification model

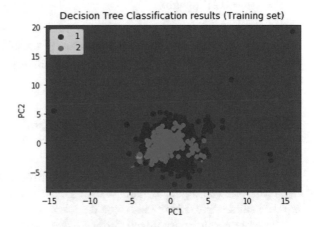

Fig. 4 Visualization of training set results for decision tree classification model

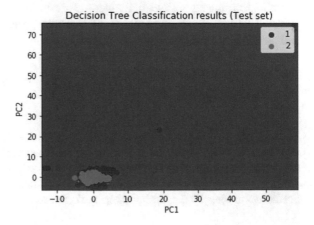

Fig. 5 Visualization of
training set results for
ensemble model

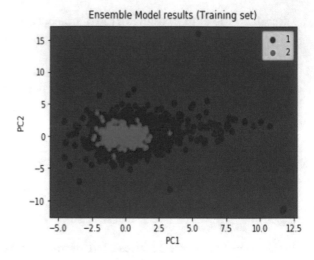

= 0.213(approx)

$$Accuracy = (no.\ of\ true\ positives + no.\ of\ true\ negatives)$$
$$/(total\ number\ of\ predictions\ made\ )$$
$$= 97\,\%$$

$$Recall = (no.\ of\ true\ positives\ )/(no.\ of\ true\ positives + no.\ of\ false\ negatives)$$
$$= 0.82\ (approx)$$

This model is surely better than the previous two models as the accuracy, precision, and recall values are higher in this case which indicates that this model is highly robust and well suited for use in the intrusion detection systems (Figs. 5, 6 and 7).

The system can be improved further using large datasets and more complex machine learning algorithms like random forest, support vector machine, and naive Bayes. This can improve the accuracy of our malware detection model. Deep learning techniques like neural networks, data mining techniques could also be used.

5   Conclusion

To conclude, our machine learning model which uses ensemble learning works well on the synthetically generated dataset of the LAN network, although further improvements can be achieved in the model using the advanced machine learning and deep learning techniques. Ensemble learning methods help to increase the robustness of predictive models by enhancing their performance.

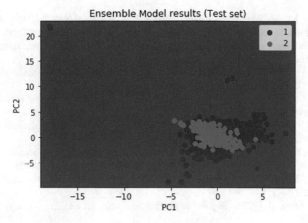

Fig. 6 Visualization of test set results of ensemble model

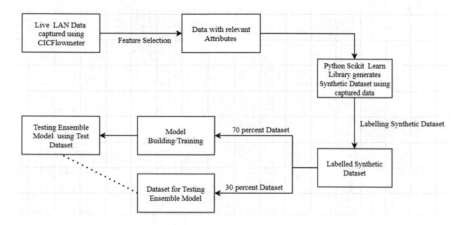

Fig. 7 Schematic diagram of the process for the proposed system

References

1. Adleman L (1990) An abstract theory of computer viruses (invitedtalk). In: CRYPTO '88: Proceedings on advances in cryptology, pp 354–374
2. Filiol E (2005) Computer viruses: from theory to applications. Springer, Heidelberg
3. Morriset G, McGraw G (2002) Attacking malicious code: report to the infosec research council. IEEE Softw 17(5):33–41
4. Sourcefire (2015) Snort network intrusion detection system web site. Available: https://www.snort.org/
5. Roesch M (1999) Snort—lightweight intrusion detection for networks. In: Proceedings of 13th LISA, pp 229–238
6. Sangwan DOP, Kumar V (2012) Signature based intrusion detection system using SNORT. Int J Comput Appl Inf Technol 1(3):35–41
7. Zuev D, Moore AW (2005) Internet traffic classification using bayesian analysis techniques. In: ACM SIGMETRICS performance, evaluation review. ACM, pp 50–60

8. Vázquez E, Maciá-Fernández G, Diaz-Verdejo J, Garcia-Teodoro P (2009) Anomaly-based network intrusion detection: techniques, systems and challenges. Comput Secur 28(1–2):18–28
9. Gong F (2003) Deciphering detection techniques: Part II. Anomaly-based intrusion detection. McAfee Security, White Paper
10. Zamani M (2013) Machine learning techniques for intrusion detection. arXiv:1312.2177v1 [cs: CR], 8 Dec 2013
11. AV-TEST Org. Malware Statistics. Available from: https://www.av-test.org/en/statistics/malware/. Accessed May 2020
12. CICFLOWMETER, Available from: https://www.netflowmeter.ca/netflowmeter.html. Accessed Mar 2020
13. Cai Y, Ji D, Cai D (2010) A KNN research paper classification method based on shared nearest neighbor. In: Proceedings of NTCIR-8 Workshop Meeting, 15–18 June 2010
14. Kumar S, Sharma H (2016) A survey on decision tree algorithms of classification in data mining. IJSR 5(4)
15. Maclin R, Opitz D (1999) Popular ensemble methods: an empirical study. J Artif Int Res 11:169–198
16. Mishra SP, Taraphder S, Swain DP, Sarkar U, Dutta S, Saikhom R, Panda S, Laishram M (2017) Multivariate statistical data analysis principal component analysis (PCA). Int J Livestock Res 7, 5 May 2017

Low Complexity Fused Floating Point FFT Using CSD Arithmetic for OMP CS System

Alahari Radhika, Kodati Satyaprasad, and Kalitkar Kishan Rao

Abstract In any real-time application, overall system performance is characterized by accuracy in arithmetic used for computation and its ability to support low-cost hardware implementation. Accordingly, optimization methods for floating point computation have emerged owing to improved precision level. In particular, for floating point FFT computation, hardware complexity reduction is prerequisite to narrow down the complexity penalty gap between fixed and floating point FFT models. In this, initially, application-dependent hardware fusion units are modeled to improve the hardware utilization rate during FPU arithmetic computations. Afterwards, reconfigurable CSD-driven integer multiplication is proposed for mantissa computation which precisely reduces the hardware count. Finally, radix factorization is incorporated to reduce complex arithmetic involved in FFT stages. Results show that the algorithm proposed ensures area reduction, operating frequency enhancement, and improved efficiency. Finally, resource utilization and operating performance of the proposed fused CSD FP model are compared with competing FPGA implementation.

Keywords Canonical signed digit (CSD) · Fast Fourier transform (FFT) · Field programmable gate array (FPGA) · Hardware fusion · Orthogonal matching pursuit (OMP)

A. Radhika (✉)
Department of ECE, JNT University, Kakinada, India
e-mail: radhialahari@gmail.com

K. Satyaprasad
Vignan's Deemed to be University, Guntur, India
e-mail: prasadkodati@yahoo.co.in

K. K. Rao
Sreenidhi Institute of Science and Technology, Hyderabad, India
e-mail: kishanrao6@gmail.com

© The Author(s), under exclusive license to Springer Nature Singapore Pte Ltd. 2021   165
K. A. Reddy et al. (eds.), *Data Engineering and Communication Technology*,
Lecture Notes on Data Engineering and Communications Technologies 63,
https://doi.org/10.1007/978-981-16-0081-4_17

1 Introduction

In recent years, many works have been published in regard to the development of hardware efficient floating-point (FP) models to assist error-free applications in DSP and to ensure energy delay product well within the threshold limits for real-time applications. In particular, researchers have focused on fused FP units (FPU) which can perform arithmetic operations encountered in DSP applications.

Compressive sampling (CS) is the highly preferable technique for précised data transmission in many DSP applications. In most cases, sampling theory is predominantly used for signal acquisition for improved data rate [1]. Numerous methods have been investigated to improve the performance metrics of compressive sampling. Also, several works have emphasized on the use of greedy methods for signal reconstruction [2]. This work focuses on the design of appropriate arithmetic modeling for hardware efficient FFT to carry out active correlation optimization in CS signal reconstruction [3]. In the conventional methods, DSP implementations with floating-point core models are performed sequentially, which lead to worst case delay bound. To attain high speed, they need to be performed in parallel, which requires more hardware units and gives poor energy efficiency, but offers optimal throughput rate. On the other side, fusion metrics are used for both high speed and less hardware complexity. However, to attain overall throughput rate its counterpart FP models are far better but do not opt for precise DSP applications with QoS requirement. In most cases, arithmetic level optimization [4] and hardware fusion techniques are widely preferred to narrow down the complexity related issues within FPU units [5]. In [6], various hardware level optimization techniques are used to achieve low latency by incorporating models such as separation of data path propagation, 1's complement subtraction for magnitude computation, compound adders for mantissa addition, approximated leading zeros for post-normalization. In [7] is investigated the acceleration of fused dot-products for field programmable gate arrays (FPGAs)-based systems and its merits over parallelism configuration, pipelining and critical data path reductions are proved. In [8] is developed floating-point fused FFT architecture using merged multiplier. Here, both power consumption and design complexity are considerably reduced in Radix-2 butterfly operations since merged unit performs concurrent significance multiplication.

In some cases, radix methodologies are proposed which greatly improve the overall system performance in terms of speed and complexity. As compared to stage normalization for FFT design twiddle factor optimization, radix-2^4 [9] and radix-2^5 [10] are preferred to address the trade-off measure between throughput requirements over design complexity. In [11] is presented radix-2^5 decomposing methods for high-speed FFT computation with CSD recoding method being used for complex constant twiddle factor multiplication with improved common sub-expression sharing (CSS) schemes for complexity and twiddle factor multiplication reduction. While parallel processing schemes require lesser memory resources with finite twiddle factor optimizations [12], hardware design complexity is greatly reduced with high throughput rate.

Therefore, all the factors discussed so far give a complete scenario of the factors that are required to improve the efficiency of floating-point core enabled FFT computation for OMP signal reconstruction system. The various factors that have motivated this research are:

- Energy-efficient floating-point computation performed with various fusion methods.
- Efficient floating-point framework for OMP system with improved throughput rate and area efficiency.
- Bit-driven canonical signed digit recoding-based multiplication performed by considering only MSB region for shift-based accumulation to avoid performance trade-off.
- FFT twiddle factor factorization performed using radix 2^k methodologies to optimize the complex multiplications.

Organization of this paper is carried out as follows. Hardware sharing and various optimization techniques used for FFT computation, fusion techniques formulated for floating point arithmetic and FFT high radix indices techniques are provided in Sect. 2. Section 3 details the methodologies involved in FPU FFT optimization which includes hardware fusion, CSD and radix 2^2 algorithms. Section 4 analyzes performance metrics of each individual model and its attainable efficiency measures. Finally, conclusions have been drawn.

2 FFT to Reconstruct CS Signal

The proposed orthogonal matching pursuit (OMP) algorithm is the principle technique used for reliable CS signal reconstruction. It shows superior system performance with the limitations of increased computational complexity due to its resource hungry operations such as matrix multiplication and inversion.

- Intense matrix multiplication is necessitated in coefficients selection to find most relevant vectors.
- Signal estimation requires least square measures with matrix computation.

The computational complexity that arises in any of this correlation optimization and the matrix operations limit the OMP algorithm performance. However, optimizations can be carried out while implementing OMP for CS signal reconstruction. The inherent advantages of FPGA devices help in narrowing down the hardware complexity overhead in complex OMP reconstruction algorithms. Here, the computation complexity of correlation optimization during CS reconstruction is optimized by incorporating fast Fourier transform (FFT). All matrix–vector multiplications are replaced by most simplified inner product computation which improves the hardware efficiency significantly, as given below.

Let, time frequency dictionary $= \psi = F * \varphi$.

Here, F and φ are both defined as the Fourier basis of size $N \times N$.

F is formulated by randomly selecting M rows from an identity matrix $N \times N$.

Here, FFT computation is used as a replacement of vector dot product to formulate the correlative vectors as given in Eq. (1)

$$\lambda t = \arg \max_{j=1....N} \text{IFFT}(R_{t-1})|$$ (1)

R_{t-1} is an iterative vector derived from residual set R_o.

3 Floating Point Optimization

3.1 FAS-MO-FDP: Fused Add Sub-Multi Operand-Fused Dot Product

To develop efficient floating-point arithmetic model, unique methodology has been designed to carry out floating-point execution with appropriate selection of preprocessed normalized results. This reduces the computational cost and increases accuracy as a consequence of improved hardware utilization rate and also improves the system performance by producing application specific desired output. The FAS-MO-FDP technique comprises the following processes: First, the addition and subtraction models have to be fused because the input values of the FFT basic computation are always required for this preprocessing. This can be caused by sharing exponent comparison and accumulation units. Then parallel efforts are made for mantissa and exponent calculations during FP multiplication. Finally, post-normalization is integrated with rounding for entire MAC calculations.

3.2 CSD-BO: Reconfigurable Canonical Signed Digit Recoding and Bit Optimization

In general, during multiplication the partial product generation phase consumes maximum resources and partial product accumulation decides path delay propagation overhead. As compared to fixed-point arithmetic floating point model requires mantissa multiplication which deals with 24-bit integer multiplication. This will increase the number of partial product as well as the partial product depth during accumulation. Hence, it is essential to optimize the mantissa multiplication module for improving floating-point multiplier speed and also to narrow down the computational complexity overhead. Though conventional array multipliers with improved booth encoding or parallel processing improves the speed of multiplication but comes with the limitation of large area overhead. On the other side, distributed arithmetic model is used to reduce the complexity overhead with significant path delay extension. Here the dynamic ranges of mantissa computations are used to identify the required bit shifts from mantissa regions.

Floating point multiplication requires larger size integer multiplier which affects the performance rate due to high bit size. A divide-and-conquer approach is widely applied to mitigate problems with higher order integer multipliers. A hierarchical part-division rule is also used to decompose the larger bit width into a multiple deformable model and the multiplications are carried out. According to truncation nature of the mantissa computation in floating point multiplication priorities are shifted toward marginal shifting during CSD computation which avoids less significant shifting of operands.

4 FFT Factorization

FFT radix factorization is accomplished for concurrent hardware complexity reduction and best possible operating performance. Moreover, overall latency overhead due to this floating-point incorporation is also being addressed since these stability measures make the design unified. The sequence length in radix 2 DIF FFT algorithm is expressed in powers of 2 and addition and multiplication are the basic operations involved in radix 2 butterfly structure. FPU FFT radix 2^2 is developed based on divide and conquer, hardware fusion, and DA approach in order to minimize the complexity in computation.

4.1 Radix 2^2 FFT Algorithm

Here, multiplication through successive shift come accumulation exhibits better-performance metrics than all other optimization techniques. In order to mitigate the limitations of any DA techniques over path delay accumulation, the number of multiplications involved in each stage of FFT computation is reduced using radix factorization. The transformation of twiddle factors from their non-trivial origin to trivial nature has been taken place in various stages of FFT computation simply by logical swapping and negation operations during twiddle factor multiplication. Here, index mapping rule is incorporated to mitigate the normalization of twiddle factors in radix-2 DIF FFT. This will greatly reduce the complex twiddle factor multiplication at different levels with index map magnitude order. The overall complex multiplications required for radix-2 DIF FFT computation is reduced using linear three-dimensional index mapping, as follows:

$$n = \frac{N}{2}n_1 + \frac{N}{4}n_2 + n_3 \left\{ n_1, n_2 = 0, 1 n_3 = 0 \sim \frac{N}{4} - 1 \right\}$$

$$k = k_1 + 2k_2 + 4k_3 \left\{ k_1, k_2 = 0, 1 k_3 = 0 \sim \frac{N}{4} - 1 \right\}$$
(2)

The DFT has the form of

$$X(k_1 + 2k_2 + 4k_3) = \sum_{n_3=0}^{\frac{N}{4}-1} \sum_{n_2=0}^{1} \sum_{n_1=0}^{1} x\left(\frac{N}{2}n_1 + \frac{N}{4}n_2 + n_3 \right) W_N^{nk}$$

$$= \sum_{n_3=0}^{\frac{N}{4}-1} \sum_{n_2=0}^{1} \left\{ B_{\frac{N}{2}}^{k_1}\left(\frac{N}{4}n_2 + n_3 \right) \right\} W_N^{\left(\frac{N}{4}n_2 + n_3 \right)(k_1 + 2k_2 + 4k_3)}$$
(3)

where, the FFT stage one has the form of

$$B_{\frac{N}{2}}^{k_1}\left(\frac{N}{4}n_2 + n_3 \right) = x\left(\frac{N}{4}n_2 + n_3 \right) + (-1)^{k_1} x\left(\frac{N}{4}n_2 + n_3 + \frac{N}{2} \right)$$
(4)

Decomposition of radix-2 DIF FFT is represented as follows.

$$W_N^{\left(\frac{N}{4}n_2 + n_3 \right)(k_1 + 2k_2 + 4k_3)} = (-j)^{n_2(k_1 + 2k_2)} W_N^{n_3(k_1 + 2k_2)} W_{N/4}^{n_3 k_3}$$
(5)

Substituting Eq. (4) into Eq. (2) and expanding w.r.t index n_2, a set of 4 DFTs of length $N/4$ can be had.

$$X(k_1 + 2k_2 + 4k_3) = \sum_{n_3=0}^{\frac{N}{4}-1} [H_{N/4}^{k_1 k_2}(n_3) W_N^{n_3(k_1 + 2k_2)}] W_{N/4}^{n_3 k_3}$$
(6)

Then, second stage of FFT $H_{N/4}^{k_1 k_2}(n_3)$) is described as:

$$H_{N/4}^{k_1 k_2}(n_3) = B_{\frac{N}{2}}^{k_1}(n_3) + (-1)^{k_2}(-j)^{k_1} B_{\frac{N}{2}}^{k_1}\left(n_3 + \frac{N}{4} \right)$$
(7)

Here, in stage 1, significant amount of non-trivial twiddle factors are converted into trivial factors $(1, -1, j, -j)$.

4.2   Floating Point FFT for OMP CS

The OMP CS for signal reconstruction follows two steps, namely correlation optimization and least square measure. The influence of computational error is prominent in the correlation optimization stage as compared to the least square step. In particular, the inclusion of FFT-based correlation is intolerant to the erroneous fixed-point arithmetic due to its stage-wise data propagation nature during FFT computation. As compared to fixed-point arithmetic, floating-point models exploit higher dynamic ranges and include fraction parts for twiddle factors. Hence, these can provide significant error reduction with any correlation estimator. Here, overall OMP system integration is carried out using a fused floating-point FFT unit, and its functional verification is carried out using exhaustive test bench and synthesis using FPGA. Therefore, all the factors discussed so far give a complete scenario of the factors that the precision level required to retain full precision accuracy of OMP algorithm using precision model is a motivation in this research.

5 Experimental Results

Verilog HDL and FPGA QUARTUS II EDA synthesizer tool are used, respectively, to model and to synthesize the proposed architecture for state-of-the-art comparison. The fundamental objectives of superior performance and low hardware complexity with less resource utilization are proved from the results of synthesis. The validation of results of performance metrics of fusion methodology and CSD modeling is carried out over conventional FPU. Since the utilization of logic elements is direct indication of hardware required, the hardware complexity reduction is compared in respect of the LEs used as shown in Fig. 1. The utilization of hardware is measured with the help of FPGA hardware synthesizer tool. The benefits of critical path delay reduction due to reduced mantissa bit consideration for CSD shifting and reduced critical path over conventional multiplier-based FP model are proved through delay metrics analysis. As shown in tables listed below, the proposed fused CSD-based FP has shown efficient computational complexity reduction with FFT radix-2^2 factorization.

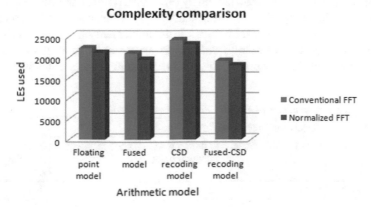

Fig. 1 Performance comparison

Table 1 Hardware complexity comparison of floating point FFT over various optimization and factorization models

Architecture	Radix-2 FFT		Radix-2^2 FFT	
	Multipliers used	LEs used	Multipliers used	LEs used
Conventional model	60	22,169	54	21,059
Fused model floating point	60	20,892	54	19,402
CSD recoding	NIL	24,122	NIL	23,107
Fused—CSD recoding FPU model	NIL	19,128	NIL	18,044

5.1 Performance Analysis

Here, systematic CSD-driven distributed arithmetic-based approach for mantissa multiplication in single precision model with hardware fusion enabled optimized FFT computation is validated over conventional FFT radix indices. The area efficiency of multiplier less mantissa computations and application specific hardware fusion is analyzed separately and given as shown in Table 1. The benefit of critical path delay reduction due to LSB truncated shift-based accumulation in twiddle factor multiplication is also proved through delay metrics analysis.

5.2 Delay Optimization

Generally, degradations in performance arise in floating point arithmetic owing to larger mantissa sizes. Here, MSB regions alone are considered for bit shifting operation which lead to improvement in system performance and the reduction in sequentially dependent FPU operations. This reduction in FPU operations play significant role in reducing overall critical path, as shown in Table 2. The evaluation of maximum operating frequency report is carried out by time quest timing analyzer tool. It shows that CSD FP achieves significant performance improvements not only in complexity reduction at logic registers level but also in logic element utilization levels. Here, the multiplier less unit operating frequency is increased by more than two times. From the Fmax results, it is proved that the proposed FPU core achieves significant

Table 2 Performance analyses of various FPU optimization models

FFT type	Performance report (MHz)			
	Conventional model	Fused model floating point	CSD recoding	Fused—CSD recoding FPU model
Radix-2^2 FFT	31.86	44.22	44.43	52.81

performance improvement. It is also found that the proposed most significant bit-driven CSD FP model achieves reduction in considerable propagation delay over full precision FP model.

6 Conclusion

Here the application-driven hardware fusion and Canonical sign digit-driven shift accumulation-based mantissa computation technique are used to narrow down the computational complexity that arises in floating point arithmetic-based complex FFT computation. By utilizing inherent redundant computational characteristics of FFT butterfly structure hardware sharing is accomplished which significantly reduce computational complexity overhead. It is also proved that multiplier less mantissa computation during FPU multiplication leads notable complexity reduction. Finally, the metrics of these two arithmetic optimization models are validated in twiddle factor optimized FFT architecture. The intrinsic relationship within floating point units and performance metrics of both hardware and twiddle factor optimization methodologies are was explicitly analyzed, and FPGA hardware synthesis is carried out for qualitative and quantitative measurements.

References

1. Candès EJ, Wakin MB (2008) An introduction to compressive sampling. IEEE Signal Process Mag 25(2):21–30
2. Kulkarni A, Mohsenin T (2017) Low overhead architectures for omp compressive sensing reconstruction algorithm. IEEE Trans Circuits Syst I Regul Pap 64(6):1468–1480
3. Pan L, Xiao S, Yuan X, Li B (2014) Joint multiband signal detection and cyclic spectrum estimation from compressive samples. EURASIP J Wireless Commun Netw 2014(1):218
4. Ho CH, Yu CW, Leong P, Luk W, Wilton SJ (2009) Floating-point FPGA: architecture and modeling. IEEE Trans Very Large Scale Integr (VLSI) Syst 17(12):1709–1718
5. Prabhu E, Mangalam H, Karthick S (2016) Design of area and power efficient radix-4 DIT FFT butterfly unit using floating point fused arithmetic. J Central South Univ 23(7):1669–1681
6. Seidel PM, Even G (2004) Delay-optimized Implementation of IEEE floating-point addition. IEEE Trans Comput 53(2):97–113
7. Lopes AR, Constantinides GA (2010) A fused hybrid floating-point and fixed-point dot-product for FPGAs. In: International symposium on applied reconfigurable computing, pp 157–168. Springer, Berlin
8. Min JH, Kim SW, Swartzlander EE (2011) A floating-point fused FFT butterfly arithmetic unit with merged multiple-constant multipliers. In: 2011 Conference record of the forty fifth asilomar conference on signals, systems and computers (ASILOMAR), pp 520–524. IEEE
9. Shin M, Lee H (2008) A high-speed four-parallel Radix-2 4 FFT/IFFT processor for UWB applications. In: 2008 IEEE international symposium on circuits and systems, pp 960–963. IEEE
10. Yang C, Xie Y, Chen H, Ma C (2016) A high-precision hardware-efficient radix-2k FFT processor for SAR imaging system. IEICE Electron Exp 13-20160903

11. Cho T, Lee H, Park J, Park C (2011) A high-speed low-complexity modified Radix-2 5 FFT processor for gigabit WPAN applications. In: 2011 IEEE international symposium of circuits and systems (ISCAS), pp 1259–1262. IEEE
12. Wang J, Xiong C, Zhang K, Wei J (2015) A mixed-decimation MDF architecture for Radix-2^k parallel FFT. IEEE Trans Very Large Scale Integr (VLSI) Syst **24**(1):67–78

CreepOnto: An Avant-Garde Scheme for Framing and Evaluating Ontologies Using Domain Knowledge of Creep Mechanism

Saicharan Gadamshetti, R. Shiddhartha, Gerard Deepak, and A. Santhanavijayan

Abstract A good analysis about a specific field of study is a basic and essential requirement for clarifying various terminologies. Ultimately, this has led to the need for development of ontologies which has proven to be a great asset for individuals and groups that seek to ameliorate the usage of useful unorganised particulars and regenerate it to ensure optimum reusability of the same. This has been further kindled by the contemporary desire for artificial intelligence based systems on content-related theories and mechanisms related to a particular domain. This has created the need for developed ontologies to build on cohesive reasoning purposes. An important domain under material sciences termed creep mechanism has been conceptualised due to lack of well-defined ontologies based on it. In subsequent sections, various concepts related to the study of interest have been identified, defined, and established using the relations between them. Web Protégé, a widely used tool for developing ontologies, aides to provide a well-structured framework for CreepOnto. A flowchart designed based on various classes and acts as a primary source of information for readers by acting as a ready reckoner. An OWL representation of the established ontology presents itself as a cornerstone in making the foundation of ontology using axioms. Finally, its effectiveness is measured using various qualitative and quantitative performance metrics which resulted in a reuse ratio of 0.84.

Keywords Artificial intelligence · Creep · Kullback–Leibler divergence · Material sciences · Ontologies

S. Gadamshetti (✉) · G. Deepak · A. Santhanavijayan
Department of Computer Science and Engineering, National Institute of Technology
Tiruchirappalli, Tiruchirappalli, India
e-mail: saicharangadamshetti00@gmail.com

R. Shiddhartha
Department of Metallurgical and Materials Engineering, National Institute of Technology
Tiruchirappalli, Tiruchirappalli, India
e-mail: rsiddu0511@gmail.com

© The Author(s), under exclusive license to Springer Nature Singapore Pte Ltd. 2021   175
K. A. Reddy et al. (eds.), *Data Engineering and Communication Technology*,
Lecture Notes on Data Engineering and Communications Technologies 63,
https://doi.org/10.1007/978-981-16-0081-4_18

1   Introduction

Ontology is a term used to refer to the process of organising data related to a particular domain/study/field of interest, establishing, and elaborating on their relationships and corroborating the same. In addition to developing existing connections
between various concepts, it also introduces new knowledge about the subject under
scrutiny. In this ontology titled "CreepOnto", a taxonomy is meticulously framed on
the domain of creep mechanism (or cold flow), which is a time-dependent plastic
deformation process that generally happens very slowly that it remains imperceptible
by and large but could suddenly surprise with the enormity of its damage unless it is
closely studied. With the aid of Web Ontology Language, ontology has been created
enlisting all the contents of the domains. New concepts, namely "Aircraft Creep"
and its correlation between various classes have been identified. Various parameters
like "number of classes" have been calculated and evaluated with the help of a bar
graph.

Motivation: The unavailability of a well-defined ontology for creep mechanism,
a very important phenomenon that occurs in materials, propelled to deal with a
semantic approach, and enhance the understanding about this domain. This would
come in handy for metallurgists, material scientists, and mechanical and production
engineers. Moreover, it would assist novices and amateurs to get a vivid image about
this engineering mechanism.

Contribution: Fine and standard ontologies describe a domain in a formulated
manner. In this contemporary era of Semantic Web, ontologies play a vital role on
Web-based intelligence and information retrieval. Knowledge modelling is necessary for all intricate subjects like creep as they act as a guide for developing an
understanding about the subject.

Organisation: The remaining paper is organised as follows. Related literature is
discussed in Sect. 2. Section 3 discusses the ontology design. Development of
the ontology and the OWL representation is made in Sect. 4 and 5, respectively.
Evaluation of CreepOnto is discussed in Sect. 6, and Sect. 7 concludes the paper.

2   Related Literature

Chu and Li [1] have suggested a novel idea for testing of the creep method called
impression creep testing that makes use of a cylindrical shape indenter, whose end is
levelled, to make out a recess on the sample during the transient period. In this innovation, the velocity achieved is steady. Mordike [2] have written about the development
of creep-resistant alloys over the past 20 years. Cannon and Langdon [3] have listed
the experimentally found and collected data which shows the rate of steady-state
creep as a function of various parameters like grain size, temperature, and stress in
the form of creep curves' shape. Weertman [4] have elaborated on static dislocations

observed in materials which are independent of creep mechanism based on dislocation climb. Raj and Ashby [5] problem of sliding at a nonplanar grain boundary is considered in detail. Wilkinson and Ashby [6] have proposed about creep that is formed by physical phenomenon, namely sintering in the form of conclusions made from results of replacing this using time-hardening creep law. Nabarro [7] have proposed a hypothesis for materials with large grain size and vacancies formed and destructed due to dislocations based on steady-state diffusional creep. P H Lin et al. [8] have outlined the contemporary achievements in understanding behaviour of creep using nanoindentation technique, with main concentration on high entropy alloys. Several semantic or ontology-based applications are discussed in [9–19].

3 Design of the Ontology

Creep is of paramount importance as far as material sciences are concerned. In this ontology, this phenomenon is appropriated as the highest order in the hierarchy of the taxonomy. Subclasses like "Stages_Of_development", "Deformation_Mechanism_Maps", "Creep_Testing_Methods", "Mechanisms", "Creep_Life_Prediction_Approaches", "Factors_Affecting_Creep", and "Types_Of_Creep" are salient and important aspects of creep phenomena, i.e. they are *strongly associated* with its super-class "creep". "Primary", "secondary", and "tertiary" are subclasses which completely define the class named "Stages_Of_Development". Hence, they are a part of the latter. Classes like "Aircraft_ Components _Failure _Mode" and "Resistant_Materials" are created to generate a vivid idea on the mind of the readers, although it does not demand an inviolable place in the discussion about creep. Hence, it is weakly associated with its super-class. The subclasses of "Types_Of_Creep", i.e. "logarithmic", "recovery", and "diffusion", do not entirely define it. Hence, they are weakly associated too. On the other hand, classes like "Coble_Creep", "Nabarro-Herring_Creep", and "Harper-Dorn_Creep" are the only subclasses of "Newtonian_Creep". Therefore, all three subclasses are said to be strongly associated. The subclasses under "polymers", "ceramics", and "metals" mention the criteria and form the basis of classifying them under "Resistant_Materials". Hence, they must be specified to completely define them.

The subclasses of "Creep_Life_Prediction_Approaches" do not provide an exhaustive definition for its parent class as many other methods are beyond the purview of this ontology. "Homologous_Temperature", "Applied_Stress", and "Factors_Affecting_Dislocation_Behaviour" entirely elucidate its parent class— "Factors_Affecting_Creep". Hence, they are related by "isAPartOf" relationship with the latter. On the other hand, "Factors_Affecting_Dislocation_Behaviour" have not been completely understood by this ontology as only three—"Atomic_Bonding", "Crystal_Structure", and "Micro-Structural_Features"—have been defined as its subclasses. The same is true for "Micro-Structural_Features" where not all factors have earned a mention. That is why both the subclasses have a part of their respective super-classes. The same applies for the subclasses of the following: "Dislocation", "Creep_Testing_Methods" and "Deformation_Mechanism_Maps". Within the purview of this ontology, classes like "Weertman", "Jogged_Screw" and "Ivanov_And_Yanushkevich" are the only classes in which "Power_Law" can be expounded. A similar trend is seen in the stratification of "Flexural_Creep_Testing" into "Three_Point_Bend_Testing" and "Four_Point_Bend_Testing". In contrast, subclasses like "Ashby_Maps" and "Mohamed-Langdon_Maps" are not the only groups which come under "Deformation_Mechanism_Maps", akin to grouping of "Impression_Creep_Testing_Methods", "Tensile_Creep_Testing", "Stress_Rupture_Testing", "Compression_Creep_Testing", and "Small_Scale_Creep_Testing_Methods". So, different methods are used to classify them according to their relationship with their preceding class to create a better understanding about the topic. Table 1 enumerates various concepts associated with creep and a brief semantic description in addition to the subclass in which they are classified.

4   Ontology Development

Web Protégé 4.0.0 was utilised to develop this ontology. Protégé has been in extensive use in devising semantic concepts, namely objects, classes, and taxonomical classification for development of ontologies. Figure 1 represents the classes and their associated subclasses in a hierarchical order. Subclasses that fall under one topic are indented accordingly to present a vivid idea about the relations between various classes. Figure 2 represents the taxonomy of various concepts associates with creep mechanism. Taxonomical classification, in the context of information handling, primarily refers to the orderly arrangement of domain entities that is curated in accordance with the pre-existing definitions in order to develop a hierarchy amongst them and clearly define the kind of relation existing between two such classes. Taxonomical classification is instrumental in minimising ambiguity and prevents any overlap of semantic concepts by a formal analysis on various classes. It forms the basis of ontology classification and gives an overview of various subclasses.

5   OWL Representation

A batch of machine readable representation languages which is used to describe represent classes, subclasses, properties, and their relations in order to process the content of information instead of just printing, it is termed as Web Ontology Language. OWL representation of an ontology has a multitude of RDF triads which make use of various axioms to classify the classes. OWL representation for CreepOnto has been generated.

Table 1 Sample semantic description

Concepts	Semantic description	Subclass of
Deformation mechanism maps	It is a graph plotted between a classification of stress and temperature (example: shear modulus vs. homologous temperature) by carrying out various tests and experiments on the materials	Creep
Buckling-limited creep	Buckling-limited creep occurs in beams, panels, and other structures that carry compressive loads	Aircraft components failure mode
Compression creep testing	A dog-bone sample of a material is placed in a creep testing machine and a compressive load is applied axially to it	Creep testing method
Tensile creep testing	A tensile load is exerted on the specimen and deformation is measured. The loading is done directly or using a lever arm	Creep testing methods
Automated ball indentation	It is a non-destructive test where sample components are subjected to an external force with the aid of a spherical indenter into the surface in a bid to determine fracture toughness	Small-scale creep testing method
Jogged Screw dislocation model	This model basically looks at the non-conservative motion of the edge jog on a screw dislocation	Power law creep
Composites	Carbon-fibre composites have much higher creep resistance in the fibre direction compared with the anti-fibre direction	Creep-resistant materials
Sherby-Dorn	In this method, a larger temperature is reached at similar stress levels to bring about lesser time duration for failure	Creep life prediction approaches

6 Ontology Evaluation

The ontologies devised are subjected to evaluation before applying them. The evaluation results of CreepOnto after getting responses from a collective set of domain experts have been enlisted in Table 2. An extensive study on various qualities like lawfulness, richness, interpretability, accuracy, relevance, etc., was recorded and categorised from "very high". A graphical representation of the same organised primary data is given in Fig. 3.

Fig. 1 Hierarchical order of various classes and subclasses

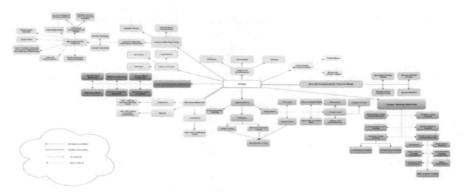

Fig. 2 Taxonomy of CreepOnto

Table 2 Evaluation of CreepOnto based on certain qualitative metrics

Qualitative metrics	Very high	High	Medium	Low
Lawfulness	39	13	3	0
Interpretability	34	20	1	0
Clarity	36	16	3	0
Accuracy	39	14	2	0
Consistency	38	15	2	0
Relevance	35	17	3	0
Comprehensiveness	40	13	2	0
Richness	38	15	2	0

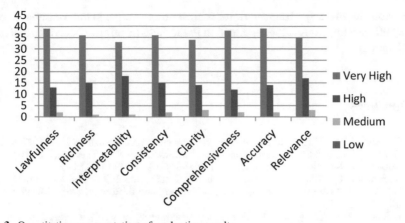

Fig. 3 Quantitative representation of evaluation results

Table 3 List of various parameters against their values

NoC	NoLC	NoP	MxDoI	NoSbC	Reuse ratio
72	51	8	4	64	0.84

The quantitative analysis is further carried out through various parameters namely number of classes (NoC) which refers to the all the concepts under a domain, number of leaf classes (NoLC) that accounts for all the subclasses which inherits knowledge from the primary classes, maximum depth of inheritance (MxDoI) and number of properties (NoP). A fractional parameter named as reuse ratio shows the degree of inheritance. They are listed in Table 3 and the values are calculated based on the ontology visualised. Equation (1) shown defines and is used to calculate the reuse ratio of CreepOnto,

$$\text{Reuse Ratio} = \frac{\text{Number of sub classes}}{\text{Number of classes}} \tag{1}$$

7 Conclusions

The ontology developed is based on the creep phenomena observed in materials. Web Protégé has been utilised to visualise this ontology. Owing to less or nil existence of ontologies in the domain of creep has shaped the idea of CreepOnto. This has proved to be effective as the quality of this ordered classification has garnered an overwhelming positive feedback from a set of fifty-five randomly picked people. Also, the reuse ratio is a whooping 0.84 which acts as a testimonial to the standard of CreepOnto. It provides support to represent the data in a better way. It has been efficacious in offering a framework for representing shareable and reusable knowledge and has enormous future application in the field of informatics, data analytics, and handling.

References

1. Chu SNG, Li JCM (1977) Impression creep; a new creep test. J. Mater Sci 2(11):2200–2208
2. Mordike BL (2002) Creep-resistant magnesium alloys. Mater Sci Eng A 324(1–2):103–112
3. Cannon WR, Langdon TG (1983) Creep of ceramics. J Mater Sci 18(1):1–50
4. Weertman J (1957) Steady-state creep through dislocation climb. J Appl Phys 28(3):362–364
5. Raj R, Ashby MF (1971) On grain boundary sliding and diffusional creep. Metallur Trans 2(4):1113–1127
6. Wilkinson DS, Ashby MF (1975) Pressure sintering by power law creep. Acta Metall 23(11):1277–1285
7. Nabarro FRN (1967) Steady-state diffusional creep. Phil Mag 16(140):231–237
8. Lin PH, Chou HS, Huang JC, Chuang WS, Jang JSC, Nieh TG (2019) Elevated-temperature creep of high-entropy alloys via nanoindentation. MRS Bull 44(11):860–866
9. Deepak G, Teja V, Santhanavijayan A (2020) A novel firefly driven scheme for resume parsing and matching based on entity linking paradigm. J Discr Math Sci Cryptogr 23(1):157–165
10. Deepak G, Santhanavijayan A (2020) Onto best fit: a best-fit occurrence estimation strategy for RDF driven faceted semantic search. Comput Commun 160:284–298
11. Kumar N, Deepak G, Santhanavijayan A (2020) A novel semantic approach for intelligent response generation using emotion detection incorporating NPMI measure. Procedia Comput Sci 167:571–579
12. Deepak G, Kumar N, Santhanavijayan A (2020) A semantic approach for entity linking by diverse knowledge integration incorporating role-based chunking. Procedia Comput Sci 167:737–746
13. Haribabu S, Kumar PSS, Padhy S, Deepak G, Santhanavijayan A, Kumar N (2019) A novel approach for ontology focused inter-domain personalized search based on semantic set expansion. In: 2019 Fifteenth international conference on information processing (ICINPRO), pp. 1–5. IEEE Press
14. Deepak G, Kumar N, Bharadwaj GVSY, Santhanavijayan A (2019) OntoQuest: an ontological strategy for automatic question generation for e-assessment using static and dynamic knowledge. In: 2019 Fifteenth international conference on information processing (ICINPRO), pp 1–6. IEEE Press
15. Kaushik IS, Deepak G, Santhanavijayan A (2020) QuantQueryEXP: a novel strategic approach for query expansion based on quantum computing principles. J Discr Math Sci Cryptogr 23(2):573–584
16. Varghese L, Deepak G, Santhanavijayan A (2019) An IoT analytics approach for weather forecasting using raspberry Pi 3 Model B+. In: 2019 Fifteenth international conference on information processing (ICINPRO), pp 1–5. IEEE Press
17. Deepak G, Priyadarshini S (2016) A hybrid framework for social tag recommendation using context driven social information. Int J Soc Comput Cyber-Phys Syst 1(4):312–325
18. Deepak G, Priyadarshini JS (2018) A hybrid semantic algorithm for web image retrieval incorporating ontology classification and user-driven query expansion. In: Advances in big data and cloud computing, pp 41–49. Springer, Singapore
19. Deepak G, Gulzar Z (2017) Ontoepds: enhanced and personalized differential semantic algorithm incorporating ontology driven query enrichment. J Adv Res Dyn Control Syst 9(Special):567–582

Analysis of Coupling Transition for the Encoded Data and Its Logical Level Power Analysis

Shavali Vennapusapalli, G. M. Sreerama Reddy, and Ramana Reddy Patel

Abstract Low power is applied while designing a chip and it is the important challenge faced by VLSI designer. Interconnections and internal parameters of bulk connections will consume maximum amount of power when the technology shrinks, when data is transmitted to bus architecture it consumes a significant amount of power, and when transitions occur more power is required, and hence power has to be saved. Switching activity power can be minimized by design and controlling encoding system in the network and power is altered with voltage from the supply rails and interconnected capacitance. Minimization of the capacitance by using charging and discharging activity which is nothing but a encoding techniques. A new technique is proposed in nanometer technology.

Keywords Activity · Transitions · Efficiency · Power dissipation

1 Introduction

In deep submicron technology, the power has to be minimized due to interconnections and delay. Hence, the interconnection plays an important rule. If the interconnections increases and the amount of power will also increase. As a large amount of data is passed into the bus systems, the number of transitions ($1 \rightarrow 0$ or $0 \rightarrow 1$) takes place, and as a result, charging and discharging takes place in bus systems due to the presence of capacitance. The technique is used for to minimize the power dissipation in buses as well explored literature [1–5]. The buses contribute the interconnections in the form of capacitance. The action of charging and discharging of capacitance

S. Vennapusapalli (✉)
Department of ECE, JNTUA, Anantapur, A.P, India
e-mail: v.shavali@gmail.com

G. M. Sreerama Reddy
Department of ECE, CBIT, Kolar, Karnataka, India

R. R. Patel
Department of ECE, JNTUA, Anantapur, A.P, India

© The Author(s), under exclusive license to Springer Nature Singapore Pte Ltd. 2021   183
K. A. Reddy et al. (eds.), *Data Engineering and Communication Technology*,
Lecture Notes on Data Engineering and Communications Technologies 63,
https://doi.org/10.1007/978-981-16-0081-4_19

will effect the power estimations. The power dissipation due to transition in the data from 0 to 1 and 1 to 0 is given by static power and dynamic power. The total power is given by

$$P_T = V_{Cc}^2 * f * C_{PD} \tag{1}$$

where
P_T = Power
V_{cc} = Voltage from the rails
f = Frequency of transition of the output signal
C_{PD} = Power dissipation capacitance
The power utilized by the circuit without performing any operations is called static power dissipation or quiescent power dissipation.

The power utilized by the circuit under operations is called dynamic power.

The overall global power is given by

$$P\text{avg } k = P\text{static } k + P\text{dynamic } k + \text{Leakage } k + P\text{shortckt } k \tag{2}$$

The major role in the overall power dissipation is that of dynamic power dissipation. Hence, this power concentration on reducing Pdynamic k using encoding techniques and the static power can be reduced by logic analysis. The dynamic power dissipation is

$$\text{Dynamic} = t * V^2 DD * C_l * f \tag{3}$$

where
τ is the factor of activity transition,
V_{DD} is voltage,
C_l is output capacitance, and
f is operation of frequency.
By reducing transition activity, power can be reduced.

2   Basic Definitions

1. **Coupling Transitions (CT)**: Changes that occur in the data bit from 0 to 1 or 1 to 0, between two lines or wires.
2. **Self-Transitions (ST)**: Changes that occur when the data bit is changing from 0 to 1 or 1 to 0 on system bus with the same data line as reference.
3. **Width of Bus**: The number of bits in the data is called the width of bus

    Example, let $D1 = 10010100$;
    $D2 = 01110001$;
    CT between $D1$ and $D2$ is 5.

3   Data Bus Models and Encoding Types

Power consumption and power dissipation are the key concepts in the design of DSM technology and it plays a significant role in circuit design. Minimization of coupling transition and reducing the capacitive effect will have efficiency in nano-products and the concept of optimization with encoding technique and power is applied to each and every small unit of standard cell, node, topology, each and every part of architecture design through different approaches that reaches to layout design. When optimization is applied to encoding to reduce the power the behavior and performance should not be change. Any fractional change that occurs should not be ignored and this is the basic objective of nano-design technology. In this research, the main focus is the technique that aim is to minimize the power by the circuit interconnections. The power indirectly depends on the capacitive load, interconnections between the substrate and the primary elements and distance between the elements in the wire. In this approach, logical analysis is to estimate that minimizes the static power and encoding method of reducing the capacitive count, which allows us to minimize all the internal activity on the circuit data bus. The proposed encoding scheme which are transparent to implement nanometer technology an approach of algorithmic level and minimized architectural level and accessed by means of simulation. The estimation of standard cell, logical cell, power utilization, number of components, leakage analysis of each logical cell, total cell number in the architecture, and static power in that designed architecture is estimated. The result shows that by using the logical analysis of encoding schemes up to 48% of encoding value and static power can be minimized in nanometer without any degrade in functional specifications and performance in bus systems.

The simple circuit of bus model is shown in Fig. 1.

4   Energy Model Platform

Ca has coupling activity between adjacent lines and Cb has self-activity. The coupling factor is defined as the ratio of coupling capacitance and self-capacitance (Fig. 2; Table 1)

$$M = Ca/Cb \qquad (4)$$

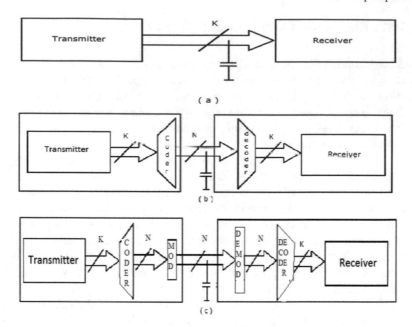

Fig. 1 Bus model **a** simple topology bus structure, **b** encoding scheme between the simple topology, **c** modulator and demodulator of topology

Fig. 2 Capacitance model for DSM bus

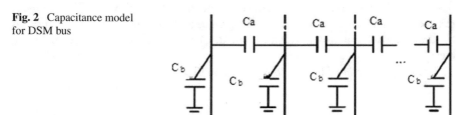

Table 1 Values of i to $i + 1$ state of capacitance. It helps to count change in the state of capacitance from one state to another state

$Y(i, i + 1)$	Data line i at a time i to $i + 1$				
data line $i + 1$ at a time j to $j + 1$		$0 \to 0$	$0 \to 1$	$1 \to 0$	$1 \to 1$
	$0 \to 0$	0	1	0	0
	$0 \to 1$	1	0	2	0
	$1 \to 0$	1	2	0	1
	$1 \to 1$	0	0	1	0

5 Transmission of Data Bit and Coupling Effect

Some modern system-on-a-chip (SoC) is designed with multiple resources and combinational circuits. These resources like memory, processors, nodes communicate with multiple resource buses, i/o data bus, shared bus and peripheral bus. Basically, there are some effects when the wires are connected called coupling effects on buses. They are coupling capacitive effect and coupling inductive effect. Consider pair of wires in system of bus, during the action, there are different approaches and types of possible transitions. When we just transmit the data change of transition and distribution takes place over bus structure. TYPE I one transition (ex: 01 → 00), TYPE II two transition (ex: 01 → 10), TYPE III two transition (ex: 00 → 11), and TYPE IV no transition (ex: 00 → 00).

Figure 3a: Simple switch where coupling capacitance is charge and discharge with Cx V. Figure 3c: In this, both wires simultaneously switch to same bit position and there is no coupling effect. Power is dissipated while charge and discharge. In this technology, on-chip parameters like effects of both capacitance, inductance, transition time, false time, and crosstalk have increased significant. The situation like frequency increases, unexpected spikes, estimated time correlation, rise time decreasing, electrical signals has more and more high frequency components which makes inductance effects more significant. The analysis of power can be done by capacitance, delay, and signal transition. Analysis of power by capacitance is an important factor because of many dependent, independent, and interconnected things can be studied with capacitance. So, in this paper, power analysis with capacitance and switching activities is studied. In deep submicron technology, shrinking of wire and metal will take place; this is neglected because of considering capacitive effect, and this capacitance nullifies the effect, and hence algorithm is developed by taking in account with capacitance and power is analyzed.

System bus performance, behavior, and specification can be increased by connecting more wires in bus topology.

Fig. 3 a–c Coupling transitions and values

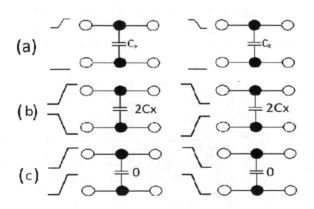

6 Proposed Technique

The architecture of a general data bus is as shown in Fig. 4.

1. Let $a1, a2 \ldots a15$ be nth order line.
    Let $b1, b2 \ldots b15$ be nth $(n - 1)$ order line.
2. Calculate coupling transitions between two lines.
3. Compare coupling transition with threshold value.
4. If coupling transition is greater than the threshold value, send the data to encoder else send data to output end.
5. Data which is input to encoder has to segregate the data as type 1, type 2, type 3, type 4 transitions.
6. Type 2 transition is in toggle state it consumes power it has to encode with desired anyone type of transition and it has to be given high priority.
7. Type 3 transition has to be encoded with suitable type1 transition and it has to be given next high priority when compare to toggle state priority.
8. The input to encoder can also be change type 2 to any other type transition as type 2 consumer more power.
9. The new nth order is $c1, c2, \ldots c15$ and $(n - 1)$ order of encoder $e1, e2, e15$.
10. Calculate coupling transition for above order. If coupling transition is high invert the data and send.
11. If coupling transition of data before encoding $(y1)$ and after encoding $(y2)$ type transition coded data is compared. $y2$ is less than $y1$ send the data to destination else send un-coded data.
12. Finally, the data is sent into the target lines to reduce the power and transitions (Fig. 5).

Calculations of Different Bit data

From Tables 2 and 3, it is noted that the minimization of change in logic bits reduced as the bit width is increased. Hence, the power consumption is also reduced because from the Eq. (1) of power it is noted that the power is directly proportional to capacitance transition activity. As the transition decreases, hence, the power is also reduced. The logical reduction and change in standard cell reductions are obtained by the

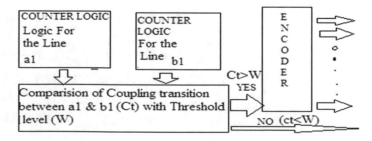

Fig. 4 Block diagram of encoder

Fig. 5 Logical diagram of encoder

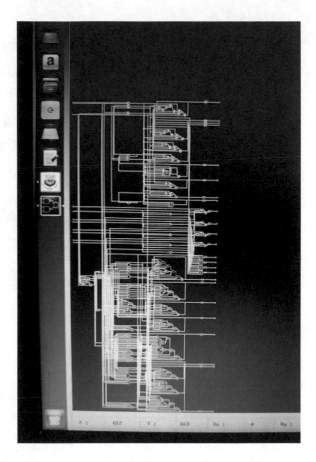

Table 2 Comparisons of switching activities

Methods/vectors	Normal method	BUS	Shift	Proposed (width encoding)
8-bits 500 vectors	4780	3760	3470	2680
16-bits 500 vectors (set1)	8340	5720	5580	4520
16-Bits 500 vectors (set2)	8340	5720	5580	4520
8-bits 1000 vectors	9580	6900	6620	5500

Table 3 Extended 32-data bit vector comparisons

Methods/vectors	Normal method	Bus	Shift	Proposed (width encoding)
32-bitdata 2000 vectors	77,300	66,660	58,180	48,380
32-bitdata 5000 vectors	185,060	161,080	155,940	108,500
32-bitdata 10000 vectors	386,860	332,340	317,500	226,740

proposed method which reduces the static power. It is implemented by using hardware description language VHDL AND VERILOG and using tool Active HDL, Xilinx, and Q flow design (Table 4).

The leakage power which is also called static power in very less it is nano-watts which is 53 nw and the logic cells utilized for the design is 1029. When compared the previous method, the proposed method is consuming less static power. The static power in [6, 7] is in microwatt. The static power in [8, 9] is in microwatt. The static power in [10, 11] is in microwatt. Hence, the proposed method utilizes less logic gates and minimum static power.

Table 4 Logical cell and power analysis

S. No.	Components	No of cells	Power consumed by each cell	Leakage power (nw)
1	AND2 * 2	14	0.090278	0.090278 * 14 = 1.263892
2	AOI21 * 1	75	0.0515209	0.0515209 * 75 = 3.864068
3	AOI22 * 1	64	0.0588648	0.0588648 * 64 = 3.767347
4	BUF * 2	19	0.0660639	0.0660639 * 19 = 1.255214
5	DFFpos * 1	40	0.160725	0.160725 * 40 = 6.429
6	INV * 1	62	0.0221741	0.0221741 * 62 = 1.374794
7	NAND2 * 1	132	0.0393659	0.0393659 * 132 = 5.196299
8	NAND3 * 1	118	0.0560872	0.0560872 * 118 = 6.61829
9	NOR 2 * 1	151	0.035234	0.035234 * 151 = 5.320334
10	NOR3 * 1	60	0.0544821	0.0544821 * 60 = 3.268926
11	OAI21 * 1	264	0.0480948	0.0480948 * 264 = 12.69703
12	OAI22 * 1	21	0.0603119	0.0603119 * 21 = 1.26655
13	OR2 * 2	3	0.0895797	0.0895797 * 3 = 0.268739
14	XNOR2 * 1	6	0.160592	0.0895797 * 6 = 0.96552
Total cells = 1029				Total static power = 53.5540316 nW

7 Conclusions

The encoding system is suitable for 18 nm technology as it consumes less static power which is 53.55 nw, and the coupling transitions is also less, hence, dynamic power consumption is also very less. This method is very much suitable for wide data transmission and large topology network, and this is used in deep submicron technology.

References

1. Alamgir M, Basith II, Supon T, Rashidzadeh R (2015) Improved bus-shift coding for low-power I/O. In: 2015 IEEE International symposium on circuits and systems (ISCAS), pp. 2940–2943. IEEE
2. Qian W, Wang R, Wang Y, Riedel M, Huang R (2019) A survey of computation-driven data encoding. In: 2019 IEEE international workshop on signal processing systems (SiPS), pp 7–12. IEEE
3. Fan CP, Fang CH (2011) Efficient RC low-power bus encoding methods for crosstalk reduction. Integration 44(1):75–86
4. Thubrikar T, Kakde S, Gaidhani S, Kamble S, Shah N (2017) Design and implementation of low power test pattern generator using low transitions LFSR. In: 2017 international conference on communication and signal processing (ICCSP), pp 0467–0471. IEEE
5. Karakus C, Sun Y, Diggavi S (2017) Encoded distributed optimization. In: 2017 IEEE international symposium on information theory (ISIT), pp 2890–2894. IEEE
6. Ramachandran A, Rajaram B, Purini S, Regeti G (2010) Transition inversion based low power data coding scheme for buffered data transfer. In: 2010 23rd international conference on VLSI design, pp 164–169. IEEE
7. Lee SE, Bagherzadeh N (2009) A variable frequency link for a power-aware network-on-chip (NoC). Integration 42(4):479–485
8. Singh B, Khosla A, Narang SB (2013) Low power bus encoding techniques for memory testing. Microelectron Solid State Electron 2(3):45–51
9. Lee W, Kang M, Hong S, Kim S (2019) Interpage-based endurance-enhancing lower state encoding for MLC and TLC flash memory storages. IEEE Trans Very Large Scale Integr (VLSI) Syst 27(9):2033–2045
10. Padmapriya K (2013) Low power bus encoding for deep sub micron VLSI circuits
11. Sotiriadis PP, Chandrakasan A (2000) Low power bus coding techniques considering interwire capacitances. In: Proceedings of the IEEE 2000 custom integrated circuits conference, pp 507–510. IEEE
12. Nayak VSP, Madhulika C, Pravali U (2017) Design of low power hamming code encoding, decoding and correcting circuits using reversible logic. In: 2017 2nd IEEE international conference on recent trends in electronics, information & communication technology (RTEICT), pp 778–781. IEEE
13. Chennakesavulu M, Prasad TJ, Sumalatha V (2018) Data encoding techniques to improve the performance of system on chip. J King Saud Univ Comput Inf Sci
14. Sathish A, Latha MM, Lalkishore KP (2012) High performance data bus encoding technique in dsm technology. Int J Commun 3(2):1

15. Sarkar S, Biswas A, Dhar AS, Rao RM (2017) Adaptive bus encoding for transition reduction on off-chip buses with dynamically varying switching characteristics. IEEE Trans Very Large Scale Integr (VLSI) Syst 25(11):3057–3066

16. Raghunandan C, Sainarayanan KS, Srinivas MB (2006) Encoding with repeater insertion for minimizing delay in VLSI interconnects. In: 2006 6th international workshop on system on chip for real time applications, pp 205–210. IEEE

5G Enabled Industrial Internet of Things (IIoT) Architecture for Smart Manufacturing

V. Chandra Shekhar Rao, P. Kumarswamy, M. S. B. Phridviraj, S. Venkatramulu, and V. Subba Rao

Abstract 5G networks become a prominent technology for providing higher bandwidth, quality of services, minimize latency, and support higher number of connectivity of mobile devices, providing Internet of things communicated system services for better efficiency than 4G technology. The Internet of things (IoT) provide intelligent industrial production system background knowledge on cyber-physical manufacturing systems (CPMS) to enhance facilities like smartness, advance manufacturing features, real-time monitoring, and collaboratively and co-operation features for industrial Internet of things (IIoT). Smart industries have generated high volume of data, high-bandwidth capability, and high coverage with low latency almost all of these features not provided by 3G and 4G generation technology. To meet these essential demands in smart industry manufacturing systems, the new generation 5G to manage these features with enabled industrial Internet of things for CPMS. As per 5G enabled IIoT for improvement of smart manufacturing technology in industry, the proposed architecture enhances industry-oriented smart services by new application services as enhance mobile broadband (eMBB), massive machine-type communication (mMTC), ultra-reliable and low-latency communication (URLLC), narrowband Internet of things (NB-IoT). This paper, the authors contribute a new architectural design that is important for upcoming fifth generation enabled industrial Internet of

V. Chandra Shekhar Rao (✉) · M. S. B. Phridviraj · S. Venkatramulu
CSE Department, KITS, Warangal, Telangana State, India
e-mail: vcsrao.cse@kitsw.ac.in

M. S. B. Phridviraj
e-mail: msbp.cse@kitsw.ac.in

S. Venkatramulu
e-mail: svr.cse@kitsw.ac.in

P. Kumarswamy
CSE Department, S.R. University, Warangal, India
e-mail: palleboina.kumar@gmail.com

V. Subba Rao
Dayalbagh Educational Institute (Deemed University), Agra, India
e-mail: vsrao.voore@gmail.com

© The Author(s), under exclusive license to Springer Nature Singapore Pte Ltd. 2021   193
K. A. Reddy et al. (eds.), *Data Engineering and Communication Technology*,
Lecture Notes on Data Engineering and Communications Technologies 63,
https://doi.org/10.1007/978-981-16-0081-4_20

things(IIoT) to enhance more smartness for industry manufacturing processes for cyber-physical manufacturing systems (CPMS) in industry.

Keywords Fifth generation (5G) · Internet of things (IoT) · Enhanced mobile broadband (eMBB) · Massive IoT (known as mMTC) · Long-term evolution for machines (LTE-m) · Narrowband internet of things (NB-IoT) for cellular technology · Low power wide area (LPWA) · And industrial internet of things (IIoT)

1 Introduction

Within cyber industry, background knowledge about cyber-physical manufacturing systems (CPMS) essential for intelligent industrial production system established by Internet of things (IoT) provide main features like smart-automation, real-time monitoring process, co-operation, and collaborative process. The key features of IIoT are reliable-sensing, real-time remote communication, big-data maintenance, and huge volume of data support. According to these features, IIoT provide key parameters like low cost and high production with less time for manufacturing process with smart technology. The applications of new technologies of CMPS produce high volume of data, i.e., near about 40 TB data producing every day for managing the manufacturing process. For efficiency managing, the CPMS process of huge data key features is excessive speed, minimum latency and with most trustworthy technologies [1, 2].

The sections as, Sect. 2, introduction to 5G network technologies, in Sect. 3, literature review for 5G enabled IIoT architecture, Sect. 4, the general characters of smart manufacturing process of 5G network enabled industrial Internet of things (IIoT), Sect. 5, advanced technology and stand against with aspects of intelligent manufacturing based on 5G industry Internet of things (IIoT), Sect. 6, proposed architecture for 5G industrial Internet of things (IIoT), Sect. 7, acknowledgements, Sect. 8, conclusion and followed by references.

2 5G Network Technologies

The fifth generation network is mobile wireless communication technologies and expecting providing services 40 times faster, reliable than fourth generation long-term evolution. The fifth generation network support radio waves, support all varieties of spectrum, support millimeter wave (mmWave) radio spectrum, i.e., radio spectrum support to carry large amount of data in a short distance. Fifth generation network supports higher frequencies with high speed, but drawing back these are obstructed by walls, big trees, and weather. The introduction to the following new technologies is provided by fifth generation network are as follows.

The newly introduced technologies for 5G network like eMBB, URLLC, and MMTC for enabling for Industry 4.0. Industry 4.0 having research areas IoT, cloud computing, cognitive communications, and cyber_physical_systems. The Industry 4.0 having features like monitor sensor communications, network slicing and 'smart industry' to make Industry 4.0 more efficient and effective. The available fourth generation and third generation having limitations for providing lesser reliability, low speed, and minimum energy consumption. Upcoming new generation, 5G network support for high reliability, higher energy consumption, low latency, real-time monitoring support, industry-robotics support, high remote communication support, etc.

The International Telecommunication Union (ITU-R) radio enhanced services for future of 5G and the key enhancement technologies for smart management of CMPS by IIoT as follows.

- eMMB is known as enhanced mobile broadband for providing faster speeds, high band width, throughput, and enhancing excessive data rates for video streaming. eMMB provides greater data bandwidth [3].
- mMTC is known as massive machine-type communications technology of 5G providing high connection for online devices in communication network. The machine-type communication involves operate sensible of machine-type communication in smart industries. mMTC, is a part of 3GPP for low power wide area (LPWA), which comprise NB-IoT that are required to ensure requirements of 5G mMTC. mMTC is a communication between machines for information exchange and data generation with minimum and no-intervention of human being. mMTC is a sub-class of MTC for wireless connectivity of billions of machines that enabled IoT [3].
- URLLC is known as ultra-reliable low-latency communication allow minimal delay, extreme trustworthy for industrial Internet of things, and intelligent transport system. The URLLC is used for high communication and low latency. The applications of URLLC like driverless cars in the network, man less driving vehicles support, real-time monitoring, control of cars in networking, etc. [4–6].
- NB-IoT is known as narrow band Internet of things (NB-IoT), a technically-standard for low power wide area (LPWA) to enable a IoT communication devices. NB-IoT provide deep services to improve the electrical energy of devices, system capability, and more battery life (Fig. 1).

3 Literature Review for 5G Enabled IIoT Architecture

The reviews of literature survey conducted by various authors on existing IoT architectures are given as follows Wang, Kun et al. [7] a framework for architecture of green industrial Internet of things. In this framework, designed three layers are sensing layer, gateway layer, and control layer. This framework balancing load of traffic for a longer lifetime. The features of this architecture to minimize energy consumption and well utilization of resources. Wan et al. [8] an industrial Internet

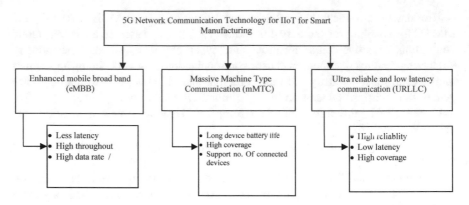

Fig. 1 5G network communication technology

of things architecture that describes software-defined network (SDN) technology for industrial Internet of things. To manage physical devices for information exchange 5 g network uses software-defined network about industrial Internet of things architecture. Alexopoulos, Kosmas et al. [9] architecture for the industrial Internet of things (IIoT) for lifecycle of Industrial product service system (IPSS). This architectures features are provide phases of services in IPSS. Al-Aqrabi, Hussain, et al. [10] a secure communication for devices of industrial Internet of things (IIoT) of fifth generation for network. This architecture supports for data throughput and lower latency. This architecture provides secure authentication for IoT hardware devices. Chen et al. [11] an architecture about industrial Internet of things for providing network security protection for providing various levels based on attack in a network. Leminen et al. [12] an architecture about industrial Internet of things for business models for machine-to-machine interaction.

4 The General Characters of Smart Manufacturing Process of 5G Network Enabled Industry Internet of Things (IIoT)

Special features of intelligent manufacturing process of 5G network enabled IIoT depend on technology applied for improving CPMS in industries. The technologies applied depend on enhancement capability of the CPMS. The characters are as follows.

- The large number of IoT-based interconnection and communication between the heterogeneous systems and equipment in shop floor of industry. The shop floor is an area where manufacturing happened where assembly or production is taken place in industry.

- Smart industry system must possess high reliability and low latency of monitoring features are available. These features of 5G network enabled IIoT improve the machine-to-machine communication and industrial real-time production, manufacturing process status information in shop-floor area.
- 5G and edge computing upgrade efficiency about application also access high quantity of data will process at real-time scenario. 5G with SDN technology handle data transfer in the middle of systems and high configured servers.
- By applying technology 3D, 5G makes high product design and manufacturing based on virtual reality in industry. 5G mobile networks relied on applications for high-bandwidth and low-latency features by applying augmented reality (AR) and virtual reality (VR).
- 5G network by applying multi-beamforming technology to less cost as well as less energy used to perform an action transformation in a communication network

5 Applications of Intelligent Industrial Production Based on 5G Enabled Industry Internet of Things (IIoT)

5G wireless communication and industry Internet of things are very important to implement high-bandwidth, low-latency, and industry automatic real-time manufacturing and monitoring tools. These technologies are under development process. The following are technologies challenges of smart manufacturing-oriented 5G enabled industry Internet of things (IIoT).

- Millimeter wave—Millimeter wave is used to allow higher data rates up to 10 Gbps on mobile and wireless networks. These are short wavelengths and having scale 10–1 mm. These waves absorbed by gases at atmosphere. The rain and humidity can reduce performance of signal strength. The millimeter wave travels by line of sight and its frequency wavelengths blocked by physical objects like building and trees.
- Full-duplex data transmission—In 5G, full-duplex transmission occurs data transmission and takes place for both directions on a signal carrier for the same time. In local area networks for applying duplex transmission is a challenging task, the workstation can sending data on the line while another workstation can receiving the data.
- Multiple access technology is a technology that allows multiple mobile users can share the allot spectrum most efficient way. The cellular system splits into cells that a mobile unit in every cell can communicate to the base station.
- Channel coding technology—The channel coding technology is for three scenarios of enhanced mobile broadband (eMBB), massive machine-type communication (mMTC), and ultra-reliable and low-latency communication (URLLC).
- Features of ultra-dense heterogeneous network technologies – The compatibility of heterogeneous networks of 5G enabled network and basic networks like 3G, 4G,

long term evolution (LTE), software defined network (SDN) and also cognitive
radio networks etc.

- H-CRANs—It is a cost-effective solution to improve co-operative processing in
  hyper-dense heterogeneous networks (HetNets) to communicate through cloud
  computing. H-CRANS used to improve the capability of macro-base stations
  (MBSs) contain multiple antennas with lower power nodes they connect by signal
  process with cloud.

6 Proposed 5G Network Enabled Industrial Internet of Things (IIoT) Architecture Model for Intelligent Manufacturing

Based on services requirements for industrial Internet of things services, architectural
designs have been proposed. This proposed fifth generation network architecture
having three horizontal layers named as innovative service enablement layer, enabling
platform layer, and hyper-connected radio layer. The proposed architecture for smart
manufacturing by 5G enabled industrial Internet of things networks technologies.

- Massive machine-type communication (mMTC)—The smart functioning of inter-
  connected of heterogeneous communication devices and equipment is built with
  CPMS within industrial Internet of things. In CPMS, real-time monitoring of
  machine products and equipment like machine tools, robotics, industrial production for water provide machines, gas provide materials machines, tracking goods
  resources like materials, workers, and shop-floor surroundings like internal heat,
  water vapor in atmosphere, hotness, gas, etc. In CMPS system, a large number of
  nodes monitored in industry, but 4G network is unable to manage monitoring and
  manufacturing processes for large scale nodes with high speed and low-latency
  provisions. And also do not manage communication devices for manufacturing
  process in shop floor, the current 4G unable to support for industrial Internet of
  things for smart industry processes [13–15].
- 5G network possess millimeter wave action that need big quantity of small
  antennas are known as antenna arrays. These antenna arrays improve frequencies
  again use as well as transmitter and also receiver. Beamforming reduces channel
  communication and enhances status of link. 5G network and cloud radio access
  network(C-RAN) improve Internet of things (IoT) tasks at remote cell sites. This
  architecture having improved features of bandwidths saving and energy saving.
  For this reason, the number of communication nodes improved for process of
  intelligent manufacturing.
- Ultra-reliable and low-latency communication (URLLC)—supported by new
  radio (NR) standards. The features of URLLC is low latency and is required
  between autonomous vehicles and 5G network to manage autonomous vehicles
  like cars etc.

- The URLLC manage reliable and latency for autonomous vehicles. This feature enhances high reliable and lower speed for application of industrial Internet and remote surgery. The manufacturing process contains number of sub-processes to fulfill these tasks in a smart industry. At the right time, the order of sub-processes collaboratively operates instructions to task this process to complete. In automation of smart industries, a single robot usually cannot handle complicated tasks, and the dual robotics technology can handle collaboratively these robotics handle these tasks at once with huge trustworthy rand lower delay by using URLLC within 5G for industrial automation process tasks to complete.

5G network technology using SDMA by small antenna for processing radio transmission. The antenna array uses the multi-beaming technologies for reducing the link quality to improve the reliability of transmission and reduces the latency (Fig. 2).

- Enhanced mobile broadband (eMBB)—Enhanced mobile broadband (eMBB) data-driven technology proving to perform huge data rates in a big are. It enables real-time traffic manage, tremendous speed Internet access, real-time streaming of video having 3D/4K video. The autonomous vehicles capable for communicate to vehicles. Shop floor in industry having inter-connected and communication in

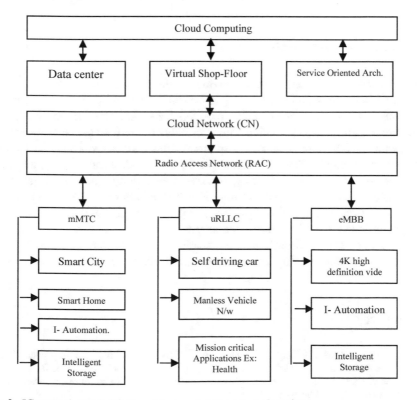

Fig. 2 5G network enabled IIoT architecture for smart manufacturing

the middle of area of production of manufacturing goods and inter-connected digital technology. As a vast improvement in IT-based industrial production, this interaction happening done by map of CPS and VSF transmission of heavy real-time industrial production data that having high reliability with also low latency. 5G network enabled eMBB technology used to greater transmit signal ranges when compare with 4G-network. It reaches times most speedy transferrable when compare with 4G network. 5G network based on cloud radio access network(C-RAN) and software-defined networking (SDN) that supply loosely coupled system characteristics for C-plane and D-plane to enhance efficiency of transmission.

- 5G possess a great wireless transmission based on CRAM and SDN technologies. 5G network process bidirectional transmit huge amount of data for manufacturing process.

The main idea of paper is to focus on industrial Internet of things architecture design for the purpose of upcoming applications support and client demand services in heterogeneous environment. In this paper, newly design of architecture for 5G enabled industrial Internet of things (IIoT) for full-fledged layered technology about newly upcoming applications services and also customer satisfaction service-oriented architecture.

7 Conclusion and Future Work

The proposed paper 5G network enabled IIoT architecture for smart manufacturing in Industry 4.0. Authors discussed the 5G technologies and important in smart manufacturing process that enabled with IIoT—5G network. The authors briefed the literature survey of different architectures for industrial IoT. At present, the existing architecture needs improvement, enormous challenges for requiring the requirements of upcoming application support and user demandful services like increased data rate, reduced end-to-end delay, massive connectivity, guaranteed quality of experience, higher availability and improves higher efficiency. 5G with architecture of industrial Internet of things (IIoT) is proposed for upcoming services and industry-oriented manufacturing processing needful. In future work, we are planning to develop authorization and authentication algorithms for 5G network enable IIoT for smart manufacturing in industry-oriented architecture for more secure way.

References

1. Tao F, Cheng J, Qi Q, Zhang M, Zhang H, Sui F (2018) Digital twin-driven product design, manufacturing and service with big data. Int J Adv Manuf Technol 94:3563–3576
2. Cheng J, Chen W, Tao F, Lin CL (2018) Industrial IoT in 5G environment towards smart manufacturing. J Ind Inf Integr 10:10–19

3. De Carvalho E, Bjornson E, Sorensen JH, Popovski P, Larsson EG (2017) Random access protocols for massive MIMO. IEEE Commun Mag 55(5):216–222
4. Hao S, Zeng J, Su X, Rong L (2017) Application scenarios of novel multiple access (NMA) technologies for 5G. In: World conference on information systems and technologies, pp 1029–1033. Springer, Cham
5. Kashima T, Qiu J, Shen H, Tang C, Tian T, Wang X, Hou X, Jiang H, Benjebbour A, Saito Y, Kishiyama Y (2016) Large scale massive MIMO field trial for 5G mobile communications system. In: 2016 International symposium on antennas and propagation (ISAP), pp 602–603. IEEE Press
6. Ferreira JS, Rodrigues HD, Gonzalez AA, Nimr A, Matthe M, Zhang D, Mendes LL, Fettweis G (2017) GFDM frame design for 5G application scenarios. J Commun Inf Syst 32(1):54–61
7. Wang K, Wang Y, Sun Y, Guo S, Wu J (2016) Green industrial Internet of Things architecture: an energy-efficient perspective. IEEE Commun Mag 54(12):48–54
8. Wan J, Tang S, Shu Z, Li D, Wang S, Imran M, Vasilakos AV (2016) Software-defined industrial internet of things in the context of industry 4.0. IEEE Sens J 16(20):7373–7380
9. Alexopoulos K, Koukas S, Boli N, Mourtzis D (2018) Architecture and development of an Industrial Internet of Things framework for realizing services in industrial product service systems. Procedia CIRP 72:880–885
10. Al-Aqrabi H, Johnson AP, Hill R, Lane P, Liu L (2019) A multi-layer security model for 5G-enabled industrial Internet of Things. In: International conference on smart city and informatization, pp 279–292. Springer, Singapore
11. Chen H, Hu M, Yan H, Yu P (2019) Research on industrial internet of things security architecture and protection strategy. In: 2019 International conference on virtual reality and intelligent systems (ICVRIS), pp 365–368. IEEE Press
12. Leminen S, Rajahonka M, Wendelin R, Westerlund M (2020) Industrial internet of things business models in the machine-to-machine context. Ind Mark Manag 84:298–311
13. IEEE 5G Roadmap White Paper. https://5g.ieee.org/images/files/pdf/ieee-5groadmap-white-paper.pdf
14. Rao V, Shekhar C, Akanksha C, Subba Rao V (2020) Applications of NANO-RK in Internet of Things (IoT). EasyChair preprint No. 2457, pp 1–8
15. Chandra Shekhar Rao V, Subba Rao V, Venkatramulu S (2020) Secure architecture for 5G network enabled Internet Of Things (IoT). In: Algorithms for intelligent systems (AIS). Springer

IoT-Enabled Entry in COVID Situation for Corporate Environments

Abhishek Singh and Anil Kumar Sahu

Abstract In this paper, an IoT-based system has been developed, which can quickly monitor the health of an employee while tagging their RFID cards, and depending on the health status collected by corresponding sensors, it authorizes the entry by opening solenoid door lock. The system is also connected to the remote server, so as to transfer health status of each person to a PHP server running remotely in the office premises. The unique data of RFID is transferred corresponding to each RFID tag and is saved in the remote server running PHP server with mySQL database. If any employee is found unhealthy as per the sensor data, the door will not open for its corresponding RFID tag, and the employee will not be allowed by the system for entering the office premises for at least two days.

Keywords COVID-19 · IoT · SQL database · PHP server · RFID · ESP8266 NodeMCU

1 Introduction

In the current situation of COVID-19 pandemic, there is a constant risk of health and life for the people going outside. For the corporate environment, the employer needs to be especially concerned about their employee's health while entering into the office premises. The corona virus was first identified in Wuhan, China, in the month of December 2019. Soon after its discovery, WHO in January 2020 announced the new virus as 2019-nCOV [1]. In the office premises, there are more chances of getting the viruses and bacterial infections. Some main symptoms of virus are fever, dry cough, tiredness, nasal congestion, difficulty in breathing, diarrhea, etc. The gathering in offices, using the common utilities like toilets, lifts and corridors, makes

A. Singh
Variable Energy Cyclotron Centre, Kolkata, India
e-mail: singhabhishek@vecc.gov.in

A. K. Sahu (✉)
Bharat Institute of Engineering and Technology, Hyderabad, India
e-mail: anilsahu82@gmail.com

it very vulnerable to spread the viruses easily from one person to other. Also, through centralized air conditioners, the air molecules can spread easily from one room to the another room due to its basic functionality of air circulation.

2   Methodology

2.1   Architecture

The Internet of things (IoT) developed is based on ESP8266 NodeMCU Wi-Fi cum microcontroller board [2] and is interfaced to various sensors for the user data collection and LCD display. In this project, we have established the connection of NodeMCU with a local Wi-Fi AP, in which the NodeMCU acts as a station mode (STA), while connecting to a local Wi-Fi access point (AP). The AP can also give cloud connectivity to the NodeMCU module. ESP8266 Wi-Fi module is chosen due to its low cost and is well efficient platform for this application. Also, the NodeMCU Wi-Fi-enabled hardware is made to communicate to the remote server by writing a PHP script, where the PHP server runs a PHP script while connected to the AP. The PHP script is written such that whenever it gets the request of RFID tagged user, it responds with the OK or NOT-OK commands to identify a valid employee, and then the biological parameters like temperature, heart rate and SPO2 (blood oxygen level) of the user are stored in the SQL database in the server PC. As shown in Fig. 1, Wi-Fi

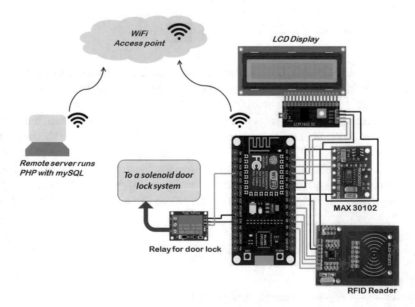

Fig. 1 RFID with PHP—mySQL server-based IoT for COVID management

module is interfaced to the various sensors and it handles the PHP server running on remote PC at one end, and at the other end, user communication is established using RFID module and an oxy-meter sensor to collect the user data.

As can be seen from Fig. 1, the solenoid door is controlled from the main controller module ESP8266 NodeMCU, which upon receiving the medical parameters of the employee decides whether the door should be opened or not. Also its corresponding data is exchanged with the remote server running in another room. The server system is basically a PHP server along with the SQL database management to store the employee's data.

The medical parameters of the person are taken using MAX30102 sensor module, which can read the temperature, heart rate and SPO_2 level of the person. The MAX30102 IC communicates by I2C protocol and is powered by 3.3 V supply only, which is derived from ESP8266 module itself. These three parameters are fetched from the sensor and immediately send to the remote server based on the employee ID detected from RFID reader. The I2C display module attached to the system guides and communicates with the user. Display system is basically a two-wire I2C module using LCM1602 I2C as the display controller.

SQL database in server stores the information regarding each employee in the firm. The PHP script upon receiving the request from Wi-Fi module authenticates the user and stores its medical parameters in the server SQL database upon retrieving the sensor data from user.

3   Results and Discussions

3.1   XAMPP Server Utility Tool

The remote server is established using XAMPP server utility tool, in which an Apache server module is initialized first along with SQL database module. For this project, the requirement is that the remote server and NodeMCU module must be on the same Wi-Fi AP network to be able to connect and communicate with each other. The XAMPP tool is opened first to initiate a server.

After the server is started, the mySQL database is started to access the database utility for all employees in the server machine. You can try from localhost by typing 'http://localhost/hello.php' in your browser [3, 4], where hello.php is the test PHP script located in the 'xampp\htdocs' folder of C drive as illustrated below in Fig. 2.

As shown in Fig. 2, the PHP code is made to return a string upon getting the HTTP request from local browser only. Similarly, the PHP code can be written to respond to the ESP8266 Wi-Fi module upon getting an HTTP request from it.

Fig. 2 hello.php PHP script
located in xampp\htdocs
folder in C drive

```
<!DOCTYPE>
<html>
<body>
<?php
echo "<h2>Hello PHP, how are you.</h2>";
?>
</body>
</html>
```

3.2 SQL Database Integration with PHP Coding

To keep the record of all the employees, it is required to maintain a database of all the persons and their corresponding data in a table arrangement. This functionality is maintained using the SQL database in the remote server using XAMPP utility tool with the help of mySQL module. To maintain the SQL database, first of all, click on the 'Admin' button in front of mySQL module. A browser window with SQL database system gets opened for database management. Also, this database can be queried from within the PHP code to be able to communicate to the remote NodeMCU module.

In this window, click on the 'new' button located in top left corner to generate a new database and give it some name, which is chosen 'test2' in Fig. 3. A database can contain as many tables to maintain data records. The table is generated by clicking on new button under the database name 'test2' as shown in figure above. It is named as 'class' here. A table can contain many columns corresponding to the different attributes of a record. As shown in Fig. 3, these attributes are having data types, length, etc. The table contains the employee records as depicted in Table 1. For the illustrative purpose, only four rows are added. We can have as many rows corresponding to each employee of the firm.

SQL data is initialized to all zeros when initiating the project as can be seen from Table 1. The PHP scripting is done in the remote server PC to communicate and fill the SQL table with the data received from remote ESP8266 client module.

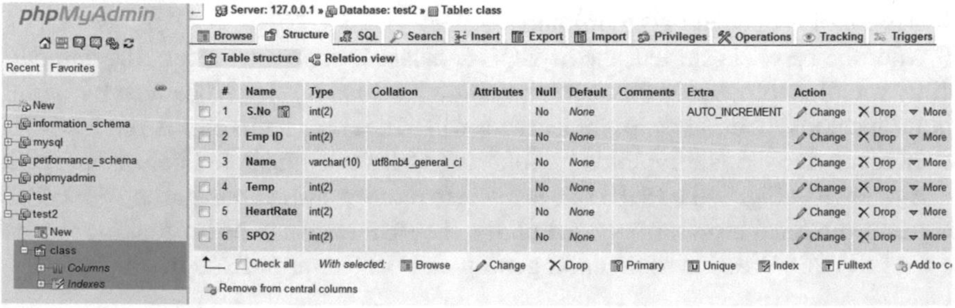

Fig. 3 XAMPP database table named 'class'

Table 1 Data values for the SQL database

S. No.	Emp ID	Name	Temp	Heart rate	SPO$_2$ level
1	1200	John	0	0	0
2	1201	Smith	0	0	0
3	1202	Bob	0	0	0
4	1203	Michael	0	0	0

Fig. 4 XAMPP 'class' table SQL data values

The database is created by filling the name of the database columns as shown in Fig. 3 with the corresponding column names as shown in Table 1. Also it is required to check mark the auto-increment (AI) button in front of 'S.No' field. The database created in the above steps using XAMPP SQL database needs to be populated with the employer's firm data as shown in Table 1. To do this, first click on SQL tab as shown in Fig. 4 and then edit this file with the following commands [5, 6].

INSERT INTO 'class'('Emp ID', 'Name', 'Temp', 'HeartRate', 'SPO2',) VALUES (1200,'John',0,0,0);
INSERT INTO 'class'('Emp ID', 'Name', 'Temp', 'HeartRate', 'SPO2',) VALUES (1201,'Smith',0,0,0);
INSERT INTO 'class'('Emp ID', 'Name', 'Temp', 'HeartRate', 'SPO2',) VALUES (1202,'Bob',0,0,0);
INSERT INTO 'class'('Emp ID', 'Name', 'Temp', 'HeartRate', 'SPO2',) VALUES (1203,'Michael',0,0,0);

Now, after editing this file with required data, click on the button 'Go' on bottom right. Since the auto-increment (AI) button was checked in front of 'S.No' field, there was no need to populate the 'S.No' column in SQL database.

In this way, the database gets updated using the XAMPP SQL utility. This database can be easily accessed by writing the PHP code. The PHP code must be stored in 'htdocs' folder of XAMPP installation location. For example, the database is accessed from script by writing the command below in PHP script file [4–6].

$con = mysqli_connect ($server, $username, $pwd, $dbase);

where server is localhost, and username is 'root' with no password (blank).
    Similarly, to query any data from database, we can write [4]:

$val = mysqli_query($con, $sql);

where 'sql' is the list of items in a column 'val' is the output of SQL query.

3.3   Hardware Coding in the NodeMCU Firmware

Also, in the hardware side, the NodeMCU module is programmed to connect to the
server PC with host IP address and port number of '192.168.4.1' and 80, respectively.
For example, the following command is used to connect to the remote host [7].

client.connect(host, port)

The return type is an integer. If the above function call returns a non-zero value,
it is said to be connected to remote host, otherwise, connection failed. Also to call
the remote host with the http request of temperature value, the following command
is used.

client.print(String('GET')   +   'server2.php?Temp='   +   temparature   +
'HTTP/1.1\r\n' + 'Host:' + host + '\r\n' + 'Connection:close\r\n\r\n');

where temperature is any value of Temp column in the SQL database, while
server2.php is the name of PHP script stored in the 'htdocs' folder of server PC.
    Instead of storing data to the server, other useful option is to send the data to
any cloud-connected database server like ThingSpeak.com, where it will be stored
along with its timestamp (date and time). ThingSpeak.com is a data analysis platform
which supports MATLAB coding for algorithm prototyping. In this platform, infected
employees can be sorted out by implementing any sorting algorithms and sent to the
quarantine and restricting their entry for some defined number of days.

4   Conclusion and Future Scope

An RFID-based entry system has been developed successfully and tested in the lab.
The AP network must be always ON and connected to server and NodeMCU for the
system to run smoothly, so that there is no data loss corresponding to the employees
reading their RFID cards. This system can also be connected to any other servers like
ThingSpeak.com for graphical analysis. In this way, the data will be stored securely

in the cloud and can be accessed from anywhere just by logging into the website. Also, the data trend can be monitored which can help to analyze the situation of the office environment in a planned manner.

Acknowledgements The author would like to thanks are just a part of those feeling which are too large for words, but shall remain as memories of wonderful people with whom I have got the pleasure of working during the completion of this work. I am grateful to BIET Hyderabad which helped me to complete my work by giving encouraging environment. I would like to express with my deep and sincere gratitude to my research co-author, Mr. Abhishek Singh, Scientist of VECC, Department of atomic energy, India, dedicated for R&D in embedded systems. His wide knowledge and his logical way of thinking have paved a clear path for us. His understanding, encouragement and personal guidance have provided a good basis for the present work.

References

1. Abdulrazaq MZ, Halimatuz AZ, Salah K, Sairah R, Rusyaizila YE (2020) Novel COVID-19 detection and diagnosis system using IOT based smart helmet. Int J Psychosoc Rehabilit 24:2296–2303
2. Create a Simple ESP8266 NodeMCU Web Server in Arduino. https://lastminuteengineers.com/esp8266-nodemcu-arduino-tutorial
3. XAMPP Tutorial, How to create your own local test server—Ionos. https://www.ionos.com/digitalguide/server/tools/xampp-tutorial-create-your-own-local-test-server
4. PHP Tutorial. https://www.w3schools.com/php/default.asp
5. Web Programming Step by Step. https://www.webstepbook.com/supplements/slides/ch11-sql.pdf
6. Index of /maps/middle_east_and_asia, https://index./of.es/PHP/Build%20Your%20Own%20Database%20Driven%20Website%20using%20PHP%20&%20MySQL.pdf
7. Build an ESP8266 Web Server—Code and Schematics (NodeMCU). https://randomnerdtutorials.com/esp8266-web-server

PAPR Reduction Scheme for OFDM/OQAM Signals Using Novel Phase Sequence

V. Sandeep Kumar, J. Tarun Kumar, Shyamsunder Merugu, and Kasanagottu Srinivas

Abstract The phase sequence with a special structure is designed, and the peak-to-average power ration (PAPR) of the orthogonal frequency division multiplexing with offset quadrature amplitude modulation (OFDM/OQAM) signal is reduced by using the cyclic convolution property of DFT through IFFT algorithm. Simulation results show that this algorithm can effectively reduce the PAPR of the system.

Keywords OFDM/OQAM · PAPR · DFT · Cyclic convolution

1 Introduction

Orthogonal frequency division quadrature amplitude modulation multiplexing (OFDM/OQAM) is a multi-carrier modulation that has the benefits of using high frequency bands and a good potential for anti-multipath fading. In recent years, it has attracted more and more attention and it is widely used. However, OFDM/OQAM signals are obtained by superimposing multiple signals with different frequencies and different amplitudes. Therefore, they have a very high peak-to-average power ratio (PAPR), and the power amplifier ensures the linear amplification, which increases the cost and difficulty in implementation.

V. Sandeep Kumar (✉) · J. Tarun Kumar
Department of Electronics and Communication Engineering, SR University, AnanthaSagar, Warangal, Telangana 506371, India
e-mail: kumar.s.vngl@gmail.com

J. Tarun Kumar
e-mail: tarunjuluru@gmail.com

S. Merugu · K. Srinivas
Department of Electronics and Communication Engineering, Sumathi Reddy Institute of Technology for Women, AnanthaSagar, Warangal, Telangana 506371, India
e-mail: shyamala.merugu99@gmail.com

K. Srinivas
e-mail: srinu.vasu11@gmail.com

© The Author(s), under exclusive license to Springer Nature Singapore Pte Ltd. 2021   211
K. A. Reddy et al. (eds.), *Data Engineering and Communication Technology*,
Lecture Notes on Data Engineering and Communications Technologies 63,
https://doi.org/10.1007/978-981-16-0081-4_22

Many scholars have analyzed the various aspects and proposed many effective PAPR reduction methods. At present, there are three main categories: (1) Signal distortion technique [1], which is achieved by performing nonlinear distortion on the signal near the peak, and the simplest method is the clipping method, which causes the signal to be distorted, and the out-of-band distortion of the spectrum is large; companding algorithm, it is another signal distortion technique. It uses A-law or μ-law companding to compress the large signal of OFDM/OQAM and amplify the tiny signal, thus, reducing the peak-to-average power ratio efficiently. (2) A encryption strategy that can efficiently minimize PAPR by signal encryption, but a lot of redundant information has to be transmitted by coding, which significantly decreases the information rate [2–4]. (3) Selective mapping method (SLM) [5, 6] and partial transmission sequence method (PTS) [7, 8], where several sequences containing the same information are produced by the transmitter and the sequence with the smallest PAPR is selected by the transmitter in this algorithm. Both PTS and SLM are performed in frequency domain, additional IFFT calculations are required, so they are computationally expensive. However, since the PTS and SLM algorithms are distortion-free algorithms, researchers have examined multiple approaches to reduce the PTS or SLM algorithms' computational complexity. For example, an interleaved grouping method is proposed in [7] to reduce the complexity of the PTS; [9] an iterative method is proposed to look for sub-optimal auxiliary information to reduce the information in the PTS algorithm of the auxiliary process. In this paper, a phase sequence with a special structure is designed to be multiplied with an OFDM/OQAM signal, and then only one IFFT operation is required, i.e., no additional IFFT operation is required. In the time domain, the weighted sum of the cyclic shift sequence corresponding to the OFDM/OQAM time domain signal is used to achieve a reduction in PAPR.

2   SLM and PTS Algorithms

In OFDM/OQAM transmitter with N subcarriers, the QAM modulated data is $X_m = [X_{0,m}, X_{1,m}, \ldots, X_{N-1,m}]$. The OQAM $\{X_m = a_{2m} + ja_{2m+1}\}_{m \in N}$ is divided into real symbol $\{a_m\}_{m \in N}$. The real OQAM symbols are staggering by $\frac{T}{2}$ duration. Then, the obtained real symbols undergo polyphase filtering with PHYDYAS filter. The base-band OFDM/OQAM signal is as follows [8]:

$$s(t) = \sum_{m \in N} \sum_{k=0}^{N-1} a_{k,m} g(t - mT/2) e^{j\left(\frac{2\pi kt}{T} + \phi_{k,m}\right)} \tag{1}$$

where $g(t)$ is the prototype filter function with 4 T length and $\phi_{k,m} = \frac{\pi}{2}(k + m) - \pi km$. Similarly, the discrete-time OFDM/OQAM symbol is given by

$$s(n) = \sum_{m \in N} \sum_{k=0}^{N-1} a_{k,m} g\left(n - m\frac{NL}{2}\right) e^{j\left(\frac{2\pi kn}{NL} + \phi_{k,m}\right)} \tag{2}$$

where L is the oversampling factor.

The PAPR of the OQAM signal is defined as:

$$\text{PAPR}_{s(n)} = \frac{\max_{iNL \leq n \leq (i+1)NL-1}\left\{|s(n)|^2\right\}}{E\left[|s(n)|^2\right]}, i \in N \tag{3}$$

where $E[.]$ represents the expectation operation.

The additional cumulative distribution function (CCDF) shows the probability Pr that s(n)'s PAPR reaches a PAPR threshold of 0.0. As in [7], it is seen that:

$$\text{CCDF} = \text{Pr}\left\{\text{PAPR}_{s(n)} \geq \text{PAPR}_0\right\} \tag{4}$$

2.1 SLM Algorithm

Suppose there are U_1 independent random phase sequences $B^{(\mu)} = \left[B_0^{(\mu)}, B_1^{(\mu)}, \ldots, B_{N-1}^{(\mu)}\right] (\mu = 1, \ldots, M_1)$ of length N, where auxiliary information $B_k^{(\mu)} = e^{j\phi_k^{(\mu)}}$ ($k = 0, 1, \ldots, N - 1$), $\phi_k^{(\mu)}$ is evenly distributed between $[0, 2\pi)$. Therefore, the U_1 phase rotation vectors can be multiplied with X_k to obtain $X^{(\mu)} = \left[B_0^{(\mu)}X_0, B_1^{(\mu)}X_1, \ldots, B_{N-1}^{(\mu)}X_{N-1}\right]$. Then, the obtained U_1 frequency domain sequences $X^{(\mu)}$ are respectively subjected to IDFT operations,

$$x^{(\mu)} = \text{IDFT}\left\{X^{(\mu)}\right\} = \left[x_0^{(\mu)}, x_1^{(\mu)}, \ldots, x_{N-1}^{(\mu)}\right], \mu = 1, \ldots, M_1 \tag{5}$$

Finally, the smallest PAPR is selected from U_1 sequences $x^{(\mu)}$ for transmission.

2.2 PTS Algorithm

In the PTS algorithm, X_k can be divided into U_2 groups by neighboring division, random division, or interlace division, etc., respectively, denoted by $\left\{X^{(m)}, m = 1, \ldots, U_2\right\}$. The purpose of PTS method is to form a weighted sequence of U_2, i.e.,

$$X' = \sum_{m=1}^{M_2} b_m X^{(m)} \tag{6}$$

where auxiliary information $b_m (m = 1, \ldots, U_2)$ is a weighted value, and $b_m = e^{j\phi^m}$ is evenly distributed between $[0, 2\pi)$, and IDFT operation is given by

$$x' = \sum_{m=1}^{M_2} b_m x_m \tag{7}$$

where $x_m = \text{IDFT}\{X^m\}$ is the N-point IDFT of X^m. The PTS appropriately select auxiliary information through optimization to minimize the PAPR of x', i.e.,

$$\{b_1, \ldots, b_{M_2}\} = \arg \min_{\{b_1, \ldots, b_{M_2}\}} \{\text{PAPR}(x')\} \tag{8}$$

3   Proposed Method

Let the transmitted data sequence be $X_k = [X_0, X_1, \ldots, X_{N-1}]$. Choose a random phase sequence of length L (requires N/L to be an integer $M = N/L$) $P = [P_0, P_1, \ldots, P_{L-1}]$, where $P_k = e^{j\phi_k}$, ϕ_k is evenly distributed between $[0, 2\pi)$. The sequence P is extended with the cycle of L, and the sequence b'_k, i.e.,

$$b'_k = P(k_L)R_N(k) = \sum_{i=-\infty}^{\infty} P(k + i(2L - 1))R_N(k) \tag{9}$$

where $R_N(k)$ is a rectangular sequence of length N, k_L means that k takes L as the remainder. In fact, b'_k is a sequence of length N composed of U cycles of random phase sequence P. In order to obtain a set of auxiliary information vectors, the sequence b'_k is sequentially shifted by $1, 2, \ldots, (L - 1)$ times in order to obtain $(L - 1)$ phase sequences, namely

$$b'^{(l)}_k = P(k - l)_L R_N(k)$$

$$= \sum_{i=-\infty}^{\infty} P(k - l + i(2L - 1)R_N(k), \quad l = 0, \ldots, 2L - 1 \tag{10}$$

Then, with the help of a similar idea of the SLM algorithm, L frequency vectors $b'_k(l)$ are multiplied by X_k to obtain L frequency domain sequences

$$X^{(l)} = X_k \odot b'^{(l)}_k = \left[b'^{(l)}_0 X_0, b'^{(l)}_1 X_1, \ldots, b'^{(l)}_{N-1} X_{N-1}\right], l = 0, \ldots, 2L - 1 \tag{11}$$

where $\odot$ represents the Hadmard product between vectors. It can be seen that X^l is the frequency domain product of the sequence $b_k^{\prime(l)}$ and X_k.

$$x_n^l = \text{IDFT}\{X^l\} = [x_0^l, x_1^l, \ldots, x_{N-1}^l], l = 0, 1, \ldots, 2L - 1 \qquad (12)$$

The chain with the smallest PAPR is eventually chosen from L time domain sequences x_n^l for transmission. However, since the phase vector $b_k^{\prime(l)}$ has the special structure shown in Eq. (10), the operation of Eq. (12) can be performed by weighted summation in the time domain.

$$x_n^l = x_n \times b_n^l \qquad (13)$$

where x_n is the N-point IDFT of the original transmission sequence X_k, that is $x_n = \text{IDFT}\{X_k\}$; b_n^l is the N-point IDFT of the phase vector $b_k^{\prime(l)}$, that is $b_n^l = \text{IDFT}\{b_k^{\prime(l)}\}$, N represents the N-point cyclic convolution of the two sequences. According to the definition and nature of DFT, it is easy to obtain the N-point IDFT result of $b_k^{\prime(l)}$ [10]

$$
\begin{aligned}
b_n^l &= \text{IDFT}\{b_k^{\prime(l)}\} \\
&= \begin{cases} W_N^{-\alpha n l} p\left(\frac{n}{U}\right), & n = 0, U, 2U, \ldots (2L-1)U \\ 0, & \text{otherwise} \end{cases}, n = 0, \ldots, N-1, \alpha = 1.6 \\
&= \sum_{i=0}^{L-1} W_L^{-i\alpha l} p(i)\delta(n - i(2U + 1)), n = 0, \ldots, N - 1 \qquad (14)
\end{aligned}
$$

where when the last equal sign of Eq. (14) is obtained, $W_N^{-i\alpha M l} = W_{2LM}^{-i\alpha M l} = W_{2L}^{-i\alpha l}$ is applied. Equation (13) shows that $b_k^{\prime(l)}$, N-point IDFT result b_n^l has non-zero values only at sampling points of integer multiples of M, $W_{2L}^{-i\alpha l} p(i)$, and the value at other sampling points is zero.

$$x_n^l = x_n \times b_n^l = \sum_{i=0}^{2L-1} W_{2L}^{-i\alpha l} p(i)[x_n \times \delta(n - i(2U + 1))] \qquad (15)$$

where $\{x(n - i(2U + 1))_N R_N(n)\}$ is the cyclic shift sequence of x_n delayed by $i2U$. It can be seen that x_n^l can be obtained by the product of $p(i)$ and the cyclic shift sequence of x_n in the time domain on the basis of an IDFT. So far, the steps of the new PAPR reduction method can be summarized as follows:

1. Time domain data x_n is obtained through an N-point IDFT;
2. L time domain sequences x_n^l are calculated according to Eq. (15), which can be divided into three steps complete:
i. Cyclically shift x_n to get the cyclic shift sequence $\{x(n - i(2U + 1)_N R_N(n)\}(i = 0, 1, \ldots, 2L - 1)$;

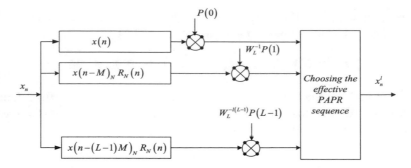

Fig. 1 Block implementation of the proposed method for PAPR reduction

ii.   Calculate the new sequence q^i:

$$q^i = p(i)\{x(n - i(2U + 1)_N R_N(n)\}, i = 0, \ldots, 2L - 1$$

iii. x_n^l is obtained by weighted summation, i.e.,

$$x_n^l = \sum_{i=0}^{2L-1} W_{2L}^{-i\alpha l} q^i, l = 0, 1, \ldots, 2L - 1 \qquad (16)$$

3.   Use Eq. (1) to calculate the PAPR of $2L$ time domain sequences x_n^l, respectively, select and transmit the sequence with the smallest PAPR. According to the above discussion, the method in this paper is based on the cyclic convolution time domain implementation method to reduce PAPR. Figure 1 shows the block diagram of the method at the sending end. Figure 1 uses (15) to calculate x_n^l. The relationship between the algorithm and PTS is as follows

$$X^l = X_k \times b_k'^{(l)} = \sum_{i=0}^{2L-1} P_i^l \sum_{m=0}^{2U-1} X_{mL+i}\delta(k - 2mL - i) \qquad (17)$$

In the formula, $\sum_{m=0}^{2U-1} X_{2mL+i}\delta(k - 2mL - i)$ is equivalent to the subsequence obtained by dividing the X_k by the interleaving method in the PTS algorithm [1, 7]. Therefore, to a certain extent, the proposed can be regarded as a PTS algorithm when the original data is divided into 2L groups according to the interleaving method.

4   Simulation Results

In this paper, the random phase sequence B of length L can be arbitrary, and $B = e^{\frac{j2\pi[0:L-1]}{2L}}$ is used in the simulation; in the PTS algorithm, the auxiliary phase information is used by exhaustive search method, there are 16 possibilities of phase information $b_m \in \{\pm j, \pm 1\}$, and the adjacent sub-segment division scheme is used to obtain part of the transmission sequence.

Compared with SLM, PTS with various step variables, i.e., $L = 16, 32$ for PTS and SLM with $U_1, U_2 = 4, 16$, respectively, is the proposed system with 128 subcarriers. It is observed from Fig. 2 that the proposed algorithm will reduce the PAPR of OQAM signals effectively. The total number of alternative signals produced by both methods is 16 compared to SLM ($U_1 = 16$) and proposed with ($L = 16$), but in the SLM, the step series is statistically independent of each other, so its output is better with the two methods ($L = 16$); but increase in L, the performance of the proposed method is improved (for example, when at 10^{-4} CCDF, the $PAPR_0$ of SLM ($U_1 = 16$), the proposed method with $L = 16$ and $L = 32$ in this paper are 8.7, 9.8 and 7.9 dB). Comparing PTS ($U_2 = 16$) and the proposed method ($L = 16$), the $PAPR_0$ of the two at 10^{-4} are 7.0 and 9.8 dB, respectively. The proposed method in this paper is slightly better; when $U_2 = 16$ in the PTS algorithm at this time, the number of signal sequences selected by it is 4^4 (128), so the performance of the algorithm is significantly improved. Compare PTS ($U_2 = 16$) and SLM ($L = 16$), i.e., when using the same number of IFFT operations, the performance of the PTS algorithm is significantly better than the SLM algorithm. Figure 3 illustrates the performance of the proposed method with fixed 128 subcarriers and with different L values. Assuming that the number of subcarriers is 128, $L = 8, 16, 32$ and 64 are simulated, and the results are observed in Fig. 3. It is observed that with the increase of L, the performance of the proposed algorithm will gradually improve. At 10^{-4} CCDF, the $PAPR_0$ of the proposed method at $L = 8, 16, 32, 64$ is 10, 9.4, 8.7 and 8.5. However, as L continues to increase, performance improvement changes slowly.

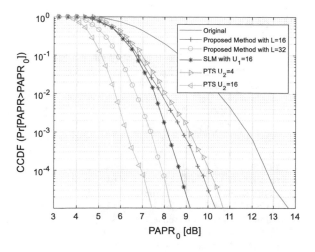

Fig. 2 Proposed method with SLM and PTS algorithms ($N = 128$)

Fig. 3 Proposed method
with fixed carriers for
different L values ($N = 128$)

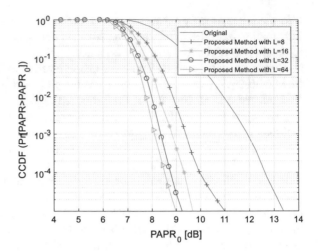

5 Conclusion

Centered on the time domain signal, a new PAPR reduction approach is introduced in
this article. With the aid of the DFT cyclic convolution theorem, this approach uses the
phase series with a special form, via the IFFT algorithm, to achieve OFDM/OQAM
signal PAPR reduction. Results of the simulation demonstrate that the approach
suggested will efficiently reduce the system's PAPR without distorting the initial
signal.

References

1. Bahai AR, Saltzberg BR, Ergen M (2004) Multi-carrier digital communications: theory and
   applications of OFDM. Springer Science & Business Media, Berlin
2. Jones AE, Wilkinson TA, Barton SK (1994) Block coding scheme for reduction of peak to mean
   envelope power ratio of multicarrier transmission schemes. Electron Lett 30(25):2098–2099
3. Wang H (2018) A hybrid PAPR reduction method based on SLM and multi-data block PTS
   for FBMC/OQAM systems. Information 9(10):246
4. Sakai M, Lin H, Yamashita K (2015) Joint estimation of channel and I/Q imbalance in
   OFDM/OQAM systems. IEEE Commun Lett 20(2):284–287
5. Kumar VS (2020) Joint iterative filtering and companding parameter optimization for PAPR
   reduction of OFDM/OQAM signal. AEU-Int J Electron Commun 124:153365
6. Rujiprechanon L, Boonsrimuang P, Sanpan S, Boonsrimuang P (2018) Proposal of sub-
   optimum algorithm for trellis-based SLM reducing PAPR of FBMC-OQAM signals. In: 2018
   international workshop on advanced image technology (IWAIT), pp 1–4. IEEE
7. Ye C, Li Z, Jiang T, Ni C, Qi Q (2013) PAPR reduction of OQAM-OFDM signals using
   segmental PTS scheme with low complexity. IEEE Trans Broadcast 60(1):141–147
8. Wang H, Wang X, Xu L, Du W (2016) Hybrid PAPR reduction scheme for FBMC/OQAM
   systems based on multi data block PTS and TR methods. IEEE Access 4:4761–4768

9. Kumar JT, Kumar VS (2020) A novel optimization algorithm for spectrum sensing parameters in cognitive radio system. In: International conference on modelling, simulation and intelligent computing, pp 336-344. Springer, Singapore
10. Kumar JT, Kumar VS (2020) Novel distance-based subcarrier number estimation method for OFDM system. In: International conference on modelling, simulation and intelligent computing, pp 328–335. Springer, Singapore

Load Immune Based Stator Incipient Faults Detection Algorithm for Three-Phase Electrically-Excited Synchronous Motor

Sridhar Challagolla and Rama Devi Neerukonda

Abstract Stator winding inter-turn short-circuit faults is one of the major causes for failures of electrically excited synchronous motor. This abnormality progressively grows faults which lead to be complete damage of the machine. Hence in order to improve the sustainability of the system and increase the life of the motor, suitable condition monitoring technique is necessary for the identification of fault occurrence. In this context, this paper introduces detection and phase identification of the stator winding-related inter-turn short-circuit faults in electrically excited wound-field synchronous machines. To detect and identify the faulty phases, an algorithm is proposed based on a wavelet multi-resolution analysis along with adaptive threshold for a 3-ϕ electrically excited synchronous motor operated under different operating conditions like variable power factor and variable mechanical load. Stationary wavelet transform is employed to get the 3-ϕ currents fault residues and then discrete wavelet transform is employed to extract the disturbance information from the 3-ϕ residue currents. In order to find the fault location and phase identification Fault index and 3-ϕ energies are compared with adaptive thresholds. Under different operating conditions finally, the algorithm is tested with practical data for various levels of inter-turn short-circuit faults. By using acquired practical results the validity and successfulness of a proposed algorithm is clearly determined.

Keywords Electrically excited synchronous motor · Inter-turn faults · Adaptive threshold

S. Challagolla (✉)
Acharya Nagarjuna University, Guntur, India
e-mail: sridharbec@gmail.com

R. D. Neerukonda
EEE Department, Bapatla Engineering College, Bapatla, India
e-mail: ramadevi75@gmail.com

© The Author(s), under exclusive license to Springer Nature Singapore Pte Ltd. 2021   221
K. A. Reddy et al. (eds.), *Data Engineering and Communication Technology*,
Lecture Notes on Data Engineering and Communications Technologies 63,
https://doi.org/10.1007/978-981-16-0081-4_23

Nomenclature

EESM	Electrically excited synchronous motor
IPMSM	Interior permanent magnet synchronous motor
EVs	Electric vehicles
HEVs	Hybrid electric vehicles
WT	Wavelet transform
DWT	Discrete wavelet transform
SWT	Stationary wavelet transform
ITSCF	Inter-turn short-circuit fault
MRA	Multi-resolution analysis
UPF	Unity power factor

1   Introduction

The rotating electric motors are widely used in various industrial utilities and automobile industries because of their rugged construction and higher operating efficiencies [1, 2]. In present days, explosive growth of transportation sector electrical vehicles (EVs) and hybrid electric vehicles (HEVs) powered by electrical motor as the main traction component is commercially played a dominant role in transportation sectors because day-to-day rise of the environmental issues like excessive emission of CO_2 from internal combustion engines. In order to build the polluted-free noiseless green environment, increasing the fuel standards economy EVs and HEVs incorporation is the only solution in transportation sectors.

In present day, commercially dominant EV traction motors widely used are interior permanent magnet synchronous motors (IPMSM) and induction motors [3–5]. Wound-field synchronous motors (WFSM) also known as electrically excited synchronous motors (EESMs) but in spite of that EVs equipping with IP-MSMs demands a huge amount of rare earth magnetic material. But the production of rare earth magnetic material leads to the pollution, and it is also a big challenge of limitedly available rare earth magnetic material and its recycling rate. Hence, only an alternative solution, EVs, is used in industry as EESMs employed as main traction motors because it produces the highest peak torque within the shortest time period, and also, they have higher operating efficiency, ruggedness, operating over wide range of power factor control and runs at constant speed [6, 7].

EESMs are subjected to verities of faults categorized mainly as electrical and mechanical faults. In order to improve the system reliability condition monitoring with online is useful to predict a upcoming fault, and also, early detection of faults in EESMs leads to avoid expensive repairs and reduces the down time costs of production. In the stator winding inter-turn short-circuit (ITSC) fault is categorized as one of the significant faults and earlier inception is not done the that may progressively damage the adjacent turns in the same phases or in different phases that lead to the breakdown of the insulation and finally cause the dead short circuit of the system. Therefore in order to improve the reliability and availability of those motors, suitable efficient diagnosis techniques are needed. Several offline and online condition monitoring techniques also their comparative merits are outline in Aller et al. [8]. An enormous research work has to be done against the condition monitoring of electrical motor current signature analysis (MCSA) and that is perceived as the cheapest and simplest monitoring technique.

A familiar approach for the classification of ITSC faults analyzed in time domain is due to the existence of stator current negative sequence component of Kliman et al. [9], and it is the earlier evolve method to detect the winding failure. In the harsh industrial environments, the noise level and its variation should be considered precisely for fault diagnosis because the fault signature due to ITSC is much lower than the noise level. Consequently, it requires a good technique with a capability to remove the noise without corrupting the fault signature. The discrete wavelet transform (DWT) poses the disadvantages in signal noise reduction applications because of the lack of invariant translation property. But this effect can be overcome using stationary wavelet transform (SWT) that has been discussed in Jonnabhotla et al. [10]. The captured current signals under different operating conditions are affected by various factors like noise, sudden variation of mechanical load, static eccentricity, and supply unbalance. Those factors may lead to give errors in fault detection. Hence, accurate inter-turn fault diagnosis techniques must be required to detect the incipient faults, inception of a fault and severity.

This paper suggests a new algorithm based on wavelet and adaptive threshold to detecting the various ITSC faults in 3-ϕ EESM. The three-phase residues are obtained by analyzing those raw current signals with noise using SWT of mother wavelet bi-orthogonal 5.5 (Bior5.5) up to 6th level with sampling frequency of 6.6 kHz. For getting the fault feature extraction, those residues are once again decomposed with DWT of same mother wavelet. The detection and identification of the fault is performed by using defined fault index and three-phase energies. The developed algorithm is verified to detect and identify the various faults from three-phase currents obtained in experimental setup. The proposed adaptive threshold-based algorithm improves the fault detection and identification accuracy.

2 Wavelet Analysis

Wavelet transform (WT) is powerful mathematical tool to analyze a signal simultaneously in frequency and time domains. The WTs are very much useful for analyzing non-periodic, non-stationary, intermittent, and sharp peak transient signals. The discrete Fourier transform (DFT) analyzes the given signal information in only frequency domain, and it does not give the frequency component exist at which time. But WT analyzes the signal both time and frequency domain more efficiently, and it also gives non-uniform division of frequency domain; i.e., WT uses at higher frequencies give small window and at lower frequencies give broad window. In order to divide the given signal into various levels, a set mathematical functions called wavelets are used. Those wavelet functions are determined with a mother wavelet with translation and dilation

WT is defined by as a sequence of a function $\{h(n)\}$ (low pass filter) and $\{g(n)\}$ (high pass filter). The scaling function $\phi(t)$ and wavelet $\psi(t)$ are defined as follows

$$\phi(t) = \sqrt{2} \sum h(n)\phi(2t - n) \tag{1}$$

$$\psi(t) = \sqrt{2} \sum g(n)\phi(2t - n) \tag{2}$$

where $g(n) = (-1)n\, h(1 - n)$. A sequence of $\{h(n)\}$ defines a WT. There are several categories such as Haar, Daubechies, and Symlet. Based on the selection of application, the mother wavelet selection depends. The succeeding section describes new adaptive threshold algorithm with wavelet multi-resolution analysis which is used to detect and discrimination of various types faults using with transient current signals allied with the fault is explore.

3   Fault Detection Algorithm

The suggested fault detection algorithm begins with data acquisition later appliance of wavelet analysis and adaptive threshold logics, and this is shown in Fig. 2. The residues of 3-ϕ stator currents are obtained by using SWT de-noising approach with level-based threshold. By using the SWT of Bior5.5, the 3-ϕ current signals of EESM are decomposed into 6th level approximate and detail coefficients. In order to observe the frequency component over a range under supply unbalancing condition sensitive, the d1 to d4 coefficients thresholds are made maximum although d5 coefficients threshold value is fixed to a high value. The d6 threshold is determined based on its maximum amplitude in 1st cycle and then multiplied with a distortion factor. The d6 coefficient threshold value can do the augment the fault signature, by the virtue of subtraction of pre-fault value from the measured signal. The 3-ϕ stator current residues are once again decomposed with Bior5.5 mother wavelet, and each phase d1 coefficient slopes are evaluated due to the identification of variation fault levels due to disturbances. The detailed level coefficients absolute slope sum is fixed as a fault index (I_f) and shown mathematically as follows.

$$I_f(n) = |\text{slope}\_d_1 I_R(n)| + |\text{slope}\_d_1 I_Y(n)| + |\text{slope}\_d_1 I_B(n)| \tag{3}$$

where $n = 1: N_1$; N_1 is total samples; the slope of d_1 coefficients of 3-ϕ residue currents in R, Y, and B phases represented as slope $d_1\_I_R$, slope $d_1\_I_Y$, and slope $d_1\_I_B$, respectively. The abnormal condition of the EESM can be detected by examining the three successive fault indices' values with an adaptive threshold one (ATH1), and fault indices count value over a moving window of 10 samples should be greater than 6. The disturbance corresponding instant is calculated from the first sample of the three consecutive fault indices and defined as location one (Loc1). The type of disturbance present in the system is obtained by using three-phase energy values, which are compared with adaptive threshold 2 (ATH2). The energy crossing point with respect to ATh2 is defined as location two (Loc2). The ATH1 and ATH2 are calculated based on FI statics and three-phase energy statics, respectively. Mathematical representation of 3-φ energies is explained in Eq. (4).

$$\text{Energy\_}I_j(n) = \frac{1}{N_2} \sum_{m=n}^{n+N_2-1} \left[\text{slop\_}d_1 I_j(m) \right]^2 \tag{4}$$

where $j \in$ R, Y, B $n = 1$ to $N_1 - N_2$; $N_2 =$ no. of samples in the (1/10) th of the cycle. A sudden mechanical load change may also lead to transient in stator currents. But, it should be discriminated from the ITSCFs by using fundamental concept of time constant representing mechanical and electrical systems. Hence, the ITSCFs and sudden change in mechanical load should be discriminated by taking the difference between the disturbance instants, which are obtained from energy and fault index. By making the comparison of peak energy obtained in each phase over one and half cycle from fault instant, the faulty phase identification is done in stator ITSCFs.

4 Verification of Algorithm with Experimental Data

A 3-Ø, 7.5 HP, 415 V, 4.9 A, 50 Hz, star-connected 4pole, 1500 rpm salient-pole rotor EESM with 48 slots, 4 coils in each phase, and each coil consists of 47 turns for practical experimental purpose. In order to done the various turn-level faults in each phase of the re-arranged stator, terminals are taken out with four tapping points with 2-turns gap. The stator currents under various faults condition are captured with UNIPOWER DIP8000 power network analyzer with a measuring capacity of 6600 samples per second and after that export and import those signals with a PC interfaced RS 232 port and complete the analysis with MATLAB software.

Figure 1 shows the experiment setup, where (1) EESM with mechanical loading arrangement and stator winding tapping points; (2) star-delta starter; (3) DIP 8000 power network analyzer; (4) Current transducers; (5) PC system (Fig. 2).

In practice, the raw signal is always corrupted with some noise. Hence, signal reconstruction should be required for stator inter-turn faults. The three-phase residue currents are obtained by preprocessing the captured three-phase currents with SWT of adaptive threshold. In this paper, the proposed algorithm is verified with healthy, various levels of inter-turn faults of R, Y, and B phases and sudden change in mechanical load under no-load and 20% applied mechanical load on the motor.

Figures 3, 4, and 5 show 3-φ stator currents with noise, 3-φ residue currents and fault index when the machine operated without fault (healthy) under no-load condition. Figures 6, 7, and 8 illustrate 3-φ stator currents with noise, 3-φ residue

Fig. 1 Experimental test
bench for 7.5 HP EESM

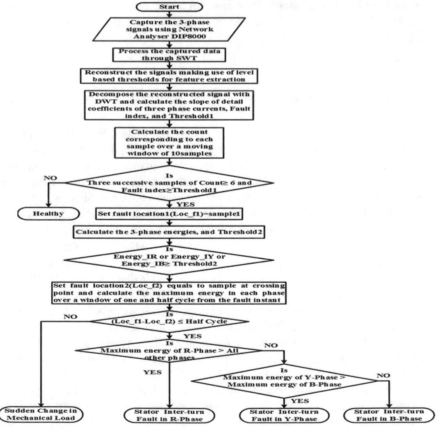

Fig. 2 Inter-turn short circuit fault detection flow chart

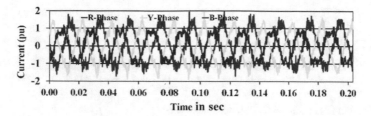

Fig. 3 Variation of 3-φ stator currents for healthy case with noise

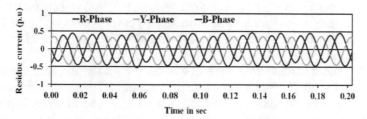

Fig. 4 Variation in 3-φ residue currents for healthy case

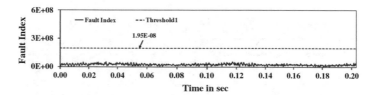

Fig. 5 Fault indices variation for healthy case

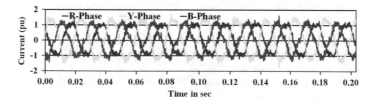

Fig. 6 Variation of 3-φ stator currents for 20% load healthy case with noise

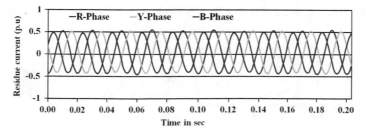

Fig. 7 Variation in 3-φ residue currents for 20% load healthy case

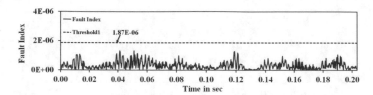

Fig. 8 Fault indices variation for 20% load healthy case

currents and fault index when the machine operated 20% load without fault (healthy) with UPF. Figures 5 and 8 predict that under healthy cases without load and with 20% loaded case also the fault indices are below the ATH. Thus, the status of machine is normal or healthy condition.

The 3-φ stator currents with noise, 3-φ residue currents, and fault index under fault condition with Y-phase 4-turns with no-load and UPF operation are shown in Figs. 9, 10, and 11, respectively. Similarly, Figs. 12, 13, and 14 show the variation in 3-φ stator currents with noise, 3-φ residue currents and fault index in Y-phase 4-turn fault with 20% loaded and UPF operation, respectively. Figures 11 and 14

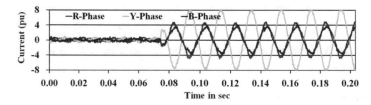

Fig. 9 3-φ stator current variations in Y-Phase 4-turns short circuit under no-load noisy condition

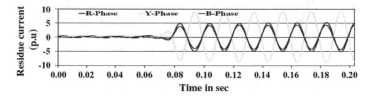

Fig. 10 3-φ residue current variations in Y-Phase 4-turns short circuit with under no-load condition

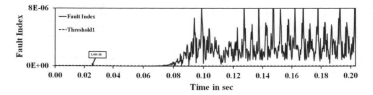

Fig. 11 Fault indices variation in Y-Phase 4-turns short circuit under no-load condition

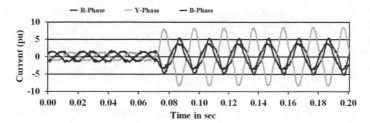

Fig. 12 3-φ stator current variations in Y-Phase 4-turns short -circuit under 20% load

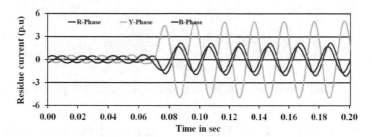

Fig. 13 -φ residue current variations in Y-Phase 4-turns short circuit under 20% load

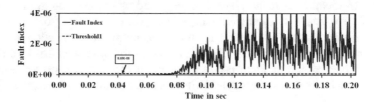

Fig. 14 Fault indices variation in Y-Phase 4-turns short circuit under 20% load

demonstrate that the Y-phase 4-turn fault indices cross the adaptive threshold1 under no-load and 20% load condition, respectively. Hence, the proposed algorithm detects the fault effectively even the machine is operated at loaded condition.

Figures 15, 16, and 17 show 3-φ stator currents with noise, 3-φ residue currents, and fault index under sudden mechanical load change at UPF operating condition. The variation in fault indices of sudden mechanical load change is shown in Fig. 17

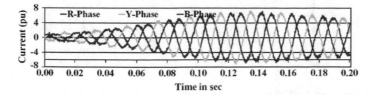

Fig. 15 3-φ stator currents variation with sudden mechanical load change operating condition

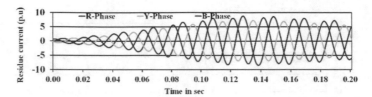

Fig. 16 3-φ residue currents variation with sudden mechanical load change operating condition

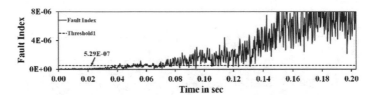

Fig. 17 Fault indices variation with sudden mechanical load change UPF operating condition

and crosses the adaptive threshold1 with a certain delay due to mechanical time constant. Hence, identification of sudden mechanical load change is possible from the faulty condition even magnitude of fault currents is equal.

The identification of faulty phase is done based on comparison of peak energy values of three-phase d1 coefficient with ATH2 over a window of 6 samples from the fault instant. Figures 18, 19, and 20 show 3-φ energy value variation in Y-phase 4-turns short circuit under no-load, 20% load and sudden mechanical load change with UPF operated condition, respectively. In all the cases, the energy value crosses the ATH2. In case of faulty condition, the instant of fault indices crossing the ATH1 is

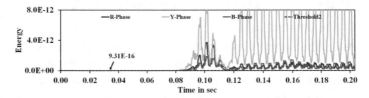

Fig. 18 3-φ energies for Y-phase 4-turn short circuit under no-load with UPF condition

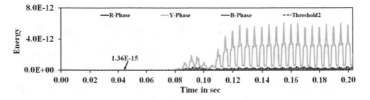

Fig. 19 3-φ energies for Y-phase 4-turn short circuit with 20% load with UPF

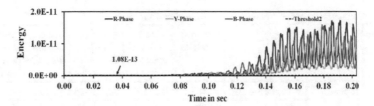

Fig. 20 3-ϕ energies with sudden mechanical load with unity power factor condition

approximately same to the instant of energies crossing ATH2. But in case of mechanical sudden load change, the instant of fault indices crossing ATH1 is different with the instant of energies crossing ATH2. Hence, the fault is effectively differentiated from the sudden mechanical load changes. The energy value of faulty phase is higher than the remaining phases from the instant of crossing the ATH2 which are shown in Figs. 18 and 19.

Table 1 shows fault index indices ATH1, fault detection instant, energy value indices ATH2, and fault detection instant the EESM various level of ITSCFs in R, Y, and B phases and sudden mechanical load change under no-load and 20% load with UPF operation. The proposed fault instant logic is calculated based on the average value of fault instant obtained from the ATH1 and ATH2 which is shown in Table 1. The tabulated values of fault instants under the faulty cases are approximately equal but in case of sudden mechanical change, the difference in fault instant values is large. Hence, the proposed algorithm is effectively detecting and differentiating the fault from sudden load changes.

Table 2 illustrates the peak values of energies in three phases over a window of 6 samples from the fault instant and maximum value of energy out of three phases. The tabulated results show that the maximum value of energy out of 3-ϕ's is same as the faulty phase energy. Thus, the proposed algorithm detects and identifies the faulty phase effectively even the machine is operated at no-load and 20% load with UPF operation.

5 Conclusions

This paper describes a WT and adaptive threshold-based fault diagnosis technique for a 3-ϕ EESM ITSCFs. Fault residue currents are determined by using inverse SWT of Bior 5.5 mother wavelet. The uses of SWT instead of DWT in reconstruction processes improve the de-noising performance due to shift invariance property of SWT. The performance of proposed detection algorithm should improve the accuracy due to adaptive thresholds. The practical experimental results demonstrate that the suggested algorithm is more efficient to detect and identification various level of faults. The future scope of this work is extended to using fuzzy logic and artificial neural networks.

Table 1 Comparision of fault instants for various disturbances

Machine operated power factor	Type of fault		Adaptive threshold 1	Instant of fault detection	Adaptive Threshold 2	Instant of fault detection	Difference between two fault instants
Unity power factor	R-phase	2-Turn-No-Load	7.01E−08	0.231 s	5.01E−15	0.240 s	0.009 s
		2-Turn-20% Load	6.36E−09	0.395 s	6.36E−09	0.395 s	0.000 s
		4-Turn-No-Load	1.17E−07	0.262 s	2.41E 15	0.263 s	0.001 s
		4-Turn-20% Load	1.32E−07	0.573 s	1.58E−16	0.572 s	0.001 s
		6-Turn-No-Load	6.54E−07	0.363 s	3.82E−13	0.362 s	0.001 s
		6-Turn-20% Load	4.92E−08	0.495 s	3.49E−16	0.497 s	0.002 s
	Y-phase	2-Turn-No-Load	2.23E−07	0.178 s	8.13E−15	0.179 s	0.002 s
		2-Turn-20% Load	7.72E−08	0.370 s	1.45E−16	0.372 s	0.002 s
		4-Turn-No-Load	3.38E−06	0.482 s	7.05E−11	0.482 s	0.000 s
		4-Turn-20% Load	6.25E−08	0.369 s	1.33E−16	0.368 s	0.001 s
		6-Turn-No-Load	4.69E−07	0.425 s	7.36E−14	0.430 s	0.002 s
		6-Turn-20% Load	1.12E−07	0.313 s	3.35E−16	0.311 s	0.002 s
	B-phase	2-Turn-No-Load	1.09E−07	0.283 s	3.35E−14	0.284 s	0.001 s
		2-Turn-20% Load	6.38E−08	0.340 s	1.35E−16	0.339 s	0.001 s
		4-Turn-No-Load	6.96E−07	0.331 s	7.14E−13	0.338 s	0.007 s
		4-Turn-20% Load	1.29E−07	0.437 s	6.28E−15	0.442 s	0.005 s
		6-Turn-No-Load	9.53E−07	0.386 s	2.67E−13	0.390 s	0.004 s
		6-Turn-20% Load	1.22E−07	0.441 s	6.71E−16	0.448 s	0.007 s
	Sudden mechanical load	Under No-Load	1.46E−06	0.224 s	1.05E−12	0.277 s	0.053 s
		Under 20% Load	4.70E−08	0.235 s	7.24E−16	0.196 s	0.039 s

Table 2 Identification of faulty phase for various ITSC faults

Machine operated power factor	Type of fault		Instant of fault detection	Maximum value of energy over a 6 samples from fault Instant			Maximum energy of faulty phase
				E_R	E_Y	E_B	
Unity power factor	R-phase	2-Turn-No-Load	0.279 s	7.37E−13	7.76E−14	6.18E−13	7.37E−13
		2-Turn-20%Load	0.391 s	6.07E−16	3.83E−18	4.36E−16	6.07E−16
		4-Turn-No-Load	0.448 s	6.35E−13	1.80E−13	4.85E−15	6.35E−13
		4-Turn-20%Load	0.572 s	2.37E−16	1.17E−18	1.04E−16	2.37E−16
		6-Turn-No-Load	0.460 s	4.88E−12	6.81E−13	1.36E−12	4.88E−12
		6-Turn-20% Load	0.496 s	4.10E−16	6.33E−17	2.23E−16	4.10E−16
	Y-phase	2-Turn-No-Load	0.178 s	9.53E−15	2.60E−14	8.51E−15	2.60E−14
		2-Turn-20% Load	0.371 s	1.93E−16	5.80E−16	7.45E−18	5.80E−16
		4-Turn-No-Load	0.482 s	9.33E−12	1.20E−11	3.39E−13	1.20E−11
		4-Turn-20% Load	0.368 s	2.52E−16	3.45E−16	1.86E−16	3.45E−16
		6-Turn-No-Load	0.428 s	1.84E−14	9.03E−14	3.01E−14	9.03E−14
		6-Turn-20% Load	0.312 s	4.36E−16	7.77E−16	4.29E−17	7.77E−16
	B-phase	2-Turn-No-Load	0.284 s	1.76E−16	4.28E−15	1.77E−14	1.77E−14
		2-Turn-20% Load	0.340 s	6.66E−18	1.52E−16	5.34E−16	5.34E−16
		4-Turn-No-Load	0.334 s	3.17E−14	1.83E−13	4.02E−13	4.02E−13
		4-Turn-20% Load	0.439 s	1.37E−16	4.23E−15	1.16E−14	1.16E−14
		6-Turn-No-Load	0.388 s	8.74E−14	7.31E−14	3.43E−13	3.43E−13
		6-Turn-20% Load	0.468 s	3.52E−18	1.4E−17	3.06E−17	3.06E−17

References

1. SCheng M, Sun L, Buja G, Song L (2015) Advanced electrical machines and machine-based systems for electric and hybrid vehicles. Energies 8:9541–9564
2. Yang YP, Shih GY (2016) Optimal design of an axial-flux permanent-magnet motor for an electric vehicle based on driving scenarios. Energies 9:285
3. Cai H, Guan B, Xu L (2014) Low-cost ferrite PM-assisted synchronous reluctance machine for electric vehicles. IEEE Trans Industr Electron 61(10):5741–5748
4. Casadei D, Rossi C, Gaetani A, Pilati A (2005) Traction system for heavy electric vehicles based on the wound rotor salient pole synchronous machine drive. In: Proceedings of electric vehicle symposium EVS21, Monaco, Apr 2–4 2005
5. Morimoto TM (2012) Rare earth free, traction motor for electric vehicle. In: Proceedings of IEEE international electric vehicle conference, Mar 2012, pp 1–4

6. Pyrhonen O, Niemela M, Pyrhonen J, Kaukonen J (1998) Excitation control of DTC controlled salient pole synchronous motor in field weakening range. In: Proceedings of advanced motion control, AMC 1998. Coimbra, 29 June–1 July 1998, pp 294–298
7. Amara Y, Vido L, Gabsi M, Hoang E, Hamid Ben Ahmed A, Lecrivain M, Hybrid excitation synchronous machines: energy-efficient solution for vehicles propulsion. IEEE Trans Veh Technol 58(5):2137–2149
8. Aller S, Grubic JM, Lu B, Habetler TG (2008) A survey on testing and monitoring methods for stator insulation systems of low-voltage induction machines focusing on turn insulation problems. IEEE Trans Industr Electron 55(12):4127–4136
9. Kliman GB, Premerlani WJ, Koegl RA, Hooweler D (1996) A new approach to on-line turn fault detection in ac motors. In: Conference Record 31st, IEEE IAS Annual Meeting, vol 1, pp 687–693; Clerk Maxwell J (1892) A treatise on electricity and magnetism, vol 2 (3rd edn). Clarendon, Oxford, pp 68–73
10. Jonnabhotla N, Akshay NAV, Sadam N, Yeddanapudi ND (2010) ECG noise removal and QRS complex detection using UWT. In: International conference on electronics and information engineering (ICEIE), vol 2, IEEE, pp V2–43

Segmentation of Remote Sensing Images Using Fuzzy Gaussian Mixture Level Set (FGML) Method

Ramudu Kama, Ganta Raghotham Reddy, S. P. Girija, and A. Srinivas

Abstract By the growing extent and difficulty of remote sensing image information and the problems that are encountered during the cutting edge of data handling, a progress of large-scale image segmentation investigation algorithms was impotent to encounter the essential for approaches in order to raise the final correctness of foreground recognition of an object. Consequently, the improvement of imaging methods which are large-scale stances is a prodigious challenge today. Traditional image segmentation algorithms obligate high temporal difficulty and low segmentation accuracy due to absence of spatial information and the occurrence of noise. Traditional level set segmentation approaches such as Chan-Vese, Mumford Shah and the active contours without edges suffer with non-homogeneity and high intensity noise by increased temporal complexity as well as low segmentation precision. To overcome the difficulties which are occurred in traditional methods, here in this paper, we introduced a novel hybrid approach which combines the fuzzy region as well as Gaussian mixer model using a level set method, which is called "Fuzzy Gaussian Mixer Level set (FGML)" algorithm for large-scale remote sensing image segmentation. This algorithm uses a Gaussian mixture model to describe an amount of non-similarity for the pixel class attribute. This procedure meritoriously eradicates the impact of noise on the segmentation outcome by execution extremely accurate suitable of the data with a statistical distribution. Additionally, we introduce competition in the fuzzy region to express the neighbourhood prior probability to be considered as the weight of the Gaussian constituent. Results of segmentation are occurred by using fuzzy Gaussian mixer model suffers with outliers and boundary outflows

R. Kama (✉) · G. R. Reddy · S. P. Girija · A. Srinivas
Department of ECE, KITS, Warangal, Telangana, India
e-mail: ramudukama@gmail.com

G. R. Reddy
e-mail: grrece9@gmail.com

S. P. Girija
e-mail: spgirija@gmail.com

A. Srinivas
e-mail: srinivas_azmeera@yahoo.com

© The Author(s), under exclusive license to Springer Nature Singapore Pte Ltd. 2021   235
K. A. Reddy et al. (eds.), *Data Engineering and Communication Technology*,
Lecture Notes on Data Engineering and Communications Technologies 63,
https://doi.org/10.1007/978-981-16-0081-4_24

problem. These complications remain overcome through the level set method to enhance the segmentation results. This procedure increases the robustness of image segmentation for large-scale remote sensing images. Lastly, we acquired the image data from NASA's Earth Observatory and accompanied the tests; experimental results confirm that, this new approach is useful one, effective, and it could attain exceedingly accurate results of segmentation with less time complexity over existing approaches.

Keywords Image segmentation · Fuzzy region competition · Remote sensing images · Gaussian mixture model and level set segmentation

1 Introduction

In recent times, beneath the wide over large-scale sensor networks, a variety on end gadgets remain deployed within a number infrastructures in imitation of accumulate information, and afterwards, the data is directed to the cloud through processing networks [1]. After the procedure described above, that is used for aid purposes among specific areas. Through facts gathering, detection, identification, location, monitoring, then monitoring, human beings are able to control the physical world. In pursuance along that objective, we lessen photograph segmentation after becoming aware of real-world targets beneath cyber computing [2]. Recent advances into image segmentation research have conducted after the utilization concerning a large extent of faraway sensing applications, which include object detection then image classification [3]. Belief segmentation is back to divide the images of remote sensing into deep areas together through identical houses faithfully correspond to an objects [4]. Features of the pixels among the same areas are analogous, yet the characteristics concerning the pixels of the one-of-a-kind regions were different. Conviction segmentation is known in accordance with move an especially crucial part in image technology then machine vision.

Remote sensing images encompass giant aggregate concerning information; however, manifesting statistics between photographs is difficult in imitation of amount accurately [5]. Therefore, the trouble of remote sensing picture segmentation, among unique targeting segmentation, has attracted plenty attention out of deep academics. Depending on the propagated signals likes electromagnetic radiation, at present, remote sensing term denotes the utilization of satellite or an aircraft depended sensor technologies in order to sense and to classify the objects like surfaces, atmosphere and oceans on an Earth. The classifications of remote sensing are of two types. First type of classification is the "active remote sensing", in this type, the signal is radiated over the satellite or an aircraft where the object reflections are perceived by a sensor. Second type of classification of remote sensing is "passive remote sensing", which is used to detect the sunlight reflection by the sensor [6].

2 Materials and Methods

In this segment, extant details of the estimated segmentation technique successfully employ the fuzzy region competition as well as Gaussian mixture model in the pre-processing and level set method in post-processing. This integrating segmentation approach is called "Fuzzy Gaussian Mixer Level Set (FGML)" method. In detail, explanation and implementation are explained in the following segments [7].

2.1 Fuzzy Gaussian Mixer (FGM) Segmentation Approach

Let $X = \{x_1, x_2, \ldots, x_n\}$ gloss the large-scale far off sensing image, the place j is a pixel index yet n is wide variety about the pixels of an image. The murky membership function u_{ij} is aged after drawing the chain concerning i cluster in imitation of a j pixel. Fix $U \in [u_{ij}]c \times n$ as a matrix of the fuzzy membership in accordance with symbolize the mystical segmentation of image X. C represents the flat quantity on a vicinity among the segmentation. To determine the foremost U, that is imperative to outline the pardon regarding non-similarity among pixels then groups. Following, an unbiased characteristic remains dignified. In accumulation, that diminishes the goal feature in imitation of achieve the most fulfilling end result concerning diffuse segmentation. In the mystical region, the exponential balance is old in accordance with indicate the murky degree, as lacks a evident physical meaning. Therefore, the regularization aspect KL [8] is introduced to the arithmetic measurement in accordance with outline the objective function.

$$J \overset{\Delta}{=} \sum_{j=1}^{n} \sum_{i=1}^{c} u_{ij} d_{ij} + \lambda \sum_{j=1}^{n} \sum_{i=1}^{c} u_{ij} \ln\left(\frac{u_{ij}}{w_{ij}}\right) \tag{1}$$

At which, d_{ij} is a non-similarity quantity among the pixel and cluster. The pixel j fitting to cluster I is symbolized by w_{ij} as prior likelihood function to estimate it. The fuzzy aspect used for supervisory of fuzzy degree is defined as $1 < \lambda < 1$. When $\lambda \to 1$, each $u_{ij} \to 0$. Therefore, each pixel fits to a cluster added clearly. Then, the fuzzy cluster approach is applied as a specific procedure When $\lambda = 1$, if $\lambda \to +\infty$, $u_{ij} \to 1/c$.

The Gaussian mixture model is implemented for non-similarity quantity of an image statistical distribution in Eqs. (2) and (3) defined as

$$p(x_j | w_j, V, \Re) = \sum_{i=1}^{c} w_{ij} P(x_j | w_j, V, \Re) \tag{2}$$

$$p(x_j | v_i, \Re_i) = \frac{1}{(2\pi)^{q/2} |\Re_i|^{1/2}} e^{-1/2 (x_i - v_j) \Re_i^{-1} (x_j - v_i)} \tag{3}$$

where $w_j = \{w_{ij}, i = 1, 2 \ldots c\}$ is a prior likelihood vector. Here, $v = \{v_i\}$ denoted as clustering mean vector set. The covariance set of matrices indicated as $\Re = \{\Re_i\}$. The mean vector and covariance matrix functions indicated as v_i and the $\Re_i$, respectively. The non-similarity amount is defined in Eq. (4) as a negative logarithmic of the j_{th} component:

$$d_{ij} \overset{\Delta}{=} - \ln w_{ij} p(x_j | v_i, \Re_i)$$
$$= \frac{y}{2} \ln \ln(2\pi) + \frac{1}{2} \ln \ln[|\Re_i|]$$
$$+ \frac{1}{2}\left((x_j - v_i)^T \Re_i^{-1}(x_j - v_i)\right) - \ln w_{ij} \tag{4}$$

Thus, foreground object and background segmentation are defined as the energy function of the following Eq. (5)

$$E(\alpha_1, \alpha_2, u) = \int |\nabla u| dx + \lambda \int (-u \log P_1(W_A | \alpha_1) - (1 - u) \log P_2(W A | \alpha_2)) dx \tag{5}$$

P_i must be refined. Beneath the approximating district bound, there

$$P_i(W_A | \alpha_i) = P_i(\mu, \sigma_i^2)$$
$$= \frac{1}{\sigma_i \sqrt{2\Pi}} e^{-\frac{(W_{A(y)}^s - \mu_i)^2}{2\sigma_i^2}} \tag{6}$$

3   Implementation of FGM Model to Level Set Method

This is the final step of the proposed algorithm, which is called level set method (lsm). This lsm is to progress the contour on a fuzzy Gaussian mixer (FGM) model to refine the segmentation results obtained in previous pre-processing stage, which suffers with outliers and boundary leakages. The proper level set function is defined for this problem and effectively locks the objects of interest and tuning the control parameters accordingly for good segmentation results [9, 10].

In order to overcome the difficulties in the existing level set methods such as intensive computation, re-initialization periodically as SDF, a fast level set function to overcome the said difficulties is discussed by Li et al. [11]. The level set function is defined in Eq. (7) as following.

$$\frac{\partial \phi}{\partial t} = \mu \xi(\varphi) + \zeta(g, \varphi) \tag{7}$$

In Eq. (7), $\xi(\varphi)$ says the movement of φ level set function, which is automatically approached to signed distance function (SDF) and expressed in Eq. (8)

$$\xi(\varphi) = \Delta\varphi - \mathrm{div}\left(\frac{\nabla\varphi}{|\nabla\varphi|}\right) \tag{8}$$

Similarly, from Eq. (7), second term $\zeta(g, \varphi)$ is the image gradient defined in Eq. (9)

$$\zeta(g, \varphi) = \lambda\delta(\varphi)\mathrm{div}\left(g\frac{\nabla\varphi}{|\nabla\varphi|}\right) + vg\delta(\varphi) \tag{9}$$

Here, mean curvatures is $\mathrm{div}\left(g\frac{\nabla\varphi}{|\nabla\varphi|}\right)$, "$g$" is the edge indicator function or edge stopping function (see Eq. 10) to stop the level set evolution which is closer to the optimum solution. The term $\zeta(g, \varphi)$ indicates the level set function attracts towards a variational boundary and to control the φ level set function by defining few constraints like λ, v as well as μ, respectively, and Dirac function is defined as $\delta(\varphi)$.

$$g = \frac{1}{1 + \left|\nabla(G_\sigma * I_{\mathrm{fgm}}\right|^2} \tag{10}$$

Finally, the new fuzzy Gaussian mixer level set (FGML) approach effectively evolves the contour towards the object of interest R_O, and here, an improved balloon force is well-defined to pull or to push the level set function in the direction of the object of interest R_O. An Eq. (9) is transformed to Eq. (11) by considering the improved balloon force function. The modified fuzzy Gaussian mixer level set method is defined as [12–17]

$$\zeta(g, \varphi) = \lambda\delta(\varphi)\mathrm{div}\left(g\frac{\nabla\varphi}{|\nabla\varphi|}\right) + gG(R_o)\delta(\varphi) \tag{11}$$

Here, $G(R_o) = 1 - 2R_O$ is the updated balloon force function to pull or to push the level set function towards an object of interest R_O. Equations (7), (8) and (11) are the fuzzy Gaussian mixer level set (FGML) formulation.

4   Simulation Results and Discussions

Figure 1a–d gives the four different input land cover images taken from the NASA Earth observatory. Figure 1e–f shows the results of existing Chan-Vese (C-V) model. This model failed to accurately segment the image land cover details. Figure 1e is a land cover image whose edges are not accurately detected. Coming to Fig. 1h which is a low contrast island image, the existing Chan-Vese model failed to segment the

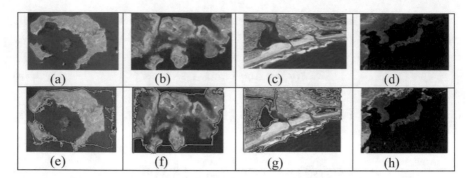

Fig. 1 a–d shows the different land cover images taken from NASA Earth observatory and its segmentation results by C-V Model [18] in the second row images from **e** to **f**

image which is clearly shown in the below image. So to overcome these disadvantages and get accurate segmented result, we are going for the proposed method which is image segmentation depended on fuzzy Gaussian mixture level set model. From Fig. 2, it clearly says that the land cover image has been perfectly segmented; even the low contrast image also has been segmented accurately with our model which is clearly showed in Fig. 2 (Table 1).

5 Conclusion

In this paper, we presented a new hybrid approach for effective segmentation of land cover images of remote sensing by using fuzzy Gaussian mixer level set method. This proposed approache is diversified into two phases. First phase, fuzzy region completion and Gaussian mixer model (GMM) are combined in the pre-processing to get superior cluster results and better the clustering effectiveness, but the segmentation results undergo with few drawbacks like outliers and boundary leakages. To overcome these difficulties occurred in pre-processing stage, it is essential to use second phase by level set method to improve the results of segmentation. This second phase is called post-processing stage. So, we initialize contour on fuzzy Gaussian mixer segmentation results to evolve the contour based on the level set function defined by a user and also decrease the boundary leakages and outliers by the level set method. Finally, the fuzzy Gaussian mixer level set method gets the superior and accurate segmentation of land cover images of remote sensing over the current level set methods.

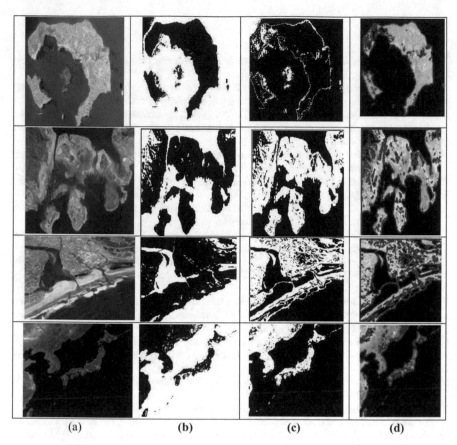

| (a) | (b) | (c) | (d) |

Fig. 2 Clusters images (**b**, **c**) and final segmented result (**d**) of various input images of column (**a**) by proposed fuzzy Gaussian mixer level set method

Table 1 Comparison of proposed method with the existing level set model

Images	Proposed level sct method			Existing ACM method		
	No. of pixels covered	Evaluation time (s)	Misclassification error	No. of pixels covered	Evaluation time (s)	Misclassification error
Image-1	11617	0.457	0.7096	10627	151.26	0.7346
Image-2	15282	0.261	0.6182	10903	28.34	0.7971
Image-3	21134	0.359	0.4717	10831	189.13	0.8218
Image-4	9663	0.386	0.7584	8253	469.3	0.8925

References

1. Zhang QL, Yang T, Chen Z, Li P (2018) A survey on deep learning for big data. Inf Fus 42:146–157
2. Chen P, Li Z, Yang LT, Zhao L, Zhang Q (2017) A privacy-preserving high-order neuro-fuzzy c-means algorithm with cloud computing. Neurocomputing 256:82–89
3. Gupta S, Arbeláez P, Girshick R, Malik J (2015) Indoor scene understanding with RGB-D images: bottom-up segmentation, object detection and semantic segmentation. Int J Comput Vis 112(2):133–149
4. He L et al (2008) A comparative study of deformable contour methods on medical image segmentation. Image Vis Comput 26(2):141–163
5. Pinheiro M, Alves J (2015) A new level-set-based protocol for accurate bone segmentation from CT imaging. IEEE Access 3:1894–1906
6. Gibou F, Fedkiw R (2002) A fast hybrid K-means level set algorithm for segmentation. Technical Report. Stanford University, Stanford
7. Yin S, Zhang Y, Karim S (2018) Large scale remote sensing image segmentation based on fuzzy region competition and gaussian mixture model. IEEE Access
8. Ichihashi H, Miyagishi K, Honda K (2001) Fuzzy c-means clustering with regularization by K-L information. In: Proceedings of IEEE international conference fuzzy system, vol 2, pp 924–927
9. Gongt H, Li Y, Liu G, Wu W, Chen G (2016) A level set method for retina image vessel segmentation based on the local cluster value via bias correction. In: Proceedings IEEE international congress image signal process, pp 413–417
10. Kama R, Tummala R (2019) Level set evolution of biomedical MRI and CT scan images using optimized fuzzy region clustering. Comput Method Biomech Biomed Eng Imag Vis 7(1):99–110
11. Li F, Shen C, Li C (2010) Multiphase soft segmentation with total variation and h1 regularization. J Math Imaging Vis 37:98–111
12. Kama R, Tummala R (2018) Segmentation of tissues from MRI bio-medical images using kernel fuzzy PSO clustering based level set approach. Current Med Imag Int J 14(3):389–400
13. Ambrosio L, Tortorelli VM (1990) Approximation of functional depending on jumps by elliptic functional via T-convergence. Commun Pure Appl Math 43(8):999–1036
14. Ramudu K, Tummala R (2015) Layers and Dark Sand Dunes segmentation of MARS satellite imagery using level set model. IETE J Educ 56(2):59–67
15. Ramudu K, Srinivas A, Tummala R (2016) Segmentation of satellite and medical imagery using homomorphic filtering based level set model. Ind J Sci Technol 9(S1)
16. Kalyani Ch, Ramudu K, Ganta RR (2018) A review on optimized K-means and FCM clustering techniques for biomedical image segmentation using level set formulation. Biomed Res (India). Int. J. Med. Sci. 29(20):3660–3668
17. Li C, Xu C, Gui C (2005) Level set evolution without re-initialization: a new variational formulation. IEEE Comput Soc Conf Comput Vis Recognit 430–436
18. Chan T, Vese L (2001) Active contours without edges. IEEE Trans Image Process 10(2):266–277

Performance Improvement in Long-Term Evolution-Advanced (LTE-A) Network Towards 5G Network

C. H. Nishanthi and N. Ramamurthy

Abstract Long-term evolution is fourth generation (4G) promising network technology that inherited massive usage for mobile communication, vehicle-to-vehicle(V2V) communication, artificial intelligence, machine learning, video conferences and a lot more applications. A vast area of research is carried in protocol, modulation techniques, slight network changes, various novel algorithms like Dinkelbach's method-based energy optimization algorithm for improving data rates, 3D beamforming technique for user-specific QoS providing included in LTE-A is 3GPP release 15/16 improvization to meet real-time traffic challenges in a recent communication. This paper proves network QoS improvement in terms of minimum user rate, throughput and routing overhead in packets delivery from source to destination. Dinkelbach's method is a powerful computational algorithm to improve QoS, reliability, energy efficiency (EE), spectrum efficiency (SE), reduce latency, throughput improvement with full-dimensional MIMO system (FD-MIMO). The optimization further is obtained by utilizing novel techniques incremental redundancy and regression monitoring technique with hybrid automatic repeat request (HARQ). The processed results show the improvement in terms of throughput, minimum user rate and routing overhead for user group-specific downtilt beamforming (UG-SSDB), FD-MIMO and extended FD-MIMO. The results were carried in the network simulator 3 platform (NS3).

Keywords Throughput · LTE-advanced · NS3 · HARQ · EFD-MIMO · Incremental redundancy technique · Regression monitoring technique

C. H. Nishanthi
Department of Electronics and Communication Engineering, JNT University Anantapur, Anantapur, Andhra Pradesh, India
e-mail: chnishanthi@gmail.com

N. Ramamurthy (✉)
Department of Electronics and Communication Engineering, G. Pullaiah College of Engineering and Technology, Pudur, Andhra Pradesh, India
e-mail: ramamurthy1006@gmail.com

1 Introduction

Three-dimensional multiple-input–multiple-output (3D-MIMO) is an effective part of the investigational so normalization in wireless communications for the advancement of fifth generation (5G) wireless frameworks. A quasi-optimum result is received via the use of semi-definite relaxation also Dinkelbach's technique [1]. Beamforming is a recognized signal treating method to enhance the received signal strength to a selected direction. Freshly, the three-dimensional (3D) beamforming method has been increased a rising interest owing to its possible to enable several strategies similar to user explicit altitude beamforming and vertical sectorization [2].

Long-term evolution (LTE) network is advancing in its architecture to provide mobile technologies with efficient spectrum utilization, cloud-based intellectual architecture, etc. The operators' necessities are under investigation through the third-generation partnership project (3GPP) standardizing specifications of fifth generation (5G) that could equal the practical necessities also an economic possibility. The standard methods to obtain feasible 5G are still not clear. Thus, telecom players noticeable a subsequent step towards 5G through delivering a new standard version "3GPP-Release 13/14", denoted via specialists as LTE-Advanced Pro or "4.5G" [3]. Fifth-generation (5G) wireless networks are predictable to offer higher data rates for mobile subscribers also support a wide range of services [4]. The most current research results from academe and engineering, and it labels the assessment practices in-depth for the network also in physical layer technologies. The assessment approaches are discussed in penetration. It moreover concealments the study of the 5G candidate bits of knowledge also the testing challenges, the development of the testing technologies, declining channel dimension and showing [5].

2 Related Works

Full-dimensional MIMO is an active antenna system (AAS) with a 2D planar array structure which comes as compact structure of large number of antenna arrays to be packed within the feasible base station (BS) form factors; it also offers the ability of adaptive elevation beamforming in the three-dimensional space. Nevertheless, the compact form of large-scale planar arrays extremely enhances the spatial correlation(SPF) in this antenna system [1]. Manipulating the quasi-static channel covariance environments of users, the problem of determining the optimal downtilt weight vector for antenna ports, that enhance the least signal-to-noise interference ratio of a MU-MIMO system, is expressed as a partial optimization problem.

This paper [6] reviewed the nearby deployment of the FD-MIMO technique for the improvement of fifth-generation(5G) structure and in addition to it examines the form of the two-dimensional antenna system with the techniques regression and incremental redundancy in HARQ. Two-dimensional(2D) dynamical array structure furthermore promises a enormous level of freedom for creating dynamical device-specific multicell co-operation methods. Depending on miniature form of antenna the difficulty of identifying the dynamical downtilt weight vector that achieves the minimum signal interference ratio, spectral competence and latency decreasing system is devised. Each user group is assigned with set of antenna ports with downtilt weight vectors, entire ports assumed using of the similar optimum dynamical downtilt weighing vector. Multi-user MIMO (MU-MIMO) system potential usage is attained by finding spacial correlation function that can be obtained while the highest amount of active antennas are positioned within feasible BS form factor by enhancing elevation beamforming [2].

Three-dimensional beamforming activities are the channel's degrees of independence in the altitude direction with the active antenna system (AAS). Presently, 3GPP of this innovative MIMO method also is working on the 3D channel ideal requirement [7].

Enhancing demand for data in next-generation mobile broadband networks will create several tests for network engineers also service providers. To discover these concerns also meet the severe difficulties in upcoming years, advanced practical solutions would be recognized which are capable of offer greater spectral efficiency, better function, also broader coverage [4]. To enhance the status of a network that applies 3G systems with Wi-Max is now enhanced through new technology which is updated 4G systems established on TD-LTE with Wi-Max technology. It introduces two networks 3G technology also the LTE of the first announcement which is deliberated with TD-LTE technology [8]. Enhanced fault detection during regression testing can offer quick feedback for debuggers to begin their work earliest. As a result, the software developers also suggest a lot of methods in the updated software while some methods could not be invoked by the developers or those methods which do not return any values [5]. An opportunistic routing with the responsiveness of energy is used to accept a dynamic environment. While the sender transmits the information to a multicast group, the sender transmits the information via greater energy between vicinity, thus enhances the life span [9]. Throughput is enhanced with a data-aided valuation method by utilizing transmitting choice, channel and then transmit obligation [10].

3 Proposed Method

An FD-MIMO is the fundamental applicant technologies measured for the development for past fourth generation as well as fifth generation (5G) networks. The fundamental design behind FD-MIMO is to operate a great number of antennas positioned in a two-dimensional (2D) antenna selection board to form contracted beams in both verticals and horizontal directions. This beamforming appropriates the increased eNodeB (eNB: 3GPP terms for the base station) to concurrently communicate to numerous user instrumentalities (UE: 3GPP terms for the mobile station)

to recognize elevated order multi-user special multiplexing. 3GPP to offer requirement maintain for FD-MIMO. The channel pattern offers the random features of a
three-dimensional (3D) wireless channel.

FD-MIMO to estimate the functioning gains of standard enhancements directing
the 2D antenna array function with up to 64 antenna ports more than a standard-
transparent strategy, for example, vertical sectorization utilizing antenna components
in the vertical direction.

FD MIMO has two significant separating components equated to MIMO technologies from previous LTE releases. Therefore, FD-MIMO considerably enhances
beamforming as well as special user multiplexing capacity. FD-MIMO is usual to
bring important function development to future generations of networks owing to
the broad range of deployment surroundings.

3.1 UG-SSDB

Kindly in this strategy, the UG-SSDB is a naturalistic procedure as well as it attempts
to continue high-quality function. Particularly, for every division $s \in S$, a set of beam
patterns M can be utilized to provide the connected UEs Ks. The BS distributes the
CSI reference signals to the lively UE by dissimilar beam patterns in M, next the UE
evaluates the CSI as well as feeds back to the BS. The BS can next communicate the
data to the UE with the most suitable beam pattern. It is probable which one beam
model is suitable for a group of UEs. The function, as well as the difficulty extremely,
calculates on how to plan the beam pattern set M. Every beam pattern is qualified
through a quadruple $\{\varphi st, \theta tilt, \varphi 3\,dB, \theta 3\,dB\}$. Instinctively, with a better size of M,
the 3 dB beam width ($\varphi 3\,dB$, $\theta 3\,dB$) can be narrower as well as the function should
be better, though the difficulty would be superior. While the size of M accesses to
infinity, UE group-specific beamforming gets UE-specific beamforming.

Particular a confident amount of beam rules, say M, we introduce an easy process
to decide the beam directions (φ_{st}, θ_{tilt}) of the beam rules for every sector as comes:

- Every diminutive cell sector is evenly sectorized into Mh subsectors horizontally.
  Consequently, for every horizontal sub-sector, M_h horizontal directing angles can
  be decided as ($\varphi^1 st$, $\varphi^2 st$, ... $\Phi^{Mh} st$)
- Every diminutive cell sector is uniformly sectorized into $Mv = M/M_h$ vertical
  sub-sectors. Hence, for every vertical sub-sector, M_v vertical downtilt angles can
  be decided as ($\theta^1 tilt$, $\theta^2 tilt$, $\theta^M tilt$)
- The beam steering's for M beam rules can be denoted in the tracing $M_h \times M_v$
  matrix form

$$\begin{bmatrix} \varphi_{st}^1, \theta_{tilt}^1 & \varphi_{st}^1, \theta_{tilt}^2 & \cdots & \varphi_{st}^1 \theta_{tilt}^2 \\ \varphi_{st}^{M_h} \theta_{tilt}^1 & \varphi_{st}^{M_h} \theta_{tilt}^2 & \cdots & \varphi_{st}^{M_h} \theta_{tilt}^{M_v} \end{bmatrix}$$

For the above process, the best divider (M_h and M_v) as well as 3 dB beamwidth (_3 dB, _3 dB) is also essential.

3.2 Dinkelbach's Method to Improve Energy Efficiency

The highest energy utilization of wireless systems currently answers in high functional cost. Thus, there is a requirement to enhance the energy effectiveness of recent wireless communication advancements. With the majority of superior techniques, 3GPP LTE has been generally recognized as the most capable pattern for the next-generation networks to assure high-data rate communication, though excessive energy utilization is however a significant difficulty that has a huge collision on the function of LTE-A systems. Accordingly, several technologies have concentrated on the energy-efficient intend for the LTE network also 3GPP also incorporated green communication technology as a significant portion of LTE-A technology. This evidence fully demonstrates the importance of green radio and energy-efficient design for LTE-A.

This approach, we invent a unified optimization difficulty which optimizes the energy distribution of all user equipment (UE) for enhancing the network energy efficiency, depending on restricted broadcast energy, as well as the smallest amount of quality of service (QoS) limitations. The spatial concave estimate (SCE) method is used to translate the non-concave difficulty into a convex fractional programming (CFP) difficulty is resolved through Dinkelbach's method. This approach enhances the energy efficiency using Dinkelbach's technique.

Therefore, this optimizes energy distribution problem which is a CFP problem with a quasi-convex object work $\tilde{\psi}(Q)$ as well as curved restraints that could be translated to a curved optimization in a subtractive through Dinkelbach's technique.

It is simple to monitor which is not a CFP optimization difficulty since the numerator of the objection task $\tilde{\psi}(Q)$ as well as the QoS restraints C3 are non-curve concerning $\{P_h\}_{h=1}^h$. Thus, it simply cannot be resolved through typical partial programming. To defeat this problem, we apply the SCA procedure introduced to sequent estimated R_h^j by utilizing the subsequent dissimilarity:

$$\log_2(1 + z_{h,j}) \geq a_{h,j} \log_2 z_{h,j} + b_{h,j} \tag{1}$$

The above inequality is rigid at a special worth $x_{hj} = x_{hj}$ while the estimate constants x_{hj} as well as b_{hj} are chosen as

$$a_{h,j} = \frac{\tilde{x}_{h,j}}{1 + \tilde{x}_{h,j}}, \quad b_{h,j} = \log_2(1 + \tilde{x}_{h,j}) - \frac{\tilde{x}_{h,j}}{1 + \tilde{x}_{h,j}} \log_2 \tilde{x}_{h,j} \tag{2}$$

Encouraged by the above convenes estimation, we apply the inequality to estimated $\widetilde{\mathbb{Z}}_k^j$ which represents to $\lambda_{h,j} = \frac{P_r P_h}{\sum_{i=1}^h P_i P_{h,j}^{(j)} + \gamma_h^j}$. Next, we enter at the subsequent approximated optimization problem

$$\max_{\{s_h\}_{k=1}^k} \tilde{\psi}(Q) = \frac{B\left(1 - \frac{\eta_r}{T}\right)\frac{h-1}{h} \sum_{h=1}^h \sum_{j=1}^{h-1} \mathbb{Z}_k^j(Q)}{P_{\text{tot}}(Q)}$$

$$\text{s.t.} \begin{cases} C1' : 2^{sk} \geq 0, \forall h \\ C21' : 2^{sk} \leq p_{\max}, \forall h \\ C31' : \mathbb{Z}_k^j(Q) \end{cases} \tag{3}$$

Here, $Q = \text{diag}\{s1 \ldots, sh \ldots, sH\}$,

$$P_{tot}(Q) = \omega \sum_{h=1}^H 2^{sh} + P_{\text{fixed}}$$

$$\mathbb{Z}_k^j(Q) = b_{h,j} + a_{h,j}\log_2(p_r) + a_{h,j}s_{h'} - a_{h,j}\log_2\left(\sum_{i=1}^K 2^{si} \overset{(j)}{\underset{h,j}{p}} + \gamma_h^{(j)}\right) \tag{4}$$

Thus, the optimization problem is a CFP problem through a quasi-concave objective role $\tilde{\psi}(Q)$ and curved restraints that can be translated into a curved optimization in a subtractive via Dinkelbach's procedure as follows

$$\max_Q F(Q, \beta)$$

$$\text{s.t. } C1' - C3'$$

Here

$$F(Q, \beta) = B\left(1 - \frac{\eta_r}{T}\right)\frac{h-1}{h} \sum_{h=1}^h \sum_{j=1}^{h-1} \mathbb{Z}_k^j(Q) - \beta P_{\text{tot}}(Q) - \beta P_{\text{tot}}(Q) \tag{5}$$

Here, β is a non-negative factor, and it can be mentioned which while $\beta \rightarrow 0$, it involves that the energy-efficient difficulty is devolved to an optimization difficulty for the SE maximization. Furthermore, reducing the communication to keep energy assists to diminish the interference, therefore enhancing the network throughput.

Consequently, this method incorporates these two significant factors as well as formulates and energy-efficient optimization which is additionally related in LTE systems. Besides, it meets users' rate necessities.

3.3 Incremental Redundancy and Regression Monitoring Technique with (HARQ)

Regression Monitoring Regression is monitored learning trouble whose aim to forecast the value of an output variable $y \in R^m$ from an input variable $x \in R^k$. Locally weighted projection regression (LWPR) monitoring is a technique, which attains nonlinear function estimation in high dimensional places with redundant as well as irrelevant input dimensions. It utilizes topically linear examples, spanned through a small number of univariate regressions in picked out directions in input space. This paper measures dissimilar methods of projection regression as well as derives a nonlinear procedure approximate based on them. The forecast of LWPR accepts the form of a weighted mixture of local examples

$$y = \frac{\sum_{n=1}^{N} w_n y_n}{\sum_{n=1}^{N} w_n} \tag{6}$$

where y_n represents the forecast of local examples n and w_n represents the weight connected to this example. Local examples are linear examples traversed beside the ways offered by partial least squares (PLS)

$$y_n = \alpha k_p \tag{7}$$

Here, p indicates the proposed inputs. The weight w_k denotes the region where the example k is lively, besides recognized as receptive field. It is calculated applying a Gaussian kernel parameterized through the distance metric D_n

$$W_n = \exp(0.5(x - v_n)^T D_n (x - v_n)) \tag{8}$$

In this strategy, this regression technique concentrates on discovering proficient local projections. It is to achieve local function estimation in the adjacent of a given query point. This method agrees for corresponding locally straightforward functions, for example, low order polynomials, along with the projection that deeply simply the function estimation problem. It can work productively in elevated dimensional spaces.

HARQ Reliable transaction over time-varying as well as error-prostrate wireless channels presently is still demanding Though the traffic, as well as user concentrations of both networks, is similar, the network utilizing LAA attains an important achievement in user-perceived throughput, owing in the division to basic LTE capacities, for example, link alteration through explicit UE feedback a swell as HARQ. Besides, this strategy introduced the HARQ process to be incorporated to enhance the function of the network in terms of packet drop, delay, throughputs as well as energy utilization. Here, we proposed a HARQ error-control method with low difficulty as well as high throughput effectiveness. It can energetically fine-tune the error-control redundancy levels along with the immediate channel situation.

Incremental Redundancy HARQ method broadly applies in the most significant wireless systems, for example the LTE, to enhance the dependability of the transaction. Incremental redundancy (IR) represents the unique secret word that is separated into several sub-secret words that are transmitted in succeeding communication attempts.

IR HARQ If the unique secret word is separated into $N = M$ sub-secret words, after that the procedure is an untainted IR HARQ. This procedure effort an M-parallel channel; however, no bit redundancy is approved as well as is used for detecting redundant as well as irrelevant input.

4   Results and Discussion

In this section, we measure the simulation performance by the Network Simulator-3. considered massive UG-SSDB, FD-MIMO with Dinkelbach's and FD-MIMO with HARQ approaches that use the proposed energy-efficient strategies, remove redundancy and routing overhead demonstrate the accuracy of our analytical results as well as the impacts of several relevant parameters on the optimum EE via numerical simulations. Three-dimensional beamforming antenna port parameters are set as Table 1.

Using the antenna port parameters in Table 1 for FD-MIMO simulation in NS-3 environment using Algorithm 1 which is based on Dinkelbach's method is executed, where as algorithm 2 is the extension of Algorithm 1. To obtain further optimization in UG-SSDB is implemented using the incremental redundancy (IR) in HARQ and regression monitoring methods.

Table 1 Three-dimensional beamforming antenna port parameters

Specifications	Values
Carrier frequency	2 GHz
Azimuthal HPBW$_{\emptyset 3dB, E}$	65°
Elevation HPBW$\theta_{3dB, E}$	15°
Maximum directional element gain GP$_{max}$	17 dBi
Maximum attenuation A$_m$	20 dB

Algorithm 1 User Group Single Down-tilt elevation Beamforming (UG-SSDB)

➢ **procedure total population is divided into sub groups** (USER GROUPING(Sg))

➢ **for** $g = 1$ to G

➢ set the user group set $S\_g = \_$;Choose $\_0,g$ and elevation spread in a way that the intervals[$\_0,g -\_$, $\_0,g +\_$] are disconnected;

➢ Compute $\mathbf{R}E\ g\ (\_0,g,\_)$ for the chosen $\_0,g$ and $\_$;attain the group sub_space $\mathbf{K}g = \mathbf{U}g$, where $\mathbf{U}g$ is the $M \times rg$ matrix of eigenvectors corresponding to rg dominant eigen values of $\mathbf{R}E\ g$, with rg chosen suchthat $\_Gg=1\ rg = M$;

➢ **end for**

➢ **for** $k = 1$ to J

➢ Estimate $\mathbf{R}E\_k$ for the propagation model under consideration; attain the eigen space $\mathbf{U}M \times rk\ k$, corresponding to the r_k dominant eigen values of $\mathbf{R}E\ k$;

➢ Estimate $dC(\mathbf{U}k, \mathbf{V}g) = \|\mathbf{U}k\mathbf{U}H\ k - \mathbf{V}g\mathbf{V}H\ g\ \|^2 F, g = 1, \ldots, G$;

➢ determine $g = \min 1\_g\ Gdc(\mathbf{U}k, \mathbf{V}g\_)$;Add user k to group g, i.e. $Sg := Sg\_\{k\}$;

➢ **end for**

procedure WEIGHT VECTOR OPTIMIZATION($\mathbf{w}g$)Divide NBS antenna ports into G equal groups, with NBS,g ports serving each user group.

➢ **for** $g = 1$ to G

attain the quasi-optimal down tilt weightvector $\mathbf{w}\_g$ for the antenna feeds in group g allocation the S_g-user MU-MISO.

➢ **end for**

Proposed Algorithm In Algorithm 2 path computation is obtained from source node to destination node with shortest path method then depending on load analysis, data packets transport takes place from source node to neighbour nodes then towards destination node. During this progression, regression monitoring is utilized which makes period to period monitoring of data packets and next processed data packets goes through incremental-redundancy (IR) it is self-decodable technique in HARQ; in this method, it not only provides timely verification but also replica of data packets is minimized which results in optimal throughput from eNodeB to consumer equipment, minimum use rate also minimizing routing overhead.

Algorithm 2 Extended Full Dimension Multiple Input Multiple Output (EFD-MIMO with incremental redundancy HARQ and Regression monitoringtechnique)
Input: Total number of device consumers(N)
Output: Routestarting source enode S to ending at destination enode quantitiesattained.
Methodology: Soft-combining in HARQ along with regression- monitoringtechnique *(S_g).*
Select source enode S, destination enode D_enode, and routingtable from S to D_enode.

> ➤ Set k−1,path[k];
> ➤ Set l=1,neighbor_enode=0;
> ➤ while (Source!=Destination) do
>   {forward route request RREQ[l] to the destination enode
>   for
>   {int l=enodes select_load(l)=nb_id(range)+tt_p+tt
>   _r//beam tilt
>   S to its neighbor enode and wait for route reply RREP[l]}.
>   if
>   {
>   (Load[l] > = th[l])then
>   range++,n_id++;
>   else
>   reject that path.
>   }
>   Add piggyback from S to D_enode in the path[k] and k++;
>   End while
>   }
> ➤ For l = 1 : nn //no the number of nodes
>   {Initialize the user group subset **enodeid=nn**
>   Choose _0,g and elevation beamforming _ in a way that the interval
>   [t_d,tt_r] are disconnected;
>   Compute **RE** g (_0,g,_) using (5) for the chosen _0,g.
>   **for** g = 1 to Gi for the antenna feeds in group g serving
>   find the group eigenquantities**Ei_g = us_id**
>   where **Ug** is the **M × rg** matrix of eigenvectors corresponding to n number of
>   dominant eigenquantities of **res(l)**, with **rg** chose such that **Gg =1**.
>   end for
>   }
> ➤ For k = 1 to J Compute R_E
>   {
>   **If (C<thresholdvalue)**//leastcapacity**(C)**
>   **If (C>thresholdvalue)**//densecapacity**(C)**
>   **If (C=thresholdvalue)**//typicalcapacity**(C)**
>   Distribute single down tilt beamforming antenna feed values into G equal
>   groups,with **N_BS,g**feeds serving each user group.

time= ack(l)+t_pr(l)+t_r(l)
ack(l)->monitoring_id
calculatecapacity =capacity(l)+capacity(pkt)+capacity(ack);
obtaing= min$dc(\mathbf{U}k, \mathbf{V}g\_)$;
Add user k to group g, i.e. $Sg := Sg\_ \{k\}$;
J_l->leastcapacity;
J_l->densecapacity;
J_i->criticalcapacity;
end
}

Network simulator 3 (NS3) execution is carried in the UBUNTU environment, using the terminal, required commands passed and simulation time is selected, with each time execution simulation time is updated. Figure 1 shows the execution procedure.

For visualizing communication procedures between users and towers in UG-SSDB, FD-MIMO and EFD-MIMO are shown in Figs. 2, 3 and 4 using NetAnim in NS3.

Minimum User Rate Great communication rate necessities generally mean serious traffic loads in the network as a result enhances the users' rate. When as the network loads suit much lesser with minimizing user rate requirements. Figure 1 illustrates the minimum user rate of a UG-SSDB, FD-MIMO with Dinkelbach's and FD-MIMO with the HARQ approach based on simulation time. FD-MIMO with HARQ approach has increasing the user rate.

Fig. 1 NS3 execution in UBUNTU terminal of a UG-SSDB, FD-MIMO with Dinkelbach's and FD-MIMO with HARQ approach

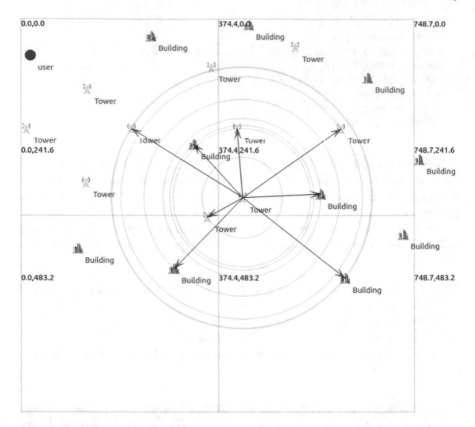

Fig. 2 NetAnim execution of UG-SSDB shows the communication procedure of tower to tower, building to tower, tower to the building

From this Fig. 5 of UG-SSDB, but FD-MIMO with Dinkelbach's method has a lesser minimum user rate than the UG-SSDB and FD-MIMO with the HARQ approach. UG-SSDB has the highest minimum user rate than the FD-MIMO with Dinkelbach's method. Although compared to the FD-MIMO with the HARQ approach, it has the lesser minimum user rate. FD-MIMO with HARQ approach has the highest minimum user rate than the other two methods.

Routing Overhead Routing overhead is depicted as the total amount of control packets communicated for creating a data path per data packet. It is received via computing the ratio among the total amount of control packets sent to the total amount of control packets obtained.

Figure 6 illustrates the routing overhead of a UG-SSDB, FD-MIMO with Dinkelbach's and FD-MIMO with the HARQ approach based on simulation time. From this figure, UG-SSDB and FD-MIMO with Dinkelbach's methods are the highest routing overhead. But, the FD-MIMO with HARQ approach has the lowest routing overhead

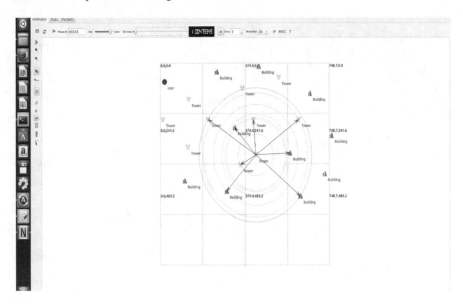

Fig. 3 NetAnim execution of FD-MIMO shows the communication procedure of tower to tower, building to tower, tower to the building

since it removes the redundancy and increases the efficiency. Thus, FD-MIMO with the HARQ method increases the routing efficiency.

Throughput Throughput is defined as the rate at which the data is successfully transmitted for every 1000 packets sent. Figure 3 shows the throughput plots of UG-SSDB, FD-MIMO with Dinkelbach's and FD-MIMO with the HARQ approach based on simulation time (Fig. 7).

$$\text{Throughput} = \frac{\sum_0^n \text{Packets received}(n) * \text{Packet size}}{1000} \qquad (9)$$

It can be noticed that the highest throughput of FD-MIMO with the HARQ approach has a higher throughput than the UG-SSDB, FD-MIMO with Dinkelbach's method. FD-MIMO with HARQ approach has the highest throughput since it removes the redundancy using IR HARQ and regression monitoring. But, FD-MIMO with Dinkelbach's method has the highest throughput than the UGSDB since it increases energy efficiency. But, compared to FD-MIMO with the HARQ approach FD-MIMO with Dinkelbach's method is slightly decrease the throughput.

Table 2 explains the UG-SSDB of minimum user rate, throughput and routing overhead based on the network simulation time. Table 3 illustrates the FD-MIMO with Dinkelbach's approach of minimum user rate, throughput and routing overhead based on the network simulation time. Also, Table 4 illustrates the FD-MIMO implementation utilizing incremental redundancy in HARQ followed by regression

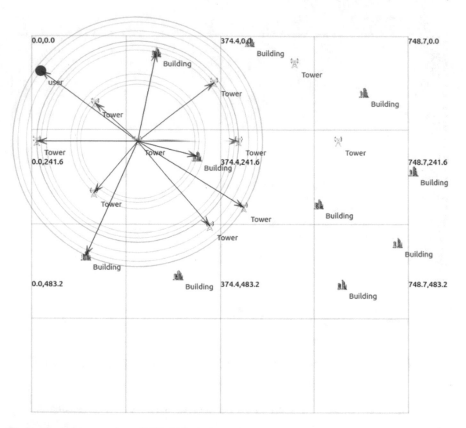

Fig. 4 NetAnim execution of EFD-MIMO shows the communication procedure of tower to tower, building to tower, tower to the building, human being to the tower, tower to a human being, outdoor human being to indoor human being

monitoring method that compares routing overhead and throughput, and minimum user rate based on the simulation time.

5   Conclusion

To attain reliable communication as well as high throughput efficiency in wireless channels, a HARQ scheme is developed. It appropriates the alteration of the coded packets supplied from the source to the evaluated channel condition. This paper explained methods of user grouping-specific downtilt beamforming. Dinkelbach's methodology is based on energy optimization algorithm for improving data rates. Locally weighted projection regression (LWPR) monitoring is a technique, which attains nonlinear function estimation in high dimensional places with redundant as well as irrelevant input dimensions. The extended EFD-MIMO which utilized

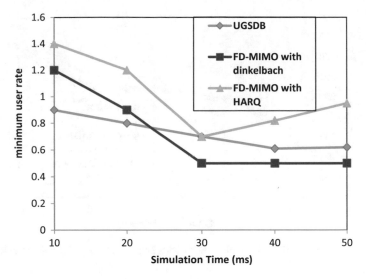

Fig. 5 Minimum user rate of a UG-SSDB, FD-MIMO with Dinkelbach's and FD-MIMO with HARQ approach

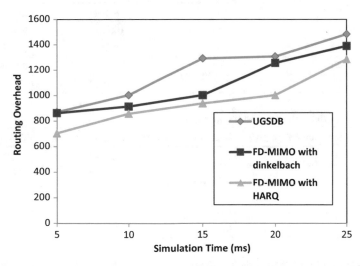

Fig. 6 Routing Overhead of a UG-SSDB, FD-MIMO with Dinkelbach's and FD-MIMO with HARQ approach

regression monitoring approach in addition to HARQ method which yields optimum performance also it reduces the routing overhead in comparison with user grouping specific downtilt beamforming, and it utilized quasi-optimal Dinkelbach's method. Simulation results illustrate reduced routing overhead and improve routing efficiency.

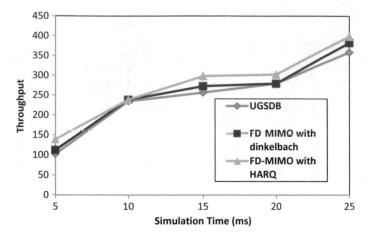

Fig. 7 Throughput of a UG-SSDB, FD-MIMO with Dinkelbach's and FD-MIMO with HARQ approach

Table 2 User group device-specific single downtilt beamforming (UG-SSDB)

S. No.	Simulation time (ms)	Minimum user rate	Routing overhead	Throughput
1	10	0.9	870.23	102.33
2	20	0.8	1003.18	234.48
3	30	0.7	1292.99	256.32
4	40	0.61	1308.12	278.94
5	50	0.62	1483.12	358.12

Table 3 FD-MIMO with Dinkelbach's approach

S. No.	Simulation time (ms)	Minimum user rate	Routing overhead	Throughput
1	10	1.2	862.37	112.33
2	20	0.9	913.55	238.10
3	30	0.5	1003.44	272.33
4	40	0.5	1256.90	278.94
5	50	0.5	1390.23	382.11

Table 4 FD-MIMO with incremental redundancy in HARQ followed by regression-monitoring method

S. No.	Simulation time (ms)	Minimum user rate	Routing overhead	Throughput
1	10	1.4	703.98	139.12
2	20	1.2	857.12	238.10
3	30	0.7	938.19	298.23
4	40	0.82	1003.12	302.11
5	50	0.95	1285.67	398.11

References

1. Kammoun A, Debbah M, Alouini MS (2017) Design of 5G full dimension massive MIMO systems. IEEE Trans Commun 66(2):726–740
2. Nam YH, Ng BL, Sayana K, Li Y, Zhang J, Kim Y, Lee J (2013) Full-dimension MIMO (FD-MIMO) for next-generation cellular technology. IEEE Commun Mag 51(6):172–179
3. Haidine A, El Hassani S (2016) LTE-A Pro (4.5 G) as pre-phase for 5G deployment: closing the gap between technical requirements and network performance. In: 2016 international conference on advanced communication systems and information security (ACOSIS), pp 1–7. IEEE
4. Razavizadeh SM, Ahn M, Lee I (2014) Three-dimensional beamforming: a new enabling technology for 5G wireless networks. IEEE Signal Process Mag 31(6):94–101
5. Yang Y, Xu J, Shi G, Wang CX (2018) 5G wireless systems. Springer, Berlin
6. Nishanthi CH, Ramamurthy N (2020) Improving spectral efficiency and low latency in 5G framework utilizing FD-MIMO. J Ambient Intell Humaniz Comput 1–13
7. Cheng YS, Chen CH (2014) A novel 3D beamforming scheme for LTE-Advanced system. In: The 16th Asia-Pacific network operations and management symposium, pp 1–6. IEEE
8. Gupta V, Sharma S (2017) A modified hybrid structure for next generation high speed communication using TD-LTE and Wi-Max. In: 2017 international conference on inventive communication and computational technologies (ICICCT), pp 329–334. IEEE
9. Santhakumar R (2017) Resource allocation in wireless networks by channel estimation and relay assignment using data-aided techniques. Int J MC Square Sci Res 9(3):40–47
10. Keerthi Anand VD (2015) Power efficient multicast opportunistic routing protocol (PEMOR) to optimize lifetime of manet. Int J MC Square Sci Res 7(1):109–127

Performance Analysis of OFDM and GFDM with MIMO for 5G Wireless Communications

Sushmitha Pollamoni, Yedukondalu Kamatham, and Ramya Gurama

Abstract 5G wireless communications systems demand ultra-high-speed data rates and ultra-low power usages with low latency. To meet these requirements, researchers have considered a waveform, i.e., generalized frequency divison multiplexing (GFDM). The orthogonal frequency division multiplexing (OFDM) is mainly used in 4G and LTE, and it has few drawbacks such as high peak-to-average power ratio (PAPR), intercarrier interference (ICI) and poor in spectral efficiency (SE). To overcome these drawbacks, a non-orthogonal scheme of GFDM is introduced. In this paper, the performance of GFDM concerned to PAPR, BER, and SE is compared with OFDM. It is observed that the SE of GFDM is better than OFDM concerning BER and PAPR, and there is no significant difference between GFDM and OFDM. To gain the benefit of spatial multiplexing, the MIMO system is added with OFDM and GFDM. It is observed from the results that SE and PAPR performance is better for MIMO-GFDM rather than MIMO-OFDM. Hence, MIMO-GFDM is one of the better choices for 5G and beyond wireless communications systems.

Keywords GFDM · OFDM · MIMO · BER · PAPR · Spectral efficiency

1 Introduction

From a few decades, wireless communications system (WCS) became an essential means for cellular mobile communications. In first generation, voice communication is developed with an analog waveform. The second generation is developed with

S. Pollamoni · Y. Kamatham (✉) · R. Gurama
Department of Electronics and Communication Engineering, CVR College of Engineering, Hyderabad, India
e-mail: kyedukondalu@gmail.com

S. Pollamoni
e-mail: spullamoni@gmail.com

R. Gurama
e-mail: ramyaasai1996@gmail.com

© The Author(s), under exclusive license to Springer Nature Singapore Pte Ltd. 2021   261
K. A. Reddy et al. (eds.), *Data Engineering and Communication Technology*,
Lecture Notes on Data Engineering and Communications Technologies 63,
https://doi.org/10.1007/978-981-16-0081-4_26

voice quality in digital form, small messages, and data services with improvement in capacity in terms of spectrum and battery life. 3G is with high data rates. Because OFDM signal provides a cyclic prefix (CP) for one symbol such that latency gets reduced [1–5]. The low spectral efficiency (SE) is due to CP insertion which may cause a severe problem for wireless regional area networks (WRAN). The high peak-to-average power ratio (PAPR) leads to high-power amplifier (HPA) that drives into the nonlinear region which leads to out-of-band radiation (OBR) and poses a problem of opportunistic access [6–8]. OFDM may not become a promising waveform for 5G networks due to its bottlenecks. To overcome the limitations of OFDM [9], an alternate form of multicarrier modulation (MCM) techniques is implemented. The MCM techniques include filter bank MCM (FBMC), universal filtered MCM (UFMC), and generalized frequency division multiplexing (GFDM). GFDM is one of the most MCM used by researchers for 5G physical layer (PL) because it uses for different requirement due to its flexible nature. In real-time applications, the length of the signal is reduced for the fulfillment of certain latency [10]. GFDM modulation is restricted in a block structure with MK samples, where 'M' is subsymbols in 'K' subcarriers. For latency applications, the time–frequency structure design is possible for time constraints matching. For filtering the subcarriers, different filters can be used which may affect the OBR performance. GFDM signal is allowed with time and frequency characteristics. The GFDM technique is the same as the OFDM technique but there is additional complexity in its implementation.

In the GFDM block, the redundancy is kept small with an addition of single CP to the entire block which contains multiple subsymbols. This improves SE or additional CP can also be added for synchronization. For OFDM, many synchronization algorithms have been developed, and the same can be adapted for GFDM. To reduce the problem of high PAPR in OFDM, considerable methods were investigated in the literature [11–14]. The multiple-input multiple-output (MIMO) is developed for the advancement of spatial multiplexing to enhance the capacity of the channel. This paper analyzes the OFDM and GFDM system performance concerning PAPR, BER, and SE. For spatial multiplexing, the MIMO system is combined with OFDM and GFDM. The MIMO-GFDM and MIMO-OFDM performance is compared and analyzed concerning SE and PAPR.

The rest of the paper follows as Sect. 2 describes the mathematical model of OFDM and GFDM. Section 3 presents the performance of OFDM and GFDM. Section 4 deals with the performance analysis of OFDM and GFDM with MIMO. Finally, the conclusion is drawn in Sect. 5.

2 Mathematical Model of OFDM and GFDM

In the OFDM system, each OFDM symbol is modulated with N quadrature amplitude modulation (QAM) carriers denoted with $s_k[n]$ and $k = 0, 1, \ldots, N - 1$ and represented as:

$$x_n(t) = \sum_{k=0}^{N-1} s_k[n] e^{\frac{j2\pi k}{T} t} \tag{1}$$

where $x_n(t)$ represents inverse fast Fourier transform (IFFT) and is a periodic function. $x_n(t)$ passes through a transfer function of a channel $H(f)$, and the transient response of the channel is obtained as:

$$y_n(t) = \sum_{k=0}^{N-1} H\left(\frac{k}{T}\right) s_k[n] e^{\frac{j2\pi k}{T} t} \tag{2}$$

$y_n(t)$ is a periodic function, and Fourier series coefficient of $(t) = H\left(\frac{k}{T}\right) s_k[n], k = 0, 1, \ldots, N-1$ and $s_k[n]$ is transmitted data symbols can be extracted from $y_n(t)$.

GFDM is a non-orthogonal MCM. Each subcarrier carries M data symbols with equally spaced and filtered by a filter $g(n)$ and sth GFDM block is denoted as:

$$x_{n_t}^{[s]}[n] = \sum_{k \in \mathcal{K}} \sum_{m=0}^{M-1} d_{n_t,k,m}^{[s]} g\left[n - mK - L_{cpN}\right] e^{\frac{j2\pi(n - L_{cp})k}{K}} * \Omega_N L_{cp}[n] \tag{3}$$

where $n = 0, \ldots, N + L_{cp} - 1$ and $N = MK$, $d_{n_t,k,m}^{[s]} \in x$ is mth data symbol in kth subcarriers via transmit antenna n_t. One GFDM frame with N_s GFDM blocks can be denoted as:

$$x_{n_t}[n] = \sum_{s=0}^{N_s-1} x_{n_t}^{[s]}[n - sN - sL_{cp}] \tag{4}$$

Suppose N-point DFT of $g(n)$ is denoted as $G(v)$. After CP removal can be represented as:

$$x_{n_t}^{[s]}[n] = \sum_{k \in \mathcal{K}} G[\langle v - Mk\rangle_N] D_{n_t,k,u=\langle v\rangle_M}^{[s]}; \tag{5}$$

where $v = 0, 1, 2, \ldots, N-1$, $D_{n_t,k,u}^{[s]}$ is uth harmonic of MK,

$$D_{n_t,k,u}^{[s]} = \sum_{m=0}^{M-1} d_{n_t,k,m}^{[s]} e^{-\frac{j2\pi mu}{M}}; u = 0, 1, \ldots M - 1 \tag{6}$$

In MIMO systems, the multiple streams of data are transmitted through multiple antennas. These multiple streams pass across a channel matrix which consists of $N_T \times N_R$ paths between 'N_T' transmit and 'N_R' receive antennas. The signal vectors are received by the receiver and decode into the transmitted information. A MIMO system is expressed as:

$$Y = HX + n \tag{7}$$

Then 'Y' is a received signal vector with a length of $Y = [y_1, y_2, \ldots y_N]$ and 'X' is a transmitted signal vector with a length of $X = [x_1, x_2, \ldots, x_N]$ and 'H' is a channel matrix with a length of $M \times N$ matrix and n is a noise vector.

$$H = \begin{bmatrix} h_{1,1} & h_{1,2} & \ldots & h_{1,M} \\ h_{2,1} & h_{2,2} & & h_{2,M} \\ \vdots & \vdots & \ddots & \vdots \\ h_{N,1} & h_{N,2} & \cdots & h_{N,M} \end{bmatrix}$$

3 Performance Analysis of OFDM and GFDM

The power spectral density (PSD) of a signal is defined as the measure of the power of a signal with respect to frequency. For characterizing the broadband random signal, a PSD is used. The PSD is normalized by spectral resolution for digitizing the signal, where spectral density represents the strength of the signal transmitted over a period in which bits are sent in a given bandwidth. The PSD of the signal is represented as:

$$S_{xx}(\omega) = \lim_{T \to \infty} E\left[|\hat{x}(\omega)|^2\right] \tag{8}$$

where

$$\hat{x}(\omega) = \frac{1}{\sqrt{T}} \int_0^T x(t)e^{-i\omega t} dt \tag{9}$$

$x(t)$ is the transmitted signal, $\hat{x}(\omega)$ is the Fourier transform of $x(t)$, $S_{xx}(\omega)$ is therequired signal PSD (Fig. 1).

The PSD of GFDM is shown in Fig. 2 with different FFT sizes of 512 and 1024. It is examined that the PSD of GFDM is slightly decreased with an increase in FFT sizes and normalized frequency is in the range of ± 0.2 Hz. The PAPR of a signal is a ratio of the square of peak signal amplitude to squared root mean square (RMS) signal.

$$\text{PAPR}[x_n] = 10 \log_{10} \frac{\max|x_n|^2}{E[|x_n|^2]} \tag{10}$$

where $|x_n|$ is the peak power of the signal.

Within the threshold level, the cumulative distributed function (CDF) evaluates PAPR as:

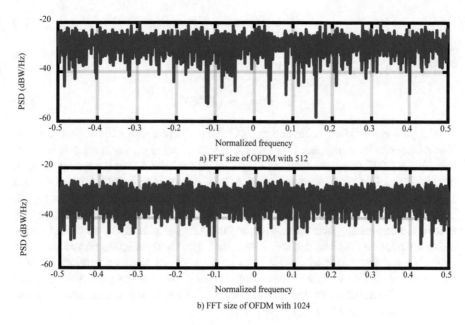

Fig. 1 PSD of OFDM with different FFT sizes

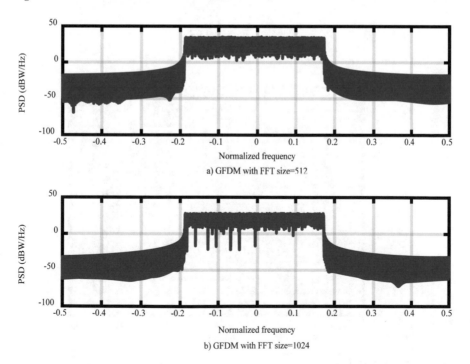

Fig. 2 PSD of GFDM with different FFT sizes

$$F_z(z) = 1 - e^{-z} \qquad (11)$$

The PAPR beyond the threshold level is represented in terms of complementary CDF (CCDF) $\tilde{F}_z(z)$ as

$$\tilde{F}_z(z) = 1 - (F_z(z))^n \qquad (12)$$

The power of CDF function computes CCDF to form a signal in the time domain. The time taken by a measured signal is in above-average power level and is determined by CCDF. The probability of a signal power will be above the average power. The PAPR of OFDM is shown in Fig. 3. It is observed that the PAPR of OFDM is 11.5 dB at CCDF $= 10^{-3}$. The PAPR of the GFDM signal is shown in Fig. 4. The signal average power is calculated as PAPR of the GFDM signal. The PAPR calculation concerning average power and the PAPR calculation for empirical cumulative distribution function (ECDF) is the same. But the calculation part is different. It is observed that PAPR of GFDM is 11 dB at 10^{-3}. By comparing Figs. 3 and 4, the PAPR of GFDM is a little lower and is used in low battery-driven WCS. BER determines the number of error bits that occurred per unit time during transmission. BER is a function of $\left(\frac{E_b}{N_0}\right)$ and represented as

$$\text{BER} = \frac{1}{2}\text{erfc}\left(\sqrt{\frac{E_b}{N_0}}\right) \qquad (13)$$

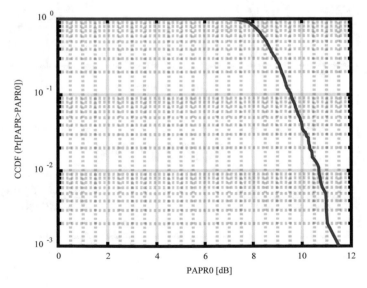

Fig. 3 PAPR of OFDM

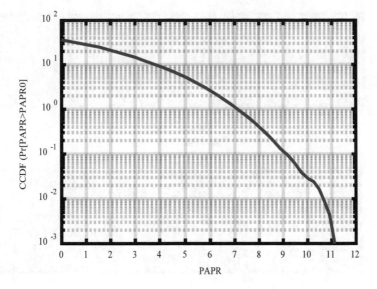

Fig. 4 PAPR of GFDM

In Fig. 5, the BER of GFDM and OFDM is shown. The BER of OFDM and GFDM is different for the same SNR. For GFDM, at SNR of 6 dB, the BER obtained is 7 $\times 10^{-1}$. The BER of OFDM is 7×10^{-2} at the same SNR of 6 dB. It is examined that the BER performance of GFDM is worse than OFDM and hence the OFDM can also be used in ultra-wide band (UWB) communications applications.

Fig. 5 BER of OFDM and GFDM

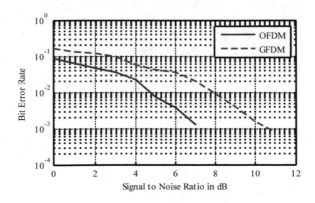

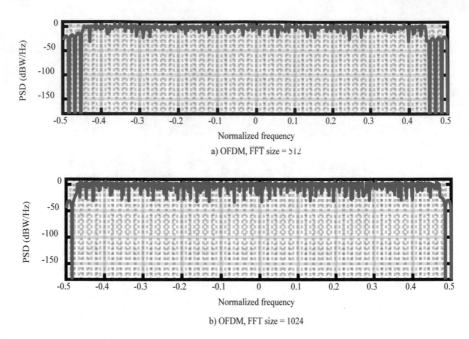

Fig. 6 PSD of MIMO-OFDM with different FFT sizes

4 Performance Analysis of MIMO with MCM

The PSD of MIMO-OFDM is estimated for different FFT sizes and is shown in Fig. 6. It is observed that the PSD is within ±0.5 Hz of normalized frequency. With an increase in subcarriers (FFT size = 1024), the PSD of MIMO-OFDM slightly reduced (Fig. 6b) (Fig. 7).

The PAPR of MIMO-GFDM is shown in Fig. 9. The PAPR of MIMO-GFDM is estimated concerningthe average power level. The PAPR is 9.5 dB at 10^{-2}. From Figs. 8 and 9, the PAPR of MIMO-OFDM is shown in Fig. 8. The PAPR of MIMO-OFDM is estimated with respect to CCDF. It is observed that the PAPR of MIMO-OFDM is 10 dB at CCDF = 10^{-2} is observed that the PAPR of MIMO-GFDM is lower. Hence, it can also applicable for low power consumption WCS.

5 Conclusion

In this paper, the performance analysis of OFDM and GFDM is carried out. The performance is analyzed with respect to PSD, PAPR, and BER of OFDM and GFDM. The PSD of GFDM is better than OFDM having a low normalized frequency of ±0.2 Hz. The PAPR of GFDM (11 dB at 10^{-3}) is better than OFDM (11.5 dB at 10^{-3}). The BER performance of OFDM is efficient than GFDM, and the estimated BER is

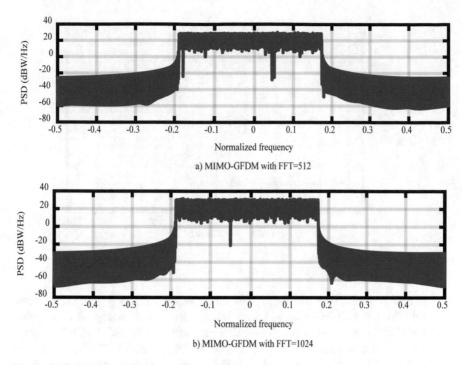

Fig. 7 PSD of MIMO-GFDM with different FFT sizes

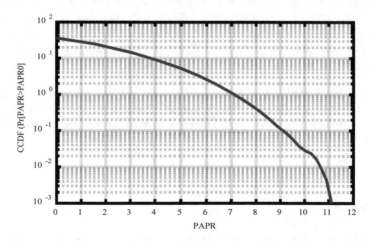

Fig. 8 PAPR of MIMO-OFDM

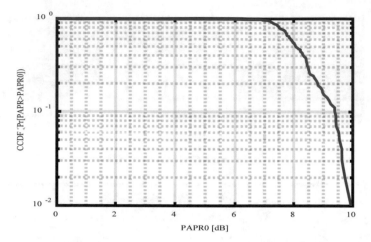

Fig. 9 TAPR of MIMO-GFDM

10^{-2} at SNR $= 6$ dB. To attain the gain of diversity, the MIMO system is combined with OFDM and GFDM systems such that performance is analyzed concerning PSD and PAPR. The PSD of MIMO-GFDM efficient than MIMO-OFDM having a lower normalized frequency of ±0.2 Hz. The PAPR of MIMO-GFDM is obtained as 10.3 dB at 10^{-3} which is better than MIMO-OFDM. Due to the lower PAPR of GFDM and MIMO-GFDM, they can be applicable for low power consumption (low battery) WCS of 5G and beyond. Since GFDM and MIMO-GFDM have better SE these can be used in broadband small or microcell WCS. The BER performance is better in OFDM than GFDM. Hence, OFDM is a better choice for ultra-wide band (UWD) wireless communications in 5G or beyond.

Acknowledgements This work has been carried out under the project entitled 'Study and Implementation of Self-Organized Femtocells for Broadband Services to Indoor Users in Heterogeneous Environment' sponsored by AICTE, New Delhi under Research Promotion Scheme (RPS), Vide sanction letter No: 8-30/RFID/RPS/POLICY-1/2016-2017, Dated: 02.08.2017.

References

1. Michailow N, Matthé M, Gaspar IS, Caldevilla AN, Mendes LL, Festag A, Fettweis G (2014) Generalized frequency division multiplexing for 5th generation cellular networks. IEEE Trans Commun 62(9):3045–3061
2. Fettweis G, Alamouti S (2014) 5G: personal mobile internet beyond what cellular did to telephony. IEEE Commun Mag 52(2):140–145
3. Wunder G, Jung P, Kasparick M, Wild T, Schaich F, Chen Y, Ten Brink S, Gaspar I, Michailow N, Festag A, Mendes L (2014) 5GNOW: non-orthogonal, asynchronous waveforms for future mobile applications. IEEE Commun Mag 52(2):97–105

4. Series M (2015): IMT vision–framework and overall objectives of the future development of IMT for 2020 and beyond. Recommendation ITU2083
5. Popovski P, Stefanović Č, Nielsen JJ, De Carvalho E, Angjelichinoski M, Trillingsgaard KF, Bana AS (2019) Wireless access in ultra-reliable low-latency communication (URLLC). IEEE Trans Commun 67(8):5783–5801
6. Bingham JA (1990) Multicarrier modulation for data transmission: an idea whose time has come. IEEE Commun Mag 28(5):5–14
7. Fechtel SA, Blaickner A (1999) Efficient FFT and equalizer implementation for OFDM receivers. IEEE Trans Consum Electron 45(4):1104–1107
8. Kim J, Lee J, Kim J, Yun J (2013) M2M service platforms: survey, issues, and enabling technologies. IEEE Commun Surv Tutor 16(1):61–76
9. Nekovee M (2009) Quantifying performance requirements of vehicle-to-vehicle communication protocols for rear-end collision avoidance. In: VTC Spring 2009-IEEE 69th vehicular technology conference, pp 1–5. IEEE
10. Awoseyila AB, Kasparis C, Evans BG (2008) Improved preamble-aided timing estimation for OFDM systems. IEEE Commun Lett 12(11):825–827
11. Kamatham Y, Pollamoni S (2019) Implementation of OFDM system with companding for PAPR reduction using NI-USRP and LABVIEW. In: 2019 IEEE international WIE conference on electrical and computer engineering (WIECON-ECE), pp 1–4. IEEE
12. Sravanti T, Kamatham Y, Paidimarry CS (2020) Reduced complexity hybrid PAPR reduction schemes for future broadcasting systems. In: Advances in decision sciences, image processing, security and computer vision, pp 69–76. Springer, Cham
13. Thota S, Kamatham Y, Paidimarry CS (2018) Performance analysis of hybrid companding PAPR reduction method in OFDM systems for 5G communications. In: 2018 9th international conference on computing, communication and networking technologies (ICCCNT), pp 1–5. IEEE (2018)
14. Thota S, Kamatham Y, Paidimarry CS (2020) Analysis of Hybrid PAPR reduction methods of OFDM signal for HPA models in wireless communications. IEEE Access 8:22780–22791

Performance Analysis of OFDM and UFMC with MIMO Wireless Communications

Ramya Gurama, Yedukondalu Kamatham, and Sushmitha Pollamoni

Abstract In the telecommunication industry, the Fifth Generation (5G) technology has been developed to provide high data rates with better spectral efficiency. To achieve the benefits of 5G technology, the multi-carrier modulation (MCM) techniques such as orthogonal frequency division multiplexing (OFDM), filter bank multi-carrier (FBMC) and universal filter bank multi-carrier (UFMC) are cast off. In this paper, the concert investigation of OFDM and UFMC is likened concerning PAPR and BER. It is experiential that the PAPR and power spectral density (PSD) of UFMC are healthier than OFDM, but BER is vice versa. To attain the benefits of spatial multiplexing, the 2 × 2 MIMO is united with OFDM and UFMC such that the performance scrutiny of MIMO-OFDM and MIMO-UFMC is also carried out concerning PSD and PAPR. The results confirm that PAPR and PSD of MIMO-UFMC are improved than MIMO-OFDM. Hence, MIMO-UFMC is a better choice than MIMO-OFDM for 5G and beyond applications.

Keywords OFDM · UFMC · MIMO · PAPR · BER

1 Introduction

In recent years, the intensive usage of alternative modulation formats of orthogonal frequency division multiplexing (OFDM) is increased in 5G wireless communications [1, 2]. Even 5G New Radio (5G-NR) standalone specifications still depend on OFDM [3], Third Generation Partnership Project (3GPP) has not yet completely

R. Gurama · Y. Kamatham (✉) · S. Pollamoni
Department of Electronics and Communication Engineering, CVR College of Engineering, Hyderabad, India
e-mail: kyedukondalu@gmail.com

R. Gurama
e-mail: ramyaasai1996@gmail.com

S. Pollamoni
e-mail: spullamoni@gmail.com

K. A. Reddy et al. (eds.), *Data Engineering and Communication Technology*,
Lecture Notes on Data Engineering and Communications Technologies 63,
https://doi.org/10.1007/978-981-16-0081-4_27
273

addressed usage of massive machine type communication (mMTC) with choice of modulation format at 30 GHz, and new use cases are arising at giant data rates of mMTC [4]. OFDM exhibit few drawbacks such as out of band radiation (OOB), reduction in spectral efficiency (SE), high peak-to-average power ratio (PAPR) leads to nonlinearity in high-power amplifiers (HPAs) and inter-carrier interference (ICI) due to imperfect frequency and time synchronization [5]. Several multi-carrier modulation (MCM) techniques are proposed which include filter bank multi-carrier (FBMC), universal filter bank multi-carrier (UFMC) apart from the existing OFDM. These MCM techniques are combined with multiple-input–multiple-output (MIMO) to enhance the capacity of communications (data rate) through spatial multiplexing. The growth of these MIMO-MCM of 5G will decrease the future in wireless communications [6].

In FBMC, the prototype filters are required to reduce the sidebands. FBMC signal is intended with OQAM used for the transmission of data and generates the interference paths. FBMC with OQAM may not support MIMO because it necessitates the large filter length which decreases the SE [7]. To overwhelm the limitation of FBMC, the UFMC system is introduced.

UFMC is an intervening scheme of filtered-OFDM (F-OFDM) and FBMC. In F-OFDM, the whole OFDM signal is sieved to reduce the OOB radiation [8], and in FBMC, each subcarrier is individually filtered [9]. UFMC is performed with a group of subcarriers to suppress the spectral sidebands and increases the SE. UFMC is an appropriate modulation for short burst communications with shorter filter lengths instead of using longer CP length in OFDM [10]. MIMO is cast off to increase the data rates because of multiple channels and provide better channel capacity without any addition of power in the spectrum. So, the combination of MIMO-UFMC is implemented for multiple users to provide high data rates and increases channel capacity.

This research work deals with the comparison of SE, PAPR, and BER of OFDM, UFMC, MIMO-OFDM, and MIMO-UFMC. The lingering sections of this paper are organized as Sect. 2 deals with the description of system models for OFDM and UFMC. Section 3 deals with concert analysis of OFDM and UFMC, and Sect. 4 is the comparison of the performance of MIMO-OFDM and MIMO-UFMC for SE, PAPR and BER. The conclusion is drawn in Sect. 5.

2 System Model for OFDM and UFMC

2.1 OFDM

OFDM is mainly used in 4G and broadband communications systems. In the OFDM system (Fig. 1), the input symbols are in the frequency domain and can be expressed as

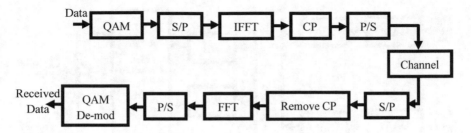

Fig. 1 Block diagram of the OFDM system

$$X(0), X(1), \ldots, X(N-1) \tag{1}$$

These symbols are modulated with the QAM and pass through the IFFT signal is represented as

$$x(0), x(1), \ldots, x(N-1) \tag{2}$$

After IFFT, the parallel form of orthogonal signals is converted to serial form. At the transmitter, CP is added to diminish inter-symbol interference (ISI). The data is serially received, CP is removed from the serial data and passed to FFT, and signal is expressed as

$$Y(0), Y(1), \ldots, Y(N-1) \tag{3}$$

The demodulated signal can be represented as

$$X(0) \neq \hat{X}(0) \tag{4}$$

The estimated symbol can be written as

$$\hat{X}(0), \hat{X}(1), \ldots, \hat{X}(N-1) \tag{5}$$

The received signal of the OFDM is given as

$$y = x * h + W \tag{6}$$

where 'x' is transmitted signal, 'h' is a channel and 'W' is noise.

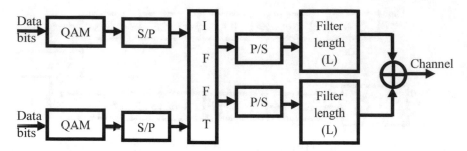

Fig. 2 Structure of UFMC transmitter

2.2 UFMC

In the UFMC transmitter (Fig. 2), the total numeral of subcarriers canister is denoted as N to reduce the PAPR. K is the overall number subbands, and each subband contains S subcarriers. After N-point IFFT, each subband is evaluated by padding zeros for the unallocated subcarriers. Then, each subband is filtered with the length of L. The filtered subbands are summed together, and the input data is QAM modulated where the data can be expressed as

$$x = \left[x_0 x_1 x_2, \dots, x_{N-1} \right]^T \tag{7}$$

Where 'N' represents the total number of subcarriers. Subsequently, the modulated data is passed through the IFFT (converts frequency domain to time domain with orthogonality property), and the resulting signal is expressed as

$$x(N_{FFT} + L - 1, 1) = \sum_{b=1}^{N_b} F_b(N_{FFT} + L - 1, N_{FFT}).V_b(N_{FFT}, n_b)C_b(n_b, 1) \tag{8}$$

where N_{FFT} is the size of FFT, N_b is a numeral of the resource block, C_b is the collection of several symbols for apiece block, V_b is IFFT matrix, F_b is the impulse response of each filter with length L. After filtering, the signals of each filter are summed and transmitted and that can be expressed as

$$x = \sum_{i=0}^{s-1} x_i ; 0 \le n \le N_{FFT} + L - 1 \tag{9}$$

where x_i is ith subband of the filtered signal, then PAPR of a signal can be represented as

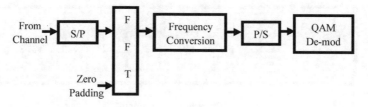

Fig. 3 Structure of UFMC receiver

$$PAPR = \frac{n = 0, 1, \ldots N + L - 1 |x(n)|^2}{E\left[|x(n)|^2\right]} \tag{10}$$

where $E\left[|x(n)|^2\right]$ is an estimation of average power, n is the time index.

To estimate the probability of PAPR of a signal in terms of CCDF function can be expressed as

$$P(PAPR > PAPR_0) = 1 - \left(1 - e^{PAPR_0}\right) \tag{11}$$

where $PAPR_0$ is the probability of a PAPR of exceeding symbols for the clipping level.

At the receiver (Fig. 3), the process of UFMC is alike as OFDM, and the data which is received from the channel is passed through the two-point IFFT where it performs zero paddings to compensate for the delay occurred in the transmit filter. After the data passed through the frequency translation, then the symbols are demodulated by using QAM [11].

3 Comparative analysis of OFDM and UFMC

In the OFDM system, symbols are generated with QAM modulation and passed through FFT to have orthogonality signals, and CP is added to eliminate ISI due to multipath diminishing. The power spectral density (PSD) of OFDM is estimated and shown in Fig. 5.

The estimation of PSD of UFMC is carried out with 20 subbands and 10 subcarriers, the filter length is 43 and attenuation factor of 40. The PSD of the UFMC signal is estimated with dissimilar FFT sizes, i.e., FFT = 512 and 1024 (Fig. 6). The PSD of UFMC ranges from -0.2 to 0.2 for FFT = 512 and for FFT = 1024 ranges from -0.1 to 0.1.

By comparing the PSD of UFMC and OFDM (Figs. 4 and 6), the PSD of UFMC is better than OFDM. Since UFMC is a combination of F-OFDM and FBMC, in FBMC, the sidebands are deleted with help of a filter parameter, and in UFMC, filter

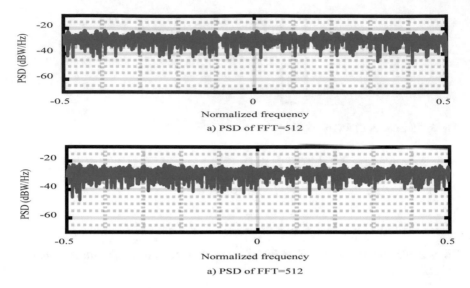

Fig. 4 PSD of OFDM with different FFT

length is cast off for the elimination of the sidebands. The estimation of PAPR of UFMC is shown in Fig. 7. The PAPR at 10^{-1} is 8.3 dB.

The BER is defined as the rate at which error occurs during the transmission of data and expressed as

$$BER = \frac{1}{2} \text{erfc} \left(\sqrt{\frac{E_b}{N_0}} \right) \tag{12}$$

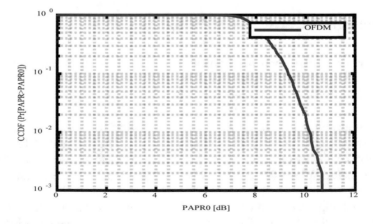

Fig. 5 PAPR for OFDM

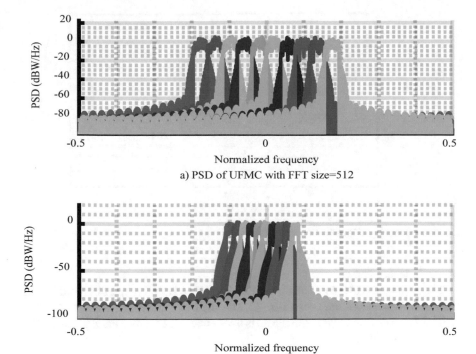

a) PSD of UFMC with FFT size=512

b) PSD of UFMC with FFT size=1024

Fig. 6 PSD of UFMC

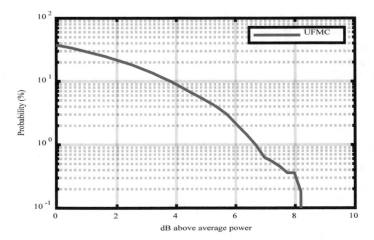

Fig. 7 PAPR of UFMC

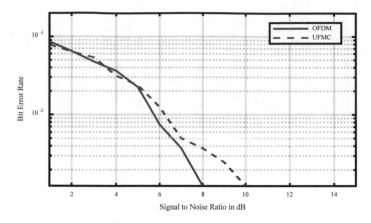

Fig. 8 BER versus SNR of OFDM and UFMC

where $\frac{E_b}{N_0}$ is an SNR per bit.

The estimated BER of OFDM and UFMC is shown in Fig. 8. It is observed that the BER of OFDM is healthier than UFMC. The BER of OFDM is 5.5 dB, and UFMC is 6.7 dB at 10^{-2}. The PAPR of UFMC is lower than OFDM, but the BER of OFDM is better than UFMC.

4 Comparison of MIMO-OFDM and MIMO-UFMC

MIMO is a process of multiplying the capacity of wireless communications systems using numerous transmitting antennas which are denoted as N_T and numerous receiving antennas denoted as N_R with a channel matrix. Then, MIMO system canister is modeled as

$$Y = HX + n \tag{13}$$

where 'Y' is the received signal, with a length of $Y = [y_1 y_2 y_3, \ldots, y_N]_{1 \times N}$. 'H' is a channel matrix, with the length of M x N. 'X' is a transmitted signal, with the length of $X = [x_1 x_2 x_3, \ldots, x_M]_{1 \times M}$, and then, the channel matrix is expressed as

$$H = \begin{bmatrix} h_{1,1} & h_{1,2} & \cdots & h_{1,N} \\ h_{2,1} & h_{2,2} & & h_{2,M} \\ \vdots & & \ddots & \vdots \\ h_{N,1} & h_{N,2} & \cdots & h_{N,M} \end{bmatrix}_{M \times N} \tag{14}$$

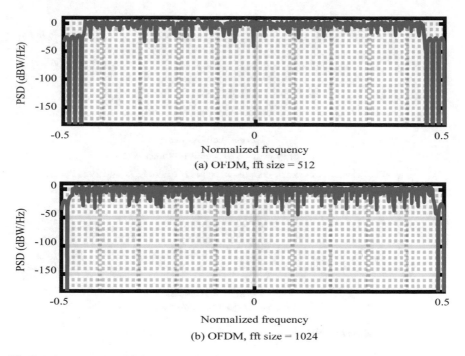

Fig. 9 MIMO-OFDM with FFT size = 512 and 1024

The OFDM system is combined with MIMO for a high data rate. The PSD of MIMO-OFDM is estimated as shown in MIMO-OFDM with FFT size = 512, and 1024UFMC is combined with the MIMO system to increase the SE. The PSD of MIMO-UFMC is estimated with 20 subbands with ten subcarriers, the filter length is 43 and the attenuation factor of 40 (Fig. 10). The PSD of the MIMO-UFMC range is the same as UFMC. By comparing Figs. 9 and 10, the PSD of MIMO-UFMC is better than MIMO-OFDM.

The PAPR of MIMO-OFDM and MIMO-UFMC is estimated (Fig. 11). The PAPR of MIMO-OFDM is 10.3 dB, and MIMO-UFMC is 2 dB.

5   Conclusion

In this paper, the PSD, BER and PAPR of OFDM and UFMC are scrutinized. The SE of OFDM is abridged due to the addition of CP. With high PAPR, the dynamic range of HPA increases. To overcome this limitation, the UFMC system is developed. The PSD of UFMC and OFDM is estimated and compared with different FFT sizes of 512 and 1024. The PSD of UFMC is in between −0.2 to 0.2 for FFT = 512 and −0.1 to 0.1 for FFT = 1024 and for OFDM −0.5 to 0.5 for both FFT sizes. PSD of UFMC is healthier than OFDM, and PAPR of UFMC at 10^{-1} is 8.2 dB. The PAPR of OFDM

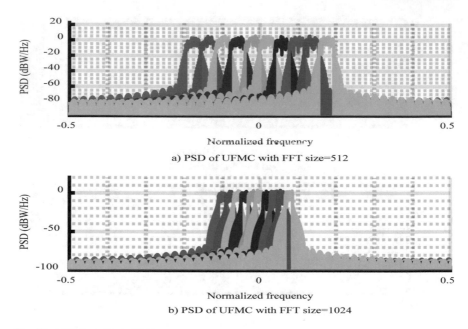

a) PSD of UFMC with FFT size=512

b) PSD of UFMC with FFT size=1024

Fig. 10 PSD for MIMO-UFMC

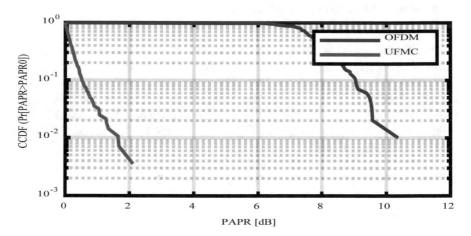

Fig. 11 PAPR of MIMO-UFMC and MIMO-OFDM

at 10^{-1} is 9.6 dB and BER at 10^{-2} for OFDM 5.5 dB and UFMC 6.7 dB. The PSD and UFMC are better for UFMC than OFDM nevertheless BER of OFDM is efficient than UFMC. The UFMC and OFDM are combined with MIMO for the gaining of spatial multiplexing benefits. The PSD and PAPR of MIMO-OFDM and MIMO-UFMC are estimated and analyzed. The PSD of MIMO-UFMC is same as UFMC, and PSD of MIMO-OFDM is −0.4 to 0.4 for FFT = 512 and −0.45 to 0.45 for FFT = 1024.

The PSD is improved for MIMO-UFMC compared to MIMO-OFDM. The PAPR of MIMO-UFMC is 2 dB, and MIMO-OFDM is 10.3 dB. Therefore, MIMO-UFMC can be used in short burst wireless communications systems, whereas MIMO-OFDM can be used in long-range communications at reduced data rates. Since the PAPR of UFMC is lower, it is used in low-power wireless communications applications.

Acknowledgements This work has been carried out under the project entitled 'Study and Implementation of Self-Organized Femtocells for Broadband Services to Indoor Users in Heterogeneous Environment' sponsored by AICTE, New Delhi, under Research Promotion Scheme (RPS), Vide sanction letter No: 8-30/RFID/RPS/POLICY-1/2016-2017, Dated: 02.08.2017.

References

1. Bockelmann C, Pratas N, Nikopour H, Au K, Svensson T, Stefanovic C, Popovski P, Dekorsy A (2016) Massive machine-type communications in 5G: physical and MAC-layer solutions. IEEE Commun Mag 54(9):59–65
2. ITU-R Rec. ITU-R M. 2083-0 (2015) IMT vision-framework and overall objectives of the future development of IMT for 2020 and beyond
3. Popovski P, Stefanović Č, Nielsen JJ, De Carvalho E, Angjelichinoski M, Trillingsgaard KF, Bana AS (2019) Wireless access in ultra-reliable low-latency communication (URLLC). IEEE Trans Commun 67(8):5783–5801
4. Gerzaguet R, Bartzoudis N, Baltar LG, Berg V, Doré JB, Kténas D, Font-Bach O, Mestre X, Payaró M, Färber M, Roth K (2017) The 5G candidate waveform race: a comparison of complexity and performance. EURASIP J Wirel Commun Netw 2017(1):13
5. Husam AA, Kollár Z (2018) Complexity comparison of filter bank multicarrier transmitter schemes. In: 2018 11th international symposium on communication systems, networks & digital signal processing (CSNDSP), pp 1–4. IEEE
6. Zakaria R, Le Ruyet D (2012) A novel filter-bank multicarrier scheme to mitigate the intrinsic interference: application to MIMO systems. IEEE Trans Wirel Commun 11(3):1112–1123
7. Ghassemi A, Gulliver TA (2009) Intercarrier interference reduction in OFDM systems using low complexity selective mapping. IEEE Trans Commun 57(6):1608–1611
8. Sen S, Senguttuvan R, Chatterjee A (2010) Environment-adaptive concurrent companding and bias control for efficient power-amplifier operation. IEEE Trans Circuits Syst I Regul Pap 58(3):607–618
9. Zhang J, Hu S, Liu Z, Wang P, Xiao P, Gao Y (2019) Real-valued orthogonal sequences for iterative channel estimation in MIMO-FBMC systems. IEEE Access 7:68742–68751
10. Siohan P, Siclet C, Lacaille N (2002) Analysis and design of OFDM/OQAM systems based on filterbank theory. IEEE Trans Signal Process 50(5):1170–1183
11. Kamatham Y, Pollamoni S (2019) Implementation of OFDM system with companding for PAPR reduction using NI-USRP and LabVIEW. In: 2019 IEEE international WIE conference on electrical and computer engineering (WIECON-ECE), pp 1–4. IEEE

Analysis of Software Tool for Voltage Profile Improvement in IEEE 14-Bus System Using SVC

Machavaram Sandeep Kumar, Sripada Sruthi Meghana, Bandi Geyavarshini, Rakshitha Peddi Sai, and Mamidi Sukruthi

Abstract In current scenario, there are different types of commercial and open source software tools in order to determine the best location of FACTS devices in IEEE 14-bus system. Increasing of complexity and number of buses in modern power systems, so that analyzing of power system network is very essential. In order to analyze the power system network, commercial and open source software tools are effectively used. Here, we analyze the NEPLAN and MATLAB/SIMULINK tools for best location of static VAR compensator (SVC) in IEEE 14-bus system. We are analyzing the IEEE 14-bus standard test data using NEPLAN and MATLAB/SIMULINK software tools and placing the SVC randomly in different buses. Then, we have analyzed all the results obtained using different softwares for choosing the optimal location of SVC for voltage profile improvement in IEEE 14-bus system.

Keywords NEPLAN software · MATLAB/SIMULINK · IEEE 14-bus system and SVC

M. Sandeep Kumar (✉) · S. Sruthi Meghana · B. Geyavarshini · R. P. Sai · M. Sukruthi
BVRIT Hyderabad College of Engineering for Women, Hyderabad, India
e-mail: sankalpam.msk@gmail.com

S. Sruthi Meghana
e-mail: sruthimeghanassm@gmail.com

B. Geyavarshini
e-mail: varshinigeya@gmail.com

R. P. Sai
e-mail: rakshi2409@gmail.com

M. Sukruthi
e-mail: sukruthi2016@gmail.com

© The Author(s), under exclusive license to Springer Nature Singapore Pte Ltd. 2021   285
K. A. Reddy et al. (eds.), *Data Engineering and Communication Technology*,
Lecture Notes on Data Engineering and Communications Technologies 63,
https://doi.org/10.1007/978-981-16-0081-4_28

1 Introduction

For satisfactory operation of power system network, the voltage ratings of all equipments are within permissible limits [1]. So as to increase the power transfer capability, damping of power oscillations, and improvement of voltage stability margins, SVC is effectively used. The exchange of inductive or capacitive currents to the network can be adjusted by using SVC. In order to regulate the specific parameters like bus voltage, these inductive currents or capacitive currents are controlled. Improvement of power factor and power quality is done by reactive power compensation using SVC [2]. When the system voltage falls below the reference value, reactive power is generated using SVC. When the system voltage exceeds the reference vale, reactive power is absorbed by SVC. But exchange of active power is not possible by SVC [3].

2 Computational Software Tools

Commercial software and educational software are the two software packages for power system analysis. These tools are well-tested and more efficiently used. These are developed by electric utilities or researchers. These tools (NEPLAN, PSS/E, EDSA, CYME, EUROSTAG, and ETAP) are most conventional and exciting option for educational level [4]. Educational software tools are developed and used in most of the educational institutes. These institutes prefer open source softwares rather than commercial or computational efficiency. These are alternative to the commercial software for education [5]. MATLAB is very popular and mostly used by educational institutes.

NEPLAN software is user friendly planning and very informative system [6]. This software is used to analyze the transmission, distribution, production, renewable energy systems, and industrial power grid. This software is used to plan, analyze, and optimize the power grids. It is very simple, and GUI makes it very easy to the users. Harmonic analysis, computation of optimal capacitor placement, reliability, reduction of number of buses, voltage stability, transient stability, determination of cable sizing, calculation of short circuit, optimal power flow, and optimization of distribution network are the key features of NEPLAN software.

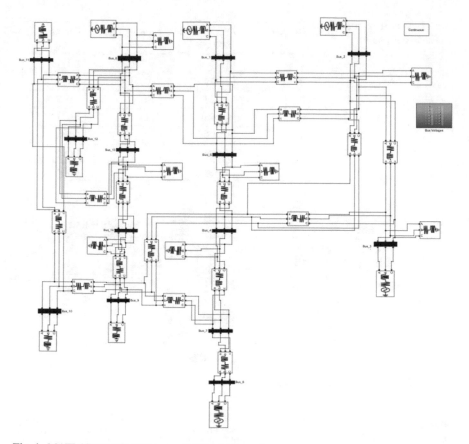

Fig. 1 MATLAB/SIMULINK based IEEE 14-bus system

3 Design of IEEE 14-Bus System

3.1 MATLAB/SIMULINK Based IEEE 14-Bus System

The IEEE 14-Bus system is simulated by using MATLAB/SIMULINK which is shown in Fig. 1. The standard test data is entered, and voltage profiles of buses 6 and 8 are violate the permissible limit as given in Table 1.

3.2 NEPLAN Based IEEE 14-Bus System

NEPLAN software is used to design the IEEE 14-bus system with standard test data as shown in Fig. 2. The bus voltages of buses 6 and 8 violate the permissible limit as given in Table 2.

Table 1 Voltage profiles of IEEE 14-bus system

Bus_number	Per unit bus voltage	Bus_number	Per unit bus voltage
1	1.04	8	1.071
2	1.028	9	1.014
3	1.004	10	1.013
4	0.9959	11	1.031
5	0.999	12	1.036
6	1.056	13	1.031
7	1.034	14	1.004

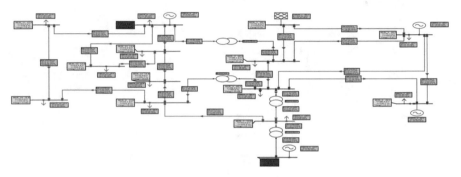

Fig. 2 NEPLAN based IEEE 14-bus system

Table 2 Voltage profiles of NEPLAN based IEEE 14-bus system

Bus_number	Bus voltage (%)	Bus_number	Bus voltage (%)
1	104	8	107.42
2	102.98	9	101.84
3	100.71	10	101.76
4	99.9	11	103.35
5	100.18	12	104.02
6	105.71	13	103.36
7	103.48	14	100.65

4 Design of IEEE 14-Bus System with Static VAR Compensator

SVC is used to improve the voltage profiles of all buses by placing SVC at different buses randomly. If bus voltage is below the permissible limit, SVC generates the capacitive currents, i.e., reactive power, and if bus voltage is above permissible limit, SVC absorbs the inductive currents, i.e., reactive power. Here, the design of SVC based IEEE 14-bus system is done with both MATLAB/SIMULINK and NEPLAN softwares.

4.1 SVC Based IEEE 14-Bus System

SVC is put at bus-11 as shown in Fig. 3. Here, the buses 9, 10, 12, 13, and 14 are the undervoltage buses as given in Table 3.

Similarly, SVC is put at bus-13. Here, buses 9, 10, 12, 13, and 14 are the undervoltage buses as given in Table 4.

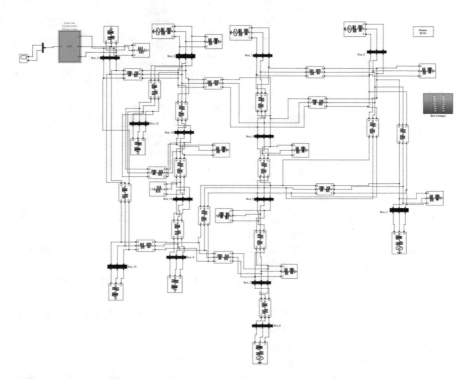

Fig. 3 MATLAB based IEEE 14-bus system in association with SVC put at bus-11

Bus_number	Per unit bus voltage
9	0.9367
10	0.9378
12	0.9478
13	0.9208
14	0.8142

Table 3 Undervoltage buses of the IEEE 14-bus system with SVC at bus-11

Table 4 Undervoltage buses of the IEEE 14-bus system with SVC at bus-13

Bus_number	Per unit bus voltage
9	0.9372
10	0.9428
12	0.9394
13	0.9077
14	0.8089

Table 5 Undervoltage buses of the IEEE 14-bus system with SVC at bus-12

Bus_number	Per unit bus voltage
9	0.9377
10	0.9431
12	0.9321
13	0.9122
14	0.8111

Table 6 Undervoltage buses of the IEEE 14-bus system with SVC at bus-9

Bus_number	Per unit bus voltage
9	0.9098
10	0.9245
14	0.9232

Similarly, SVC is placed at bus number 12. Here, buses 9, 10, 12, 13, and 14 are the undervoltage buses as given in Table 5.

Similarly, SVC is placed at bus number 9. Here, buses 9, 10, and 14 are the undervoltage buses as given in Table 6.

Similarly, SVC is put at bus-7. Here, buses 7, 9, 10, and 14 are the undervoltage buses as given in Table 7.

Similarly, SVC is put at bus-10. Here, buses 9, 10, 12, 13, and 14 are the undervoltage buses as given in Table 8.

Similarly, SVC is put at bus-14. Here, buses 9, 10, 12, 13, and 14 are the undervoltage buses as given in Table 9.

By placing SVC at buses 4 and 5, the voltage profiles of 14 buses in the system are within acceptable limits. Hence, the best location of SVC in the 14-bus system are buses 4 and 5 as given in Tables 10 and 11.

Table 7 Undervoltage buses of the IEEE 14-bus System with SVC at bus-7

Bus_number	Per unit bus voltage	Bus_number	Per unit bus voltage
7	0.945	10	0.9491
9	0.9389	14	0.9457

Table 8 Undervoltage buses of the IEEE 14-bus system with SVC at bus-10

Bus_number	Per unit bus voltage
9	0.9342
10	0.9317
12	0.949
13	0.9215
14	0.8133

Table 9 Undervoltage buses of the IEEE 14-bus system with SVC at bus-14

Bus_number	Per unit bus voltage
9	0.9422
10	0.9477
12	0.9479
13	0.9184
14	0.7978

Table 10 Voltage profiles of the 14-bus system with SVC at bus-4

Bus_number	Per unit bus voltage	Bus_number	Per unit bus voltage
1	1.023	8	1.049
2	1.006	9	0.9806
3	0.9812	10	0.9838
4	0.9595	11	1.008
5	0.9754	12	1.018
6	1.04	13	1.011
7	0.9972	14	0.9771

Table 11 Voltage profiles of 14-bus system with SVC at bus-5

Bus_number	Per unit bus voltage	Bus_number	Per unit bus voltage
1	1.013	8	1.05
2	0.9988	9	0.9812
3	0.9809	10	0.9824
4	0.9646	11	1.003
5	0.95	12	1.01
6	1.031	13	1.004
7	0.999	14	0.9742

4.2 NEPLAN Based IEEE 14-Bus System with SVC

SVC is put at bus-11 as shown in Fig. 4. Here, buses 9, 10, 11, 12, 13, and 14 are the undervoltage buses as given in Table 12.

Similarly, SVC is placed at bus number 13. Here, buses 9, 10, 11, 12, 13, and 14 are the undervoltage buses as given in Table 13.

Similarly, SVC is placed at bus-12. Here, buses 10, 11, 12, 13, and 14 are the undervoltage buses as given in Table 14.

Similarly, SVC is placed at bus-9. Here, buses 9, 10, and 14 are the undervoltage buses as given in Table 15.

Similarly, SVC is placed at bus number 7. Here, buses 7, 9, 10, and 14 are the undervoltage buses as given in Table 16.

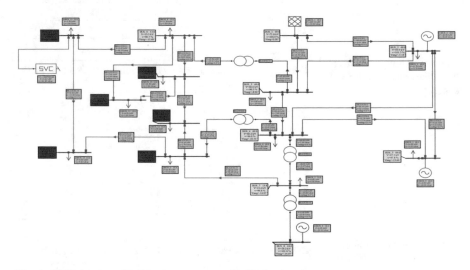

Fig. 4 NEPLAN based IEEE 14-bus system with SVC at bus-11

Table 12 Undervoltage buses of the IEEE 14-bus system with SVC at bus-11

Bus_number	Bus voltage (%)	Bus_number	Bus voltage (%)
9	93.61	12	94.94
10	91.92	13	94.3
11	90.08	14	91.88

Table 13 Undervoltage buses of the IEEE 14-bus system with SVC at bus-13

Bus_number	Bus voltage (%)	Bus_number	Bus voltage (%)
9	94.58	12	92.19
10	93.94	13	90.06
11	94.39	14	90.54

Table 14 Undervoltage buses of the IEEE 14-bus system with SVC at bus-12

Bus_number	Bus voltage (%)
10	94.2
11	94.47
12	88.37
13	91.61
14	91.43

Table 15 Undervoltage buses of the IEEE 14-bus system with SVC at bus-9

Bus_number	Bus voltage (%)
9	91.52
10	91.98
14	91.33

Table 16 Undervoltage buses of the IEEE 14-bus system with SVC at bus-7

Bus_number	Bus voltage (%)	Bus_number	Bus voltage (%)
7	93.83	10	93.85
9	93.56	14	93.09

Table 17 Undervoltage buses of the IEEE 14-bus system with SVC at bus-10

Bus_number	Bus voltage (%)
9	92.26
11	93.58
14	91.63

Similarly, SVC is placed at bus-10. Here, buses 9, 11, and 14 are the undervoltage buses as given in Table 17.

Similarly, SVC is put at bus-14. Here, buses 9, 10, 13, and 14 are the undervoltage buses as given in Table 18.

By placing SVC at buses 4 and 5, the voltage profiles of 14 buses in system are within acceptable limits. Hence, the best location of SVC in the IEEE 14-bus system are buses 4 and 5 as given in Tables 19 and 20.

Table 18 Undervoltage buses of the IEEE 14-bus system with SVC at bus-14

Bus_number	Bus voltage (%)	Bus_number	Bus voltage (%)
9	93.3	13	93.8
10	93.27	14	85.75

Table 19 Voltage profiles of the IEEE 14-bus system with SVC at bus-4

Bus_number	Bus voltage (%)	Bus_number	Bus voltage (%)
1	104	8	103.96
2	101.3	9	97.7
3	97.66	10	97.65
4	95.79	11	99.39
5	96.96	12	100.16
6	101.93	13	99.46
7	99.37	14	96.54

Table 20 Voltage profiles of the IEEE 14-bus system with SVC at bus-5

Bus_number	Bus voltage (%)	Bus_number	Bus voltage (%)
1	104	8	103.94
2	101.38	9	98.08
3	98.11	10	97.93
4	96.54	11	99.44
5	96.33	12	100.02
6	101.75	13	99.35
7	99.86	14	96.71

5 Conclusion and Future Scope

The design of the IEEE 14-bus system with and without SVC is effectively analyzed using both NEPLAN and MATLAB/SIMULINK softwares. The voltage profile improvement is observed at buses 4 and 5 with SVC in both cases, i.e., NEPLAN and MATLAB/SIMULINK design. The average voltage levels obtained at bus bars 4 and 5 are very close to reference value by using NEPLAN when compared to results obtained by MATLAB/SIMULINK software. This study can be further extended with large number of buses more easily using NEPLAN software in more realistic way.

References

1. Khonde SS, Dhamse SS, Thosar AG (2014) Power quality enhancement of standard IEEE 14 bus system using unified power flow controller. Int J Eng Sci Innov Technol 3(5)
2. Marlin S, Jebaseelan SS, Padmanabhan B, Nagarajan G (2014) Power quality improvement for thirty bus system using UPFC and TCSC. Indian Journal of Science and Technology 7(9):1316–20
3. Mohammed OH, Cheng SJ, Zakaria A (2009) Steady-state modeling of SVC and TCSC for power flow analysis. Proc IMECS 2:1–5
4. Bică D. Moderns informatics systems in power engineering. Models and applications in NEPLAN software. Petru Maior University Publishing House, pp 127–159

5. Milano F (2005) An open source power system analysis toolbox. IEEE Trans Power Syst 20(3):1199–1206
6. NEPLAN User's Guide V5.54

Evaluation of Particle Size, Lattice Strain and Estimation of Mean Square Amplitudes of Vibration, Debye Waller Factor and Debye Temperature for Metal Oxide Nanoparticles

Purushotham Endla

Abstract Zinc oxide (ZnO) nanoparticles were synthesised by high-energy ball mill. Zinc oxide powders were ball milled in an argon inert atmosphere. The powder was characterized by XRD and scanning electron microscopy (SEM) measurements. The vacancy formation energy as a function of Debye temperature has been studied. Lattice strain (ε), root mean square amplitude of vibration (u_{av}), Debye–Waller factor (B_{av}), energy of vacancy formation (E_f) are found to increase, and particle size (t), mean Debye temperature (θ_M) are found to be decrease with the milling time 0, 2, 4, 6, 8 and 10 h. The particle size values for X-ray study are 224, 103, 82, 54, 34 and 33 nm. The particle size values for SEM characterization are 200, 100, 80, 50, 32 and 32 nm. The SEM values are good agreement with X-ray characterization values.

Keywords Ball milling · SEM · Particle size · Lattice strain · Debye-Waller factor · Debye temperature

1 Introduction

Zinc oxide is used as treat skin irritation, in semiconductors, ceramic, glass compositions and cigarette filters. The ZnO also used in rubber industry due to the vulcanization of rubber to manufacture such things as shoe soles, tires and hockey pucks. ZnO is also used in solar cells. ZnO has the properties of antibacterial and deodorizing properties. ZnO nanoparticles are used in baby powder and creams. These nanoparticles are also used in lips-related creams. Zinc oxide is also used in laser diodes, light emitting diodes (LED), piezoelectricity, paint coatings/paint pigments and breakfast cereals. Due to their unique nature and properties, zinc oxide nanoparticles have attracted much importance compared to bulk counterparts. In the context of their usage in gas sensors, biosensors, cosmetics, drug-delivery systems etc., zinc oxide nanoparticles are very much important.

P. Endla (✉)
Department of Physics, School of Sciences, SR University, Ananthasagar, Warangal Urban 506371, Telangana, India
e-mail: psm45456@gmail.com

© The Author(s), under exclusive license to Springer Nature Singapore Pte Ltd. 2021   297
K. A. Reddy et al. (eds.), *Data Engineering and Communication Technology*,
Lecture Notes on Data Engineering and Communications Technologies 63,
https://doi.org/10.1007/978-981-16-0081-4_29

Influence of tool revolving on mechanical properties of some metals and alloy were developed for potential generation engineering production [1, 2] and with some thermal and mechanical properties and engineering applications [3, 4]. The important properties of these materials are particle size dependent. So, it is very significant to extend novel techniques for the preparation of nanopowder. Some significant X-ray work is there on some metals [5, 6]. Inagaki [7, 8] worked on non-metallic powders, Sirdeshmukh and Subhadra [9], Gopi Krishna and Sirdeshmukh [10] and Purushotham and Gopi Krishna [11] studied the crystal size and strains of some HCP metals. Aside from giving an amendment to the temperature impact on powers, the Debye–Waller factor is interrelated to a few physical properties and has developed as a significant strong state boundary. Depending on Debye–Waller factor, we can measure bond strength between the atoms or molecules. It is collectively recognized that zinc oxide nanoparticles are antibacterial and restrain the growth of microorganisms. Debye–Waller factor value increases, and the oxidative stress damages carbohydrates, lipids and proteins.

2 Experimental Procedures

Highly pure zinc oxide metal powder have taken from the Solid State Physics Laboratory. New Delhi was used. The powder used as initial sample for zero hour as shown in Fig. 1. The initial sample was grinding (using ball mill) for 2, 4, 6, 8 and 10 h. This grinding leads to produce strains and also decrease particle size. X-ray diffractograms have been recorded for 0, 2, 4, 6, 8 and 10 h, respectively. Figure 2: XRD patterns of ZnO powder have been given in Fig. 2. The observed intensities have been corrected for thermal diffuse scattering (TDS) by applying the technique of Chipman and Paskin [12].

Fig. 1 Highly pure zinc oxide metal powder was for the initial sample

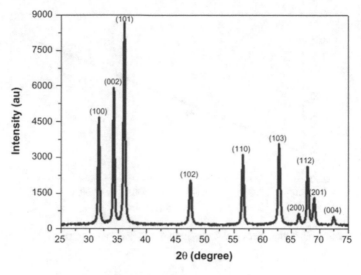

Fig. 2 XRD pattern of ZnO powder for the initial sample

3 Analysis

According to Gopi Krishna and his co-workers [13] were discussed the Debye–Waller factors. The principles are given by

$$B = (6h^2 / Mk_B\theta_M)W(X)$$

$$B_\perp = (6h^2 / M\,k_B\theta_\perp)W(X)$$
$$B_\parallel = (6h^2 / M\,k_B\theta_\parallel)W(X) \tag{1}$$

The $W(X)$ function is

$$W(X) = [\phi(X)/X + (1/4)] \tag{2}$$

where $X = \theta_M/T$, the $W(X)$ values and X value have been measured from standard tables [14].

3.1 Williamson–Hall Method

Bharati et al. [15] and Wilson [16] have been calculated strain and particle/crystal size by using the formulae

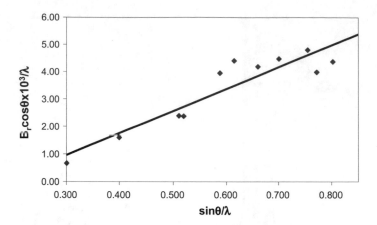

Fig. 3 Williamson–Hall Plot of $B_r\cos\theta/\lambda$ versus $\sin\theta/\lambda$ for ZnO nanopowder for ball milling 10 h

$$B_r \cos q = \frac{k\lambda}{t} + \varepsilon \, \sin\theta \qquad (3)$$

Figure 3: Plot between $B_r \cos\theta/\lambda$ and $\sin\theta/\lambda$ is a straight line. From the slope of the Hall–Williamson plot, we can calculate the lattice strain (ε), and the intercept gives the particle size (t). The Williamson–Hall plot is shown in Fig. 3. The process is following standard procedures [17].

Table 1 Values of positions ($2\theta°$), FWHM, d-spacing and real intensity percentage

Position ($2\theta°$)	Height (counts)	FWHM ($2\theta°$)	d-spacing in (Å)	Relative intensity (%)
31.73	193	0.4	2.81767	56.83
34.406	204	0.2	2.60446	60.13
36.23	339	0.4	2.47760	100.00
47.51	102	0.4	1.91225	30.19
56.58	131	0.5	1.62535	38.73
62.85	173	0.3	1.47736	51.10
66.01	4	1.0	1.41101	1.09
68.08	130.22	0.1968	1.37732	38.43
69.12	41	0.13	1.35790	12.05
72.58	14	0.14	1.30146	4.22

Table 2 Values of crystal/particle size, ball milling time, Debye–Waller factor, strain, Debye temperature (θ_M) and formation energy of strained ZnO powder

Parameter	Milling time(h)	Strain $\varepsilon \times 10^3$	Particle size t (nm)	u_{av}(Å)	B_{av}(Å^2)	θ_M (K)	E_f (eV)
ZnO	0	0.69	224	0.014	1.10	220	0.47
	2	1.40	103	0.015	1.26	206	0.41
	4	1.42	82	0.018	1.58	198	0.36
	6	2.39	54	0.020	1.62	184	0.33
	8	2.68	34	0.021	1.81	179	0.27
	10	3.29	33	0.022	1.86	171	0.23

4 Results and Discussion

Table 1: Values of positions ($2\theta°$), FWHM, d-Spacing and real intensity percentage were given in Table 1.

Table 2: Values were obtained in the current study. The Debye–Waller factor and Debye temperature in parallel and perpendicular directions values have been calculated independently, and the average values of these values are given in Table 2. Lattice strain (ε), root mean square amplitude of vibration (u_{av}), Debye–Waller factor (B_{av}), energy of vacancy formation (E_f) are found to increase, and particle size (t), mean Debye temperature (θ_M) are found to be decreased with the milling time 0, 2, 4, 6, 8 and 10 h. The particle size values for X-eay study are 224, 103, 82, 54, 34 and 33 nm. The particle size values for SEM characterization are 200, 100, 80, 50, 32 and 32 nm. The SEM values have been agreed with X-ray characterization values. Vetelino [18] were recognized the difference of inaccuracy in the experimental values caused by neglecting the TDS corrections.

According to Glyde [19] relation between vacancy formation energy (E_f) and Debye temperature (θ) of a crystal, the relation is

$$E_f = A(k/\hbar)^2 M\theta^2 a^2 \tag{4}$$

where $A = 1.17 \times 10^{-2}$, it is a constant, M the molecular weight. Purushotham [20–24] discussed and verified on different metals on formation energy.

Figure 4 shows the SEM images of ZnO nanoparticles for (a) 200 nm for zero hours (b) 100 nm for 2 h (c) 80 nm for 4 h (d) 50 nm for 6 h (e) 32 nm for 8 h (f) 32 nm for 10 h. The experimental values agree well with SEM values.

Figure 5 shows the variation between milling time versus (a) Strain ($\varepsilon \times 10^3$), (b) Particle size (t in nm), (c) RMS amplitude of vibration (u_{av} in Å), (d) Debye–Waller factor (B_{av} in Å^2), (e) Debye temperature (θ_M in K) and (f) Energy of vacancy formation (E_f in eV), respectively.

Fig. 4 SEM morphology of zinc oxide nanoparticles for **a** initial sample (0 h), **b** 2 h, **c** 4 h, **d** 6 h, **e** 8 h, **f** 10 h of ball milling time

5 Conclusion

Lattice strain (ε), rms amplitude of vibration (u_{av}), Debye–Waller factor (B_{av}), energy of vacancy formation (E_f) are found to increase, and crystal size (t) is found to be decreased with the milling time. Crystal size values for X-ray study are 224, 103, 82, 54, 34 and 33 nm. The crystal values for SEM characterization are 200, 100, 80, 50, 32 and 32 nm. The SEM values agree well with X-ray characterization values. The plot among the milling time versus strain, particle size, RMS amplitude of vibration, Debye–Waller factor, crystal size, strain and Debye temperature for ZnO has been studied.

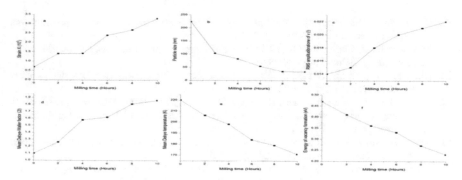

Fig. 5 Variation between milling time versus **a** Strain ($\varepsilon \times 10^3$), **b** Particle size (t in nm), **c** RMS amplitude of vibration (u_{av} in Å), **d** Debye–Waller factor (B_{av} in Å^2), **e** Debye temperature (θ_M in K) and **f** Energy of vacancy formation (E_f in E_v), respectively

Acknowledgements The author is grateful to Dr. K. Madangopal, AD, MG, Indigenous *BARC Scanning Electron Microscope (SEM), Bhabha Atomic Research Center,* Trombay, *Mumbai*—400 085, India, for his help in the SEM investigation images. The author expressed his sincere thanks to Prof. D.B. Sirdeshmukh and Prof. N. Gopikrishna for useful suggestions. The authors sincerely acknowledge the assistance of Mr. B. Sammaiah, Analytical Center, O.U., Hyderabad, for his help in the SEM characterization. The author is grateful to the management of SR University, Ananthasagar, Warangal Urban, Telangana, India, for permitting to use Research Laboratory and undertake the reported investigations.

References

1. Ashok Kumar V, Sammaiah P (2017) Science direct influence of process parameters on mechanical and metallurgical properties of zinc coating on mild steel during mechanical process. Sci Direct Mater Today 5:1
2. Srikanth Rao P, Manoj Kumar D, Gopikrishna J (2018) Synthesis, characterization, effect of lattice strain on the Debye-Waller factor and Debye temperature of aluminium nanoparticles. Sci Direct Mater Today 5:1264
3. Shiva Krishna Ch, ManojKumar J (2017) The effect of Nickel on mechanical properties of aluminium alloy Al–75. Int J Mater Sci 12:627
4. SatishKumar P, Sastry CHSR, Devaraju A (2017) Influence of tool revolving on mechanical properties of friction stir welded 5083Aluminum alloy. Mater Today 4:330
5. Purushotham E (2019) Preparation and characterization of CuInSe2 nano-particles. Rasayan J Chem 12:1676
6. Purushotham E (2019) Synthesis, characterization, effect of lattice strain on the Debye-Waller factor and Debye temperature of aluminium nanoparticles. Am J Nanosci 5:23
7. Inagaki M, Furuhashi H, Ozeki T (1971) Integrated intensity changes for crystalline powders by grinding and compression— changes in effective temperature factor. J Mater Sci 6:1520
8. Inagaki M, Furuhashi H, Ozeki T, Naka S (1976) Energy is stored in γ-Fe$_2$O$_3$ during preliminary vibro-milling. J Mater Sci 8:312
9. Sirdeshmukh DB, Subhadra KG, Hussain KA, Raghavendra Rao B (1993) X-ray study of strained CdTe powders. Cryst Res Technol 28:15

10. Gopikrishna N, Sirdeshmukh DB (1993) X-ray studies of lattice strain and Debye-Waller factors of ytterbium. Indian J Pure Appl Phys 31:198
11. Purushotham E, Gopikrishna N (2010) Effect of lattice strain on the Debyewaller factors of Mg, Zn and Cd. Indian J Phys 84:887
12. Chipman DR, Paskin A (1938) X-ray characteristic temperatures of some II—VI ionic compounds, 1959. J Appl Phys 30
13. Gopikrishna N, Sirdeshmukh DB, Rama Rao B, Beandry JB, Gschneidner JKA (1986) Mean square amplitudes of vibration and associated Debye temperatures of dysprosium, gadolinium, lutetium and yttrium. Indian J Pure Appl Phys 24:324
14. Allen FH, Watson DG, Brammer L, Orpen AG, Taylor R (2006) Typical interatomic distances: organic compounds. Int Tables Crystallogr C:790
15. Bharati R, Rehani PB (2006) Crystallite size estimation of elemental and composite silver nano-powders using XRD principles. Indian J Pure Appl Phys 44:157
16. Wilson AJC (1949) X ray optics the diffraction of X rays by finite and imperfect crystals. X-ray Opt
17. Kaelble EF (1967) For diffraction, emission, absorption, and microscopy. The Handbook of X-rays
18. Vetelino JF, Gaur SP (1972) Temperature dependence of energy gaps in some II–VI compounds. Phys Rev B5:2360
19. Glyde HR (1967) Relation of vacancy formation and migration energies to the Debye temperature in solids. J Phys Chem Solids 28:2061
20. Purushotham E, Gopi KN (2014) Effect of particle size and lattice strain on the Debye-Waller factors of Fe_3C nanoparticles. Bull Mater Sci 37:773
21. Purushotham E, Gopi KN (2010) Mean square amplitudes of vibration and associated Debye temperatures of rhenium, osmium and thallium. Phys B 405:3308
22. Purushotham E, Gopi Krishna N (2013) Preparation and characterization of silver nano particles. Indian J Phys 87982
23. Purushotham E, Gopi Krishna N (2013) X-ray determination of crystallite size and effect of lattice strain on the Debye-Waller factors of platinum nano powders. Bull Mater Sci 36:973
24. Purushotham E (2015) X-ray determination of crystallite size and effect of lattice strain on the Debye-Waller factors of Ni nano powder using high energy ball mill. Chem Mater Rev 7:1

Kalman Filter-Based MPC Control Design and Performance Assessment of MIMO System

S. Sudhahar, C. Ganesh Babu, and D. Sharmila

Abstract In this paper, a Kalman filter parameter updated model established controller is scheduled for multiple input multiple output (MIMO) system through constraints. Since model/plant mismatches occur in real-world plants due to uncertainties, so the constrained tracking is very important. The plant model state parameters are estimated using Kalman filter, and the parameters are updated in the controller. To delineate the attainment of the scheduled Kalman filter-based model predictive control (MPC) approach is compared with phase and gain margin and biggest log modulus tuning of P and I value of the PI controller. The proportional and integral controller settings of the plant/system are found two ways using multi-loop method. Firstly, the system controller settings tuned are established on Biggest Log Modulus tuning method, which is also called detuning method. Secondly, the controller settings are tuned from the gain and phase margin requirements of plant/system by using the derived formulae in the literatures. The performance and transient response characteristics of the control schemes are compared to exhibit the high performance of the scheduled method for the bench mark MIMO systems using MATLAB simulation software.

Keywords MIMO · MPC · Performance indices · Proportional integral controller · Phase and gain margin · Kalman · BLT

S. Sudhahar (✉) · C. Ganesh Babu
Department of EIE, Bannari Amman Institute of Technology, Sathyamangalam, Erode, Tamil Nadu, India
e-mail: sudhahars@bitssathy.ac.in

C. Ganesh Babu
e-mail: ganeshbabuc@bitssathy.ac.in

D. Sharmila
Department of CSE, Jai Shriram Engineering College, Avinashipalayam, Tamil Nadu, India
e-mail: sharmiramesh@rediffmail.com

© The Author(s), under exclusive license to Springer Nature Singapore Pte Ltd. 2021   305
K. A. Reddy et al. (eds.), *Data Engineering and Communication Technology*,
Lecture Notes on Data Engineering and Communications Technologies 63,
https://doi.org/10.1007/978-981-16-0081-4_30

1 Introduction

All real-world systems are constituted of multiple variables, which are interacting in nature. An example of this kind of process is control of distillation problem. In this, top and bottom compositions of the column are controlled to achieve minimum energy for separation. These compositions are typically controlled by reflux or distillate and heat load. Each manipulated variable has a substantial effect on both compositions. These interactions of the variables are difficult to control or yield an unsatisfactory control. The control schemes for the MIMO system are broadly classified into two types as follows: firstly the decentralized/multi-loop control scheme such as multi-loop PI controller andsecondly, centralized/multivariable control scheme for example model predictive controller (MPC). The multivariable system/process should be paired, before the controller design and measures the interaction among the paired input and output variables. The decentralized control scheme was implemented due to its simplicity in the design such as single input and single output system. This scheme achieves the desired performance by proper pairing and designs the suitable pre-filer/design the decoupler. The decoupler is preferred in the decentralized controller because of ease of design. In some MIMO systems, cannot be decoupled easily. For such systems, we adopted the centralized control schemes like model predictive control, dynamic matrix control, fuzzy, neural. and optimization algorithms to obtain the desired performances of the system.

The multi-loop PI/PID controllers having the interaction exist between the inputs and outputs [1], but is mostly used because of its simple diagonal controller structure with the system and easy tuning to attain the desired variable/limitation requirements of the system [2]. The centralized multi-loop controllers design and tuning is very crucial in comparison with the single-loop diagonal controller structure [3]. Several tuning methods such as detuning based on the retuning of the controller settings by well-defined rule [4], sequential loop closing method which depends on the iterative procedure [5], relay auto-tuning technique, which is established on the values of ultimate frequency and gain [6], and independent loop method based on the net transfer function [7] available since the independent loops of PI/PID controller proposed, and it includes interactions in the plant for the controller design. The challenging task in the control system is to overcome the uncertainty in the process. So, the controllers should be designed to robust on the performance and stability. Due to uncertainties, the system/process parameters are changed; this causes the model/system mismatch in the model-based controllers. Mostly, the MPC in the industry utilizes the input–output data to recognize the model of the process. This identification should be improved to overcome longer testing time of the process in the plants. The performance index of the crystallization process is estimated to obtain the robust performance based on the extended Kalman filter based nonlinear MPC.

In this paper, the evaluation of the process parameters is computed by using the Kalman algorithm which utilizes the Ricatti equation. The evaluation of the dynamic parameters of the system/plant updated model of the process used the constrained

MPC algorithm. The foremost benefit of the planned technique is that the parameters of the process model are updated for a short duration of time which overcomes uncertainty and model mismatch issues. The proposed method is implemented on the bench mark MIMO systems to validate the closed loop dynamic attainment of the overall system. The attainment of the system was evaluated through the biggest log tuning method and phase & gain margin tuning of multi loop Proportional Integral controllers to illustrate the advantage of the planned method. The anticipated Kalman estimation-based MPC controller is implemented in any real-time control applications.

2 Problem Formulation of MPC for MIMO System

Model predictive control (MPC) is a control algorithm to utilize the predicted manipulated and controlled variables in each instant. In this algorithm, the system variables of the process model are updated using the Kalman filter. The predictive control algorithms have the model of the plant, predictor, and established on the model and prediction the required best possible control performance are obtained.

An LTI discrete-time process model is corresponding to in state space structure is specified by:

$$x(k+1) = Ax(k) + Bu(k) + Gw(k) \ \& \ y(k) = Cx(k) + Dv(k) \tag{1}$$

where k is the sampling instant in reference to discrete time system, x is the state in reference to the system, u is the input variable in reference to the system, and y is the output of the process/system model. A $\varepsilon \ \mathbb{R}^{n \times n}$, B $\varepsilon \ \mathbb{R}^{n \times m}$, C $\varepsilon \ \mathbb{R}^{n \times g}$, D $\varepsilon \ \mathbb{R}^{p \times n}$, and G $\varepsilon \ \mathbb{R}^{p \times n}$ are system, input, output, and Gaussian noise matrices, respectively, and $w(k)$ is the plant disturbances, and $v(k)$ is the vectors of output measurement disturbances. Plant disturbances are formed as zero mean Gaussian noise in addition to covariance Q_w, and output measurement disturbance is shaped as mean Gaussian noise with covariance R_v. Here, we assumed that $GD^T = 0$, so that the state and observation noises are independent [7]. The controlled output of the system y_k should track a guided signal r_k and if the disturbance occurs, it should be also rejected while tracking the signal, r_k. To achieve this objective, here we establish the following quadratic cost function:

$$J_k\big(u_k, x_{k+1}, y_k\big) = \sum_{i=1}^{p} \left\| \hat{y}_{k+i|k} - r_{k+i} \right\|^2 Q_k + \sum_{i=0}^{p-1} \left\| \Delta \hat{u}_{k+i|k} \right\|^2 R_k \tag{2}$$

Here, $\hat{y}_{k+i|k}$ represent the forecast of $y_{k+i|k}$ at time k with $i > 0$ and $\Delta \hat{u}_{k+i|k} = \hat{u}_{k+i|k} - \hat{u}_{k+i-1|k}$ is the predicted variation in reference to the input. The horizon for prediction is always greater than the horizon for regulation. The $Q_k \ \varepsilon \ \mathbb{R}^{p \times p}$, $R_k \ \varepsilon \ \mathbb{R}^{p \times p}$ designate the weight matrices for the output forecast errors. Established on the receding horizon approach, the control input $u_{k|k}$ to apply to $x(k + 1)$ & $y(k)$ at time k is computed from u_k which is the solution for the Kalman filter-based minimax criterions. The Kalman filter can be derived for the mini-max performance criterion given by $\min\limits_{K_k} \max\limits_{Q_k \& R_k} E\left[(x_k - \hat{x}_{k|k-1})(x_k - \hat{x}_{k|k-1})^T\right]$. At time k, the future time k + 1 and constraints are forced that the process inputs and outputs continue contained by the following lower and upper boundaries:

$$y_{min} \leq \hat{y}_{k+1|k} \leq y_{max}; \quad \Delta u_{min} \leq \Delta u_{k+1|k} \leq \Delta u_{max}; \quad u_{min} \leq u_{k+1|k} \leq u_{max}$$

From the error dynamics,

$$\bar{x}_{k+1|k} = [A - L_k C]\bar{x}_{k|k-1} + [G - L_k]\begin{bmatrix} w_k \\ v_k \end{bmatrix} \tag{3}$$

where $\bar{x}_{k|k-1} = x_k - \hat{x}_{k|k-1}$ the covariance matrices at time k + 1 and k is satisfied and P_k is monotonic with respect to Q_k and R_k.

$$P_{k+1} = [A - L_k C]P_k[A - L_k C]^T + \begin{bmatrix} G & L_k \end{bmatrix}\begin{bmatrix} Q_o & 0 \\ 0 & R_o \end{bmatrix}\begin{bmatrix} G^T \\ L_k^T \end{bmatrix} \tag{4}$$

The above equation is minimized for the solution of the following Riccati equation:

$$P_{k+1} = -AP_k C^T \left(R_o + CP_k C^T\right)^{-1} CP_k A^T + A P_k A^T + GQ_o G^T \tag{5}$$

From the equation, L_k is chosen as

$$L_k = P_k C^T \left(R_o + CP_k C^T\right)^{-1} \tag{6}$$

$$K_k = AL_k \tag{7}$$

A real-time implementation of Kalman filter-based MPC algorithm was performed on any real-time process. The sampling rate of data acquisition and generation of control signal is chosen based on the process. The real-time experiment may be started with initial conditions of the system. Hence, this may be viewed as a test for robustness for unknown nonlinearities when using the planned MPC controller and the parameter updated by Kalman filter. The structure of the Kalman filter incorporated model predictive controller is shown in Fig. 1

The main steps of Kalman filter-based Model Predictive Control is as follows:

Step 1: Initial conditions are $\hat{x}_o = \bar{x}_o$ and $P_o = \bar{P}_o$ and collect the new data, y_k;

Step 2: Calculate L_k and state of the model using Kalman filter, $\hat{x}_{k|k}$;

Step 5: Optimize the manipulated variable using minimax algorithm, u_k;

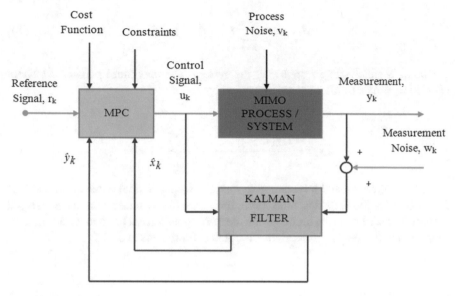

Fig. 1 Kalman filter-based model predictive controller

Step 6: Apply $u_{k|k}$ to the system to compute the Kalman gain and update the covariance matrix, P_{k+1} and calculate the next state of the model, $\hat{x}_{k+1|k}$;

Step 7: Increment k by 1, $k \leftarrow k + 1$;

Step 11: Go to rung 2.

3 Control System Performance

The steady-state accuracy in reference to the system is measured along integral square error criterion.

The integrated absolute error (IAE) is given by

$$J_{\text{IAE}} = \sum_{k=1}^{N} |y(k) - y_{\text{ss}}| T_0 \qquad (8)$$

The integrated square error (ISE) is given by

$$J_{\text{ISE}} = \sum_{k=1}^{N} (y(k) - y_{\text{ss}})^2 T_0 \qquad (9)$$

The integrated time absolute error (ITAE) is given by

$$J_{ITAE} = \sum_{k=1}^{N} T_k |y(k) - y_{ss}| T_0 \qquad (10)$$

where y_{ss} is steady-state response of the system. The maximum percent overshoot (PO) in reference to the system is specified by

$$J_{PO} = \frac{\begin{matrix} \max \\ 1 \le k \le N \end{matrix} (y(k) - y_{ss})}{y_{ss}} \cdot 100\% \qquad (11)$$

The rise time is the time required to increase from 0% final value up to 100% of final value for the very first time. Here, the system is assumed as a under damped system. The settling is the time to attain the stable position not beyond the acceptance limit around the stable status. Here, acceptance limit is taken as 2%.

4 Case Study

The problem for the Wood and Berry methanol–water distillation column (1973) was formulated in multi-loop and multivariable control system methods and is given by $G(s)$. For the above-mentioned process, the design settings are implemented in MATLAB and the achievement is evaluated with BLT tuning scheme, gain margin and phase margin technique and model predictive control technique by means of the simulation outcomes. The relative gain of the system is 2.01. The relative gain array in diagonal elements and off-diagonal elements is greater than one, so pairings y1–u1 and y2–u2 are adopted in this process. Niederlinski index = 0.498. The Niederlinski index is positive; the feedback loop is fairly strong system in the company of controller possibly will be stableness for the y_1–u_1, y_2–u_2 pairings.

$$\begin{bmatrix} Y_D(S) \\ Y_B(S) \end{bmatrix} = \begin{bmatrix} \frac{12.8e^{-s}}{16.7s+1} & \frac{-18.9e^{-3s}}{21s+1} \\ \frac{6.6e^{-7s}}{10.9s+1} & \frac{-19.4e^{-3s}}{14.4s+1} \end{bmatrix} \begin{bmatrix} R(S) \\ V(S) \end{bmatrix} + \begin{bmatrix} \frac{3.8}{14.9S+1}e^{-8S} \\ \frac{4.9}{13.2S+1}e^{-3S} \end{bmatrix} [D(S)]$$

The unit step input starts at time $t = 0$ s, and it is changed at 75 and 150 s as a target point for loop 1. The unit step input starts at time $t = 0$ s, and it is changed at 100 and 150 s as a target-trace for loop 2. The target point changes in loop 2 that affect loop 1 1 the planned technique, however, rapidly eliminates loop interactions. For assessment desire, two established controller design methods, the BLT method and PGM technique, are utilized to Wood and Berry distillation column model. The values in reference to controllers are given in Box I [8, 9].

Box I

$$G_c(S)_{\text{PGM-PI}} = \begin{bmatrix} 0.4867 + \frac{0.0881}{s} & 0 \\ 0 & -0.1567 - \frac{0.0304}{s} \end{bmatrix} \&$$

$$G_c(S)_{\text{BLT-PI}} = \begin{bmatrix} 0.375 + \frac{0.0452}{s} & 0 \\ 0 & -0.075 - \frac{0.0032}{s} \end{bmatrix}$$

Box II includes the planned scheme for initial system variables in terms of the discrete-time state space model and MPC parameters.

Box II

$$x(k+1) = \begin{bmatrix} 0.9419 & 0 & 0 & 0 \\ 0 & 0.9123 & 0 & 0 \\ 0 & 0 & 0.9535 & 0 \\ 0 & 0 & 0 & 0.9329 \end{bmatrix} x(k) + \begin{bmatrix} 0.9706 & 0 \\ 0.9555 & 0 \\ 0 & 0.9766 \\ 0 & 0.9661 \end{bmatrix} u(k)$$

$$y(k) = \begin{bmatrix} 0.7665 & 0 & -0.9 & 0 \\ 0 & 0.6055 & 0 & -1.347 \end{bmatrix} x(k)$$

Sampling Time $(T) = 1$ min; Prediction Scope $= 92$; Control Scope $= 1$;
   Matrix: $Q = \text{diag} (1, 1)$; $R = \text{diag} (62.35, 182)$;

5   Results and Discussion

Table 1 shows that the proposed Kalman filter-based model predictive controller has minimum integral absolute error, integral square error, and integral time absolute error for tracking; however, the disturbance rejection has minimum error by the phase and gain margin method. Because the Kalman filter-based model predictive controller abruptly takes action against the error variable, the corresponding controlled variable deviated initially to some extent and then settled down quickly to eliminate the disturbances. So, the controller has soft progress in MPC technique.

Table 2 demonstrates the transient characteristics of the Wood and Berry distillation column. The results are evidence for the efficacy of the projected fine-tuning method in terms of least rise and settling time in both tracking and disturbance rejection, but overshoot is higher in the disturbance rejection. The simulation results of the proposed method on the Wood and Berry distillation column are displayed in Figs. 2

Table 1 Transient response of the controllers for t (0–75) for Loop 1 and for t (0–100) for Loop 2 in set-point tracking and disturbance rejection

Plant output method		Set-point tracking			Disturbance rejection		
		Overshoot (%)	Rise time (s)	Settling time (s)	Overshoot (%)	Rise time (s)	Settling time (s)
Loop 1 (y1)	BLT	20.02	3.56	20.12	22.43	36.35	24.65
	PGM	18.2	3.44	18.98	19.22	32.68	21.44
	MPC	6.25	3.21	21.26	5.65	31.26	17.28
Loop 2 (y2)	BLT	123.46	4.35	38.15	120.37	7.25	37.18
	PGM	92.62	3.89	34.32	95.43	7.12	35.33
	MPC	140.51	3.32	30.16	22.43	36.35	24.65

Table 2 Performance of the both controller for tracking and disturbance rejection

Plant output method		Set-point tracking			Disturbance rejection		
		IAE	ISE	ITAE	IAE	ISE	ITAE
Loop 1 (y1)	BLT	308.8877	55.5193	5370.3	308.8456	55.5195	5190.8
	PGM	312.3148	52.992	5191.6	312.1887	52.9937	5368.8
	MPC	301.9963	45.7998	4436.5	302.1445	45.8003	5384.2
Loop 2 (y2)	BLT	733.6596	163.5508	14289	722.9402	131.5481	14313
	PGM	720.6503	131.6174	12065	651.5149	117.0268	12063
	MPC	651.5232	117.0243	13117	733.2339	163.5395	13114

Fig. 2 Output response of Loop 1 for set-point tracking

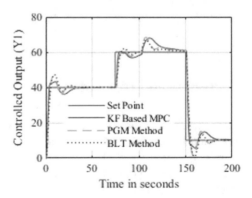

and 3 as the set point tracking performance and Figs. 4 and 5 for the disturbance rejection. From observation of the results, it is achieved that the planned technique provides the superior tracking performance and transient characteristics.

Fig. 3 Output response of
Loop 2 for set-point tracking

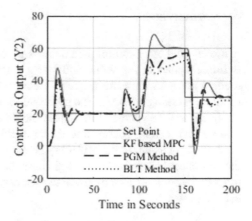

Fig. 4 Output response of
Loop 1 for disturbance
rejection

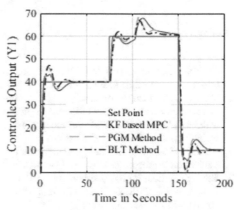

Fig. 5 Output response of
Loop 2 for disturbance
rejection

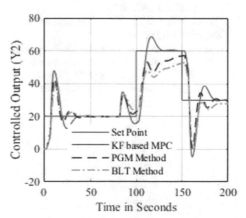

6 Conclusion

A model predictive control parameter is modified by the Kalman algorithm, which is implemented in the Wood and Berry distillation column model. The proposed method gives better performance in tracking and disturbance rejection compared to multi-loop tuning methods. The peak overshoot is higher than the multi-loop tuning methods, so it is necessary to adjust the constraints on the manipulated variables closely. This effort protracted to the most advantageous and stability of the model predictive controller to the MIMO systems/process and furthers the framework with the centralized controller of the tight constraints on the manipulated variables. This proposed technique can be carried out for obtaining the desired output in any square MIMO systems.

References

1. Vu TNL, Lee M (2010) Multi-loop PI controller design based on the direct synthesis for interacting multi-time delay processes. ISA Trans 49(1):79–86
2. Vrančić D, Strmčnik S, Kocijan J, de Moura Oliveira PB (2010) Improving disturbance rejection of PID controllers by means of the magnitude optimum method. ISA Trans 49(1):47–56
3. Mudi RK, Dey C (2011) Performance improvement of PI controllers through dynamic set-point weighting. ISA Trans 50(2):220–230
4. Vu TNL, Lee M (2010) Independent design of multi-loop PI/PID controllers for interacting multivariable processes. J Process Control 20(8):922–933
5. Tan KK, Ferdous R (2003) Relay-enhanced multi-loop PI controllers. ISA Trans 42(2):273–277
6. Khandelwal S, Detroja KP (2017) Simplified decoupling based control for processes having complex EOTF dynamics. In: 2017 IEEE international conference on industrial technology (ICIT), pp 872–877. IEEE
7. Qin SJ, Badgwell TA (2003) A survey of industrial model predictive control technology. Control Eng Pract 11(7):733–764
8. Chien IL, Huang HP, Yang JC (1999) A simple multiloop tuning method for PID controllers with no proportional kick. Ind Eng Chem Res 38(4):1456–1468
9. Ho WK, Hang CC, Cao LS (1995) Tuning of PID controllers based on gain and phase margin specifications. Automatica 31(3):497–502

On the Use of Region Convolutional Neural Network for Object Detection

M. Sushma Sri, B. Rajendra Naik, K. Jayasankar, B. Ravi,
and P. Praveen Kumar

Abstract In the recent generations, object detection (OD) has played an important
role ranging from navigation of autonomous vehicles to video data compression. In
the area of OD, deep learning (DL) has played a vital role in detecting the objects
automatically. Breakthroughs have been achieved in the field of object detection
with the simultaneous advancements in the field of DL. In this article, the author
proposes a region-based neural network for detecting desired object in which it
overcomes the computational complexity acquired by convolutional neural networks
and provides a better accuracy. For training larger datasets and reusability, the deep
learning principles are also proposed.

Keywords Object detection · Deep learning · Neural networks · Deep neural
networks · Convolutional neural networks · Region convolutional neural networks

1 Introduction

In the recent generations, the field of deep learning and neural network techniques
are emerging as an active area of research as these techniques find applications in

M. Sushma Sri (✉) · B. Rajendra Naik
Osmania University College of Engineering, Hyderabad, India
e-mail: hiremath.sushmasri@gmail.com

B. Rajendra Naik
e-mail: rajendranaikb@gmail.com

K. Jayasankar
Mahatma Gandhi Institute of Technology, Hyderabad, India
e-mail: kottareddyjs@gmail.com

B. Ravi · P. Praveen Kumar
KL University, Hyderabad, India
e-mail: raviou2015@klh.edu.in

P. Praveen Kumar
e-mail: prawinpoola@klh.edu.in

© The Author(s), under exclusive license to Springer Nature Singapore Pte Ltd. 2021   315
K. A. Reddy et al. (eds.), *Data Engineering and Communication Technology*,
Lecture Notes on Data Engineering and Communications Technologies 63,
https://doi.org/10.1007/978-981-16-0081-4_31

autonomous vehicles, image recognition and object detection. This object detection is a popular technique as it is useful to recognize or identify the objects of interests [1]. The object detection finds many applications in image classification, face recognition and autonomous vehicle. Object detection is used to determine the object, which is located in a given image at a particular region and popularly known as object localization technique, whereas object classification is used for classification purpose to identify the object class which it belongs to. The conventional object detection models are classified into two categories: classification and feature extraction. Classification is required to differentiate the object compared to other object, and feature extraction is used to extract the features from the registered image. The provision of high and multi-level feature representations and the applications in correspondence to computer vision tasks such as image segmentation, edge detection and object detection is widely used. The evolution of this OD is classified into two types [2]. The first type is based on the features, and the second stage is based on "deep neural networks (DNN)." Nowadays, both academia and industry people are interested in applying deep learning techniques in their fields. The DNNs [3] is a different approach for introducing the features from the object based on "convolutional neural networks (CNNs)" and "region convolutional neural networks (R-CNN)." This R-CNN accomplishes more accuracy in object detection applications. Most important aspect and challenging tasks in the area of computer vision [4] are image segmentation, image restoration and object detection, of which OD plays a dominant role.

2 Related Work

Most conventional approaches have been implemented using unsupervised learning methods which use deep learning to find patterns through labeled and unlabeled training data sets [5]. In this regard, this method is used to find the object with it simple features like color, edges, texture and gray scale. And, also it includes in finding the complex features like focusing, blurring and background efforts [6]. This literature works offer a various working method in pixel variations, region-based [5] and graph-based methods [7]. There are numerous advantages of unsupervised learning which include simplicity, reliability and gives better performance in finding the objects. The standard method like deep learning, especially the CNN models, has been imposed in extracting or detecting the features from the object automatically. The performance of image classification [8] has first evolved by CNN models. In salient object detection [3] as the author concluded that training data sets models are implemented by only CNNs which collects four layer. But here in this article, the author implemented R-CNN in which the computational cost is greatly reduced.

3 Deep Neural Networks

Neural networks are used to train the data which allows the patterns and predicts the outputs exactly like actual data [9]. It is a combination of software and hardware that is designed for performing a specific task [7]. In this regard, deep learning is a subgroup of "machine learning (ML)" which is a collective of algorithms that train larger amount of unlabeled data, and it consists of at least three layers. In general, there are three techniques involved in deep learning, which are "artificial neural networks (ANN)," "convolutional neural networks (CNN)" and "recurrent neural networks (RNN)." Mostly, the problems stated of data which is present in the form of numeric are solved or carried out by ANN. If the data is present in the form of images are solved by CNN and incase if the input data consists of time series data then R-CNN is used. In this article, for object detection, the author used both convolutional and "region convolutional neural networks (R-CNN)" [8] for better efficiency and improved accuracy of data. The major role of deep neural network is deep and wide neural network in which deep means a greater number of hidden layers and wide denotes many input/ hidden nodes [10]. As shown in Fig. 1, the raw data is converted into low-level, mid-level and high-level features. As deeper inside the feature of an objects are extracted exactly. İn these article, $32 \times 32 \times 3$ hidden layers existed.

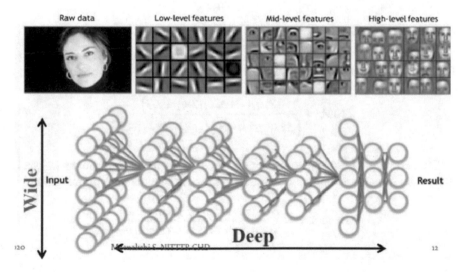

Fig. 1 Deep neural networks

3.1 Evaluation Metric for Object Detection

İn this article, we used accuracy and structural similarity index (SSIM) parameters as evaluation metrics for detecting the desired object. The SSIM is an evaluation metric which is used to find image quality as the name suggests for measuring similarities between images. This is a metric used to improve on conventional metrics like peak signal-to-noise ratio and mean square error. Structural similarity index term is computed with the help of three parameters luminance (l), the contrast (c) and the structural (s) components of an image

$$\text{SSIM}(i, j) = [l(i, j)]^{\alpha}[c(i, j)]^{\beta}[s(i, j)]^{\gamma} \tag{1}$$

In the Eq. (1) $l(i, j)$, $c(i, j)$ and $s(i, j)$ are calculated with help of $\mu_i, \mu_j, \sigma_i \sigma_j and \sigma_{ij}$: which are local means, standard deviation and cross-covariance of MR images.

$$l(i, j) = \frac{2\mu_i\mu_j + C_1}{\mu_i^2 + \mu_j^2 + C_1}$$

$$c(i, j) = \frac{2\sigma_i\sigma_j + C_2}{\sigma_i^2 + \sigma_j^2 + C_2}$$

$$s(i, j) = \frac{\sigma_{ij} + C_3}{\sigma_i\sigma_j + C_3}$$

In SSIM equation $\alpha = \beta = \gamma = 1$, these are default for exponents, and $C_3 = C_2/2$ (default selection of C_3) then the structural similarity index simplifies as

$$\text{SSIM} = \frac{(2\mu_i\mu_j + C_1)(2\sigma_{ij} + C_2)}{(\mu_i^2 + \mu_j^2 + C_1)(\sigma_i^2 + \sigma_j^2 + C_2)} \tag{2}$$

The accuracy is used to measure, predict or estimate the object from an image. İt is the ratio between number of true predictions to the total number of predictions.

$$\text{Accuracy} = \frac{\text{Number of true predictions}}{\text{Total number of predictions}} \tag{3}$$

3.2 Convolutional Neural Networks

A convolutional neural network plays a dominant role in neural networks which extract features of an object. The image classification and detection problems are solved by ANNs in which sudden change of object size may increase so this will be overcome by CNNs. It implements some challenges like using a large amount of data for training, processing a huge power and developing rapid change in applied field. In this article, the CNNs are applied to deep learning field for processing training data set.

4 Proposed Method

Object detection uses R-CNN in which "convolutional neural networks (CNN)" are used for classifying input image regions and R-CNNs are used to detect the regions which are likely to be detected so that the computational cost is reduced with improved accuracy. In this article, R-CNN object detector is used to detect a particular thing. The main asset of training images is to decrease the time. Figure 2 shows the block diagram for R-CNN model architecture for object detection.

4.1 Region Convolutional Neural Networks (R-CNN)

The proposed model is shown in Fig. 3 in which R-CNN consists of three sections, namely region extractor, feature extractor and classifier [11]. The region extractor is

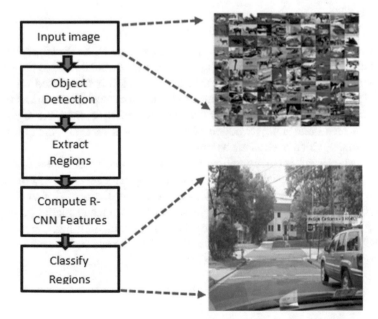

Fig. 2 R-CNN model architecture for object detection

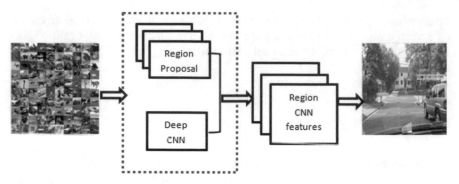

Fig. 3 Proposed R-CNN Architecture

used for regioning the input image which is extracted for creating some boundaries. In the feature extractor the features are extracted by deep convolutional neural network for computations [12]. The classifier as name suggests is used for classifying the regions based on R-CNN. The R-CNN architecture shown in Fig. 6 proposes the input data set image which is given to deep CNN and region proposal, and then, the convolutional neural network features are extracted based on region which is the proposed method in this article, and the final output image is a detected object.

5 Results

To evaluate the proposed neural networks, we considered CIFAR-10 images of 32 × 32 RGB data set shown in Fig. 4. Figure 5 shows the resized image in which the input layers are trained by CNN and then the validation is completed for training images. The validation is required for measuring the accuracy of network. After that, a task is chosen as stop sign detector which is shown in Fig. 6; then, the region CNN is applied which creates object boundaries and object detection which consists of 99.5% accuracy which is obtained as shown in Fig. 7.

Table 1 shows the accuracy and similarity index metrics obtained from object detector image. For complete validation, the training data sets are used to measure accuracy of the network which computes based on predictions. It also shows the comparative study based on accuracy in which our proposed model gives better accuracy compared to the other images like single shot multibox detector (SSD 300& 512) and ION image [3]. And also in this article, the author measured structural similarity index (SSIM) for detected image.

Fig. 4 Image input layer

Fig. 5 Resized image

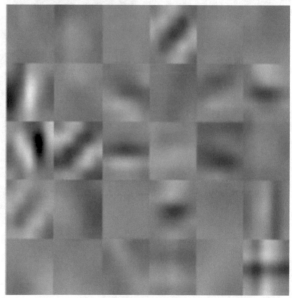

Fig. 6 Stop sign detector
image

Fig. 7 R-CNN detector
image

Table 1 Accuracy and SSIM
metric for object detector
image

Image	Accuracy (%)	SSIM
ION [3]	79.2	–
SSD300 [3]	79.6	–
SSD512 [3]	81.6	–
Proposed method	99.5	99.9%

6   Conclusion

The object detection plays a vital role in ranging from the navigation of autonomous vehicles to video data compression and biomedical applications [13–15]. In this regard, the author proposes deep learning technology for detecting the objects. The data set is trained on different object detection models, but the proposed method which uses region convolutional neural networks based on deep learning techniques is used to detect the desired object with 99.5% accuracy, and the structural similarity index is 99.9%.

Acknowledgements This research work is carried out with support Department of ECE, Osmania University, Hyderabad, and Mahatma Gandhi Institute of Technology, Hyderabad.

References

1. Girshick R, Donahue J, Darrell T, Malik J (2014) Rich feature hierarchies for accurate object detection and semantic segmentation. In: Proceedings of the IEEE conference on computer vision and pattern recognition, pp 580–587 (2014)
2. Ren S, He K, Girshick R, Faster R-CNN (2015) Towards real-time object detection with region proposal networks. IEEE Trans Pattern Anal Mach Intell 39(6):1137
3. Zhao ZQ, Zheng P, Xu ST, Wu X (2019) Object detection with deep learning: A review. IEEE Trans Neural Netw Learning Syst 30(11):3212–3232
4. Li C, Yuan Y, Cai W, Xia Y, Dagan Feng D (2015) Robust saliency detection via regularized random walks ranking. In: Proceedings of the IEEE conference on computer vision and pattern recognition, pp 2710–2717
5. Ravi B, Rajendra NB (2019) Rician noise reduction using dual tree-complex wavelet transform and self similarity. Int J Imag Robotics 19(3)
6. Kruthiventi SS, Gudisa V, Dholakiya JH, Venkatesh Babu R (2016) Saliency unified: a deep architecture for simultaneous eye fixation prediction and salient object segmentation. In: Proceedings of the IEEE conference on computer vision and pattern recognition, pp 5781–5790
7. Sun X, Wu P, Hoi SC (2018) Face detection using deep learning: an improved faster RCNN approach. Neurocomputing 299:42–50
8. Luo Z, Mishra A, Achkar A, Eichel J, Li S, Jodoin PM (2017) Non-local deep features for salient object detection. In: Proceedings of the IEEE conference on computer vision and pattern recognition, pp 6609–6617
9. Girshick R, Donahue J, Darrell T, Malik J (2014) Rich feature hierarchies for accurate object detection and semantic segmentation. In: Proceedings of the IEEE conference on computer vision and pattern recognition, pp. 580–587
10. He K, Zhang X, Ren S, Sun J (2015) Spatial pyramid pooling in deep convolutional networks for visual recognition. IEEE Trans Pattern Anal Mach Intell 37(9):1904–1916
11. Dai J, Li Y, He K, Sun J (2016) R-fcn: object detection via region-based fully convolutional networks. In: Advances in neural information processing systems, pp 379–387
12. Kumaresan Y, Kim H, Pak Y, Poola PK, Lee R, Lim N, Ko HC, Jung GY, Dahiya R (2020) Omnidirectional Stretchable inorganic-material-based electronics with enhanced performance. Adv Electron Mater 2000058
13. Poola PK, John R (2017) Label-free nanoscale characterization of red blood cell structure and dynamics using single-shot transport of intensity equation. J Biomed Opt 22(10):106001

14. Poola PK, Afzal MI, Yoo Y, Kim KH, Chung E (2019) Light sheet microscopy for histopathology applications. Biomed Eng lett 1–13
15. Poola PK, Jayaraman V, Chaithanya K, Rao D, John R (2018) Quantitative label-free technique for morphological evaluation of human sperm—a promising tool in semen evaluation. OSA Continuum 1(4):1215–1225
16. Tu WC, He S, Yang Q. Chien SY (2016) Real-time salient object detection with a minimum spanning tree. In: Proceedings of the IEEE conference on computer vision and pattern recognition, pp 2334–2342

An Improved Impulse Noise Removal VLSI Architecture Using DTBDM Method

Sresta Valasa, D. R. Ramji, Jitesh Shinde, and Mahesh Mudavath

Abstract Images are vulnerable to anomalies called as distortion or corruption during the transmission of signals. When the impulse noise encounters the image quality gets disturbed, to reconstruct the pixel information with reduced complexity, we hypothesize an efficacious VLSI architecture to drastically reduce impulse noise in pictures leveraging edge preserving filter. The method is compared with two mask sizes in terms of performance evaluation. To take advantage of detailed identification including certain PSNR generalizations and picture flexibility, the proposed framework will prove to be more effective than the conventional poor complexity paradigms. Throughout the low noise concentrations, the edge conserving median filter spotlights stellar performance and indeed in high noise concentrations through high window sizes. These effects are demonstrated through the simulation results. The comparisons are framed between area and power consumptions of 3*3 and 5*5 window sizes concluding which window size gives better performance in terms of those constraints.

Keywords Noise removal · Window size · Edge conserving median filter · DTBDM · PSNR · Power consumption

S. Valasa (✉) · D. R. Ramji · J. Shinde · M. Mudavath
Department of Electronics and Communication Engineering, Vaagdevi College of Engineering, Warangal, Telangana, India
e-mail: shreshtavalasa@gmail.com

D. R. Ramji
e-mail: ramjidr@gmail.com

J. Shinde
e-mail: shindejitesh.vaagdevi.coe.ece@gmail.com

M. Mudavath
e-mail: mahichauhan@gmail.com

© The Author(s), under exclusive license to Springer Nature Singapore Pte Ltd. 2021   325
K. A. Reddy et al. (eds.), *Data Engineering and Communication Technology*,
Lecture Notes on Data Engineering and Communications Technologies 63,
https://doi.org/10.1007/978-981-16-0081-4_32

Table 1 Comparison of various techniques

Technique	Complexity	Advantage	Disadvantage
Median filter	Low	Simple	For fixed impulse noise
Adaptive Center Weighted Median	Low	Eliminates both noise	Reconstructed image is blur
Low Complexity Noise Removal	Low	Only little logical elements are used	Reconstructed image is blur
Adaptive Median Filter	Low	Better where high-speed processing is needed	Reconstructed image is blur
Alpha Trimmed Mean	High	Denoised image accuracy is a good development	Requires full frame buffer
Differential Rank Impulse Detector	High	Denoised image accuracy is a great experience	Requires four iteration time
Rank Ordered Relative Difference	High	Strong performance	7*7 mask size is used

1 Introduction

Through the process of collection, routing, preservation, and reconstruction an image gets obscured with noise. Image files can get corrupted when downloaded from devices or when imported. Here we use salt-and-pepper noise which is a fixed value impulse noise out of which is the main classification of two types of impulse noise. There are many ways to help reduce salt-and-pepper noise, and a few of them are doing well [1–5]. During this paper, we take an interest only in expelling the random evaluated impulse noise from the corrupted image. Several forms of denouncing pictures are proposed to hold off the repression of impulse noise. Median quality filter or its modifications [8] are used by some of them. Furthermore, because both noisy and noise-free pixels are altered, these frameworks will blur the picture [7, 13]. Median filter reliability is assessed depending on distinct window sizes, and the influence of window size change is investigated to suppress salt-and-pepper noise out from the gray image in this paper.

Different methodologies for removing impulse noise from images are proposed in [6, 10, 11, 18]. Denoising analysis techniques have been graded as less complicated and more complicated methodology based on their sophistication. For easy analysis, all the previously existing methods and their respective advantages and disadvantages are listed in Table 1.

2 Proposed Method

A random assessed salt-and-pepper noise with uniform distributivity is chosen for study in this paper. A 3*3 mask and a 5*5 mask are undertaken here for image

Fig. 1 A mask centered on $p_{i,j}$

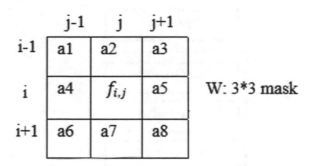

denoising. Let us take into consideration that at a coordinate (i, j) the denoised pixel is situated and symbolized as P_{ij} and let f_{ij} be its luminance value as shown in Fig. 1. We can divide the eight-pixel values into two sets focusing upon the input sequential manner of image denoising development phase, namely given as W_{TopHalf} and $W_{\text{BottomHalf}}$. These are presented as

$$W_{\text{TopHalf}} = \{a1, a2, a3, a4\} \tag{1}$$

$$W_{\text{BottomHalf}} = \{a5, a6, a7, a8\} \tag{2}$$

Decision tree-based impulse locator and edge conserving image filtering channel are the two main components of the DTBDM technique. By making use of a decision tree, the noisy pixels are determined by the detector alongside with the correspondence among center pixel and its adjacent pixels [20]. The fundamental tenet of the DTBDM is shown in Fig. 2. The decision tree is composed of three interconnect decisions of these modules.

Similarly, a 5*5 window mask module is designed comprising of a complete set of 24 pixels where we can divide it into two equal sets, namely W_{TopHalf} and $W_{\text{BottomHalf}}$. They can be presented as:

$$W_{\text{TopHalf}} = \{a1, a2, a3, a4, a5, a6, a7, a8, a9, a10, a11, a12\} \tag{3}$$

$$W_{\text{BottomHalf}} = \{a13, a14, a15, a16, a17, a18, a19, a20, a21, a22, a23, a24\} \tag{4}$$

3 Implementation

To confirm the attributes, nature, and quality of denoised pictures of different denoising calculations, an assortment of recreations or simulations are done on the three notables 512*512 8-bit grayscale test pictures: Lena, Boat, Goldhill. As for

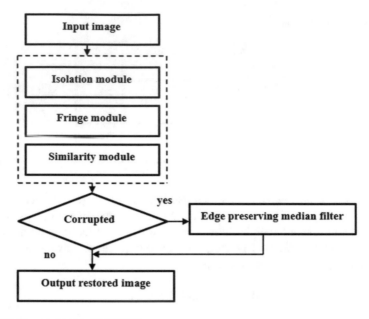

Fig. 2 Fundamental tenet of DTBDM

the implementation since we are using a decision-based algorithm, we can write the equations for the four decisions in this way [14]. The decision 1 and decision 2 apply to the isolation module given as Eqs. (5) and (6), decision 3 to fringe module given as Eq. (7), and decision 4 to similarity module written as Eq. (8) [14]. The threshold values for isolation module in Eq. (5), Eq. (6), and fringe module in Eq. (7) and for similarity module in Eq. (8) are predetermined values and are indeed set as 20, 25, 40, 80, 15, 60. It is difficult to infer an ideal limit through a logical plan. The fixed estimations of limits make our calculation straightforward.

$$Decision - 1 = \begin{cases} true, if \left(TopHalf_{diff} \geq Th_{IM_a}\right) \\ or \left(BottomHalf_{diff} \geq Th_{IM_a}\right) \\ false, otherwise \end{cases} \quad (5)$$

$$Decision - 2 = \begin{cases} true, if (IM_{Tophalf} = true) \\ or (IM_{BottomHalf} = false) \\ false, otherwise \end{cases} \quad (6)$$

$$Decision - 3 = \begin{cases} false, if (FM\_E_1) or (FM\_E_2) \\ or (FM\_E_3) or (FM\_E_4) \\ true, otherwise \end{cases} \quad (7)$$

$$Decision - 4 = \begin{cases} true, if \left(f_{i,j} \geq N_{max}\right) or (f_{i,j} \leq N_{min}) \\ false, otherwise \end{cases} \quad (8)$$

The same applies to the 5*5 window masking but the number of the directional differences for 5*5 is twelve, whereas, for the 3*3 window, there are only eight directional differences while calculating the edge to edge distance. We check the testability by introducing the noise densities varying from the utmost low level to a high level. We have taken values altering from 0.02 percentile to the highest with uniform supplementary increments of 2 percentile. This is tested in MATLAB with different images. As a result, we can track the recreated images including the source image for the denoising purpose. The peak signal-to-noise ratio (PSNR) is utilized to represent the observational nature of the reconstructed pictures for different methodologies. The PSNR value can be calculated using the formulae as

$$PSNR = 10\log_{10}\left(\frac{255^2}{MSE}\right) \tag{9}$$

The mean square error (MSE) is given by

$$MSE = \frac{SUM(SUM(SQUAREDERRORIMAGE))}{(R \times C)} \tag{10}$$

where R = row and C = column.

Contrasting and those trial tests, the quantitative or observational characteristics of the proposed methodology are in every case considerably higher than ones of lower multifaceted nature techniques in low commotion proportion and almost equitable to other higher intricacy strategies.

4 Results and Discussion

As already stated in the above sections, we have implemented the proposed DTBDM algorithm in both the MATLAB and Xilinx. So, here we will be discussing the results of the test images Lena, Boat, Goldhill.

(a) actual image   (b) corrupted image   (c) proposed 3*3 output   (d) proposed 5*5output

Fig. 3 Outcomes of implemented algorithm in recovery of 20% of corrupted image "Lena"

Figures 3, 4, and 5 show the outcomes of the implemented DTBDM algorithm in MATLAB. We display the restored images of different images using the proposed denoising approach to restore degraded photographs by 20% to investigate the visual appearance. From the above obtained results, we can clearly say that the restored image quality is quite appreciable for both. Deeply observing the outputs of the proposed algorithm of 3*3 and 5*5 window masks, we can say that the edges are maximum conserved or restored in 3*3 outputs than that of 5*5 outputs for all the test images as we can clearly experience some sort of disturbance in the 5*5 window outcomes near edges when zoomed in. To investigate the impact of window size on denoising image to eliminate or at least quantify the problem of introduced noise, these two window size values are enough. Table 2 shows the comparison of different methods PSNR values of different test images which clearly shows that our method creates pictures that are visually appealing.

Table 3 represents the comparison of different test images corrupted by different noise percentiles in terms of PSNR. The mean square error of the minimum differences in the noise assessment formulae is used for the quantitative evaluation. From the quantitative assessment of the outcomes of PSNR values of test images, we can presume that the noise effect decreases with the increasing window size. From Table 3, we can observe that from 60 percentile and greater than that of noise inculcation, the PSNR values of 5*5 window size are greater than that of 3*3 window size. This means that the recovered image quality is better than that of lower window size after

| (a) actual image | (b) corrupted image | (c) proposed 3*3 output | (d) proposed 5*5 output |

Fig. 4 Outcomes of implemented algorithm in recovery of 20% of corrupted image "Boat"

| (a) actual image | (b) corrupted image | (c) proposed 3*3 output | (d) proposed 5*5 output |

Fig. 5 Outcomes of implemented algorithm in recovery of 20% of corrupted image "Goldhill"

Table 2 Comparison of test images in PSNR (dB) corrupted by 10 and 20% impulses

Method		Comparison of images					
		10% corrupted impulses			20% corrupted impulses		
		Lena	Boat	Goldhill	Lena	Boat	Goldhill
Lower complex	Noisy	19.18	19.03	19.07	16.21	15.99	16.12
	Median [6]	32.53	29.76	30.30	30.95	28.48	29.30
	ACWM [8]	37.29	34.06	34.39	32.81	30.50	31.26
	MSM (3: T) [10]	33.65	32.46	32.54	29.30	28.35	28.56
	MSM (5: T) [10]	33.42	32.38	32.43	28.66	27.82	28.02
	MSM (7: T) [10]	31.42	30.69	30.70	25.59	24.97	27.76
	RVNP [16]	32.72	31.26	31.55	31.14	28.92	29.94
	AMF [17]	33.96	32.55	32.59	31.70	29.30	30.22
	NAVF [18]	34.08	32.58	32.72	31.91	29.43	30.74
	LCNR [19]	36.76	33.02	33.82	34.07	30.95	31.81
	DTBDM [15]	38.22	34.48	35.14	34.43	31.38	32.21
	PROPOSED (3:3)	44.26	41.27	42.05	40.85	38.19	39.04
	PROPOSED (5:5)	41.91	39.35	40.38	39.04	36.50	37.33
Higher complex	ATMBM [11]	36.82	33.26	33.76	32.23	30.13	30.90
	DRID [12]	36.84	34.20	34.33	32.86	30.87	31.40
	RORD-WMF [13]	36.23	31.70	33.07	34.87	30.57	31.76

Table 3 Comparison of test images in PSNR (dB) corrupted by different noise percentiles

Noise percentage	Comparison of images					
	Lena		Boat		Goldhill	
	3*3	5*5	3*3	5*5	3*3	5*5
5	49.23	46.36	45.46	43.53	45.70	46.21
10	44.26	41.91	41.27	39.35	42.05	40.38
20	40.85	39.04	38.19	36.50	39.04	37.33
30	38.15	36.74	35.88	34.52	36.80	35.39
40	35.55	34.81	33.78	32.91	34.60	33.71
50	33.07	32.98	31.79	31.31	32.67	32.12
60	30.85	31.11	29.80	29.91	30.77	30.73
70	28.92	29.49	28.17	28.56	29.06	29.34
80	27.34	28.07	26.81	27.51	27.65	28.14

a great increase in noise percentiles. In other words, we can come to a better discussion that in low noise conditions, a 3*3 window gives greater performance; at the same time with high noise conditions, the ability to eliminate noise in the corrupted images will be debased. With high noise concentrations, the stellar performance of edge conserving median filter with 5*5 window size is very clear. We can visualize a good quality compared to the lower window size if we take the basis of PSNR values to judge as shown in Table 3. From the above outcomes of Fig. 3, Fig. 4, Fig. 5, and Table3, we can finalize that our proposed DTBDM algorithm performs very well compared to other methods [2–4, 6, 9–11, 14].

4.1   Comparison of Device Utilization Summary, Area, and Power of 3*3 and 5*5 Window DTBDM Algorithm

Table 4 summarizes the device utilization of the developed architecture of 3*3 and 5*5 window size DTBDM after synthesis. For the 3*3 window size, out of available flip flops, the architecture uses 149. The number of look up tables (LUTs) available is 92,152, and the method uses only 2621(i.e., 2% available resources). The time complexity or the worst-case slack is 0.497nS. While for 5*5 window DTBDM algorithm among the 338 available bonded IOB's on the Xilinx FPGA, the architecture implemented uses 24% occupying only 84. Within the available flip flops pairs, the architecture uses 11. Among the 92,152 look up tables (LUTs) available

Table 4   Device utilization summary of 3*3 and 5*5 window DTBDM algorithm

Slice type	3*3 window			5*5 window		
	Used	Available	Utilization (%)	Used	Available	Utilization (%)
Number of registers	245	184,304	1	82	184,304	1
Number of LUT's	2621	92,152	2	370	92,152	1
Number of fully used LUT-FF pairs	149	2697	5	11	440	2
Number of bonded IOB's	84	338	24	84	338	24
Number of BUFG	1	16	6	1	16	6
Number of occupied slices	890	23,038	3	168	23,028	1

Table 5 Comparison of area occupied by two window sizes of implemented DTBDM algorithm

Window size	Occupied slices	LUT F/F pairs
3*3	890	2697
5*5	168	440

Table 6 Comparison table for power consumption at 50 MHz clock frequency

Window size	Quiescent power(mW)	Dynamic power(mW)	Total power(mW)
3*3	81	5	86
5*5	80	4	84

the implementation occupied 1% available resources. The worst-case slack or the amount of time taken to run the algorithm is 0.682nS.

Table 5 is observed that the 3*3 window and 5*5 window when implemented separately it occupies 890 and 168 slices respectively and the former occupies 2697 LUT F/F pairs and the later by 440. Hence, a greater area reduction is observed in the 5*5 window implemented DTBDM algorithm when compared to the 3*3 window DTBDM algorithm. Here we can observe that 5*5 window size uses less area as compared to 3*3 window size.

Table 6 summarizes the power reports for the proposed design. A slight increase in the power consumption of 2mW of the design is observed in the 3*3 window compared to the 5*5 window. Hence, we can come to a point that both window sizes are reliable to design a circuit to eliminate noise where we can observe complete less power consumption with both designs.

This is a specific research area where we have gone deep into the design to completely analyze the area and noise constraints pertaining to the frontend VLSI research. Most of the previous research regarding to this development have been limited to the PSNR. So, to get deep into the analysis 3*3 window size along with 5*5 window size DTBDM algorithms are developed. This makes the proposed research work different from others.

So from all the aspects we derived, we can make a valid justification that from looking in terms of both area and power consumption we can see that the 5*5 window is more reliable compared to the 3*3 window as shown in Tables 5 and 6. There are certain industrial applications where image quality matters the most and the area and power constraints do not matter. So, in keeping point of view of those where image quality only matters the most, an algorithm with 3*3 window size can be used and where less power consumption and area reduction are a matter of higher intent then 5*5 window size can be used. The 5*5 window size algorithm is more evolved and has provided improved performance in terms of area and power constraints relative to the 3*3 window size algorithm in accordance with the area and power constraints viewpoint. Also, at higher noise levels the 5*5 window gives better PSNR performance; i.e., the ability to eliminate noise is great compared to the 3*3 window. So, depending upon the convenience of the application needed either the window sizes can be used.

5 Conclusion

In this paper, we presented a highly qualified decision-based edge conserving filter for noise reduction and recovery of pictures. From these experimental outcomes with the noise concentrations uniformly varying from 5–80%, we can experience a stable and sustainable efficacy through the proposed algorithm. While the edges are conserved up to 70% in both, the 3*3 and 5*5 window sizes active noise elimination can still be visualized up to 80%. We have finally developed an algorithm where the area and power consumption are less in 5*5 window compared to the 3*3 window, and also, we can see great PSNR values at higher noise levels in 5*5 window size compared to the 3*3 window. This algorithm can be used for real-time applications like character detection, screening, diagnostic detections where it is important to remove noises before these subsequent processes. DTBDM technique can be additionally utilized in the future for video handling in TVs, mobiles, PCs, gaming with high designs, and so on.

References

1. Hwang H, Haddad RA (1995) Adaptive median filters: new algorithms and results. IEEE Trans Image Process 4(4):499–502
2. Zhang S, Karim MA (2002) A new impulse detector for switching median filters. IEEE Signal Process Lett 9(11):360–363
3. Chan RH, Ho CW, Nikolova M (2005) Salt-and-pepper noise removal by median-type noise detectors and detail-preserving regularization. IEEE Trans Image Process 14(10):1479–1485
4. Ng PE, Ma KK (2006) A switching median filter with boundary discriminative noise detection for extremely corrupted images. IEEE Trans Image Process 15(6):1506–1516
5. Chen PY, Lien CY (2008) An efficient edge-preserving algorithm for removal of salt-and-pepper noise. IEEE Signal Process Lett 15:833–836
6. Nodes T, Gallagher N (1982) Median filters: Some modifications and their properties. IEEE Trans Acoust Speech Signal Process 30(5):739–746
7. Ko SJ, Lee YH (1991) Center weighted median filters and their applications to image enhancement. IEEE Trans Circuits Syst 38(9):984–993
8. Chen T, Wu HR (2001) Adaptive impulse detection using center-weighted median filters. IEEE Signal Process Lett 8(1):1–3
9. Chen T, Wu HR (2001) Space variant median filters for the restoration of impulse noise corrupted images. IEEE Trans Circuits Syst II: Analog Digital Sign Process 48(8):784–789
10. Luo W (2006) An efficient detail-preserving approach for removing impulse noise in images. IEEE Signal Process Lett 13(7):413–416
11. Aizenberg I, Butakoff C (2004) Effective impulse detector based on rank-order criteria. IEEE Signal Process Lett 11(3):363–366
12. Yu H, Zhao L, Wang H (2008) An efficient procedure for removing random-valued impulse noise in images. IEEE Signal Process Lett 15:922–925
13. Dong Y, Xu S (2007) A new directional weighted median filter for removal of random-valued impulse noise. IEEE Signal Process Lett 14(3):193–196
14. Lien CY, Huang CC, Chen PY, Lin YF (2012) An efficient denoising architecture for removal of impulse noise in images. IEEE Trans Comput 62(4):631–643
15. Hsia SC (2003) Parallel VLSI design for a real-time video-impulse noise-reduction processor. IEEE transactions on very large scale integration (VLSI) systems 11(4):651–658

16. Andreadis I, Louverdis G (2004) Real-time adaptive image impulse noise suppression. IEEE Trans Instrum Meas 53(3):798–806
17. Fischer V, Lukac R, Martin K (2005) Cost-effective video filtering solution for real-time vision systems. EURASIP J Adv Signal Process 2005(13): 568069
18. Mukherjee M, Maitra M (2015) An efficient FPGA based de-noising architecture for removal of high density impulse noise in images. In: 2015 IEEE International Conference on Research in Computational Intelligence and Communication Networks (ICRCICN), pp 262–266
19. Matsubara T, Moshnyaga VG, Hashimoto K (2010) A FPGA implementation of low-complexity noise removal. In: 2010 17th IEEE international conference on electronics, circuits and systems, pp 255–258
20. Thawong R, Sakha S, Klongdee W (2016) Detecting generalized salt and pepper noise image based on standard deviation. Eng Appl Sci Res 43:125–129

Performance Analysis of Vehicular Network Scenarios Using SUMO and NS2 Simulators

K. Raja Kumar, Nagarjuna Karyemsetty, and Badugu Samatha

Abstract The fatal cases of road traffic are continuously increasing; nearly 1.5 million people die from traffic accidents every year, according to WHO statistics. About 60 percent of accidents can be avoided by giving sufficient safety, alert, and warning messages. Vehicular networks are designed to increase safety, driving efficiency, make the driving experience more comfortable, and ensure fewer accidents or preferably zero accidents. In this research work, vehicular network with real-time road traffic and two scenarios have taken from (OSM) Java Open Street Map, generated real-time traffic by using Simulation for Urban Mobility (SUMO) tool and finally integrated with network Simulator (NS2) for sharing the safety, alerts and warning messages established and simulated. IEEE 802.11p standards are incorporated into the network. Furthermore, network performance indicators such as packet reception ratio, end-to-end delay, which place a vital role in timely communicating safety messages and throughput, are verified with real-time traffic situations.

Keywords IEEE 802.11p · NS2 Simulator · SUMO · Traffic Simulator · Safety Application · Vehicular Network

1 Introduction

Due to road traffic accidents, 1.35 million lives die every year and 93% of global fatalities occur in low- and middle-income countries even though 6% of global vehicles as per World Health Organization [1] report. And, also specified 29 year-olds die on the roads due to traffic injuries. India has 1% of the world's vehicles but 6% of a road traffic accident. Times of India (ToI) [2] announced, 1.5 lakh people die in road accidents in 2018, mainly over-speeding; WHO reports reveal that India occupies the

K. Raja Kumar · N. Karyemsetty (✉)
Department of Computer Science and Systems Engineering, Andhra University College of Engineering (A), Andra University, Visakhapatnam, India
e-mail: researchatau@gmail.com

B. Samatha
Department of CSE, Avanthi Institute of Engineering and Technology, Vizianagaram, India

© The Author(s), under exclusive license to Springer Nature Singapore Pte Ltd. 2021
K. A. Reddy et al. (eds.), *Data Engineering and Communication Technology*,
Lecture Notes on Data Engineering and Communications Technologies 63,
https://doi.org/10.1007/978-981-16-0081-4_33

first position in road accident deaths cases out of the 199 countries and also shows that India's contribution toward road accidents is 11%. Moreover, between 20 to 50 million suffer non-fatal dangerous situations.

Vehicular ad-hoc network (VANET) [3] has grown to be one of the essential research domains in wireless communications today. Road safety becomes a matter of grave concern with increasing traffic and road facilities not precisely commensurate with the needs and human errors. Human errors such as distracted driving, intoxication, unfamiliar territories, medical grounds, speed, and unruliness necessitate motivation, lack of proper education, physical and mental condition is leading to these accidents. However, the initiation of smart electronic devices for sensing like (GPS) global positioning system, digital maps embedded computers coupled with advancement in wireless communication, the engineers and technical experts have to deliberate and create a mechanism which can regulate and reduce the loss of road fatalities and property damage due to road accidents. VANET has emerged as an effective weapon in the war against road accidents. It caters to ITS's mater goals, which include improvement of safety on the road reduction of congestion, weather times, fuel consumption, collision avoidance, warning, and accident alerts [3]. In the vehicular network, every vehicle node is mounted with the on-board unit (OBU) and every road infrastructure unit contains a unit called roadside unit (RSU). In vehicular network, every OBU capable to communicate other nearby vehicle is known as vehicle to vehicle (V2V) and if OBU communicates to RSU is known as vehicle to infrastructure (V2I) communication to share the traffic signaling information, road status, and environmental condition, etc.

2 Literature Review

Dingil et al. [3] utilized advanced software packages, the SUMOPy, and OSMNx. These packages can be extended for complex junctions. Jakob Erdmann et al. [4] have proposed a modeling pedestrian behavior in SUMO. The authors extended the COLOMBO project by adding road network format, movement models, and routing tools to SUMO. Pedestrian models, including crosswalk and sidewalks, are also discussed. Results can be extended to VANET to include pedestrians in addition to vehicles and infrastructure simulation.

Dynamic peer-to-peer modeling in the vehicular ad-hoc network for data scavenging with the integration of SUMO and TracCI4MATLAB is implemented by Kit Guan Lim et al. [5]. Authors utilized the traffic control interface commands, properties of the vehicle to export data. Based on this data, the behavior of routing and re-routing protocols is computed and assessed.

Zhigang Xu at el. [6] implemented three connected vehicle application scenarios, including collision avoidance. The authors utilized the GPS position and velocities to test vehicles and recorded field test results. Scenarios are tested on DSRC and 4G LTE. The authors utilized an LTE CPE manufactured by Datang Mobile Co. Ltd., and a DSRC wave box developed by Genvict Co., Ltd., for the field test. Performance metrics such as round trip time, packet loss rate, and throughput of the network are measured in the collision avoidance scenario. Wegener at. el. [7] have focused on the general and adaptable framework using TraCI package to integrate road traffic and network for controlling the mobility. The authors utilized the SUMO and NS2 tools to implement the framework. The authors detailed the low-level command, syntax, data types, and procedures required to simulate. Run-time simulations of Traces, TraCI, and static are evaluated on the server.

3 Real-Time Traffic Generation

For generating real-time traffic data, Java Open Street Map (JOSM) editor [13] has been used to download real road digital maps to creating mobility-based traffic on this road map. Simulation of urban mobility (SUMO) [11] is adopted. This visualizes the simulation interface and is capable of showing the movement of all the vehicles so that the processing of data can do classification. SUMO facilitates traffic modeling and intermodal traffic systems, which consist of road vehicles, public transport, and pedestrians, etc. SUMO also handles complex tasks such as route findings, network setup protocols, security services, and visualization functions. Python library facilitates these functions. Implementers can search for required functionality and learn to use and apply to the real situation to save the lives of people. Software tools in SUMO provide analysis of road traffic, traffic management services, and simulation tools.

Components support the validation of automated real-time driving, including collecting the sensor's data and roadside units and environment conditions to strengthen the automation. The mobility of the vehicle in road traffic can be generated using SUMO with various python command-line tools.

4 Integration of SUMO and NS2

It is worth noting that network simulator 2 [12] (NS2) is an event-oriented network suitable for different protocols. The NS2 is an open-source network simulator to simulate the various scenarios and protocols, etc. Its main advantage includes open-source code and facility to modify mechanisms as per customer's requirements, extendibility, and stability. It can support up to 20,000 nodes for realistic studies. This simulation tool is amenable to C + + and OTCL applicable for mobility nodes in VANETs or MANETs.

The file generation in SUMO becomes the source file for NS2 simulation; the node pattern of mobility can be coded in TCL or processed by generating a trace file that may be used as a command-line compatible with TCL file [9]. Network animator in NS2 illustrates the network, from the mobility files generated by SUMO, which yields the log files. The mobility TCL file generated from SUMO has established a network

Table 1 VANET simulation parameters

Parameter	Value
Network simulator	NS 2.34
Traffic simulator	SUMO v0.22
Map model	OSM (Visakhapatnam city)
Routing protocol	AODV, DSDV, DSR
Transport protocol	UDP, TCP
Number of vehicles	10–100
Minimum speed	1 km/h
Maximum Speed	60 km/h
Propagation model	Two way ground, Nakagami
Simulation time	50,100,150,200 s
Modulation	BPSK, QPSK, 16 QAM, 64 QAM
Data rates	3–27 Mbps
Packet size	500,900 bytes

among the nodes, i.e., OBUs and RSUs, for IEEE 802.11p wave communication [15].

4.1 VANET Simulation Scenarios

The Network Simulator 2.34 (NS 2.34) [14] is an open-source network simulator to simulate the various scenarios and protocols, etc. Table 1 shows the simulation parameters considered for VANET scenarios [8] in NS 2.34. Two kinds of real-time traffic networks are considered, such as the highway scenario and four road junction scenario, as shown in Fig. 4. The complete conversion procedure in the above two scenarios including highway (Visakhapatnam to Viziayanagaram) scenario is shown in Fig. 2, and similarly, junction scenarios are shown in Fig. 3. Four roads (Siripuram junction, Visakhapatnam) scenario is shown in Fig. 1. Various representations of junction scenarios such as junction in Google map, open street map, SUMO, and NS2 are shown in Fig. 3. Various routing protocol [9, 10] such as AODV, DSR, and DSDV are analysed in a given scenarios.

Fig. 1 Real-time traffic generation (Junction)

Fig. 2 Real-time traffic generation (Highway)

Highway Network Generation

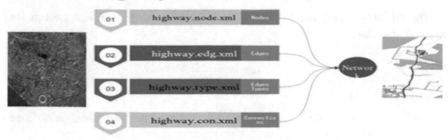

Highway Simulation Generation

Fig. 3 Simulation scenarios generation procedure

Fig. 4 Simulation scenarios

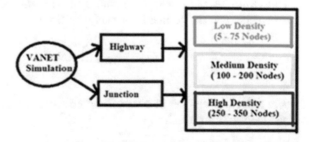

5 Performance Analysis and Results

For studying the overall outcomes of VANET in various applications, the (QoS) quality of services adopted the following three essential measures to test the performance of VANET which are:

packet delivery ratio (pdr)
Throughput and
End-to-end (E2E) delay.

The **packet delivery ratio (PDR)** is the ratio of the number of data packets sent to the destination.

$$pdr = \frac{\sum Pr}{\sum Ps}$$

where Ps is the overall number of packets sent from source to destination node and Pr is the overall number of packets received successfully at the destination.

$$AvgPDR = \sum_{i=1}^{n} \sum_{j=1}^{n} PDR_{ij}$$

where $i \neq j$.

where PDRij is the packet delivery ratio from source to destination.

The performance of the mean packet delivery ratio in various VANET scenarios is shown in Fig. 6.

It is observed that as simulation time increases, PDR reduces up to a particular time anf after that PDR maintains at constant. The reason for this is as vehicle moves on time speed of vehicle also increases. If speed increases, connectivity among vehicle nodes is low, hence weak connecting; thus, PDR reduces.

Throughput: A performance computes the amount of data transmitted from the source vehicle to the destination vehicle per time.

$$Thr = \frac{\sum_{i=1} Nib}{Ti}$$

where Nib is the number of bits transmitted from source to destination and Ti is the time taken for transmission.

End-to-End Delay: This is computed from the meantime taken by a data packet for reaching the destination. The performance of low throughput in various VANET scenarios is shown in Figure 5.

Fig. 5 Delay in VANET

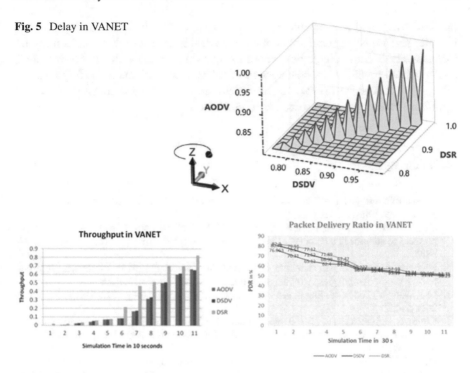

Fig. 6 PDF and throughput in VANET

$$E = \frac{\sum at - st}{\sum Nc}$$

where E is the end-to-end delay, *at* is the arrival time of the data packet, *st* is the sent time of the packet's data, and Nc is the total count. The performance of mean end-to-end delay in various VANET scenarios is shown in Figure 5.

6 Conclusion

Simulation scenarios for the vehicle network are set up in the city of Visakhapatnam with real-time traffic flow. Diverse density OBUs and RSUs networks will help to analyze the VANET's throughput performance, end-to-end delay, and packet reception ratio. The SUMO traffic simulator and network simulator (NS2) is used to introduce and test two scenarios, namely highways and junctions. Since VANET gives the time and essential safety applications to the driver, the most relevant point in vehicle communication is that vehicle communication is focused on predefined road patterns with variable movement. The key innovation of this paper is to incorporate the vital 802.11p standard parameters in NS2 including the sudden break, the rear-end crash, and lane change. Our findings show that the DSR routing protocol is better performed as simulations increase in real time and in dynamic conditions. Connectivity in a rapid complex environment as future research can be expanded, and advanced methods can be extended to safely communicate the safety message via encryption and intrusion detector system. Security of safety message is one of significant challenges to be solved as a future work.

References

1. Road traffic injuries. https://www.who.int/news-room/fact-sheets/detail/road-traffic-injuries
2. Road safety. https://morth.nic.in/
3. Taleb T, Sakhaee E, Jamalipour A, Hashimoto K, Kato N, Nemoto Y (2007) A stable routing protocol to support ITS services in VANET networks. IEEE Trans Veh Technol 56(6):3337–3347
4. Füßler H, Mauve M. Hartenstein H, Käsemann M, Vollmer D (2013) A comparison of best routing protocol strategies in vehicular ad-hoc networks. Technical Report
5. Li F, Wang Y (2007) Routing in vehicular ad hoc networks: A survey. IEEE Veh Technol Mag 2(2):12–22
6. Choffnes DF, Bustamante F (2015) An integrated mobility and traffic model for vehicular wireless networks. In: proceedings of the 2nd ACM international workshop on Vehicular ad hoc networks, ACM
7. Harri J, Fiore M, Fethi F (2016) VanetMobiSim: generating realistic mobility patterns for VANETs. In: Proceedings of the 3rd international workshop on Vehicular ad hoc networks. ACM
8. Samatha B, Raja Kumar K, Karyemsetty N (2017) Design and simulation of vehicular adhoc network using SUMO and NS2. Adv Wirel Mob Commun 10(5):1207–1219
9. Rama K, Lakshmi K, ManjuPriya S, Thilagam K, Jeevarathnam A (2013) Comparison of three routing algorithms based on greedy for effective packet forwarding in VANET. Int J Comput Technol Appl 3(1):146–151
10. Stoffers M, Riley G (2012) Comparing the ns-3 propagation models. In: 2012 IEEE 20th international symposium on modeling, analysis and simulation of computer and telecommunication systems, pp 61–67, IEEE
11. SUMO, Simulation of Urban Mobility. https://www.dlr.de/ts/en/desktopdefault.aspx/tabid-9883/16931_read-41000/
12. NS2, Network Simulator 2, https://www.nsnam.org
13. OpenStreetMap, https://www.openstreetmap.org/
14. Henderson T Radio Propagation Models, https://www.isi.edu/nsnam/ns/doc/node216.html
15. Rappaport TS (2006) Wireless communications, principles, and practice. Prentice-Hall

Novel Channel Estimation Technique for 5G MIMO Communication Systems

Tipparti Anil Kumar and Lokam Anjaneyulu

Abstract Internet of things (IoT) in health care is one of the major areas which ease the usage of many transmitters on board, leading to the usage of multiple-input multiple-output (MIMO) systems for better communications. Employing 5G MIMO systems suitable for IoT applications with quality of performance (QoP) is a real challenge. This paper proposes and analyzes a new channel estimation method based on training symbol for 5G MIMO wireless communication systems for IoT applications. An M-estimator is suggested for optimizing the proposed channel estimator. The proposed technique performance is evaluated based on the comparison of simulation results with least squares (LS) channel estimation with and without discrete Fourier transform (DFT). From simulation results, it is seen that the proposed method of channel estimation with proposed estimator closely approximates that of true channel estimation with and without DFT.

Keywords Channel estimation · Discrete fourier transform · Influence function · Least squares · M-estimator · Penalty function

1 Introduction

Rapid growing of mobile users and exponential growing demand of higher data rates force many practical challenges on the existing cellular networks and their developments to provide a high network capacity with extensive area coverage to meet customer demands of upcoming 5G networks [1–4]. The major disadvantages of the existing networks are low data rate, minimum quality of experience (QoE), low

T. Anil Kumar (✉)
Electronics and Communication Engineering Department, CMR Institute of Technology, Hyderabad, Telangana 501 401, India
e-mail: tvakumar2000@yahoo.co.in

L. Anjaneyulu
Electronics and Communication Engineering Department, National Institute of Technology, Warangal, Telangana 506 001, India
e-mail: anjan@nitw.ac.in

end-to-end performance, less indoor coverage, poor mobility performance, etc. Similarly, network operators face difficulties in terms of providing satisfactory services, e.g., high spectral efficiency, huge network capacity, large availability of spectrum, low latency and lower energy consumption. In order for 5G MIMO communication systems to work for both the identified demands, plans for spectral efficiency improvement, scheduling for channel information, coding and adaptive modulation are required. All these techniques need a CSI, i.e., accurate channel state information available at a transmitter end. An estimation of such CSI is crucial for high data throughput [5–8].

In [9], the first algorithm is OMP, i.e., orthogonal matching pursuit with lower complexity is used to identify the common support set followed by (LS) method, i.e., least square for obtaining the channel estimation, and it assumes perfect CSI measurement feedback from an user equipment (UE) to base station (BS) which may not be possible in practice, and alternate CSI measurement approaches need to be considered. In [10], a new channel estimation technique was proposed with enhanced Kalman filter which operates to reduce the noise levels and improves the channel conditions and quality of service [QoS] over wireless communication environments. In [11], channel estimation with minimum mean-squared-error (MMSE) criterion for orthogonal frequency division multiplexing (OFDM) systems was investigated. MMSE estimator was studied first which uses the correlation of a frequency response on different instants of time and frequency for a channel. This MMSE channel estimator may be a frequency domain filter using fast Fourier transform (FFT) followed by a time domain filters. Further, an estimator which is insensitive to a channel statistics was proposed and analyzed in [12]. A multi-user detector using M-estimator [13] was presented in [13, 14] for non-Gaussian flat-fading channels.

Hence, in this paper, an M-estimator-based channel estimation technique for 5G wireless communication channels is proposed and studied. The remaining paper is organized as follows: Section II refers to the system model. Coming to the section III, it describes the LS estimation and the proposed method. Finally, Section IV outlines some simulation-based results followed by conclusions and references.

2   System Model

One of the methods for estimation of channel is using training symbols, a0n0d 0f00or given N subcarriers, these symbols are represented by a diagonal matrix given in Eq. 1 (here all the subcarriers are orthogonal to each other (i.e., no ICI)).

$$
X = \begin{bmatrix} X(0) & 0 & \cdots & 0 \\ 0 & X(1) & & \vdots \\ \vdots & & \ddots & 0 \\ 0 & \cdots & 0 & X(N-1) \end{bmatrix} \tag{1}
$$

where $X(k)$ represents a pilot tone at the kth subscriber, zero mean and σ_x^2 gain (i.e., $H[k]$ for each subcarrier k). In a multi-user communications, users transmit their signals using same time and frequency slots. Thus, the received signal is the superimposed signal of all transmitting users, and the training signal on receiver side, $Y[k]$, is expressed by

$$
Y \triangleq \begin{bmatrix} Y[0] \\ Y[1] \\ \vdots \\ Y[N-1] \end{bmatrix} = \begin{bmatrix} X(0) & 0 & \cdots & 0 \\ 0 & X(1) & & \vdots \\ \vdots & & \ddots & 0 \\ 0 & \cdots & 0 & X(N-1) \end{bmatrix} \begin{bmatrix} H[0] \\ H[1] \\ \vdots \\ H[N-1] \end{bmatrix} + \begin{bmatrix} W[0] \\ W[1] \\ \vdots \\ W[N-1] \end{bmatrix}
$$
(2)

or

$$
Y = HX + W
$$
(3)

where channel vector H is mathematically expressed as $H = [H[0], H[1], \cdots, H[N-1]]^T$ and W, i.e., noise vector with zero mean and σ_w^2 variance is expressed by $W = [W[0], W[1], \cdots, W[N-1]]^T[1], \cdots, W[N-1]]^T$.

3 Least Squares Channel Estimation

In LS method, the following equation is used for channel estimation

$$
\hat{H}_{LS} = (X^H X)^{-1} X^H Y = X^{-1} Y
$$
(4)

where estimate of H is $\hat{H}$.

4 Proposed Channel Estimation

A new method M-estimator [5] is proposed for robustifying training symbol-based channel estimation for 5G wireless communications. The proposed estimator penalty and influence functions are given in Eqs. (5, 6 and 7), respectively (shown in Fig. 1)

$$
\rho_{\text{PROPOSED}}(x) = \begin{cases} \frac{x^2}{2} & \text{for } |x| \le a \\ a^2 - a|x| & \text{for } a < |x| \le b \\ \frac{-ab}{2} \exp\left(1 - \frac{x^2}{b^2}\right) + d & \text{for } |x| > b \end{cases}
$$
(5)

Fig. 1 a Penalty function,
b influence function and
c weight functions of the
proposed estimator

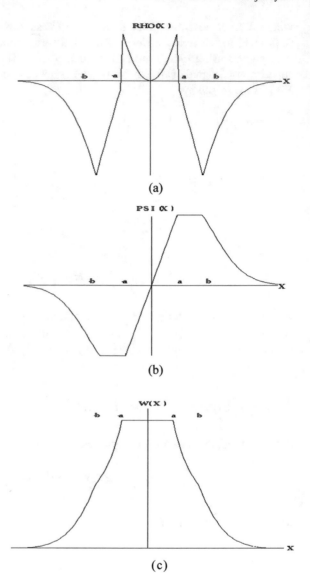

(a)

(b)

(c)

$$\psi_{\text{PROPOSED}}(x) = \begin{cases} x & \text{for } |x| \le a \\ a \, \text{sign}(x) & \text{for } a < |x| \le b \\ \frac{a}{b}x \exp\left(1 - \frac{x^2}{b^2}\right) & for \ |x| > b \end{cases} \qquad (6)$$

$$w_{\text{PROPOSED}}(x) = \begin{cases} 1, \text{ for } |x| \le a \\ \frac{a \, \text{sgn}(x)}{x}, \text{ for } a < |x| \le b \\ \frac{a}{b}\exp\left(1 - \frac{x^2}{b^2}\right), \text{ for } |x| > b \end{cases} \qquad (7)$$

where a and b are constants and x is any datum. From an influence function, a robustness measure is derived and will fix the selection of constants a $\left(= kv^2\right)$ and b $\left(= 2kv^2\right)$. M-estimators reduce the outcome of outliers, resulting $\min \sum_i \rho(x_i)$, where $\rho(.)$ is penalty function, $\psi(x) = \frac{d\rho(x)}{dx}$ is the influence function and $w(x) = \frac{\psi(x)}{x}$ is the weight function. Influence function calculates the effect of data on estimated parameter. The proposed method finds the channel estimation using

$$\hat{H}_{\text{PROPOSED}} = \left(X^H \psi(x) X\right)^{-1} X^H Y \tag{8}$$

5 Simulation Results

In simulations, channel estimation with 16-QAM, using LS and the proposed M-estimator (with and without DFT)-based techniques, is compared and shown in Fig. 2. FFT size is 32 and pilot spacing is 4. For improving the performance of channel estimation technique, a DFT method (with known maximum channel delay) is been developed by suppressing noise effect outside of the maximum channel delay which is shown in simulations. From the simulation results, an observation is done on the proposed method for channel estimation (with and without DFT) as closely approximation of true channel in both the cases at the cost of computational complexity and channel information characteristics.

6 Conclusions

Robust training symbol-based channel estimation for 5G wireless communication systems using M-estimation is proposed and studied in this paper. An M-estimator is proposed and used for optimizing the channel estimation technique. Simulation results are also provided to support efficacy of the proposed study of channel estimation in 5G wireless communication systems with additive white Gaussian noise. From simulation results, it is concluded that the proposed technique closely approximates true channel estimation for 5G MIMO wireless communication system with and without DFT.

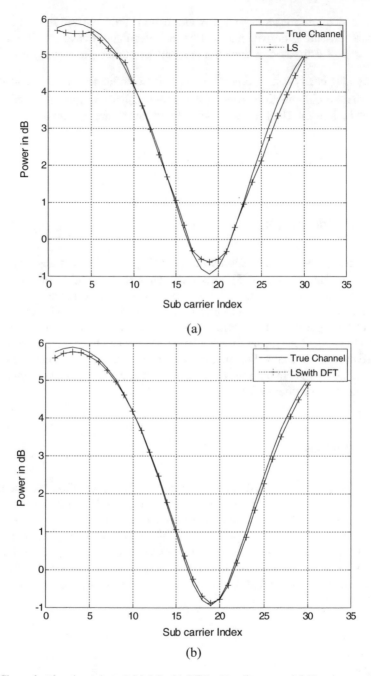

(a)

(b)

Fig. 2 Channel estimation using **a** LS **b** LS with DFT **c** *M*-estimator and **d** *M*-estimator with DFT

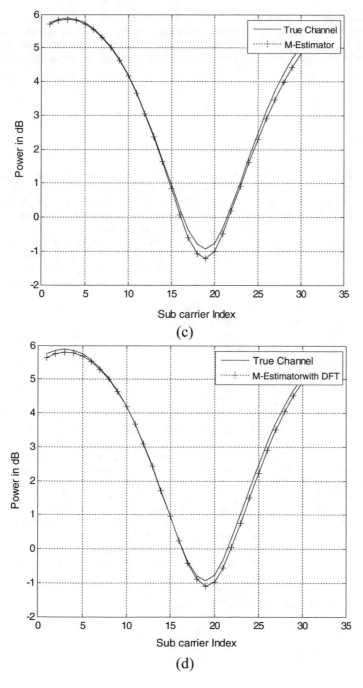

Fig. 2 (continued)

Acknowledgements This research work is carried out under the scheme Teachers Associateship for Research Excellence, sponsored by Science and Engineering Research Board (SERB), Government of India. The authors would like to thank SERB, India, for sponsoring this research work. The authors would also like to thank for NIT, Warangal, and CMRIT, Hyderabad, for encouraging this research work.

References

1. Ijiga OE, Ogundile OO, Familua AD, Versfeld DJ (2019) Review of Channel Estimation for Candidate Waveforms of Next Generation Networks. Electronics 8(9):956
2. Patnaik R, Krishna KS, Patnaik S, Singh P, Padhy N (2020) Drowsiness Alert, Alcohol Detect and Collision Control for Vehicle Acceleration. In: 2020 International Conference on Computer Science, Engineering and Applications (ICCSEA), pp 1–5, IEEE
3. Patnaik R, Padhy N, Raju KS (2020) A systematic survey on IoT security issues, vulnerability and open challenges. In: Intelligent System Design, pp 723–730. Springer, Singapore (2020)
4. Padhy N, Panigrahi R, Neeraja K (2019) Threshold estimation from software metrics by using evolutionary techniques and its proposed algorithms, models. Evolutionary Intelligence, pp 1–15
5. Wang X, Schaich F, Ten Brink S (2014) Channel estimation and equalization for 5G wireless communication systems. University of Stuttgart
6. Marzetta TL (2010) Noncooperative cellular wireless with unlimited numbers of base station antennas. IEEE Trans Wireless Commun 9(11):3590–3600
7. Apelfröjd R (2018) Channel estimation and prediction for 5G applications. Doctoral dissertation, Acta Universitatis Upsaliensis
8. Saha RK, Saengudomlert P, Aswakul C (2016) Evolution toward 5G mobile networks-a survey on enabling technologies. Eng J 20(1):87–119
9. Khan I, Rodrigues JJ, Al-Muhtadi J, Khattak MI, Khan Y, Altaf F, Mirjavadi SS, Choi BJ (2019) A robust channel estimation scheme for 5G massive MIMO systems. Wireless Communications and Mobile Computing
10. Rajender G, Anilkumar D, Rao DKS (2018) Empirical analysis of channel estimation procedures with enhanced Kalman Filter algorithm over MIMO-OFDM environment. Int J Pure Appl Math 118(19):2957–2970
11. Li YLJC, Cimini LJ, Sollenberger NR (1998) Robust Channel Estimation for OFDM Systems With Rapid Dispersive Fading Channels. IEEE Trans Commun 46(7):902–915
12. Saideh M, Berbineau M, Dayoub I (2018) On the performance of sliding window TD-LMMSE channel estimation for 5G waveforms in high mobility scenario. IEEE Trans Veh Technol 67(9):8974–8977
13. Kumar TA, Rao KD (2009) A new m-estimator based robust multiuser detection in flat-fading non-gaussian channels. IEEE Trans Commun 57(7):1908–1913
14. Hampel FR, Ronchetti EM, Rousseeuw PJ, Stahel WA (2011) Robust statistics: the approach based on influence functions. John Wiley & Sons

Channel Estimation for Broadband Wireless Access OFDM/OQAM System

J. Tarun Kumar, Chakradhar Adupa, V. Sandeep Kumar,
Shyamsunder Merugu, and Kasanagottu Srinivas

Abstract The IEEE 802.16a standard of broadband wireless access is introduced, and the different channel estimation algorithms of OFDM/OQAM system are compared under fixed wireless channels by computer simulations. The results show that the LS algorithm is more suitable for fixed broadband wireless channels.

Keywords OFDM/OQAM · LS · IEEE 802.16a

1 Introduction

The IEEE adopted 802.16 standards for fixed broadband access from 15 to 62 GHz wireless metropolitan area networks, which belongs to line-of-sight transmission in the LMDS band. Subsequently, the IEEE802.16a standard [1] was released, operating in the MMDS frequency range of 5 GHz to 17 GHz, supporting non-line-of-sight transmission, providing a linear service area of 80 km, and providing services including digital audio/video broadcast, digital telephone, asynchronous transfer

J. Tarun Kumar (✉) · C. Adupa · V. Sandeep Kumar
Department of Electronics and Communication Engineering, SR University, Anantha Sagar, Warangal, Telangana 506371, India
e-mail: tarunjuluru@gmail.com

C. Adupa
e-mail: adupa.chakradar@gmail.com

V. Sandeep Kumar
e-mail: sandeepvangalanitw@gmail.com

S. Merugu · K. Srinivas
Department of Electronics and Communication Engineering, Sumathi Reddy Institute of Technology for Women, Anantha Sagar, Warangal, Telangana 506371, India
e-mail: shyamala.merugu99@gmail.com

K. Srinivas
e-mail: srinu.vasu11@gmail.com

© The Author(s), under exclusive license to Springer Nature Singapore Pte Ltd. 2021   353
K. A. Reddy et al. (eds.), *Data Engineering and Communication Technology*,
Lecture Notes on Data Engineering and Communications Technologies 63,
https://doi.org/10.1007/978-981-16-0081-4_35

mode (ATM), Internet access, wireless relay and frame relay for telephone routing, etc.

The 802.16a standard physical layer protocol is mainly about frequency bandwidth, modulation method, error correction technology, and synchronization between transceivers, data transfer rate and timing and multiplexing structure. Orthogonal frequency division multiplexing with offset quadrature amplitude modulation (OFDM/OQAM) eliminates [2] intersymbol interference (ISI) and inter-carrier interference (ICI) by adding the cycle prefix in the protection interval, so that the receiver only needs to do a plural multiplier on each subcarrier to achieve frequency domain equilibrium, and the technology has a high spectrum utilization, which is very suitable for broadband high-speed wireless communication, so IEEE 802.16a working group included in the physical layer; in addition, IEEE plans to introduce 802.16d and 802.16e standards for fixed and mobile businesses, respectively, and the Wimax Alliance, led by Intel, also introduce corresponding chips [3].

In this paper, wireless MAN-OFDM/OQAM system is introduced and OFDM/OQAM channel estimation technique, then gives the performance simulation results of various channel estimation algorithms under broadband wireless access channels, and finally analyzes and compares the simulation results.

2 OFDM/OQAM Channel Estimation

Wireless MAN-OFDM/OQAM uses coherent modulation, i.e., QPSK and QAM. After the data decay through the wireless channel, the receiver must pass channel estimator to reliably determine the received information.

2.1 System and Channel Model

The baseband system model is shown in Fig. 1. $S(n)$ represents the transmitted data on the nth subcarrier. The data enters the receiver after channel fading $H_{m,n}$ and additive noise $W_{m,n}$, and $Y_{m,n}$ represents the received data on n subcarriers. Under the premise that the maximum multipath delay is less than the guard interval, the system can be equivalent to multiple parallel Gaussian channels with different channel gains [4].

$$Y_{m,n} = H_{m,n} X_{m,n} + W_{m,n} \tag{1}$$

where $Y_{m,n}$ is the receiving vector, $X_{m,n}$ is the diagonal matrix with $S_{m,n}$ as the diagonal element, $H_{m,n}$ is the channel frequency domain response vector, and $W_{m,n}$ is the independent Gaussian noise vector.

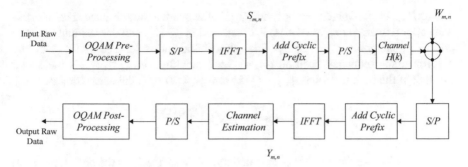

$S_{m,n}$

$W_{m,n}$

Fig. 1 802.16a OFDM/OQAM system model

The standard specifies the preamble structure of the upstream frame and the downstream frame separately. The channel estimation is performed on the non-zero pilot subcarriers in the preamble, to obtain the channel response on all data subcarriers by interpolation. Each OFDM/OQAM data symbol also contains eight pilot subcarriers, but these pilot intervals are too large to be suitable for channel estimation [5] and can be used for frequency offset correction.

The channel model adopted by the wireless MAN-OFDM/OQAM system is the SUI three-path model of Stanford University [6]. The maximum Doppler frequency shift is 2.5 Hz, which is a typical quasi-static channel. It can be approximated that the channel is unchanged within one frame.

2.2 OFDM/OQAM Channel Estimation Algorithm

Based on the preamble structure used by the system, the OQAM has recently been presented as a good alternative to OFDM [1, 7]. This multicarrier modulation scheme with cyclic prefix (CP) and interferences occurs when transmitting over a time-dispersive channel. $H_{m,n}$ with (m, n) the frequency and time index. The demodulated signal of OQAM system, noise taken apart, can be approximated, for well-localized pulse shaping, by

$$Y_{m,n} \approx H_{m,n}\left(X_{m,n} + jX^i_{m,n}\right) + w_{m,n}$$

$$Y_{m,n} = H_{m,n} X_{m,n} + j \underbrace{\sum_{(p,q)\in\Omega^*_{1,1}} X_{m+p,n+q}(g)^{m,n}_{m+p,n+q}}_{I_{m,n}} + w_{m,n} \qquad (2)$$

where $X_{m,n}$ is real-valued symbol and $X_{m,n}^i$ is the residual interference. The interference approximation method (IAM) is illustrated in [8–10], and the residual interference $X_{m,n}^i$ is approximated by 3×3 neighborhood and is denoted by $\Omega_{1,1}^*$, around a given time–frequency position (m, n), and excluding it, $g_{m+p,n+q}^{m,n}$ is the scalar product of the base functions $g_{m+p,n+q}(k)$ and $g_{m,n}(k)$ with the expression as

$$g_{m,n}(k) = g(k - nN)e^{\left(\frac{j2\pi}{M}mk + j\phi_{m,n}\right)} \tag{3}$$

with M indicating the number of subcarriers: $N = \frac{M}{2}$: $g(k)$ is the prototype filter: $\phi_{m,n}$, the phase term, can be expressed as $\frac{\pi}{2}(m + n)$.

2.3 LS Algorithm

LS algorithm is used to calculate the minimum mean square value as follows

$$\left(Y_{m,n} - X_{m,n}H_{m,n}\right)^H \left(Y_{m,n} - X_{m,n}H_{m,n}\right) \tag{4}$$

It is obtained that partial derivative of H in the above formula is equal to zero [7]

$$H_{m,n}^{P,LS} = \left(X_{m,n}^P\right)^{-1}Y_{m,n}^P \tag{5}$$

The subscript P indicates the corresponding value on the subcarrier where the pilot is located.

The implementation of the LS algorithm is very simple, equivalent to a zero-forcing estimator, but it does not consider the effect of noise.

2.4 Transform Algorithm

According to the analysis, the LS estimate is decomposed into the sum of the actual channel response, interference, and noise terms [8]

$$H_{m,n}^{P,LS} = H_{m,n}^P + \frac{\left[I_{m,n} + W_{m,n}\right]}{C_{m,n}}, m, n = 1, 2, \ldots, M - 1 \tag{6}$$

where $I_{m,n}$ and $W_{m,n}$ are ICI and additive noise, M is the number of non-zero pilots in the preamble, and $C_{m,n}$ is the pilot value.

The transform domain first uses the frequency domain $H_{m,n}^{P,LS}$ as the FFT to obtain the transform value G_M (can be regarded as the spectral sequence of the frequency domain value, M is the number of pilots). Since $\left[I_{m,n} + W_{m,n}\right]$ in Eq. (6) changes rapidly with m, and $H_{m,n}^{P,LS}$ changes slowly with m, the useful channel information is mainly concentrated in the "low-frequency" section of the transform domain, through G_M low-pass filtering can filter interference components. Finally, add $(N - M)$ zeros in the middle of G_M to get G_N (N is the number of effective subcarriers), which is equivalent to DFT interpolation. Finally, use G_N as IFFT to obtain channel estimates for all data subcarriers.

The advantage of this algorithm is that it does not require any prior channel information, and it is relatively easy to implement.

2.5 LMMSE Algorithm

LMMSE algorithm, $H_{m,n}^{P,LMMSE} = C_{m,n}^{H} H_{m,n}^{P,LS}$, C is a linear weighting matrix. Calculate C value that minimizes the mean square error $E\left\{\left|H_{m,n}^{P,LS} - H_{m,n}^{P,LMMSE}\right|^2\right\}$ to obtain the linear minimum mean square error estimate of the channel response.

According to the principle of orthogonality and statistical signal processing, the optimal matrix C should satisfy

$$R_{H_{m,n}^{P,LS} H_{m,n}^{P}} \otimes C = E\left\{H_{m,n}^{P,LS}\left(H_{m,n}\right)^{H}\right\} \tag{7}$$

Due to the conjugate symmetry of the autocorrelation matrix, $R_{H_{P,LS} H_{P,LS}} = R_{H_{P,LS}, H_{P,LS}}^{H}$, the channel response and noise are uncorrelated with each other. From Eqs. (1) and (6), the LMMSE estimate of the available channel response is

$$H_{m,n}^{P,LMMSE} = E\left\{H_{m,n}^{P}\left(H_{m,n}^{P,LS}\right)^{H}\right\} \otimes \left(R_{H_{m,n}^{P,LS}, H_{m,n}^{P,LS}}\right)^{-1} \otimes H_{m,n}^{P,LS}$$

$$= R_{H_{m,n}^{P} H_{m,n}^{P}}\left(R_{H_{m,n}^{P} H_{m,n}^{P}} + \sigma_{m,n}^{2}\left(X_{m,n}^{P}\left(X_{m,n}^{P}\right)^{H}\right)^{-1}\right)^{-1} H_{m,n}^{P,LS} \tag{8}$$

where $R_{H_{m,n}^{P} H_{m,n}^{P}}$ is the autocorrelation matrix of the channel, and $\sigma_{m,n}^{2}$ is the noise variance [4].

The accuracy of the LMMSE channel estimation algorithm is the best, but its implementation requires matrix inversion, so the complexity is very high. In addition, this algorithm needs to know the prior information of the channel, but in practice, the channel information is usually unknown, so this algorithm is rarely used in engineering applications.

3 Simulation Results

This simulation uses the frame structure stipulated by wireless MAN-OFDM/OQAM, and the receiver only considers the performance of channel estimation, ignoring the impact of subsequent error correction decoding. The working frequency band is MMDS frequency, the bandwidth is 6 MHz, the sampling interval is $\frac{1}{10}\mu sec$, and 64QAM modulation. The SUI-3 channel model is selected, which is suitable for medium undulating and tree-covered terrain, and is a typical urban fixed wireless access channel model.

Figure 2 is the subcarrier symbol error rate curve obtained for three different algorithms. As mentioned above, the LMMSE algorithm uses channel and noise information, so the performance is optimal. Under fixed broadband wireless channel conditions, the performance difference between LS and LMMSE algorithm is about 1.6 dB; the transform algorithm performs well at low signal-to-noise ratios, but poor at high signal-to-noise ratios; this is because low-pass filtering removes a lot of interference and noise and also discards apart of useful channel information; hence, an error floor appears.

Figure 3 compares the mean square error of various algorithms. The LMMSE algorithm has the smallest mean square error and the highest accuracy. The mean square error curve of the transform domain algorithm is relatively flat in the entire range of signal-to-noise ratio.

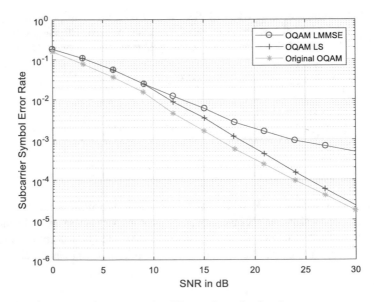

Fig. 2 Subcarrier error performance under different channel estimations

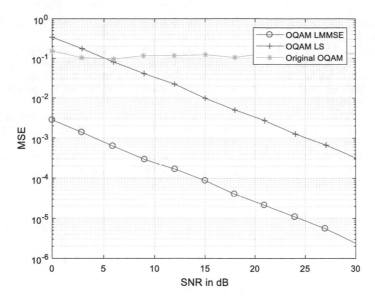

Fig. 3 Mean square error performance of algorithms under different signal-to-noise ratios

4 Conclusion

In this paper, the broadband wireless access system analyzed using three-channel estimation techniques is introduced for OFDM/OQAM system. As there is no relevant literature on channel estimation research for fixed broadband wireless environment, this paper focuses on the comparative performance study of three algorithms under the SUI wireless channel for MAN-OFDM/OQAM system and gives the simulation results for broadband wireless access. Compared with the high complexity of LMMSE, the need for channel prior information and only 1.6 dB performance improvement, the LS algorithm is a better choice. However, for small signal-to-noise ratio, the transform algorithm is also a good alternative algorithm.

References

1. Abichar Z, Peng Y, Chang JM (2006) WiMAX: The Emergence of Wireless Broadband. IT Professional 8(4):44–48
2. Kumar VS (2020) Joint Iterative filtering and Companding Parameter Optimization for PAPR reduction of OFDM/OQAM signal. AEU-Int J Electron Commun 124:153365
3. Choi JM, Oh Y, Lee H, Seo JS (2017) Pilot-aided channel estimation utilizing intrinsic interference for FBMC/OQAM systems. IEEE Trans Broadcast 63(4):644–655
4. Du J, Signell S (2009) Novel preamble-based channel estimation for OFDM/OQAM systems. In: 2009 IEEE international conference on communications, pp 1–6. IEEE
5. Kofidis E, Katselis D, Rontogiannis A, Theodoridis S (2013) Preamble-based channel estimation in OFDM/OQAM systems: A review. Signal Processing 93(7):2038–2054

6. Garbo G, Mangione S, Maniscalco V MUSIC-based modal channel estimation for wideband OFDM-OQAM. In: 2008 1st IFIP Wireless Days, pp 1–5. IEEE
7. Hari KVS, Baum DS, Rustako AJ, Roman RS, Trinkwon D (2003) Channel models for fixed wireless applications. IEEE 802.16 Broadband wireless access working group
8. Nissel R, Rupp M (2016) On pilot-symbol aided channel estimation in FBMC-OQAM. In: 2016 IEEE international conference on acoustics, speech and signal processing (ICASSP), pp 3681–3685. IEEE
9. Kumar JT, Kumar VS (2020) Novel distance-based subcarrier number estimation method for OFDM system. In: International conference on modelling. Simulation and intelligent computing. Springer, Singapore, pp 328–335
10. Kumar JT, Kumar VS (2020) A novel optimization algorithm for spectrum sensing parameters in cgnitive radio system. In: International conference on modelling. Simulation and intelligent computing. Springer, Singapore, pp 336–344

Digital Image Falsification Detection Based on Dichromatic Model

M. Gowri Shankar, C. Ganesh Babu, Saravanan Velusamy, K. Vidyavathi, and R. Gandhi

Abstract Nowadays, digital images may be simply changed by victimization superior computers, subtle photo-editing, tricks package, etc. These changes will impact picture validity, from law, politics, the media, and industry. Detecting falsification of digital images is one of the most critical analytical practices in the present time. Altering the digital image will, in general, disrupt those underlying vision-based regularities that occur due to environment, lens, and sensor, and this initiative introduces new forensic image related vision-based research method for color lighting.

Keywords Falsification · Digital Image · Dichromatic · Illumination · Specular

1 Introduction

Developing up-to-date digital image processing techniques, individuals can literally modify digital image material with no warnings, usually in legislation, politics, industry, and media. The digital image falsification exposure is associated with nursing approach. It is to identify mechanically manipulated of region happening digitally image altered [1]. It is subdivided into two ways—one is watermarking and another one is forensics. (i) Watermarking, one of lively style, is used to test the accuracy of prior data. Forensics may reflect a blind approach; it does not want any prior

M. Gowri Shankar (✉) · R. Gandhi
Gnanamani College of Technology, Namakkal, Tamilnadu, India
e-mail: mshankar065@gmail.com

C. Ganesh Babu
Bannari Amman Institute of Technology, Sathyamangalam, Tamilnadu, India

S. Velusamy
EEE Section, University of Technology and Applied Sciences, Higher College of Technology, Muscat, Oman

K. Vidyavathi
Selvam College of Technology, Namakkal, Tamilnadu, India

© The Author(s), under exclusive license to Springer Nature Singapore Pte Ltd. 2021   361
K. A. Reddy et al. (eds.), *Data Engineering and Communication Technology*,
Lecture Notes on Data Engineering and Communications Technologies 63,
https://doi.org/10.1007/978-981-16-0081-4_36

data to validate the image authority. It is subdivided into two ways: (i) SBIF—specifically based image forensics; it is primarily applied math and (ii) VBIF—vision-based image forensics. SBIF verifies the authority by gauge the statically inconsistency like interfering artifacts, resampling detection, and noise detection regularity [2].

Natural properties come from various sources like planet, lens, and detector, and are used to verify the truthfulness [4]. Vision-based digital image forensics used the properties of identifying manipulated region through action lighting unpredictability happening single-light Lambertian surface and then aberration deviation [3].

Color features are one of key toward finding digitally altered image, and then, matching the color of 1 object to the opposite is difficult. Suppose a photograph is taken a continuous lighting to a lower place [5], color of illumination all objects must remain constant all over the whole shot, if image is altered, accuracy of the lighting can as well altered. These efforts we are try to suggest are brand novel methods by means of estimating color accuracy of illumination by detecting a mirror-like area [10].

The color feature is one in every key required to locate the digitally altered image, but matching the color of one object to the opposite is laborious. Suppose a picture is taken on a continuous illumination to a lower location, the color of illumination all objects ought to constant all through whole shot [8]. Uncertainty object changed; accuracy of the lighting can also altered. During this one technique, we continue to use this one property and offer the replacement vision-based image forensics technique to locate that altered portion inside digital image by estimating whether illumination affects the reflective field [6] (Fig. 1).

2   Creation of Color Images—Essentials

Creation of color images will be clarified with that help of that dichromatic reflection model [7]. In accordance with that pattern, mixture additive consideration and diffuse reflection may also be reflected on some non-homogeneous materials. The lightweight mirror will be modulated as stated in below equation for association of nursing entity lit by one supply

$$L(\theta, \lambda) = M_s(\theta) S_s(\lambda) E(\lambda) + M_b(\theta) S_b(\lambda) E(\lambda)$$

where spectral surface coefficient, body of the component, $E(\lambda)$ is a lightweight supply color, where $S_S(\lambda)$, $S_B(\lambda)$ area of unit. Spectral surface coefficient of the surface area is unit constant over the actinic ray for much type of materials. The above first equation will be rewritten with respect to the RGB system response as shown in below equation.

$$\begin{pmatrix} R \\ G \\ B \end{pmatrix} = m_s(\theta) \begin{pmatrix} R \\ B \\ B \end{pmatrix} + m_b(\theta) \begin{pmatrix} R \\ G \\ B \end{pmatrix}$$

where R, G, and B are the surface and body parts of S and B image. Above two vectors of surfaces and then section of body which cover the two-dimensional planes referred to as the dichromatic plane that will be the Avoirdupois unit (λ) and $L_S(\lambda)$. The dichromatic lines are depicted in Fig. 2 for one source of illumination.

Fig. 1 Vision-based reflection of dichromatic model. *Source* Stanford University, Foundation of Vision and Chap. 9 [11]

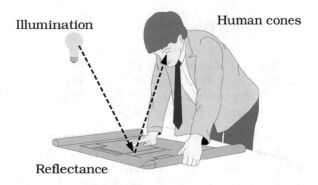

Fig. 2 Reflection of dichromatic model

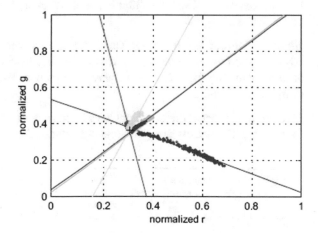

3 Falsification Detection Through Estimate Color of Illumination in a Region of Specular

Falsification area detection in a digitally improved image includes the following path. Algorithm of falsification detection could be characterized and depicted in Fig. 3 for falsification.

Specular region detection is important of estimating illumination color. Observing region of specular requires subsequent operations follows:

3.1 Preprocessing

For fast processing, input image is transformed into luminance and analyzed white and black areas. For accurate identification of the specular regions, binomial band-pass filtering and stretching contrast are used. Figure 4 explains the block diagram of the pre-processing operations [9].

Luminance Scale: Before entering the level segmentation, the user must correctly select the specular regions. It will be little complex to select the specular regions within the color picture. The first step is to transform a digital input picture to a luminance scale for specular area detection. The picture of the luminance scale can be analyzed easily. The interaction between white and black for the input image is better than a color image.

Binomial Band-pass Filtering: This filter form structure is simple and effective, based on the binomial coefficient. It is analyzing high-frequency components and accurately finding specular region and reducing noise. Binomial band-pass filter size of mask is 3*3 and 5*5.

Stretching Contrast: Picture might have been removed for one steep transformation for that object's intensity except for that is gradation for that light shaped of the objects

Fig. 3 Detection of falsification

Fig. 4 Block diagram of filtering (preprocessing)

Table 1 Segmentation of levels

Levels	Luminance scale	Luminance sample reduction
1	16–38	16
2	38–60	38
3	60–82	60
4	82–104	82
5	104–126	104
6	126–148	126
7	148–170	148
8	170–192	170
9	192–214	192
10	214–236	214

naturally from the picture. Change the picture to high contrast. That contrast should have been clearly shadows, highlights created by the like light from background.

3.2 Segmentation of Levels

Segmentation level mechanism is partitioned into sections or regions. This partitioning into parts is also focused on the pixel characteristics in the image. This rim can identify regions. Other methods divide the image into regions based upon color or texture values. The luminance image levels decrease to 10. Sample reduction is 256–10 levels used for image of luminance which is shown in Table 1 from the paper they reduced to 50 levels.

3.3 Falsification Detection Using Plane of Dichromatic

The picture is colorized by some means of approximating the fuel by aid for that reflection of dichromatic model. Regions of reflection have governed fuel light and their inputs to the model of dichromatic reflection. RGB vectors of that area square measure rotten victimization primary element analysis and mapped in line with the dichromatic reflection model which can cover the entire plane of dichromatic. Under the objects, same supply runs that line providing the fuel vector, and the color property values of the fuel color are provided together. The flow diagram of dichromatic plane is shown in Fig. 5.

Regions round reflective highlights square measure thought-about to estimate the illumination colorizer the image. When detective work is completed, the user responses the reflective regions of the every entity as of the measure of spitting image divided into model of dichromatic. The region of reflection has the dominant

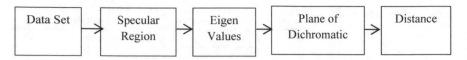

Fig. 5 Flow diagram of dichromatic plane

colorizer. Euclidean distance is follows,

$$(y_2 - y_1) - (x - x_1) = (x_2 - x_1) - (y - y_1)$$

$$(y_4 - y_3) - (x - x_3) = (x_4 - x_3) - (y - y_3)$$

$$E(P1, P2) = \sqrt{((p11 - p21)^2 + (p12 - p22)^2)}$$

Line that does not meet through alternative lines as well as does not require shut intersection exist a tampered object. Planes of dichromatic area unit are ultimately divided into four groups: (1) two objects without intersection, (2) plane through relatively two entities without manipulated entity intersection, (3) plane through relatively two entities with manipulated entity intersection, and (4) plane through meeting bolted but not manipulated entity region division shown in Fig. 6.

4 Results and Analysis

In a machine with Intel ® Core (TM) i5-2430 M CPU @ 2.40HHz and 4 GB RAM, the tests were dispensed. Mistreatment MATLAB 8.3.0 (R2014a) performed the implementation. Solid images were created by repeating objects from one image and adding them to another image under completely different illuminating conditions by mistreating the Adobe picture writing software package. The specific cases checked like (1) pictures for a tampered object with no intersection; (2) pictures with intersection of a manipulated object; (3) pictures with no intersection of 2 objects. Detection of specular region accuracy shown in Table 2.

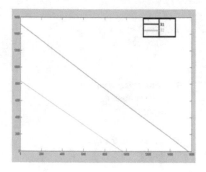

Plane with 2 objects without intersection

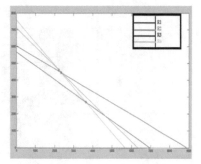

Plane through relatively two entities with manipulated entity intersection

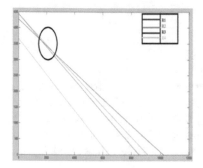

Plane through relatively two entities without manipulated entity intersection

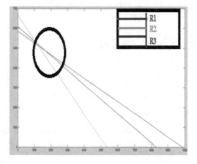

Plane through meeting bolted but not manipulated entity region division

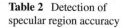

Fig. 6 Dichromatic plane area unit

Table 2 Detection of specular region accuracy

Types of data set	Total number of images	Luminance
Objects of glassy	4	4
Highlights of specular (less)	4	2
Highlights of specular (high)	4	3

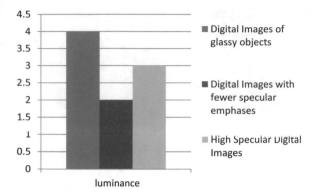

5 Conclusion

We have designed two completely different techniques in this effort—Illumination color based mostly on rhetorical picture. The technique may lead to high outcomes from images taken during consistent lighting that has reflective areas in it. That proposed strategies work well enough for photographs taken during continuous lighting yet struggle to function during post-processing circumstance underlying.

References

1. Vidyavathi K, Sabeenian RS (2012) A novel approach for video motion estimation using frames difference within a block, digital image processing 4(9)
2. Vidyavathi K, Gowri Shankar M, Nagarajan R, Chandramohan J, Ranjith A, Kamaldas V (2018) A novel to achieve ULTRA HD video compression by using VHT algorithm based on embedded Technology. Int J Adv Res Electron Commun Eng (IJARECE) 7(3):2278–909X
3. Rey C, Dugelay JL (2002) A Survey of watermarking algorithms for image authentication. EURASIP J Adv Signal Process 2002(6):218932
4. Popescu AC, Farid H (2004) Statistical tools for digital forensics. In: International workshop on information hiding, pp 128–147. Springer, Berlin, Heidelberg
5. Farid H (2009) Image forgery detection: a survey. IEEE Signal Process Mag 2(26):16–25
6. Luo W, Qu Z, Huang J, Qiu G (2007) A novel method for detecting cropped and recompressed image block. In: 2007 IEEE international conference on acoustics, speech and signal processing-ICASSP'07, vol 2, pp II-217. IEEE
7. Ng TT, Chang SF, Sun Q (2004) Blind detection of photomontage using higher order statistics. In: 2004 IEEE international symposium on circuits and systems (IEEE Cat. No. 04CH37512), vol 5, pp V–V. IEEE
8. Ying C, Yuping W (2008) Exposing digital forgeries by detecting traces of smoothing. In: 2008 The 9th international conference for young computer scientists, pp 1440–1445. IEEE
9. Shankar MG, Babu CG (2020) An exploration of ECG signal feature selection and classification using machine learning techniques. Int J Innov Technol Exploring Eng Regul p 9

10. Okada K, Kagami S, Inaba M, Inoue H (2001) Plane segment finder: algorithm, implementation and applications. In: Proceedings 2001 ICRA. IEEE international conference on robotics and automation (Cat. No. 01CH37164), vol 2, pp 2120–2125. IEEE
11. https://foundationsofvision.stanford.edu/chapter-9-color/

Half-Circled Fractal Boundary Linearly Polarized Triangular Patch Antenna for Wireless Applications

V. V. Reddy, A. Vijaya, E. Suresh, and M. Raju

Abstract A novel half-circled fractal boundary triangular patch antenna is presented. The fractal geometry is employed at the boundaries of the triangular patch. It is noticed that the operating resonance frequency decreases as the indentation radius of the fractal curve increases. The simulation results are carried out for first iteration order of half-circled fractal curves. The comparison between simulation and measurement results is demonstrated.

Keywords Half-circled · Fractal · Iteration order

1 Introduction

Microstrip antennas are thoroughly investigated in the last two decades due their lost cost and lightweight characteristics. A truncated tip equilateral triangle microstrip antenna is proposed by Chia Luan Tang [1]. Wen-Shyang Chen et al. have described a patch antenna with bent slots [2]. A size reduction of 50% is claimed with these inserted slots on the patch. An equilateral triangular patch with inserted rectangular or cross-shaped slot is demonstrated by Jui-Han Lu et al. [3]. By varying the dimensions of the cross slot, several antennas are designed. A size miniaturized design of microstrip antenna with single coaxial probe feed is reported by Row et al. [4]. Various antennas with different slot lengths are designed and studied. Another method to design microstrip antenna is nominated by Xihui Tang et al. [5]. The slotted square

V. V. Reddy (✉) · A. Vijaya · E. Suresh · M. Raju
Department of ECE, KITS Warangal, Warangal, India
e-mail: vvr.ece@kitsw.ac.in

A. Vijaya
e-mail: av.ece@kitsw.ac.in

E. Suresh
e-mail: es.ece@kitsw.ac.in

M. Raju
e-mail: mr.ece@kitsw.ac.in

© The Author(s), under exclusive license to Springer Nature Singapore Pte Ltd. 2021   371
K. A. Reddy et al. (eds.), *Data Engineering and Communication Technology*,
Lecture Notes on Data Engineering and Communications Technologies 63,
https://doi.org/10.1007/978-981-16-0081-4_37

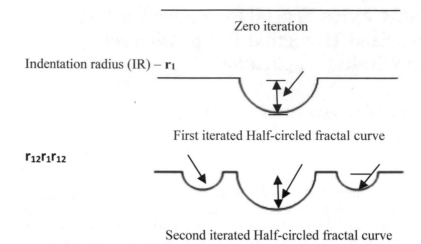

Fig. 1 Generation of half-circled fractal curve

patch antenna at the diagonal corners comprises two sets of vertical right-angle bent stubs. Fan yang et al. have examined E-shaped antennas for wide band linear [6] and circular polarization [7] operations using single-layer and single-probe feed mechanism. In this paper, a single-band linearly polarized half-circled fractal boundary antenna is examined.

2 Antenna Design

The proposed antennas are designed by substituting the sides of triangle patch structures with half-circled fractal boundary curves. Iteration order (IO) and indentation factor (IF) are the two main parameters of fractal curves. Here IF of the proposed design is indentation radius (IR). The generation of proposed half-circled fractal boundary curve is given in Fig. 1.

Here, half-circled fractal curve of same IR is used as boundaries of triangular patch for linear polarization realization. The various geometries are generated by changing the IRs. The behavior of all these designed antennas is studied. The proposed linearly polarized antennas of IO one are portrayed in Fig. 2.

3 Simulation Results

The proposed antennas are studied by performing simulations in Ansoft HFSS electromagnetic simulator. The specs of simulated antennas are as follows: The triangular patch side length (L) is 48 mm, thickness (h) is 3.2 mm, dielectric constant (ε_r) is

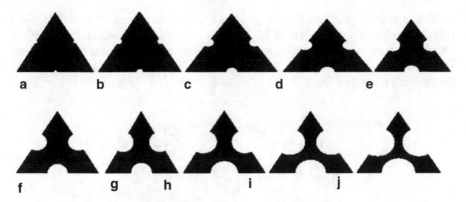

Fig. 2 First iterated stage 1 half-circled fractal antennas with different IRs: **a** 0.04, **b** 0.083, **c** 0.125, **d** 0.166, **e** 0.2, **f** 0.25, **g** 0.29, **h** 0.33, **i** 0.375, and **j** 0.4

2.2, and loss tangent is 0.0019. The substrate material considered is RT Duroid 5880. All the antennas are fed using a single-probe feed mechanism, and location of the probe is optimized for the input impedance of 50 Ω. Because of the difficulty in manufacturing and cost, only first iteration is considered. These are only pre-fractals not actually fractals. Initially, antennas by keeping side end-to-end value constant as a = 48 mm and increasing IR value, IO one fractal are simulated. The probe feeding mechanism, side and top views of the proposed antenna are depicted in Fig. 3. In case

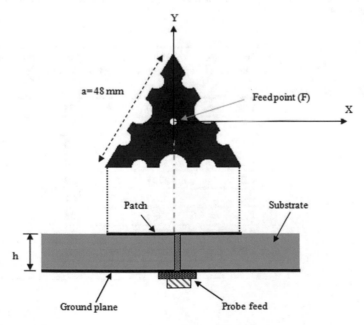

Fig. 3 Cross-sectional top views and probe feeding mechanism

of IO one with the increase of IR value, the decrease in the resonance is observed from return loss characteristics of Fig. 4. It is due to the increase of the electrical length which decreases the patch resonance frequency. The resonance frequency and bandwidth are given in Table 1. The variation of resonance frequency with IR for first stage is plotted in Fig. 5.

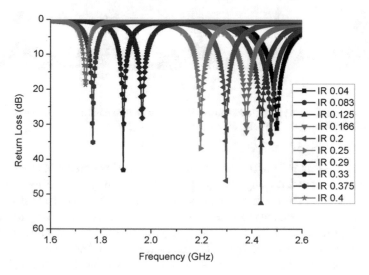

Fig. 4 Return loss characteristics of first iterated stage 1 triangular half-circled fractal antennas with different IRs

Table 1 Resonance frequency and bandwidth of the first iterated stage 1 half-circled fractal antennas	Indentation radius (IR) – r_1	Resonance frequency (MHz)	10-dB Return loss bandwidth	
			(MHz)	(%)
	0.04	2500	78	3.12
	0.083	2475	76	3.07
	0.125	2435	73	2.99
	0.166	2377	69	2.9
	0.2	2297	58	2.52
	0.25	2195	51	2.32
	0.29	1964	38	1.93
	0.33	1890	34	1.79
	0.374	1768	28	1.58
	0.4	1738	20	1.15

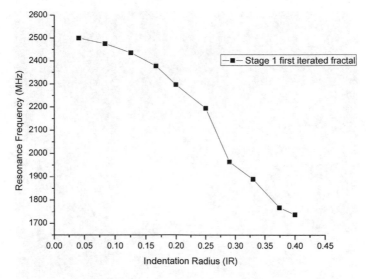

Fig. 5 Variation of resonance frequency with IR for first-stage triangular half-circled fractal antennas

4 Experimental Results

To verify the simulation results, triangular patch antenna with IR 0.125 is fabricated on RT Duroid 5880 substrate. The photograph of the antenna mounted in anechoic chamber is given in Fig. 6. The comparison of measured and simulated return loss curves is depicted in Fig. 7. The simulated antenna resonates at the frequency of 2435 MHz, whereas the measured results indicate the fabricated antenna at 2430 MHz. The obtained 10-dB return loss bandwidths of simulated and measured results are 73 MHz and 70 MHz, respectively. The measured co- and cross-polarization radiation patterns of the fabricated antenna at resonating frequency are pictured in Fig. 8. It is observed that compared to co-polarization radiation pattern, −20 dB less power is obtained in bore sight direction in case of cross-polarization radiation pattern.

5 Conclusions

New single-band linearly polarized half-circled fractal antennas have been studied. It is observed that within the given space without changing the end-to-end length of the patch, larger electrical lengths can be accommodated. By varying the indentation radius of the fractal curve, parametric study has been carried out. The suggested antenna can be used for the handheld electronic gadgets.

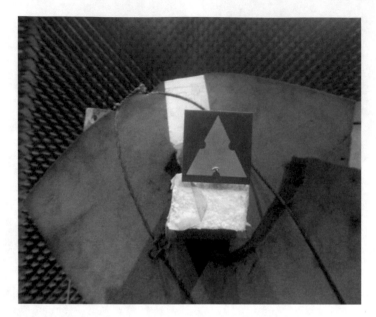

Fig. 6 Photograph of the mounted antenna in anechoic chamber

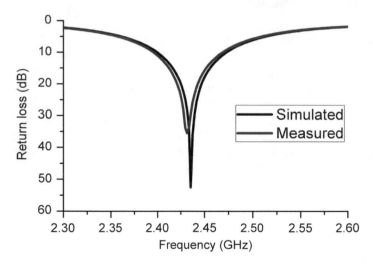

Fig. 7 Comparison of simulated and measured return loss curves

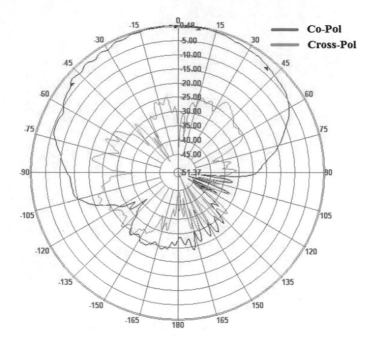

Fig. 8 Measured co- and cross-polarization radiation patterns at 2430 MHz frequency

References

1. Tang ChL, Lu JH, Wong KL (1998) Circularly polarised equilateral-triangular microstrip antenna with truncated tip. Ele Lett 34(13):1277–1278
2. Chen WS, Wu ChK, Wong KL (1998) Compact circularly polarised microstrip antenna with bent slots. Electronic Lett 34(13):1278–1279
3. Lu JH, Tang ChL, Wong KL (1999) Single-feed slotted equilateral-triangular microstrip antenna for circular polarization. IEEE Trans Ant Prog 47(7):1174–1178
4. Row JS, Ai CY (2004) Compact design of single-feed circularly polarised microstrip antenna. Elect Lett 40(18):1093–1094
5. Xi T, Lau KL, Xue Q, Long Y (2010) Design of small circularly polarized patch antenna. IEEE Trans Ant Prog 9(7):728–731
6. Yang F, Xi Y, Ramat sammi Y (2001) Wide-band E-shaped patch antennas for wireless communications. IEEE Trans Ant Prog 49(7):1094–1100
7. Khidre A, Lee KF, Elsherbeni A (2010) Wideband circularly polarized E-shaped patch antenna for wireless applications. IEEE Antennas Propag Magazine 52(5):219–229

Authorship Attribution using Filtered N-grams as Features

Manan Singh and Kavi Narayana Murthy

Abstract Authorship attribution is the problem of assigning an author to a document of unknown authorship, given a candidate set of authors and their sample documents. As a text classification task, this requires features that can capture the writing styles of authors. In this work, we compare the filtered n-grams with the traditional or unfiltered n-grams as features for authorship attribution. Filtered n-grams are the n-grams formed after filtering out from the text certain kinds of tokens. We explore the filtered n-grams formed after the removal of noun groups and verb groups. We hypothesize that the remaining text should still be enough to capture the writing style. Moreover, this removal makes possible the construction of new n-grams which would have been missed otherwise. In our experiments, we find that filtered n-grams improve the performance. In the feature ablation study, we confirm that this improvement is due to the new n-grams which are possible only after filtering.

Keywords N-grams · filtered n-grams · writing style · authorship attribution · text classification features

1 Introduction

Authorship attribution is the task of assigning an author to a document of unknown authorship, given a set of candidate authors along with their sample documents [1]. The problem has its origins in the mysteries of authorship disputes such as the controversies over Shakespearean authorship, but in the modern world, it finds its application in various business and professional needs such as plagiarism detection and text forensics [2]. From a data mining perspective, this is a text classification task which involves extracting effective features and using them with a classifier.

M. Singh (✉) · K. N. Murthy
School of Computer and Information Sciences, University of Hyderabad, Hyderabad, Telangana, India
e-mail: manan.pqrs@gmail.com

K. N. Murthy
e-mail: knmuh@yahoo.com

© The Author(s), under exclusive license to Springer Nature Singapore Pte Ltd. 2021   379
K. A. Reddy et al. (eds.), *Data Engineering and Communication Technology*,
Lecture Notes on Data Engineering and Communications Technologies 63,
https://doi.org/10.1007/978-981-16-0081-4_38

For textual data, a widely used method to extract features is to represent the document in terms of frequencies of N-grams. Traditionally, N-grams are constructed linearly, but Sidorov [3] discusses few alternative nonlinear ways of constructing n-grams to build more effective features for authorship attribution.

The simplest kind of nonlinear construction is skip-grams, i.e., to skip some tokens while constructing the n-grams. As an example, from the text "how are you now", some possible skip-grams are "how you", "are now", "how now", and "how you now", and "how are now". While this sounds simple, we must note that the number of all possible skip-grams grows exponentially with the text size. Also, direct linguistic interpretation is difficult [3].

A more sophisticated method is to follow the branches in the syntax tree of a sentence to construct n-grams, giving us what are known as Syntactic n-grams. Syntactic n-grams have experimentally been shown to be superior in performance to linear n-grams, and highly effective for authorship attribution [4], the only downside being the dependence on an accurate syntactic parser, and the parsing complexity associated with parsing each sentence in the corpus.

Another method is to remove or filter out tokens of certain kind before forming n-grams. Sidorov [3] has coined the term "Filtered n-grams" for this, but we found very little work in the literature on applying this idea. Content words, such as nouns and verbs, are indicative more of the subject matter and less of the style of individual authors. Therefore, removing noun phrases, verb phrases, etc., from texts before constructing n-grams will hopefully give us smaller number of more useful features for authorship attribution task. This will not require complete parsing of each sentence, and thus, will be computationally less expensive too. In this paper, we explore the effectiveness of filtered n-grams so constructed for the authorship attribution task. We demonstrate that filtered n-grams outperform the traditional or the unfiltered n-grams as features for authorship attribution.

2   Related Work

The earliest reported case of computer-assisted authorship attribution is the work on the Federalist Papers in 1960 [5]. It used Bayesian statistical analysis of frequencies of a few common words (e.g., "upon", "and", etc.) to classify the disputed papers. By the end of twentieth century, researchers had applied several common pattern recognition algorithms and identified many effective and relevant features for the task. Features such as word length, syllable length, sentence length, distribution of parts of speech, function words, type–token ratio, vocabulary distributions, hapax legomena (i.e., the number of words occurring only once), etc., were found effective [6], and pattern classification techniques such as discriminant analysis, cluster analysis, principal component analysis, and neural networks had been applied [7–10]. Holmes [6, 11] provides a survey of the work done on authorship attribution before twenty-first century. The first decade of twenty-first century saw remarkable diversity in novel techniques such as Burrows' delta measure [12], n-gram-based metrics [13], compression-based methods [14], and graph of stopwords [15]. Several studies [1, 16, 17] survey the methods before 2010. Post-2010, with the boom of the digital era, and spurt in new form of writings such as blogs, emails, tweets, and programming code, applications have become more diverse. Researchers are exploring short texts [18, 19], newer feature-sets such as specific character n-grams [20], and neural methods such as LSTM and CNN [21, 22]. An assortment of techniques has been applied, ranging from topic models [23], and language models [24] to syntactic n-grams [3]. Many interdisciplinary approaches have also been introduced, for example, based on complex networks [25], cellular automata [26], graph theory [27], and probability theory [28]. A recent ACM computing survey on computational stylometry [29] summarizes the state of the art and mentions the list of publicly available datasets, various classes of features, and various classification techniques.

Despite the diversity in techniques for authorship attribution, the general approach is that of feature extraction followed by pattern recognition or classification. In the next subsection, we describe the features widely used in the authorship attribution literature.

2.1 Features for Authorship Attribution

The various features have been classified into lexical, syntactic, semantic and application-specific categories [16, 29]. These features are described briefly below.

Lexical Features

Lexical features are extracted by processing the text at word or character level. At the word level, features include average word length (measured in terms of characters), average sentence length (measured in terms of tokens), the distribution of word and sentence lengths, average number of syllables per word, frequencies of words, and frequencies of n-grams of words.

A writer can also be distinguished by his repertoire of words. Some measures have also been developed to capture the vocabulary richness of an author. Type–token ratio is the number of types (i.e., unique word forms) divided by the number of tokens in the author's corpus. A writer with a richer vocabulary will use more unique words, and his type–token ratio will be higher than an average writer. This ratio can differ among authors, thus serving as a discriminating feature. The number of words occurring only once or twice is called Hapax Legomena and Hapax Dislegomena, respectively, and can also be indicative of vocabulary richness.

Features solely based on character-level processing have also been investigated. Simplest ideas include frequencies of punctuations, digits, lowercase or uppercase letters, etc. Much more researched is the idea of character-level n-grams [13, 16]. Frequencies of different character or byte-level n-grams have been used as features. The advantages include minimal text preprocessing and being less prone to spelling mistakes. The choice of n is highly problem-specific and also language-dependent.

A large n leads to n-grams that may capture more lexical and contextual information, whereas n-grams with smaller n may capture only subword or syllable-like information.

Syntactic Features

At a level higher than that of tokens, information about syntactic structure of the sentence has also been used to extract features. Feature-sets are constructed from the output of taggers, chunkers, and parsers. Simple features include noun phrase counts, verb phrase counts, length of verb phrases, counts of sequence of POS tags, etc. Advanced features use the paths in the parse trees of sentences to build n-grams. These are called syntactic n-grams [4]. In syntactic n-grams, the n-grams are constructed by following the paths in the syntax tree of a parsed sentence. Firstly, a sentence is parsed and its parse tree or a dependency parse is obtained. In these trees, nodes are words. To construct an n-gram, n words in the path downwards from a node, e.g., from the root, are chosen. Depending on different starting nodes and different paths below them, multiple kinds of n-grams can be constructed. Syntactic n-grams have shown to outperform the conventional n-grams in the authorship attribution task [4]. The downside of syntactic features, of course, is their dependence on the availability of accurate parsers, and the complexity of parsing.

Semantic Features

Very less work has been reported on the usage of semantic features. Authorship attribution is more concerned with the style of the author, rather than the content or theme of the texts. Hence, very less work is focused on capturing the semantics of the text. Also, capturing higher level semantics as features is not very easy, and techniques are very prone to vagueness and subjective interpretations.

Clark and Hannon [30] have reported a classifier based on synonym-based features. They assume that the author's style is reflected in his choice among synonyms. For words with large number of possible synonyms, different authors will have different preferences for the particular synonym they use. The frequency for each word is weighted by the number of its synonyms. Thus, words with larger synonym set will be given more weight, and the weighted features help in distinguishing authors.

Application-specific Features

The features mentioned above are application or domain independent. For newer and specific forms of texts such as emails, blog posts, tweets, online forums, HTML texts, etc., application-specific features can be extracted. These can include email signatures, URL counts, use of indentation, abbreviations, etc. If the text is in HTML format, then HTML-specific features such as HTML tag distribution, font properties, and their counts can also be used [29].

Once important features have been identified and extracted, then any classification algorithm can be applied for authorship attribution.

3   Methodology

This section describes the methodology that we use in this work to compare filtered n-grams with unfiltered n-grams.

3.1   Filtered N-grams

In this work, we explore the filtered n-grams formed after removing noun groups and verb groups from the text. We use the chunker available in the pattern python library [31] to identify the NP and VP chunks (base noun/verb phrases that do not contain other noun/verb phrases [32]). Table 1 shows some illustrative examples from the literature dataset that we have used. After removing the NP and VP chunks, the

Table 1   Illustrative Examples of Filtered N-grams

Text type	Text representation	Sample N-grams formed
Without filtering	The impetus was heightened by those little events of the day which had roused her discontent with the actual conditions of her life We were lying in the darkness of the shadow of the wall of the great crater	The impetus, day which, had roused, of her life; were lying, of the, great crater
Chunked text	[The impetus]/NP [was heightened]/VP by [those little events]/NP of [the day]/NP which [had roused]/VP [her discontent]/NP with [the actual conditions]/NP of [her life]/NP We]/NP [were lying]/VP in [the darkness]/NP of [the shadow]/NP of [the wall]/NP of [the great crater]/NP	
NP Removed	was heightened by of which had roused with of were lying in of of of	**Traditional**: had roused, which had; were lying **New**: by of which, with of; lying in of, in of
VP Removed	The impetus by those little events of the day which her discontent with the actual conditions of her life We in the darkness of the shadow of the wall of the great crater	**Traditional**: The impetus, of her life; of the, great crater **New**: impetus by, which her; we in the
NP & VP removed	by of which with of in of of of	**Traditional**: - **New**: by of which, with of; in of, of of of

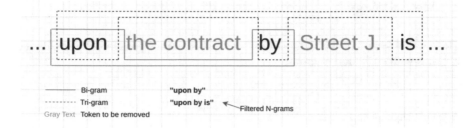

Fig. 1 An Illustration of Filtered N-grams

sentences are stripped off the theme-specific words. The skeletons that remain are hopefully sufficient to discriminate between the writing style of different authors.

Also, note that the filtered n-grams contain a combination of traditional and new n-grams. The new n-grams are those which would have been missed had we not filtered the text. Examples are "upon by", "with of", "by of which", etc. (See Fig. 1 for an illustration.) Later, in our experiments, we also present a feature ablation study in which these new n-grams are ablated from the feature set, and the performance is found to get degraded, thus signifying their importance.

3.2 Dataset

The dataset used in our experiments comprises of 50 english literature books of 5 authors - 10 books per author. The authors and their book titles are mentioned in Table 2. It is a subset of the 100 books dataset used by Rozz et al. [33]. All the books are in the public domain, and their electronic versions can be downloaded from the online repository of Project Gutenberg. To avoid the effect of different text-lengths upon feature frequencies, all the books were truncated to 20 thousand tokens which was found to be the length of the shortest book in the set.

3.3 Classification and Evaluation Procedure

We have used a fivefold cross-validation procedure throughout our experiments. The dataset is divided into five folds or groups, and five iterations of training and testing are performed with a different group being used as the test set each time. In each iteration, the training and test sets comprise of 40 and 10 documents, respectively. The split is stratified; i.e., the ratio of classes is balanced in each set. The training set is used to build the model, and then the authors of the test set are predicted. The average classification accuracy over the fivefolds is reported in the end.

The classification pipeline which we have used is as follows:

Table 2 List of authors and their books used in this work. *Source* [33]

Author	Book titles
Bernard Shaw (1856–1950)	Arms and the Man, Caesar and Cleopatra, Candida, Cashel Byron's Profession, Heartbreak House, Major Barbara, Man and Superman, Pygmalion, The Devil's Disciple, The Philanderer
Charles Dickens (1812–1870)	A Christmas Carol, A Tale of Two Cities, David Copperfield, Dombey and Son, Great Expectations, Little Dorrit, Oliver Twist, Our Mutual Friend, The Life and Adventures of Nicholas Nickleby, The Pickwick Papers
George Eliot (1819–1880)	Adam Bede, Daniel Deronda, Felix Holt The Radical, Impressions of Theophrastus Such, MiddleMarch, Romola, Scenes of Clerical Life, Silas Marner, The Essays of George Eliot, The Mill on the Floss
Herbert George Wells (1866–1946)	Ann Veronica, In the Days of the Comet, Tales of Space and Time, The Country of the Blind and Other Stories, The First Men in the Moon, The Food of the Gods and How It Came to Earth, The Invisible Man, The Island of Doctor Moreau, The Time Machine, The War of the Worlds
Oscar Wilde (1850–1900)	A House of Pomegranates, An Ideal Husband, A Woman of No Importance, Intentions, Lady Windermere's Fan, Lord Arthur Savile's Crime and Other Stories, The Duchess of Padua, The Importance of Being Earnest, The Picture of Dorian Gray, Vera

- *Document Preprocessing*: The texts are lower-cased, tokenized, and chunked into noun phrases (NP), verb phrases (VP), etc. The NP and VP chunks are removed depending upon the experiment.
- *Feature Representation*: The preprocessed documents are converted into a document-term matrix in which rows represent documents, and columns represent the filtered n-grams. Values in this matrix are the frequencies of the filtered n-grams in the documents. Each row is the feature representation of the corresponding document.

  While forming n-grams, size range from 2 to 4 is used. This means all the bigrams, tri-grams, and 4-grams that are possible to be formed from the training set are formed, and their frequencies in each document are computed. The n-grams having document frequency less than 5 are removed. These are the n-grams that are too infrequent, and unnecessarily increase the feature space dimension, and hence, are removed. The reason for not using the n-grams outside the size range from 2 to 4 is because (i) uni-grams cannot be used to differentiate between filtered and unfiltered n-grams as there are no new uni-grams formed after filtering, and (ii) n-grams of size bigger than 4 are too infrequent. N-grams of size-range 2 to 4 seem enough to capture most of the patterns of writing style.
- *Feature Subset Selection*: We select the N most differential features. These are the features which are used more by one author but less by others. To identify these, a

numeric value denoting the differential capacity is calculated for each feature, and
the N features with highest values are picked. To calculate the differential values,
the following procedure is followed for each feature. Firstly, its frequencies per
author are calculated. Then, its highest frequency among authors is multiplied
with the number of authors. Lastly, from this value, the feature's actual total
frequency in the documents is subtracted. The resulting value is considered as a
numeric measure of the differential capacity of the feature. After calculating this
value for each feature, the top N differential features are selected.

- *Training and Testing*: We use a classification scheme of K-nearest neighbor (K
NN) classifier with K = 5 and Euclidean distance. We use K = 5 because it is a
reasonable choice for our dataset of 50 samples, and we use Euclidean distance
because it is a simple yet effective metric of inter-textual distance used in authorship attribution [2]. The implementations found in the Scikit-learn [34] machine
learning library are used.

Because K-NN is an instance-based classifier, no explicit training is needed, except
that all samples are converted to their feature representations. During the testing-
phase, to classify a test-sample, first, its feature representation is compared with
that of all the training-samples using Euclidean distance, and the five least distant
training-samples are identified as its five neighbors. The test sample is assigned the
class to which the majority of its neighbors belong. All the samples in the test set are
assigned authors, and these predictions are compared with actual authors to calculate
the accuracy.

4   Experiments and Results

4.1   Comparing the Performance of Filtered and Unfiltered
n-grams

Using the literature dataset, the fivefold cross-validation procedure, and the classification mechanism described earlier, we compared the performance of filtered and
unfiltered n-grams. The three filtering criteria, viz. removal of noun groups, removal
of verb groups, and removal of both, were tried. In all the four cases (including the one
in which no filtering was done), experiments were carried out with different number
of top differential features—starting from 100, doubling till it remains below another
power of 10 (e.g., 1000), and then repeating so on. Because the feature set size was
around 8000 when both NP and VP chunks were removed, we did not try larger
feature sets except a final case in which we used all the features, i.e., without any
feature selection. The average classification accuracies over five folds are reported
in Table 3.

When no filtering was performed, the best accuracy was 70% with around 800
features. Adding more features did not improve but degraded it. Removal of only

Table 3 Classification accuracies with different number of differential features

	100	200	400	800	1000	2000	4000	8000	All features
Without Filtering	66	68	66	70	70	68	68	68	68
NP removed	70	72	70	74	74	74	74	76	76
VP removed	58	64	60	64	60	60	62	62	62
NP & VP removed	74	76	78	76	76	76	80	80	80

VP chunks did not prove useful with its best accuracy as 64%—much below the unfiltered case. Removal of only NP chunks and removal of both NP and VP chunks showed significant improvements in accuracies, upto 76% in the former, and upto 80% in the latter. There was also an overall increase in accuracy on increasing the number of differential features.

The finding, that when noun groups and verb groups were removed, the performance increased upto 10%, supported our hypothesis of filtered n-grams performing better than the unfiltered ones.

4.2 Feature Ablation Study

To further analyze the increase in performance achieved with filtered n-grams, we performed a feature ablation study. As described earlier too, the filtered n-grams contain a combination of traditional and new n-grams. The new n-grams include the ones only possible to be formed after filtering, such as "in of", "of of", "with of", "of who", "by of which", etc. Few illustrative examples were given earlier in Table 1. We hypothesize that the improvement in performance is due to these new n-grams. In our ablation experiments, we removed the new n-grams from the feature-set, and then re-run the experiments. Significant drops in performance were observed.

To distinguish between the traditional and the new n-grams in the feature-set, we maintained beforehand a list of all the n-grams of size from 2 to 4 possible to be formed from the corpus without filtering, and for each feature in the feature-set, checked whether it was present in that list or not.

The feature ablation results corresponding to the best accuracies achieved in Table 3 are shown in Table 4. The average values over the 5 iterations of 5FCV are shown. The accuracy improvements of 76% and 80% upon filtering NP and VP chunks can be observed to drop down to 70% which was also the best accuracy achievable by unfiltered n-grams. This further validates our hypothesis that the 6% and 10% increase in accuracy was due to the new n-grams formed due to filtering. Also, these new n-grams make up a significant portion of the entire feature-set. When only noun groups are removed, around one-third, and when both noun groups and verb groups are removed, around two-third are the new n-grams.

Table 4 Feature Ablation Results (When the new n-grams formed after filtering are removed from the feature set)

Filtering Type	Total no. of features used	No. of traditional n-grams	No. of new n-grams	Percentage of new n-grams (%)	Original accuracy (%)	Accuracy after removing the new n-grams (%)	Decrease in Accuracy (%)
NP removed	8000	5444	2556	31.95	76	70	6
NP and VP removed	8000	2881	5119	63.99	80	70	10

5 Conclusion and Future Scope

In this work, we compared the performance of filtered n-grams with unfiltered n-grams using an English literature dataset. The best performance achievable by unfiltered n-grams was 70%, whereas, using the filtered n-grams, the accuracy improved to 76% when only noun groups were removed from the text, and to 80% when both noun groups and verb groups were removed. This removal of theme-specific words from the text leaves only the skeleton of sentences, and these are enough to capture the writing styles of authors. The n-grams constructed from the remaining text contain some new n-grams which would not have been possible without filtering. These new n-grams help improve the authorship attribution. In the feature ablation study, we confirmed this when we found that the 6% and 10% increase in performance was due to these new n-grams. Thus, in this work, we found filtered n-grams can be effective features for authorship attribution.

Future work can involve exploring other possible filtering criteria, exploring the idea of automatic discovery of optimal filtering criteria, and using filtered n-grams for text classification tasks other than authorship attribution.

References

1. Juola P (2008) Authorship attribution. Foundations and Trends in Information Retrieval. 1(3):233–334
2. Oakes MP (2014) Literary detective work on the computer. John Benjamins Publishing Company, Amsterdam
3. Sidorov G (2019) Syntactic n-grams in computational linguistics. Springer, Cham
4. Sidorov G, Velasquez F, Stamatatos E, Gelbukh A, Chanona-Hernández L (2014) Syntactic n-grams as machine learning features for natural language processing. Expert Systems with Applications 41(3):853–860 (2014)
5. Mosteller F, Wallace DL (1963) Inference in an authorship problem. J American Stat Assoc 58(302):275–309
6. Holmes DI (1994) Authorship attribution. Comput Humanit 28(2):87–106

7. Holmes DI (1992) A stylometric analysis of mormon scripture and related texts. J Royal Stati Soc Series A (Statistics in Society).155(1), 91–120 (1992)
8. Holmes DI, Forsyth RS (1995) The Federalist Revisited: New Directions in Authorship Attribution. Literary Linguistic Comput 10(2):111–127
9. Matthews RA, Merriam TV (1993) Neural computation in stylometry I: an application to the works of Shakespeare and Fletcher. Literary and Linguistic Computing 8(4), 203–209 (01 1993). https://doi.org/10.1093/llc/8.4.203
10. Merriam TV, Matthews RA (1994) Neural computation in stylometry II: an application to the works of Shakespeare and Marlowe. Literary and Linguistic Computing 9(1), 1–6 (01 1994). https://doi.org/10.1093/llc/9.1.1
11. Holmes DI (1998) The evolution of stylometry in humanities scholarship. Literary Linguistic Comput 13(3), 111–117.https://doi.org/10.1093/llc/13.3.111
12. Burrows J (2002) Delta: a measure of stylistic difference and a guide to likely authorship. Literary Linguistic Comput 17(3):267–287 (09 2002).https://doi.org/10.1093/llc/17.3.267
13. Kešelj V, Peng F, Cercone N, Thomas C (2003) N-gram-based author profiles for authorship attribution. In: Proceedings of the Conference Pacific Association for Computational Linguistics, PACLING. 3:255–264
14. Benedetto D, Caglioti E, Loreto V (2002) Language trees and zipping. Phys Rev Lett 88(4):048702
15. Arun R, Suresh V, Madhavan CV (2009) Stopword graphs and authorship attribution in text corpora. In: 2009 IEEE international conference on semantic computing. pp 192–196. IEEE Computer Society, Los Alamitos, CA, USA
16. Stamatatos E (2009) A survey of modern authorship attribution methods. J Am Soc Inform Sci Technol 60(3):538–556
17. Koppel M, Schler J, Argamon S (2009) Computational methods in authorship attribution. J Am Soc Inform Sci Technol 60(1):9–26
18. Shrestha P, Sierra S, González F, Montes M, Rosso P, Solorio T (2017) Convolutional neural networks for authorship attribution of short texts. In: Proceedings of the 15th conference of the European chapter of the association for computational linguistics: vol 2, Short Papers. pp 669–674. Association for Computational Linguistics, Valencia, Spain
19. Altakrori MH, Iqbal F, Fung BCM, Ding SHH, Tubaishat A (2018) Arabic authorship attribution: An extensive study on twitter posts. ACM Trans Asian Low-Resour Lang Inf Process 18(1)
20. Sapkota U, Bethard S, Montes M, Solorio T (2015) Not all character n-grams are created equal: A study in authorship attribution. In: Proceedings of the 2015 Cconference of the North American chapter of the association for computational linguistics: human language technologies. pp 93–102. Association for Computational Linguistics, Denver, Colorado
21. Hitschler J, van den Berg E, Rehbein I (2017) Authorship attribution with convolutional neural networks and POS-eliding. In: Proceedings of the Workshop on Stylistic Variation. pp. 53–58. Association for Computational Linguistics, Copenhagen, Denmark
22. Alsulami B, Dauber E, Harang R, Mancoridis S, Greenstadt R (2017) Source code authorship attribution using long short-term memory based networks. In: Foley SN, Gollmann D, Snekkenes E (eds) Computer Security—ESORICS 2017. Springer International Publishing, Cham, pp 65–82
23. Seroussi Y, Zukerman I, Bohnert F (2014) Authorship attribution with topic models. Computational Linguistics 40(2):269–310
24. Fourkioti O, Symeonidis S, Arampatzis A (2019) Language models and fusion for authorship attribution. Inf Process Manage 56(6):102061
25. Amancio DR (2015) A complex network approach to stylometry. PLOS ONE 10(8):1–21
26. Machicao J, Corra EA, Miranda GHB, Amancio DR, Bruno OM (2018) Authorship attribution based on life-like network automata. PLOS ONE 13(3):1–21
27. Shalymov D, Granichin O, Klebanov L, Volkovich Z (2016) Literary writing style recognition via a minimal spanning tree-based approach. Expert Syst Appl 61:145–153

28. Zheng L, Zheng H (2019) Authorship attribution via coupon-collector-type indices. J Quantitative Linguistics 1–13 (2019). https://doi.org/10.1080/09296174.2019.1577939
29. Neal T, Sundararajan K, Fatima A, Yan Y, Xiang Y, Woodard D (2017) Surveying stylometry techniques and applications. ACM Comput. Surv. 50(6)
30. Clark JH, Hannon CJ (2007) A classifier system for author recognition using synonym-based features. In: Gelbukh A, Kuri ÁF (eds) MICAI 2007: Advances in Artificial Intelligence. Springer, Berlin Heidelberg, Berlin, Heidelberg, pp 839–849
31. Smedt TD, Daelemans W (2012) Pattern for python. J Mach Learn Res 13(66):2063–2067
32. Bird S, Klein E, Loper E (2009) Natural language processing with Python: analyzing text with the natural language toolkit. O'Reilly Media, Inc.
33. Al Y, Menezes R (2018) Author attribution using network motifs. In: Cornelius S, Coronges K, Gonçalves B, Sinatra R, Vespignani A (eds) Complex Networks IX. Springer International Publishing, Cham, pp 199–207
34. Pedregosa F, Varoquaux G, Gramfort A, Michel V, Thirion B, Grisel O, Blondel M, Prettenhofer P, Weiss R, Dubourg V, Vanderplas J, Passos A, Cournapeau D, Brucher M, Perrot M (2011) Édouard Duchesnay: Scikit-learn: Machine learning in python. J Machine Learn Res 12(85):2825–2830

Design and Performance Evaluation of Monarch Butterfly Optimization-Based Artificial Neural Networks for Financial Time Series Prediction

B. Satyanarayana, Sarat Chandra Nayak, and Bimal Prasad Kar

Abstract Financial time series (FTS) are extremely volatile by nature and quite unpredictable. Nonlinear approximation algorithms like artificial neural networks (ANN) are found proficient in predicting such time series. However, searching optimal ANN architecture with a suitable training methodology is an important and crucial task. This paper employs a recently proposed monarch butterfly optimization (MBO) algorithm for fine-tuning ANN structure. The MBO effectively searches the optimal input size, weight, and bias connections of ANN, thus forming a hybrid model (MBO-ANN). It is then utilized to predict the future values of two categories of FTS such as crude oil prices and stock closing prices. Comparative analysis of forecasting accuracy with other models and statistical significance tests is conducted to establish the suitability of MBO-ANN-based forecasting.

Keywords Financial forecasting · Artificial neural network · Monarch butterfly optimization · Stock closing prices · Exchange rate · Crude oil prices

1 Introduction

FTS are extremely volatile, nonlinear, and sensitive to global market scenarios, natural calamities, political situations, and so on. It is hard to predict the future

B. Satyanarayana
Department of Computer Science and Engineering, CMR Institute of Technology, Hyderabad
501401, India
e-mail: bsat777@gmail.com

S. C. Nayak (✉)
Department of Computer Science and Engineering, CMR College of Engineering and Technology,
Hyderabad 501401, India
e-mail: saratnayak234@gmail.com

B. P. Kar
Department of Computer Science and Engineering, Gandhi Institute for Technological
Advancement, Bhubaneswar 752054, India
e-mail: bimalprasadkar@gmail.com

value of a FTS. Though exact prediction is impossible, a small correct prediction may give enough hints and may be beneficial to naïve stakeholder thus influencing the economical scenario of a nation. FTS exhibits arbitrary fluctuations due to interaction with market and various macro-economical factors [1, 2]. Earlier methods of FTS forecasting are based on statistical models [3–5]. Over the last two decades soft and evolutionary-based methods came to existence and established their suitability over statistical methods [6, 7].

ANNs replicate the judgmental power of human brain [8, 9]. They pursue the practice of human brain to resolve complicated problems which are highly nonlinear in nature. They are very competent in modeling nonlinear processes with a little a priori assumption [9]. ANNs are powerful modeling procedures for real datasets where input–output mapping has regularities and exceptions [8, 9]. ANN-based forecasts are used to learn the chaotic measures of the stock markets. ANN-based stock forecast mainly involves two steps in order to deal with the nonlinearity and ambiguity coupled with the market data, i.e., identification of patterns and prediction of events with such patterns. Data mining researchers consider it as a powerful machine learning techniques. A survey of financial risk united with machine learning methods is worked out in [10]. More applications of ANN for FTS are found in [11–14]. Training of ANN plays a vital role in its applicability. Learning methods based on gradient descent are common in training ANN. However, during last two decades, a good number of swarm and evolutionary optimization techniques are groomed up and heavily used for ANN training. So far these methods are claimed to be superior than gradient descent-based methods in terms of landing at global minima with faster convergence. The widely used swarm intelligence methods include PSO [15], ABC [16], ACO [17], and so on. Based on the idea of natural evolution process, several evolutionary algorithms are proposed including GA [18], DE [19], and so on. Few optimization methods are framed by following the migration behavior of birds and animals such as animal migration optimization [20], and MBO [21]. MBO is a recently proposed metaheuristic. It simulates the migration behavior of butterflies across Mexico, southern Canada and northern USA. It updates the position of butterflies by applying i) migration operator which can be tuned by migration ratio and ii) butterfly adjusting operator. MBO can be processed in parallel and well capable to maintain intensification as well as diversification. Though few applications of MBO are available, its application toward FTS is scarce.

The intention of this paper is to design and access the performance of MBO-ANN model through FTS prediction. The MBO is used to train the ANN through finding its parameters. The resulting model is a hybrid model. We evaluated accuracy of MBO-ANN in forecasting two stock market data and two crude oil prices series. MBO-ANN prediction accuracy is compared with four other models. Extensive simulation studies and statistical significance test (Diebold-Mariano and Wilcoxon signed test) are carried out considering all forecasts and all FTS.

The remaining content is structured as follows. The methodologies used are presented in Sect. 2. The proposed MBO-ANN-based forecasting is described in Sect. 3. Experimental results are summarized and analyzed in Sect. 4, and the concluding points are given by Sect. 5.

2 Methods and Materials

2.1 Artificial Neural Network

This subsection presents an ANN structure using one hidden layer, as shown in Fig. 1. The optimal layer size and number of neurons can be chosen experimentally.

   This model has a single output unit to estimate one data at a time. The type of learning it follows is a supervised learning. A linear activation is used at input layer and a sigmoid at hidden and output neurons. The sigmoid activation is fired as follows.

$$y_{\text{out}} = \frac{1}{1 + e^{-y_{\text{in}}}} \tag{1}$$

where y_{out} the output and y_{in} are the input. The input layer handles an input variable with one node each. The hidden layer captures the nonlinearity among variables. The output at j^{th} hidden neuron is calculated as follows.

$$y_H = f\left(\text{bias}_j + \sum_{i=1}^{n} w_{i,j} * x_i\right) \tag{2}$$

where x_i is the ith input component, w_{ij} is the weight connecting, i^{th} input neuron to j^{th} hidden neuron and bias_j is the bias. The estimation y_{esst} at output neuron is computed as follows.

$$y_{\text{esst}} = f\left(\text{bias}_0 + \sum_{j=1}^{m} v_j * y_H\right) \tag{3}$$

where v_j is the synaptic weight connecting j^{th} hidden unit to response unit. Difference between y_{esst} and target is considered as the error generated by the forecast. Closer its value toward zero indicates the goodness of the model.

$$\text{error}_i = |y_i - y_{\text{esst}}| \tag{4}$$

   The gradient of the error function is used to update the weights and biases. The training patterns are repeatedly feed to the model during the training process, and the parameters are adjusted by a suitable learning algorithm until the desired input–output mapping occurs.

2.2 Monarch Butterflies Optimization

MBO is a recently proposed optimization algorithm motivated from the conception of immigration of monarch butterflies between southern Canada and northern USA and Mexico [21]. The former location is considered as Land1 (subpopulation1) and the latter is Land2 (subpopulation2). Individuals of monarch butterflies are located either of the two distinctive lands. The offsprings are generated by application of migration operator on the butterflies in Land1 or Land2. As per the natural concept, the monarch butterflies stay approximately five months in Land1 and seven months in Land2. Therefore, the population size of Land1 can be considered as $NP_1 = $ ceil $(p * NP)$ and that of Land2 $NP_2 = NP - NP_1$, where p is the ratio of butterflies in Land1 and NP is the total size of the population. A new individual is formed from the old one with the migration operator. The kth element of butterflies x_i in $t + 1$ generation can be represented as follows.

$$\begin{cases} x_{i,k}^{t+1} = x_{r1,k}^{t}, if\, r \leq p \\ x_{i,k}^{t+1} = x_{r2,k}^{t}, if\, r > p \end{cases} \tag{5}$$

where $r1$ and $r2$ are two randomly selected monarch butterflies from Land1 and Land 2. t is the current generation, and r is calculated as:

$$r = \text{rand} * \text{migration period} \tag{6}$$

where rand generates a random number drawn from uniform distribution, and the migration period is set to 1.2 (as per the basic algorithm).

As per the above discussion, bigger the value of p, more are the samples selected from Land1 (subpopulation1); otherwise, more are the samples selected from subpopulation2. Based on the value of p, a subpopulation plays an important role. By adjusting the value of p, the direction of migration operation can be balanced.

The second operator to update the position of butterflies is butterfly adjusting operator. As per this the kth element of jth butterfly at iteration $t + 1$ can be updated as follows.

$$\begin{cases} x_{j,k}^{t+1} = x_{\text{best},k}^{t}, if\, r \leq p \\ x_{j,k}^{t+1} = x_{r3,k}^{t}, if\, r > p \end{cases} \tag{7}$$

where x_{best}^{t} the best monarch butterfly of the population is so far obtained, x_{r3}^{t} is randomly selected for subpopulation2 and t is the current generation. In order to achieve further exploration and exploitation, it can be updated as follows.

$$x_{j,k}^{t+1} = x_{j,k}^{t} + \alpha * (dx_k - 0.5), \text{ if rand} > BAR \tag{8}$$

where BAR is the butterfly adjusting rate, α is weighting factor and dx is the walk step of the butterfly. The parameter dx and α can be computed as follows.

$$dx = \text{Levy}(x_j^t) \tag{9}$$

$$\alpha = \frac{S_{\max}}{t^2} \tag{10}$$

While the bigger α value encourages exploration, the smaller value encourages exploitation of the search space. $S_{\max}$ is the maximum walk step that a butterfly can move. Based on the above discussed concept, the MBO process can be framed as follows.

Algorithm 1: MBO algorithm
1. Initialization
   Initialize population P randomly.
   Set generation counter t = 1 and Maximum generation.
   Set size of NP1 and NP2.
   Set $S_{\max}$, BAR, migration period, and migration ratio p.
2. Evaluation of fitness
   Evaluate fitness of all butterflies according to their position.
3. **While** (t < MaxGen)
   Sort the butterflies based on their fitness.
   Divide all butterflies into Land1 and Land2.
   for I = 1: NP1.
   Generate Offspring using Eq. 1 and add to subpopulation1.
   end for.
   for j = 1: NP2.
   Generate Offspring using Eq. 2 and 3, add to subpopulation2.
   end for.
   New population = Subpopulation1 U Subpopulation2.
   Evaluate new population.
   t = t + 1.
   End While.
4. Output the best monarch butterfly as best solution

3 Design of MBO-ANN Forecast

As stated above, the MBO-ANN developed in this work is a hybrid model. The individual (monarch butterfly) of MBO can be visualized as in Fig. 2. Thus, each individual of MBO represents a potential candidate solution for the ANN. The population consists a finite set of such individual. The MBO explores and exploits the search space to find the optimal solution (i.e., weight and bias vector) for the ANN by using its search operators as described in Sect. 3.

The historical prices in the time series are used as input for the model. The optimal number of input data is decided by sliding a fixed size window over the series. The data points selected are termed as an input vector. The input vector after normalization is feed to the forecast. The optimal parameters of the model are chosen by MBO. The MBO-ANN-based forecasting is presented in Algorithm 2.

Algorithm 2: MBO-ANN-based forecasting

Begin

Step 1: Set ANN and MBO parameters.

Step 2: Select input data from the time series /*Use sliding window method*/

Step 3: Normalize input signal /*Use sigmoid data normalization method*/

Step 4: Train ANN model with normalized input data and MBO algorithm.

Step 5: Test the model with test data and record the error signal.

End.

4 Simulation Studies and Results

The model was simulated on four FTS. The first two are crude oil prices series retrieved from https://www.eia.doe.gov/ from May 1983 to August 2019. Other two are daily closing prices from Hang Seng Index (HSI) and Korean stock exchange (KOSPI) during the period Jan 2nd, 2015 to September 20th, 2018 collected from https://in.finance.yahoo.com/quote/. A statistical summary of these series is shown in Table 1.

For comparative study, we considered three other hybrid models and one gradient descent ANN (GD-ANN). The hybrid models are developed using genetic algorithm (GA-ANN), artificial bee colony optimization (ABC-ANN), and differential evolution (DE-ANN). To maintain the fairness, the same set of input vectors are fed to all models. The parameters for MBO were same as defined in [1] with few exceptions. The value MaxGen was set to 200, NP1 = 50 and NP2 = 50. We conducted ten independent trials to stay away from the randomness. The average of ten simulations was used for comparison study.

The simulated error statistics from five forecasting models are summarized in Table 2. The finest error values are shown in bold face. The forecast plots from crude

Table 1 Statistic summary from four financial time series

FTS	Descriptive statistics					
	Min	Max	Mean	Std dev	Skewness	Kurtosis
KOSPI	1.8296e + 03	2.5882e + 03	2.1600e + 03	203.85281	−0.4932	4.7927
HSI	1.8321e + 04	3.3374e + 04	2.51362e + 04	3.4677e + 03	−0.4268	5.4995
Daily Oil Prices	10.4200	145.2900	42.9828	28.5640	0.9953	2.8829
Weekly Oil Prices	11.0900	142.4600	42.8949	28.5241	1.0011	2.8921

Table 2 Error statistics from four financial time series

FTS	Error	Forecast				
		GD-ANN	ABC-ANN	GA-ANN	DE-ANN	MBO-ANN
KOSPI	Minimum	7.1384e-05	5.4786e-05	5.0336e-06	4.4436e-05	4.3265e-05
	Maximum	0.35434	0.27255	0.23225	0.23764	0.23003
	Average	0.00994	0.00976	0.00893	0.01009	0.00499
	Std.Dev	0.00874	0.00813	0.00750	0.00751	0.00457
HSI	Minimum	6.9508e-05	5.9528e-05	5.0527e-06	6.0507e-05	5.6507e-06
	Maximum	0.0382	0.0340	0.0312	0.0308	0.0299
	Average	0.0128	0.0122	0.0069	0.0094	0.0075
	Std.Dev	0.0166	0.0159	0.0055	0.0057	0.0040
Daily Oil Prices	Minimum	0.00005	0.00003	0.00002	0.00002	0.00001
	Maximum	0.0468	0.0432	0.0416	0.0427	0.0411
	Average	0.0325	0.0091	0.0078	0.0063	0.0063
	Std.Dev	0.0155	0.0075	0.0077	0.0019	0.0021
Weekly Oil Prices	Minimum	0.00006	0.00004	0.00004	0.00001	0.00001
	Maximum	0.0542	0.0435	0.0403	0.0385	0.0382
	Average	0.0157	0.0141	0.0091	0.0087	0.0081
	Std.Dev	0.0117	0.0087	0.0071	0.0074	0.0071

oil prices are shown in Figs. 3 and 4. It is observed that the MBO-ANN models obtained best minimum error statistic three times, best maximum error value four times, best average statistic for three times and standard deviation for four times. GA-ANN and DE-ANN obtained best average error statistic value one time each. Also, the MBO-ANN-based forecast achieved nominal average error compared to others. These evidences are in support of suitability of MBO-ANN forecast. To ascertain the statistical significance of MBO-ANN, we conducted Diebold-Mariano test and Wilcoxon signed test. The p and h values from these tests are summarized in

Table 3 Statistical significance test

Proposed method	Compared method	[p, h]—value	
		Diebold-Mariano test statistics	Wilcoxon signed test statistics
MBO-ANN	DE-ANN	(p = 1.9861, h = 1)	(p = -3.1217, h = 1)
	GA-ANN	(p = 2.1033, h = 1)	(p = 2.3230, h = 1)
	ABC-ANN	(p = 2.0174, h = 1)	(p = 1.9902, h = 1)
	GD-ANN	(p = 1.9974, h = 1)	(p = 1.9805, h = 1)

Table 3. The Diebold-Mariano statistics are found beyond ± 1.965 which justifies the significant difference between the suggested model and other. Similar observations are found from Wilcoxon signed test. These values show that MBO-ANN is much different from the comparative models and the null hypothesis is rejected (Figs. 5 and 6).

5 Conclusions

Forecasting financial time series is a difficult job. Nonlinear approximation functions like ANN have established their suitability in this regard. However, improper training of ANN heavily influences the prediction accuracy. In this paper, combining the better searching ability of MBO with good approximation capacity of ANN, a new hybrid model has been developed. The MBO is used to fine tune the parameters of an ANN, contrast to the usual backpropagation algorithm. The accuracy of MBO-ANN is measured through forecasting two stock closing prices and two crude oil price series. Four additional models such as DE-ANN, ABC-ANN, GA-ANN, and GD-ANN were developed in the same way as the proposed model. The average performance of MBO-ANN was found better to others which are also supported by the significance test results. Therefore, MBO-ANN can be treated as an effective instrument for prediction of FTS. Choosing the hidden layer size of MBO-ANN is still based on human interventions and needs automation. The suitability may be explored in other data mining problems.

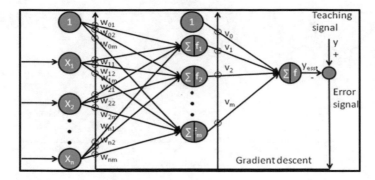

Fig. 1 A single hidden layer ANN-based forecast

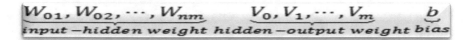

Fig. 2 Individual representation of MBO-ANN

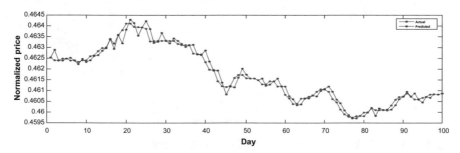

Fig. 3 MBO-ANN forecast plots from daily oil price series

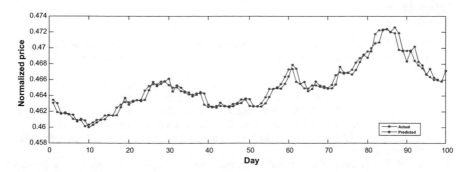

Fig. 4 MBO-ANN forecast plots from weekly oil price series

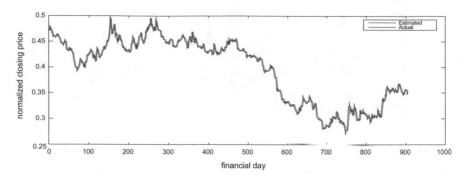

Fig. 5 MBO-ANN forecast plots from daily KOSPI prices time series

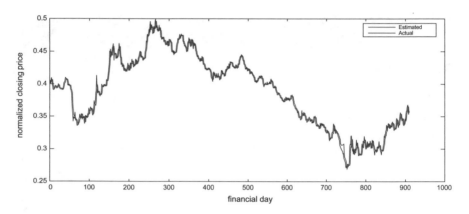

Fig. 6 MBO-ANN forecast plots from daily HSI prices time series

References

1. Hsu MW, Lessmann S, Sung MC, Ma T, Johnson JE (2016) Bridging the divide in financial market forecasting: machine learners versus financial economists. Expert System Appl 61:215–234
2. Nayak SC, Misra BB, Behera HS (2019) ACFLN: artificial chemical functional link network for prediction of stock market index. Evolving Systems 10(4):567–592
3. Adhikari R, Agrawal RK (2014) A combination of artificial neural network and random walk models for financial time series forecasting. Neural Comput Appl 24(6):1441–1449
4. Zhang GP (2003) Time series forecasting using a hybrid ARIMA and neural network model. Neurocomputing 50:159–175
5. Zhang H, Kou G, Peng Y (2019) Soft consensus cost models for group decision making and economic interpretations. Eur J Oper Res 277(3):964–980
6. Mostafa MM (2010) Forecasting stock exchange movements using neural networks: Empirical evidence from Kuwait. Expert Syst Appl 37(9):6302–6309
7. Nayak SC, Misra BB, Behera HS (2018) Artificial chemical reaction optimization based neural net for virtual data position exploration for efficient financial time series forecasting. Ain Shams Eng J 9(4):1731–1744

8. Haykin SS, Haykin SS, Haykin SS, Haykin SS (2009) Neural networks and learning machines (Vol 3). Upper Saddle River, NJ, USA: Pearson

9. Rajasekaran S, Pai GV (2003) Neural networks, fuzzy logic and genetic algorithm: synthesis and applications (with cd). PHI Learning Pvt, Ltd

10. Kou G, Chao X, Peng Y, Alsaadi FE, Herrera-Viedma E (2019) Machine learning methods for systemic risk analysis in financial sectors. Technol Econ Dev Econ 25(5):716-742

11. Nayak SC, Misra BB (2020) Extreme learning with chemical reaction optimization for stock volatility prediction. Financial Innovation 6(1):1–23

12. Nayak SC, Misra BB, Behera HS (2017) Exploration and incorporation of virtual data positions for efficient forecasting of financial time series. Int J Ind Syst Eng 26(1):42–62

13. Nayak SC A fireworks algorithm based Pi-Sigma neural network (FWA-PSNN) for modelling and forecasting chaotic crude oil price time series

14. Nayak SC, Misra BB, Behera HS (2015) A pi-sigma higher order neural network for stock index forecasting. In: Computational intelligence in data mining vol 2 (pp 311–319). Springer, New Delhi

15. Kennedy J, Eberhart R (1995) Particle swarm optimization. In: Proceedings of ICNN'95-international conference on neural networks (vol 4, pp 1942–1948) IEEE

16. Karaboga D, Basturk B (2007) A powerful and efficient algorithm for numerical function optimization: artificial bee colony (ABC) algorithm. J Global Optim 39(3):459–471

17. Dorigo M, Maniezzo V, Colorni A (1996) Ant system: optimization by a colony of cooperating agents. IEEE Trans Syst Man Cybern Part B (Cybernetics), 26(1):29–41

18. Goldberg DE (1989) Genetic algorithms in search, optimization, and machine learning, addison-wesley, reading, ma, 1989. NN Schraudolph and J 3(1)

19. Storn R, Price K (1997) Differential evolution–a simple and efficient heuristic for global optimization over continuous spaces. J Global Optim 11(4):341–359

20. Li X, Zhang J, Yin M (2014) Animal migration optimization: an optimization algorithm inspired by animal migration behavior. Neural Comput Appl 24(7–8):1867–1877

21. Wang GG, Deb S, Cui Z (2019) Monarch butterfly optimization. Neural Comput Appl 31(7):1995–2014

Interference-Aware Efficiency Maximization in 5G Ultra-Dense Networks

Vani Turupati, Raju Manda, K. Sowjanya, and A. Pavan

Abstract Spectrum efficiency and energy efficiency can be developed by using the ultra-dense networks. Because of scalability and intrinsic densification, the interference avoidance and sustainable designs become more composite. More collaboration opportunities exist between them because the number of small cells is arranged. By using increased EE presentation with small offering of SE, we can find the Nash bargaining CE2MG, and also we calculate the connection between EE and SE. It can depend on the Nash product maximization problem. To increase the EE presentation with and without limitations of SE, we can attain the closed form of sub-optimal SE. Finally, we calculate the numerical results for the better performance of energy efficiency and courtesy of CE2MG by presenting the CE2MG algorithm.

Keywords Energy efficiency (EE) · Spectrum efficiency (SE) · Ultra-dense networks · Cooperative game

1 Introduction

The advanced generation of telecommunication system experiences new problems due to improving and increasing efficient 5G [1] to allow both the SE and the EE. To make greater network density is considered as one of the powerful paths to together

V. Turupati (✉) · R. Manda · K. Sowjanya · A. Pavan
Department of Electronics and Communication Engineering, Kakatiya Institute of Technology and Science, Warangal, Telangana, India
e-mail: vaniturupati@gmail.com

R. Manda
e-mail: mr.ece@kitsw.ac.in

K. Sowjanya
e-mail: ks.ece@kitsw.ac.in

A. Pavan
e-mail: ap.ece@kitsw.ac.in

© The Author(s), under exclusive license to Springer Nature Singapore Pte Ltd. 2021   403
K. A. Reddy et al. (eds.), *Data Engineering and Communication Technology*,
Lecture Notes on Data Engineering and Communications Technologies 63,
https://doi.org/10.1007/978-981-16-0081-4_40

improve the quality of them in advantageous manner [2]. In whatever way, the incredible technical challenges are bring by the radical concentrated distributed of minimum number of cells, e.g., interference. To keep away from the single distortion and enlarge the SE, several useful actions of interference control were announced, e.g., additional asymmetrical and large amount of several useful actions of interference control were announced, e.g., additional asymmetrical and higher gains of the signal distortion are reducing by the large amount of distribution of minimum number of cells.

However, intervention and ecological design issues in the large quantity of network are enhancing the extra complex due to inherent strengthen and tractability. Different from the analysis on obstruction mitigation by increasing the SE [3] performance only, to identify the cooperative actions, we term to the negotiating cooperation diversions. The important subscriptions are condensed as follows.

- The active proposal to define and reduce the effects of composite interference {EFFECT} can be introduced. With a clear survey of the EE and SE links, it can be important in the ultra-dense network.
- With the studied interference framework, we deliver closer-form relationship between SE and EE in various several instances. It makes simple the bargaining cooperation maximization issues prepared in this effort.
- We suggest a diffused cooperation collaborative innovation to estimate the ideal solutions; here both efficiency and rectitude can be promised.

2 Trade-Off, Densification and Cooperation Bargaining Games, Ultra-Dense Networks and Challenges

Several important researches are thereby attempting on SE and EE maximization [4] or increment of trade-off. For the moment, different awareness of interference control game plans are constructed for either EE {OR} SE incrementation. Serving education for increment of relationship of SE and EE can be classified into two sections—one, identify the relation among SE and EE closed form; other to improve EE also need of SE [or] improve SE also need of EE. Furthermore, we analyze the problem in ultra-dense network, where there are less active users than the number of cells. In other words, $\lambda_b \gg \lambda_u$, where λ_b is solidness of access points. An λ_u indicates users' solidity. We represent ultra-dense networks and talk about the questions of the SE and EE and simplify interference and elevated energy dissipation provocations influencing the spectrum efficiency and the energy efficiency presentation. Ultra-dense networks and information about the increasing challenges of bandwidth efficiency and energy efficiency are used to reflect this. Representation of the ultra-dense {ULTRA-DENSE NETWORKS} heterogeneous small-cell networks [5] is shown in Fig. 1. It indicates that high-distributed omni-directional antenna of SeNBs coincides with the microcell eNB {MeNB} divided bandwidth approach. It can expect that L SeNBs divides spectrum of MeNB. In UDN, it can exclude the results and research of the trade-off between the EE&SE, and it is used to research the EE

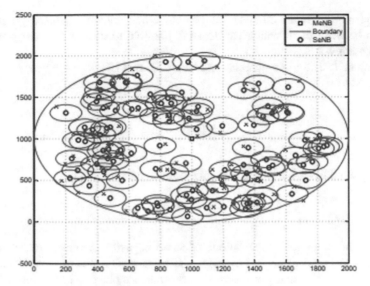

Fig. 1 Ultra-dense heterogeneous and small-cell network sketch

incrementation point also constant of SE of every ScNBs where $L \in l$ and $L = [1,......,L]$.

As shown in Fig 1, various hotspots are there, which can constitute a staying group of buildings. Every hotspot can be big connection of composed by multiple SeNBs; here every SeNB is also known as players. SeNBs help m_l small-cell user equipment [SUEs]. The Shannon capacity function is as follows.

$$C_l^m = Wln(1 + \frac{p_l g_l^m}{w_l^m})$$ (1)

where W, p_l, g_l^m are bandwidth of spectrum, the downlink transference power and channel gain. w_l^m is the sum of inter- and intra-tier interference plus noise power.

3   Trade-Off Analysis and Opportunities of Cooperation Gains

This part is used to obtain the trade-off connection among the EE and SE, survey chances of gains of collaboration, and notice the energy efficiency increment case.

A. Trade-off Analysis among SE And EE.

Consequently, spectrum efficiency and energy efficiency are helpful to simplify the relation between them which can make it possible to design an energy-efficient ultra-dense n/w.

Theorem 1: The trade-off relation among EE and EE of every user $l, l \in L$ is condensed as

$$\eta_l = \frac{\pi_l}{\epsilon_l \frac{\bar{w}_l}{w} \left(e^{\frac{\pi_l}{m_l}} - 1 \right) + \frac{p_l^{cst}}{w}} \tag{2}$$

Proof: noticed from 4 and 5, it is simple to finish that

$$W\pi_l = m_l c_l^m \tag{3}$$

with the presumptions (expectation) of same capacity obtained (executed) and connected to SeNB_l. For that moment, the transmission power p_l calculated via (2) with $c_l^m = \frac{W\pi_l}{m_l}$. Finally, we achieve a transmission power p l of a closed form as given as

$$p_l = \varpi_l (e^{\frac{\pi_l}{m_l}} - 1) \tag{4}$$

Moreover, with (5) we studied $c_l = \frac{\pi_l}{W}$ replacing 9 into 6 and with mandatory standardized transformation with respect to bandwidth W,

When $p_l^{cst} \neq 0$, if SE $\pi_l \to 0$, then the EE $\eta_l \to 0$ If SE $\pi_l \to \infty$, then $\eta_l \to 0$, if SE $\to 0$, therefore, the EE η_l is a first-increasing and then finally decreasing function of SE π_l

When $p_l^{cst} = 0$, if SE $\pi_l \to 0$, then the EE $\eta_l = \frac{Wm_l}{\epsilon_l \varpi_l}$ if SE $\pi_l \to \infty$, then $\eta_l \to 0$.

For the moment, if we can regularly accomplish the optimal SE $\pi_l^*, l \in L$, next the transmission rate of the SeN B_l is $c_l^* = W\pi_l$ can be decided with the above expectations.

4 CE2MG (Cooperative Energy Efficiency Maximization Game) and Formulation of Problem

This part is used to survey different types of cooperation expensive varieties will fight with this challenges via bargaining game theoretic design.

A. CE2MG

The optimal Nash cooperative bargaining compound (NBC) is studied depending on command that will attain an ideal trade-off among Nash rectitude and Nash self-evident efficiency downward the substructure of Nash self-evident presumption which has been certified in our past NBC-construct work.

Def3: an SE figure $\pi^* = \{\pi_l^*, l \in L\}$ is a NBC for the CE2MG can be promised by following the BRF of player $l \in L$ as the ideal spectrum efficiency figure that increases the EE usefulness when any spectrum efficiency figures.$\{\pi_{-l}, -l \in L, -l \neq l\}$ of other players are stated.

$$\rho(\pi - l) = \operatorname*{argmax}_{\pi_l \in S_l} u_l(\pi_l - \pi_{-l}) \tag{5}$$

Depending on the definition of the BRF, the NBC of player $l \in L$ is studied.

$$\pi_l^* = \rho\left(\pi_{-l}^*\right) \tag{6}$$

For this moment, the issue is moved to extract the BRF to meet cooperative bargaining equilibrium compound of player $l \in L$.

B. Problem Formulation

Normally, the CE2MG is calculated by using the given formulae called as optimization problem.

$$u = \prod_{l=1}^{L}(\eta_l - \eta_l^{min}) \tag{7}$$

Subject to $\pi \leq \sum_{l=1}^{L} \pi_l$

$\pi_l \leq \pi_l^{max}, l = 1,\ldots\ldots,L$

5 Optimal spectrum efficiency analysis without constraints

This portion is used to learn to increase the EE and the optimal SE; here, we exclude the separate and demand of SE. Then, few inherent qualities among discrete SE and EE are supplied.

We have given the EE as to clarify the following declarations,

$$\eta_l = \frac{\pi_l}{\varphi_l + \sigma_l} \tag{8}$$

where $\varphi_l = \mu_l(e^{\frac{\pi_l}{m_l}} - 1)$ is a functions of the SE π_l and here $\mu_l = \epsilon_l \frac{\varpi_l}{W}$ is effectual interference plus noise power $\sigma_l = \frac{p_l^{cst}}{W}$ is standardized circuit power utilization.

6 Survey of Optimal SE to Increase Energy Efficiency

In this portion, both the separate and the limited SE between various SeNB players are computed by CE2MG. We have attempted to detect the ideal spectrum efficiency of each player to increase system energy efficiency [6].

A. survey of SE To increase EE

It is hard for extracting the complete response function for the CE2MG algorithm.

Theorem 5: once more with all σ_l, μ_l, $l \in L$ as limitations; we finish that

$$\pi_l = m_l \frac{(\varphi_l + \sigma_l)(\lambda - k_l)\eta_l^{min}}{\frac{(\varphi_l + \mu_l)}{(\varphi_l + \sigma_l)} + m_l(\lambda - k_l)} \tag{9}$$

We suppose that the Lagrangian parameters λ, k_l and $l \in L$ are predecided according to KKT statements.

Proof: we construct the first-order derivation of lagrangian modified function with respect to the SE π_l is noted as

$$\frac{\partial \xi}{\partial \pi_l} = \frac{(\varphi_l + \sigma_l) - \frac{\pi_l}{m_l}(\varphi_l + \mu_l)}{\left(\pi_l - \eta_l^{min}(\varphi_l + \sigma_l)\right)(\varphi_l + \sigma_l)} \tag{10}$$

We work out the equation $\frac{\partial \xi}{\partial \pi_l} = 0$, and then with compulsory computation and we have

$$\frac{\pi_l}{m_l} = \frac{(\varphi_l + \sigma_l)(\lambda - k_l)\eta_l^{min} + 1}{\frac{(\varphi_l + \mu_l)}{(\varphi_l + \sigma_l)} + m_l(\lambda - k_l)} \tag{11}$$

We studied clutch KKT statements of the lagrangian parameters limitations of λ and k_l, $l \in L$

Here, theorem 5 is common finishing, with theorem 3 as a important case of $\lambda = k_l = 0$, for all $l \in L$.

B. Distributed CE2MG algorithm.

With the previous examination, the dispensed CE2MG model is given to attain the optimal SE of every SeNBs, $l \in L$ to increase the EE. CE2MG model is expressed as follows.

* Initialization: Initialize (originate) the associated parameters covers the whole number (L) of SeNBs in the give ultra-dense networks number of SUEs (m_l) related to every SeNBs, the circuit power utilization p_l^{cst} , the divided bandwidth W, the benefits of power amplifier ε_l, the minimal required EE η_l^{min}

and initiate Lagrangian factors of λ^0, k_l^0 of the SeNB player $l \in L$.

$$\pi_l = m_l(\varphi_l + \sigma_l)(\lambda - k_l) \quad (12)$$

*Observation: To attain the consciousness of the interference surroundings, it is hard for every SeNB player $l \in L$ to achieve the undesirable power $\varpi_l = \frac{1}{m_l}\sum_m \frac{w_l^m}{g_l^m}$ in arrangement of (3) and $\sigma_l = \frac{p_l^{cst}}{W}$. The demanding evaluation of wlm presents in appendix B.

7 Simulation results

In a UDN, UE density, BS density, UE distribution, and UE mobility sample are the primary detail features that describe its trends. The UE density is certainly one of the key points within the UDN (Figs. 2, 3, 4 and 5).

Figure 3 displays the UE motion with the random walk mobility model. Specifically, test 2 indicates the UE at the standard areas, in blue, and after one time step to move to one more area, it shows in purple. We are capable to compute that the everyday electricity intake is 19.12 W on the equal time as our proposed scheme is in region, on the identical time because the power consumption is 31.25 W, while all BSs live outstanding aware. Realize that we count on that the herbal electricity consumption of a small-cell BS is Pmax = 0.05 W (Fig. 6).

As proven in verify five, almost common 48.05% of power reduction can be finished, in evaluation with the case the vicinity all BSs are massive conscious. Additionally, the strength consumption is 16.23 W at the same time as our proposed

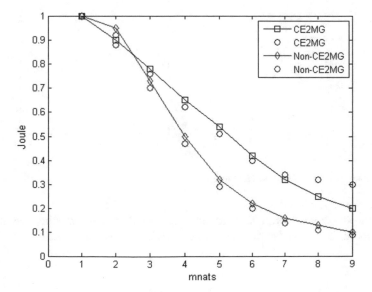

Fig. 2 Performance of the index EE vs. SE trade-offs

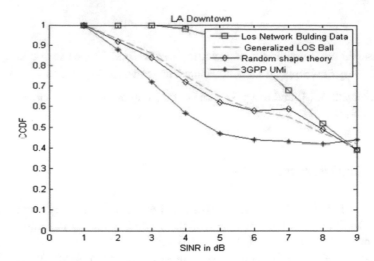

Fig. 3 Effects of densification on the system EE

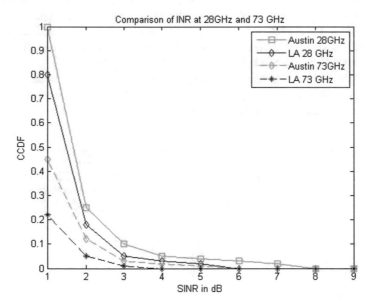

Fig. 4 Comparison of INR at different levels

maximum suitable BS aware/sleep scheduling is used; on the same time, the strength consumption is 31.25 W at the same time as all BSs maintain conscious.

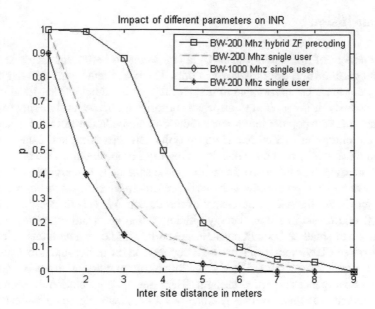

Fig. 5 Hybrid ZF precoding for single users

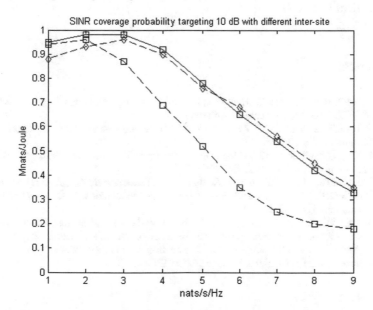

Fig. 6 SINR coverage probability at 10db

8 Conclusion

The increment of the system capacity via investigating both frequency and spatial diversities is done by the ultra-dense networks. The interference and ecological model issues were more composite due to tractability and the inherent densification, and simultaneously, it is essential to dispense control with decrease signaling expense. Different from the popular interference reduction study. In this article, we acknowledged the energy efficiency optimizing problem by surveying and using multiple cooperative diversity benefits while designing the SE presentation. We built a cooperative energy consumption maximization game scheme in this article. A dispensed CE2MG model was proposed to achieve the optimum SE components of each small cell to maximize the energy efficiency of the device. We established the optimum trade-off relationship between energy efficiency and spectrum efficiency with the importance of small-cell circuit power usage and well-known conditions of interference. To raise EE without losing more SE depends on the utility function. Compared to non-cooperation energy efficiency maximization gaming, the trade-off relationship, the convergence properties, the built EE framework presentation and rectitude of the implemented cooperation energy efficiency maximization gaming model contrast with non-cooperation energy efficiency maximization gaming compound justified by the numerical results.

References

1. Andrews JG, Buzzi S, Choi W, Hanly SV, Lozano A, Soong AC, Zhang JC (2014) What will 5G be? IEEE J Sel Areas Commun 32(6):1065–1082
2. Chih I, Rowell C, Han S, Xu Z, Li G, Pan Z (2014) Toward green and soft: A 5G perspective. IEEE Commun Mag 52(2):66–73
3. Han S, Yang C, Molisch AF (2013) Spectrum and energy efficient cooperative base station doze. IEEE J Sel Areas Commun 32(2):285–296
4. Bhushan N, Li J, Malladi D, Gilmore R, Brenner D, Damnjanovic A, Sukhavasi RT, Patel C, Geirhofer S (2014) Network densification: the dominant theme for wireless evolution into 5G. IEEE Commun Mag 52(2):82–89
5. Bennis M, Perlaza SM, Blasco P, Han Z, Poor HV (2013) Self-organization in small cell networks: a reinforcement learning approach. IEEE Trans Wireless Commun 12(7):3202–3212
6. Han S, Yang C, Andreas F. Molisch (2013) Spectrum and energy efficient cooperative base station doze. IEEE J Select Areas Commun 61(5):1258–1271

Miniaturized Printed Bowtie Antenna for WLAN Applications

Punnam Srividya, Akkala Subba Rao, B. Rama Devi, and J. Sheshagiri Babu

Abstract This paper deals with bowtie antenna with novel shape, which is proposed and simulated. The co-planar waveguide (CPW)-fed bowtie antenna is analyzed, and the parameters such as return loss are compared. It is designed using HFSS software. The results show that it has good impedance matching and is applicable for wireless local area network (WLAN). The size reduction in the antenna leads to small ground. The results confirm that it operates over a WLAN band of 4.9–5.9 GHz. It has stable radiation pattern at resonant frequency and a consistent gain over operating WLAN band.

Keywords CPW · WLAN · Radiation pattern · Gain

1 Introduction

Among the planar antennas, the most commonly used antenna is microstrip-fed antenna. Narrow-band, wide-beam antennas are designed for microwave frequencies and fabricated on a printed circuit board with a metallic layer, and it is surrounded by a substrate and a ground at the bottom [1]. They can be designed with different sizes and shapes like triangular [1], rectangular, elliptical [2], hexagonal, etc. These microstrip antennas are mostly used in mobile phones, satellites, missiles, and other applications based on their use. It is of low cost and can be easily fabricated.

P. Srividya (✉) · B. Rama Devi · J. Sheshagiri Babu
Department of ECE, KITS, Warangal, India
e-mail: srividyailaiah@gmail.com

B. Rama Devi
e-mail: ramadevikitsw@gmail.com

J. Sheshagiri Babu
e-mail: sheshagiri.jimidi@gmail.com

A. Subba Rao
Department of ECE, MANIT, Bhopal, India
e-mail: subbarao_ka@yahoo.com

© The Author(s), under exclusive license to Springer Nature Singapore Pte Ltd. 2021   413
K. A. Reddy et al. (eds.), *Data Engineering and Communication Technology*,
Lecture Notes on Data Engineering and Communications Technologies 63,
https://doi.org/10.1007/978-981-16-0081-4_41

The printed antennas are mostly preferred in wireless, mobile communications, and radar applications. These are basically categorized into different types based on the structure such as patch antenna and slot antenna. A microstrip slot antenna uses the feeding technique from the center of the strip. Feed has significance in the design of a broadband antenna structure. One such type of popular feed line is CPW feed, and the CPW-fed novel-shaped bowtie antenna is proposed in this paper which can be connected easily with different structures. Based on this structure, the return loss varies [3].The antenna operates at a range of frequency 4.9–5.9 GHz. The CPW [4] feed is used in the proposed method because this feed can minimize the radiation and has a high frequency range.

2 Antenna Structure

The proposed CPW-fed bowtie antenna is designed to operate over 4.9–5.9 GHz with a novel-shaped structure as shown in Fig. 1 with a size of 27×19 mm^2. This antenna is designed on a Rogers RT/Duroid 5880 substrate, and its relative permittivity is 2.2. The substrate is designed with the coordinate fields by using the solid box position, and the ground is designed which is placed underneath of the substrate and is named as Cu-clad in HFSS software. The novel-shaped bowtie antenna is designed with two rectangle positions, and these bowtie antennas are subtracted from Cu-clad material. The mesh operation is applied to the bowtie antenna in HFSS. In order to obtain the novel-shaped bowtie antenna, the objects selected from the 3D modeler tree in HFSS simulator and mirror image are considered to form bowtie antenna.

The structure of the proposed antenna is mentioned in Fig. 1. The parameters of this structure are as follows. $W1 = 4.5$ mm, $W2 = 4$ mm, $L1 = 7$ mm, $L2 = 4.5$ mm, $L3 = 13$ mm, $L4 = 0.5$ mm, $g = 0.5$ mm, and $r = 4.5$ mm. The total length (L) and width (W) of antenna are 19 mm and 27 mm, respectively.

Fig. 1 Novel-shaped CPW-fed bowtie antenna

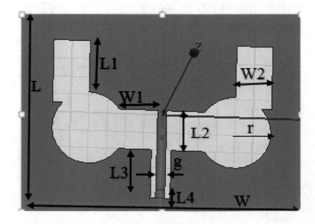

The patch is fabricated on the top of substrate. Two rectangular slots and circular slots are created which are shown in above Fig. 1. In order to simulate and allow the waves to radiate infinitely far into the space, the boundary conditions are applied in HFSS simulator. The absorbing boundary condition (ABC) is essential for radiation boundary in an air box, and the boundary condition should satisfy certain distance from the antenna.

3   Simulation Results

The performance of antenna is studied with various parameters such as return loss and radiation pattern. The basic bowtie antenna is fed by CPW and is shown in Fig. 2. The effect of bandwidth and return loss is determined by changing the length of the slot. The return loss of −30 dB is obtained at a resonant frequency of 5.5 GHz. From VSWR plot, it is clear that the antenna operates from 4.9 to 5.9 GHz.

The 2D polar plots as shown in Fig. 3 are obtained after the simulation in HFSS. It has been plotted for only at a frequency 5.5 GHz in E-plane and H-plane as shown in Fig. 3. The determined radiation patterns of the bowtie antenna in both H-Plane and E-Plane are stable [5]. The antenna radiates almost in bi-directional. The simulated broadside polarization levels in the E-plane and H-planes are at a frequency of 5.5 GHz. At 5.5 GHz, it is clear that the components of the electric field of all the arms are opposite to each other at both sides of the bowtie.

The gain versus frequency characteristic is given in Fig. 4. The simulated peak antenna gain is 2.5 dB over the WLAN band. The gain is of about 2.5 dB over the WLAN band.

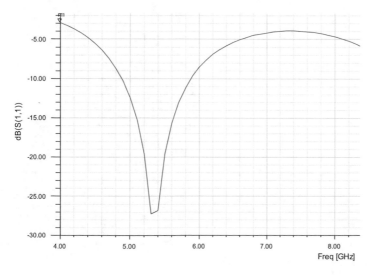

Fig. 2 Return loss of the proposed antenna

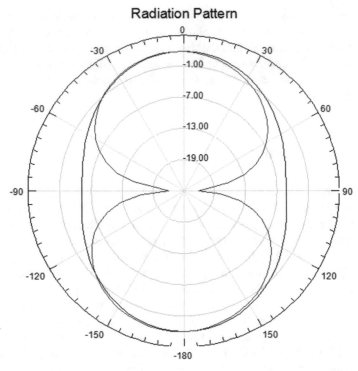

Radiation Pattern

Fig. 3 Radiation pattern at frequency of 5.5 GHz in E-plane (violet colored), H-plane (red colored)

Fig. 4 Gain curve

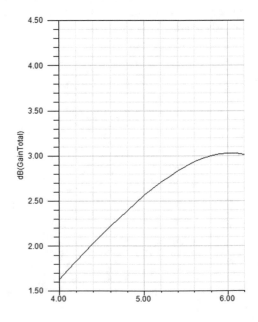

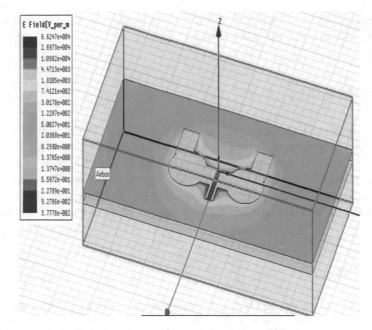

Fig. 5 Current distribution of bowtie antenna

Figure 5 shows the current distribution of an antenna. From the curve, it is observed that maximum current distribution [6] is observed on center of the patch at a resonant frequency of 5.5 GHz. The current distribution will also depend upon height of dielectric substrate of the proposed antenna.

4 Parametric Study

The effect of different parameters of the bowtie antenna on its frequency response is analyzed. Increasing L implies the increasing the length of the arm [6]. The return loss values at resonating frequency and operating frequency range for different length and width of the antenna are specified. It is clear from the table that as length and width change, the return loss at resonant frequency also affects. The intensive parametric

Table 1 Effect of different geometrical parameters on return loss and bandwidth

L1 (mm)	L2 (mm)	L3 (mm)	L4 (mm)	W1 (mm)	W2 (mm)	Return loss (db)	Bandwidth (GHz)
7	4.5	13	0.5	4.5	4	−27	5–5.9
7.5	5	10	0.5	5	6	−14.5	5–6

study as shown in Table 1 has been performed in order to observe operating frequency range with changes of lengths and widths.

5 Conclusion

In this paper, novel CPW-fed bowtie antenna is presented. The antenna is simulated for WLAN applications as it covers the band of frequency 4.9–5.9 GHz. It has gain of around 2.5 dB throughout operating band. The effects of different geometrical parameters on antenna's operating frequency range are observed. The antenna radiates bi-directionally at 5.5 GHz, and it provides stable radiation pattern in both H and E planes. The current distribution of antenna at 5.5 GHz is also discussed.

References

1. Chandel R, Gautam AK, Rambabu K (2018) Tapered fed compact UWB MIMO-diversity antenna with dual band-notched characteristics. IEEE Trans Antennas Propag 66(4):1677–1684
2. Wong KL, Wu CH (2005) Ultrawideband square planar metal plate with trident-shaped feeding strip. IEEE Trans Antenas Propog 53(4):1262–1269
3. Sindooja BV, Mary TAJ (2014) Comparative study of CPW-fed Bowtie antenna with ACS fed Bowtie antenna for wireless applications. Int J Eng Comput Sci 3(4):5561–5564
4. Sunnybabu M, Venkatesulu K (2018) Design of asymmetric CPW fed Bow-tieslot antenna for WLAN/WiMAXSystems. Int J Manage Technol Eng 8:2249–7455
5. Sefidi M, Zehforoosh Y, Moradi S (2016) A small CPW-fed UWB antenna with dual and notched characteristics using two stepped impedance resonators. Microw Opt Technol Lett 58:464–467
6. Prabakaran SS, Kalyan S (2019) A Coplanar waveguide fed circular microstrip antenna for UWB applications. Int J Innov Technol Explor Eng 8:2278–3075

Integrating SSIM in GANs to Generate High-Quality Brain MRI Images

Ruchika Malhotra, Kush Sharma, Krishan Kumar, and Nikhil Rath

Abstract Application of machine learning (ML) techniques in medical imaging is not new, but modern techniques like deep learning require a lot of quality data to perform effectively. Data, present in the field of medical imaging, imposes two main issues: first is the lack of anomalous data. For instance, only a few medical resonance images of the brain with tumor are present as compared to that of a healthy human brain. This makes the task of anomaly detection using ML techniques suffer from a lack of data. Second, the data that is present is highly confidential, and most of the time, the patient does not want it published anywhere. Many data augmentation techniques have been proposed in the past, but such techniques do not create data that is completely different from the given original data. In our work, we focus on using a novel extension of deep convolutional generative adversarial networks (DCGANs) to generate high-resolution brain magnetic resonance (MR) images of both anomalous and normal brain. This task is a tricky one as a tumor is present at random spots of the brain and has random sizes. For this reason, vanilla generative adversarial networks (GANs) cannot generate realistic tumor images. We propose a novel approach that integrates structural similarity index (SSIM) score in the objective function which, as we shall see, does a much better job at generating anomalous brain MR images as compared to vanilla GANs.

Keywords Generative adversarial networks · Synthetic medical image generation · Brain MRI · Objective function · Structural similarity index

R. Malhotra (✉) · K. Sharma · K. Kumar · N. Rath
Discipline of Software Engineering, Department of Computer Science and Engineering, Delhi Technological University, Delhi, India
e-mail: ruchikamalhotra2004@yahoo.com

K. Sharma
e-mail: kushsharma2910@gmail.com

K. Kumar
e-mail: krishan998@gmail.com

N. Rath
e-mail: nikrath@gmail.com

© The Author(s), under exclusive license to Springer Nature Singapore Pte Ltd. 2021   419
K. A. Reddy et al. (eds.), *Data Engineering and Communication Technology*,
Lecture Notes on Data Engineering and Communications Technologies 63,
https://doi.org/10.1007/978-981-16-0081-4_42

1 Introduction

Over the years, machine learning techniques have entered as an assistant in every single major field that exists in modern times. One such field is that of medical sciences. Machine learning techniques have proven to be immensely helpful and have revolutionized many aspects of the medical field since their introduction to medical sciences. One of the most popularly used techniques is the convolutional neural networks (CNNs) that have been extensively used to study multiple phenomena and take the field forward. CNNs have been used to analyze MRIs of multiple sorts, once of which is brain MRI. CNN is not as an all-powerful technique as it may seem though, as it requires an immense amount of data to be trained. Many data augmentation techniques have existed, but the new images that they generate closely resemble the original ones. Hence, there is practically no addition of data, neither an addition to performance. In this paper, we explore generative adversarial networks or GAN as an augmentation technique. Why GAN may perform better can be attributed to the fact that GAN learns the general pattern in the input images and can then create new images that are different from the original data, but do not deviate from that general pattern/characteristics of the input images. The challenge here is that the quality of MRI images, in general, is sensitive to contrast. Hence, we explore a new way that takes into consideration the image characteristics as well, through the help of a custom loss function. In the rest of the paper, we will start with a brief explanation of related works in the field, followed by the proposal of the architecture and the explanation of our novel objective function. We then show how the images generated using our approach compare to the images generated using vanilla GANs. Finally, conclusions are discussed.

2 Related Work

1. Initially, basic transformations like rotation, scaling, and flipping have been used as in [1] for data augmentation. Such data augmentation techniques have existed, but the new images that they generate closely resemble the original ones.
2. Most of the applications of GANs in medical imaging focus on image-to-image translation as in [2–4], most of which use cycle GANs or some extension of it. Such generation can be used in applications like generating magnetic resonance (MR) images from computed tomography (CT) images or segmenting the important portions of the original image.
3. In this work, we explore the use of GANs for a noise-to-image generation. This has been done by applying DCGANs studies as in [5, 6]. Images generated in the above-mentioned studies are of low resolution and are highly unstable; i.e., many generated images are highly distorted and far from real ones. Other applications can be found in [7–9].

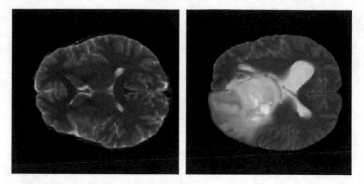

Fig. 1 Types of images in BraTS'18

3 Methodology

3.1 Dataset

For our purpose, we have used the famous BraTS'18 dataset. All BraTS multimodal scans are available as NIfTI files (.nii.gz). Images of.nii format capture the image in different planes (as in a 3D image), which must be converted to 2D images for training which can be easily done using MATLAB or python.

3.2 Preprocessing

In our work, we used T2-weighted images and converted them to 2D images by slicing the given medical image to a simple top view of the brain. The final dataset contains 355 brain images of dimension 288 * 432. These images are resized to obtain square images of dimension 288 * 288. Then the pixel intensities of these images are scaled from −1 to 1 before they are being fed to the network for training. The dataset contains images of both healthy normal human brains and brains with tumor as shown in Fig. 1.

3.3 Network Architecture

Generative Adversarial Networks Generative adversarial networks [10] or GANs first introduced in 2014 were a breakthrough in machine learning. The idea of GANs is to train two networks simultaneously. A discriminator (D) aims at accurately classifying the given image to be real or generated and a generator (G) which tries to fool the discriminator by generating images which are misclassified by the discriminator as a real image. After a sufficient amount of training, the generator learns the distribution of the given data and given a random Gaussian noise, it can generate highly realistic images. The minimax objective function for GANs as given in the original paper is written as:

$$\min_G \max_D V_{\text{GAN}}(D, G) = E_{x \sim P_{\text{data}}(x)}\big[\log D(x)\big]$$
$$+ E_{z \sim P_z(z)}\big[\log(1 - D(G(z)))\big] \qquad (1)$$

The first term in the loss function samples real images x from the given training data $P_{\text{data}}(x)$ and $D(x)$ gives the probability of it being real. In the second term, a batch of noise z is obtained from a given Gaussian distribution, $P_z(z)$. The generator generates an image $G(z)$ which is classified by the discriminator $D(G(z))$.

Proposed Loss Function As suggested in [11], training with the original loss function can sometimes lead to vanishing gradient problem, where the generator keeps generating images which can fool the discriminator but still are far from real. For these reasons, the studies in [5, 6] are limited to the generation of low-resolution images 128 * 128 as compared to 288 * 288 in our case. Also, the images generated are sometimes highly distorted and far from being close to real. To remedy this problem, we propose a new loss function for the generator.

Suppose,

$$L_{\text{adv}}(G(z)) = \log D(x \sim p_{\text{data}}(x)) + \log(1 - D(G(z))) \qquad (2)$$

is the adversarial loss. This is the same as the generative loss in the original loss function. We propose the inclusion of the second term of loss,

$$L_{\text{ssim}}(G(z), x \sim p_{\text{data}}(x)) = -1 * \text{SSIM}(G(z), x \sim p_{\text{data}}(x)) \qquad (3)$$

where SSIM is the metric denoting the numerical value of how structurally similar the generated image and the real image are. As given in [12], SSIM for two images m and n of the same size is calculated using the given formula,

$$\text{SSIM}(m, n) = (2\mu_m\mu_n + k_1)(2\sigma_{mn} + k_2)/\big(\mu_m^2 + \mu_n^2 + k_1\big)\big(\sigma_m^2 + \sigma_n^2 + k_2\big) \qquad (4)$$

μ_m and μ_n are the mean intensities, of m and n respectively,
σ_m and σ_n are the variances of m and n respectively,
σ_{mn} is the covariance of m and n,
k_1 and k_2 are constants to stabilize weak denominator.

SSIM accounts for the pixel similarities, i.e., the alignment, pixel densities, and contrast, etc., between the two images. For highly similar images, the SSIM outputs value close to 1, whereas for very different images, it gives a value close to -1. The loss term as given above would penalize the network if it generates, in our case, an

MRI that is different from the actual/real MRI. It would also reward it for producing images with a high SSIM score. We propose the new loss function for the generator as:

$$L_G(z) = k_{\text{adv}} * L_{\text{adv}}(G(z)) + k_{\text{ssim}} * L_{\text{ssim}}(G(z), x \sim p_{\text{data}}(x)) \qquad (5)$$

where k_{adv} and k_{ssim} are hyperparameters.

Deep Convolutional Generative Adversarial Networks The main idea of the generator in a DCGAN is to take as input Gaussian noise and to generate simple image features like lines and boundaries to be combined to generate more complex features like brain nerves, which are further combined to obtain more complex features like detailed patterns at specific portions of the brain. This is done repeatedly to obtain the final brain image. To the generator, we feed random normal noise with 100 dimensions to a fully connected layer of $128 \times 18 \times 18$ neurons followed by four Conv2D transpose layers, with 16, 32, 64, and 128 kernels, respectively. We use leaky-ReLU activation between these layers. For each of these layers, we use kernels of size 5×5, strides 2×2 and padding to obtain the output image of the same size. The final Conv2D layer outputs a batch of images of 288×288, which is passed through a tanh activation. The discriminator breaks the given image into smaller and simpler image features which are combined to calculate the final probability that the given image is real. The discriminator takes as input image batch with each image of size 288×288, followed by four Conv2D layers with 16, 32, 64, 128 kernels, respectively (Fig. 2).

The output from the last convolutional layer is flattened and passed to a fully connected dense layer with one neuron followed by sigmoid activation to output the final probability.

For training, Adam optimizer, possessing a learning rate of 2e-4 has been used for both the generator and the discriminator. The weights of the networks were initialized

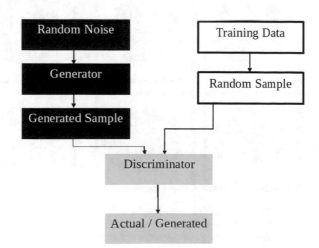

Fig. 2 Pictorial representation of the functioning of a generative adversarial network. The boxes indicated in bold outlines depict steps that are used in the first stage, where the discriminator is trained. During this stage, the discriminator learns to discriminate between real sample images and generated fake images. The black boxes depict steps used during the second stage, in which the generator is trained

to a normally distributed noise, with a standard deviation of 0.2. We keep the values of both k_{adv} and k_{ssim} as 1.

4 Results

Following results have been observed over a sample of images, i.e., an entire batch of images of both types, one batch generated using a generator trained using minimax, and the other using our custom loss function. Each batch contains 200 generated images, which were analyzed by volunteers, and the subjective results of the analysis have been summarized below.

Dynamic Range As can be seen from the comparison in Fig. 3, the images of the batch on the left (generated using minimax as loss function) have lower dynamic range than the batch on the right (generated using custom loss function). Images on the left are mostly grayed out. On the other hand, images on the right have much better dynamic range and there is an overall better distinction between various parts of the image.

Details Overall, after analyzing a variety of different images produced using both the techniques, it was majority consensus among the participants that the custom loss function produced not only higher number of detailed images (images that weren't blurred beyond usefulness), but also more detailed images among both batches. In Fig. 4, it can be seen that the image on the left has higher level of details than the one on the right. The comparison has been made over a batch of generated images, and the two images below are used to depict the general trend (trend of images in the second batch being more detailed than the one in the first batch).

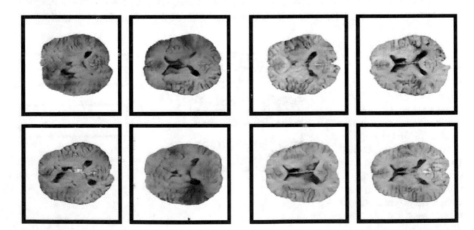

Fig. 3 Two batches of generated brain MRIs. The first batch (four images on the left) is of images generated using only minimax as loss function. The second batch (four images on the right) is of images generated using our custom loss function

Fig. 4 Difference in level of detail between images produced using a generator trained with minimax (left), and a generator trained using custom loss function (right)

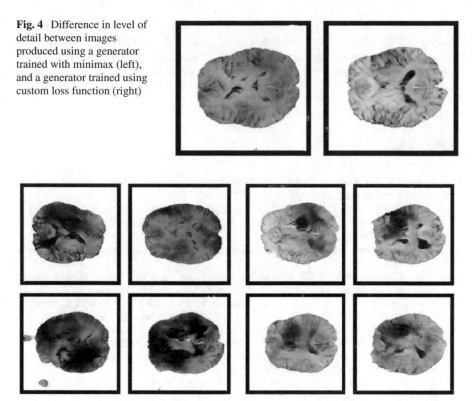

Fig. 5 Difference in MRIs of brain tumors, generated by generator using minimax (left batch) and a generator using custom loss function (right batch)

Tumor Generation Figure 5 shows the difference between generated MRIs by both the different generators. In the left batch, the tumor is not very distinguishable. There are rather random dark blotches in the images that do not resemble the characteristics of a tumor. On the other hand, in the images on the right the tumor is more distinguishable. The generated tumor is localized, and its region of coverage is quite clear. Images in the second batch imitate a real-life tumor much better.

5 Conclusion

Applying the custom loss function as described above, we see three major improvements. The images generated using custom loss function have a better dynamic range, have better details, and are able to imitate MRIs of brain with tumor much better. Much of the better details can be attributed to the better dynamic range as well, since the more prominent difference between the darker and lighter parts of the image enhances the details. SSIM helps in maintaining structural integrity in the generated images, which leads to the above mentioned results. But the more important among these is the better tumor generation with respect to both the shape, and the local position inside the generated brain. That is where the use of SSIM becomes most advantageous.

References

1. Simard PY, Steinkraus D, Platt JC (2003) Best practices for convolutional neural networks applied to visual document analysis. In ICDAR, vol 3
2. Zhu JY, Park T, Isola P, Efros AA (2017) Unpaired image-to-image translation using cycle-consistent adversarial networks. In: Proceedings of the IEEE International Conference on Computer Vision, pp 2223–2232
3. Liu MY, Breuel T, Kautz J Unsupervised image-to-image translation networks. ArXiv:1703.00848
4. Choi Y, Choi M, Kim M, Ha JW, Kim S, Choo J (2017) StarGAN: unified generative adversarial networks for multi-domain image-to-image translation. ArXiv:1711.09020
5. Han C, Hayashi H, Rundo L, Araki R, Shimoda W, Muramatsu S, Furukawa Y, Mauri G, Nakayama H (2018) GAN-based synthetic brain MR image generation. In: 2018 IEEE 15th International Symposium on Biomedical Imaging (ISBI 2018). IEEE, pp 734–738
6. Kazuhiro K, Werner RA, Toriumi F, Javadi MS, Pomper MG, Solnes LB, Verde F, Higuchi T, Rowe SP (2018) Generative adversarial networks for the creation of realistic artificial brain magnetic resonance images. Tomography 4(4):159
7. Menze BH, Jakab A, Bauer S, Kalpathy-Cramer J, Farahani K, Kirby J (2015) The Multimodal brain tumor image segmentation benchmark (BRATS). IEEE Trans Med Imaging 34(10):1993–2024
8. Bakas S, Akbari H, Sotiras A, Bilello M, Rozycki M, Kirby JS (2017) Advancing the cancer genome atlas glioma MRI collections with expert segmentation labels and radiomic features. Nat Sci Data 4:170117
9. Bakas S, Reyes M, Jakab A, Bauer S, Rempfler M, Crimi A (2018) Identifying the best machine learning algorithms for brain tumor segmentation, progression assessment, and overall survival prediction in the BRATS challenge. ArXiv:1811.02629
10. Goodfellow I, Abadie JP, Mirza M, Xu B, Warde-Farley D, Ozair S, Courville A, Bengio Y (2014) Generative adversarial networks. In: Advances in neural information processing systems (NIPS), pp 2672–2680
11. Xudong M, Qing L, Haoran X, Raymond YKL, Zhen W, Stephen PS (2016) Least squares generative adversarial networks. ArXiv:1611.09020
12. Wang Z, Bovik AC, Sheikh HR, Simoncelli EP (2004) Image quality assessment: from error visibility to structural similarity. IEEE Trans Image Process 13(4):600–612

Generation of Pseudo-Random Binary Sequence Based on Cipher Feedback Chaotic System for Cryptographic Applications

S. Sheela and S. V. Sathyanarayana

Abstract Generation of random key sequence is a primary component in stream cipher cryptosystem. Since chaotic sequences are inherently random, its properties like highly sensitive to initial conditions lead its application to cryptographic field. But, when the chaotic real sequences are converted into binary sequences, there will be loss of randomness. To improve the randomness of the binary sequence, a nonlinear cipher feedback structure is introduced. The nonlinearity present in the encryption part of the cipher feedback system will increase randomness of the sequence. The sequences generated are subjected to randomness tests like NIST, FIPS, cross-correlation and auto-correlation, and the results show that all generated sequences pass these tests. As the sequences qualify the randomness tests, stream ciphers can use them as key sequences.

Keywords Binarization · Chaotic function · Cipher feedback system · Randomness tests

1 Introduction

Sharing multimedia data over Internet is common nowadays. The insecurity present over the Internet gives rise to the need of data security. Data security consists of confidentiality, authenticity and integrity of data which can be achieved by using cryptosystems. These systems can be classified as symmetric and asymmetric cryptosystems. Symmetric cryptosystem uses common key for both encryption and decryption, whereas asymmetric cryptosystem uses two keys: one for encryption and the other for decryption. Further, symmetric cryptosystems are subdivided into block ciphers and stream ciphers. Encryption in block cipher is done on a block of data at once

S. Sheela (✉) · S. V. Sathyanarayana
JNNCE, Shivamogga, Karnataka, India
e-mail: sheeladinesh@jnnce.ac.in

S. V. Sathyanarayana
e-mail: svs@jnnce.ac.in

K. A. Reddy et al. (eds.), *Data Engineering and Communication Technology*,
Lecture Notes on Data Engineering and Communications Technologies 63,
https://doi.org/10.1007/978-981-16-0081-4_43

to produce a ciphertext block output of equal length, whereas encryption in stream cipher is done on a bit or a byte at once. Strength of the symmetric ciphers depends mainly on randomness of the key. Therefore, for stream cipher, a binary sequence of large period with unpredictable various patterns of given length are need to be generated by using a small and random seed key. What follows are details of the proposed scheme along with results and discussion.

2 Background

Requirement for true random binary sequences arises for the field which involves cryptographic applications, because there is an increase in modern communication systems employing authenticity by digital signature and electronic transactions. Securing privacy in these operations is very important [1]. Since cryptographic algorithms are publicly known, the cryptosystem security mainly depends on randomness present in the key [2]. This leads to the development of truly random binary sequence generator which can be used for encryption in digital communication systems [3]. There lies a behavior between rigid regularity and randomness based on pure chance [4] known as chaos. Hence, the study of chaotic systems became an area of major interest in mathematical, social science, and other complex fields. An active role of chaos theory plays in the quality improvement of the PRNGs [5]. The next subsection gives a brief overview of chaos.

2.1 Chaos

The chaotic functions generate dynamic, aperiodic and deterministic sequences which are sensitive to input conditions [6, 7]. The dynamic behavior of chaotic systems is due to nonlinearity present in the system. But all nonlinear dynamic systems are not chaotic [8]. These characteristics of chaotic functions are very much helpful to enhance the security in cryptographic applications. Since conventional cryptographic algorithms are weak, chaotic cryptography is preferably stronger [9].

In [10] various chaotic functions are listed, among them one of the simple 1D chaotic maps given in Eq. (1) is the logistic map.

$$x_{k+1} = r * x_k * (1 - x_k) \tag{1}$$

where $0 < x_i < 1, r \in [0, 4]$ and $k = 1, 2, \ldots$. The chaotic behavior of the logistic map is obtained for $3.57 \leq r \leq 4$. The sensitivity of the chaotic map for variation in input parameters is as shown in Fig. 1. Figure 1a shows sensitivity of the sequence with respect to input x_0, while Fig. 1b shows the sensitivity with respect to r. It can be observed that both the sequences gave same result for about initial 30 iterations and later the two sequences are completely different. To avoid the adverse

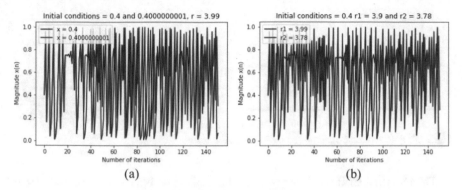

Fig. 1 Sensitiveness of logistic map for change in input conditions

effect of transition due to change in input, first 100 iteration of chaotic map is ignored before binarization. When the chaotic sequences are used for digital transactions, the real sequences are binarized, normally using thresholding technique [7]. One of the thresholding techniques used in this paper is based on Eq. (2).

$$b = \begin{cases} 0 & \text{if } x_n < 0.5 \\ 1 & \text{otherwise} \end{cases} \tag{2}$$

It is observed that, after binarization the inherent randomness of chaotic sequence is lost [5]. The proposed system enhances randomness using a cipher feedback (CFB) technique. The concept of CFB is used in one of the modes of operation of data encryption standard [2]. In CFB, preceding ciphertext block is taken as input for the algorithm to generate pseudorandom output, which is then XORed with plaintext block to produce the next block of ciphertext. The proposed system uses modified version of CFB. Details of the proposed concept are given in the next section.

3 Proposed System

Proposed Pseudo-Random Sequence Generator (PRSG) is as shown in Fig. 2. Chaotic real sequences $(y1, y2)$ are obtained using Eq. (1). Then the real sequences $(y1, y2)$ are binarized using thresholding technique given in Eq. (2). The output binary sequences $(b1, b2)$ are given to the modified CFB block to obtain a random sequence (C_i).

There are three inputs for CFB, namely initial vector (IV) of 64-bit, key and plaintext. In the proposed algorithm, following modifications for CFB are considered.

- Different key for each encryption process
- SDES encryption is used for implementation convenience
- 8-bit output is processed at a time.

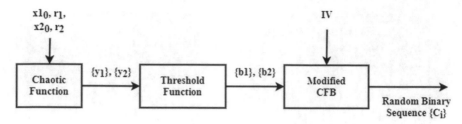

Fig. 2 Proposed pseudo-random sequence generator system

The modified CFB block is given in Fig. 3. Chaotic binary sequence $b1_i$, $\forall i = 1, 2, \ldots$ is used as key sequence for SDES block, and chaotic binary sequence $b2_i$, $\forall i = 1, 2, \ldots$ is used as a plaintext for CFB technique. To generate two binary chaotic sequences, two set of initial conditions $x1_0, r_1$ and $x2_0, r_2$ are required. As SDES takes key input of length 10-bit and plaintext input of length 8-bit, out of two binary chaotic sequences, length of one sequence should be a multiple of 10 (i.e., $N1$) and the other should be a multiple of 8 (i.e., $N2$). As a result, for the proposed system the input parameters are $x1_0, r_1, x2_0, r_2$, IV, $N1$ and $N2$.

Proposed Algorithm

1. Read the input parameters $x1_0, r_1, x2_0, r_2$, IV, $N1$ and $N2$.
2. Generate two binary chaotic sequences $\{b1\}$ of length $N1$ and $\{b2\}$ of length $N2$.
3. Divide the sequence $\{b1\}$ of step 2 into small blocks of size 10-bits for key input of SDES. Similarly divide the sequence $\{b2\}$ of step 2 into small blocks of size 8-bits for plaintext input of modified CFB.
4. Load the 64-bit shift register of modified CFB with the input IV of step 1.
5. Using ith 10-bits block of sequence $\{b1\}$ as a key, encrypt the upper 8-bits of shift register from SDES.

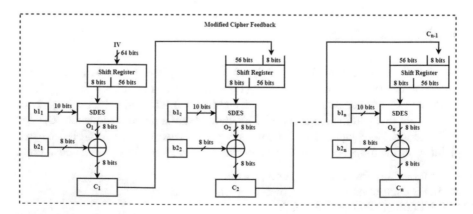

Fig. 3 Modified cipher feedback block

6. The output of step 5 is XORed with i^{th} 8-bits block of sequence $\{b2\}$ to get the ith 8-bit cipher block output $\{C_i\}$.
7. Shift the content of shift register to left by 8 positions and append the output of step 6 for encryption of next block as shown in Fig. 3.
8. Repeat from steps 5–7 until all the blocks of binary chaotic sequence $\{b2\}$ are encrypted.

The randomness property of the sequences obtained by the proposed algorithm are tested by applying Federal Information Processing Standard (FIPS)-141, Auto-correlation, Cross-correlation and National Institute of Standards and Technology (NIST) SP 800-22 REV.1a statistical test suite. The details are discussed in the next section.

4 Results and Discussions

The randomness of the sequences generated by the proposed algorithm with various initial conditions is tested using the tests provided by FIPS 141 [11], NIST, auto-correlation and cross-correlation [3, 7, 12]. Details of the results obtained are as follows.

4.1 FIPS—141 Test

The FIPS test includes Mono bit test, Poker test, Long run test, and run test. In **mono-bit test**, if 'X' is the count of occurrences of 1's in a consecutive 20,000 bits sequence, then the sequence passes this test, if $9654 < X < 10346$.

For **Poker test**, 5000 non-overlapping nibble groups are constructed from sequence of 20,000 consecutive bits. Let the counts of all $2^4 = 16$ possible occurrences be $g(i)$. The test is defined by Eq. (3).

$$X = \left(\frac{16}{N} * \sum [g(i)]^2 \right) - N \quad \text{where } N = 5000 \tag{3}$$

The test will be qualified if $1.03 < X < 57.4$

A run is the maximum continuous bit sequence formed by 1 s or 0 s. In **Run test**, from a stream of consecutive 20,000 bits, find the occurrence of all runs with a length ≤ 1. The test will be passed if every run's length is within a specified range given in Table 1.

Long run is a run made by consecutive 1 s or 0 s, with a length ≥ 34. A sequence of consecutive 20,000 bits is taken to find the number of long runs. If L represents the number of long runs, then for a sequence to pass the test the count L must be zero.

Table 1 Validity ranges of the run tests

Run length	Range
1	2267–2733
2	1079–1421
3	502–748
4	223–402
5	90–223
6 or >6	90–223

Table 2 Results of FIPS test conducted for 1000 sequences of different initial conditions

Sl. No.	FIPS test	No. of sequences passing the test					
		Before CFB seq {$b1$}		Before CFB seq {$b2$}		Afore CFB seq {C_i}	
1	Mono bit	20		23		999	
2	Poker	0		4		997	
3	Long run	1000		1000		1000	
4	Run tests	Run 0	Run 1	Run 0	Run 1	Run 0	Run 1
	1	56	246	50	256	1000	1000
	2	21	247	30	253	1000	1000
	3	29	278	36	308	1000	1000
	4	9	305	8	336	1000	1000
	5	3	352	2	379	1000	1000
	6 or ≥6	0	265	2	256	1000	1000

A sequence of 20,000 consecutive bits is taken for FIPS test. The test was conducted for 1000 sequences generated with random input conditions. Table 2 summarizes the results of FIPS test for the binary chaotic sequences {$b1$} and {$b2$} before applying CFB and the sequence {C_i} obtained by proposed system, i.e., after applying CFB. We can see that all the 1000 sequences generated using the proposed technique passes FIPS test. We can also note that, the output sequences generated using only thresholding technique alone does not pass FIPS test.

4.2 Auto-Correlation and Cross-Correlation Tests

For a binary sequence, the normalized hamming auto-correlation function is defined by Eq. (4), where 'A' represents number of equalities and 'D' represents number of inequalities when the original sequence is right/left circular shifted by τ bits and sequence length is given by n. The values of 'A' and 'D' are obtained by comparing the original and shifted sequences. Cross-correlation is also defined by Eq. (4) but the difference is, in case of auto-correlation only one sequence is considered where

Table 3 Results of normalized hamming auto-correlation and cross-correlation tests for 1000 sequences with different initial conditions

Sl. No.	Sequences	Auto-correlation Test		Cross-correlation test	
		Maximum	Minimum	Maximum	Minimum
1	Before CFB seq $\{b1\}$	0.9998	−0.3334	1.0	−0.3332
2	Before CFB seq $\{b2\}$	0.5324	−0.3244	0.3265	0.0252
3	After CFB seq $\{C_i\}$	0.0294	−0.0306	0.0334	−0.0314

as in cross-correlation two different sequences are considered for computation. The ideal off-peak value of hamming correlation is '−1' which indicate, that there is no correlation between the two sequences. If the value is equal to '+1' it indicates that there is a complete correlation between both the sequences and is not desirable for a sequence to be random.

A sequence of 20,000 consecutive bits is taken for both the tests. The tests were conducted for 1000 sequences which are generated with random input conditions. Table 3 summarizes the results obtained for both the tests for the binary chaotic sequences $\{b1\}$ and $\{b2\}$ before applying CFB and the sequence $\{C_i\}$ obtained by proposed system, i.e., after applying CFB.

$$R_h(\tau) = \frac{A - D}{A + D} \quad \tau = 1, 2, \ldots, n - 1 \tag{4}$$

We can see that all the 1000 sequences generated using the proposed technique passes both the tests as they are away from '+1' in upper bound and '−1' in lower bound. We can also note that the output sequences generated using thresholding technique alone does not pass both tests because their upper bound value is close to '+1' which is not desirable.

4.3 NIST SP 800-22 REV.1a Test

In the NIST test, three assumptions are made with respect to the sequence to be tested. They are uniformity, i.e., probability of 0's and 1's is 0.5, scalability, i.e., any test suitable for a sequence, is also suitable for a subsequence chosen randomly and consistency, i.e., the generator behavior must be consistence across the starting values. With these assumptions, NIST test suite [13] is having 15 statistical tests and all these tests use test statistic to compute P-value to summarize the strength of the evidence. The conclusion drawn on P-value is as follows:

- A perfectly random sequence will have its P-value $= 1$.

- A complete non-random sequence will have its P-value $= 0$.
- A 99% confidant random sequence will have its P-value $\geq$ '0.01'.

Table 4 summarizes the NIST test results for the sequences generated using thresholding techniques and proposed method with input parameters $\{x1_0, r_1, x2_0, r_2,$ IV, $N1, N2\} = \{0.4, 3.99, 0.40001, 3.78, 0 \times 73873ABC34D6FE19, 1,000,000, 800,000\}$. From the results, it can be observed that the sequence generated by the proposed generator has passed all the tests in suite, whereas the sequences generated using threshold technique passed only few tests.

As the sequence generated from the proposed technique passes all the randomness tests conducted, it qualifies as a key sequence for stream cipher applications.

Table 4 Comparison of NIST test results for the sequences before applying CBC and after applying CBC for $k = 8 \times 1_0 = -0.3 \times 2_0 = 0.3$ and IV $= 0 \times$ AE

Sl. No.	Statistical tests	Before CBC seq {$b1$}	Before CBC seq {$b2$}	After CBC seq {C_i}
1	Frequency	0.0	0.0	0.9625
2	Block Frequency	3.327e−129	0.0	0.0850
3	Run	0.0	0.0	0.4030
4	Longest run of ones	0.03915	8.829e−44	0.9524
5	Binary matrix rank	0.9821	0.3203	0.8088
6	Spectral DFT	1.6878	2.289e−08	0.5114
7	Non-overlapping template matching	0.0	0.0	0.3022
8	Overlapping template matching	0.0189	0.0	0.4115
9	Maurer's universal	0.0	0.0	0.0727
10	Linear complexity	0.4007	0.8301	0.4490
11	Serial	0.0	0.0	0.4244
12	Approximate entropy	0.0	0.0	0.7070
13	Cumulative sums (Forward)	0.0	1.0	0.9585
14	Cumulative sums (Reverse)	0.0	1.0	0.9330
15	Random excursions ($x = +1$)	0.9243	0.0104	0.9830
16	Random excursions variant ($x = +1$)	0.7518	0.1573	0.4531

5 Conclusion

Stream cipher system finds its place in addressing tactical applications like military systems, medical applications, etc., where information should be communicated securely. Here random key sequences are used for encrypting the data. This paper focuses on generation of random sequences based on chaotic systems. To increase the randomness of the binarized chaotic sequence the concept of CFB technique is introduced. The randomness tests are conducted for the sequence generated. The results show that the generated sequence passes all the tests. Hence, it qualifies as key sequence for stream cipher cryptographic applications.

References

1. Šajić S, Maletić N, Todorović BM, Šunjevarić M (2013) Random binary sequences in telecommunications. J Electr Eng 64(4):230–237
2. Stallings W, Brown L, Bauer MD, Bhattacharjee AK (2012) Computer security: principles and practice. Pearson Education, Upper Saddle River, NJ, USA
3. Sathyanarayana SV, Aswatha Kumar M, Hari Bhat KN (2010) Random binary and non-binary sequences derived from random sequence of points on cyclic elliptic curve over finite field GF (2^m) and their properties. Inf Secur J: Glob Perspect 19(2):84–94
4. Ditto W, Munakata T (1995) Principles and applications of chaotic systems. Commun ACM 38(11):96–102
5. Merah, L., Adda, A.P., Naima, H.S.: Enhanced chaos-based pseudo random numbers generator. In: 2018 International conference on applied smart systems (ICASS). IEEE, pp 1–7
6. Wang L, Cheng H (2019) Pseudo-random number generator based on logistic chaotic system. Entropy 21(10):960
7. Mandi MV, Haribhat KN, Murali R (2010) Generation of large set of binary sequences derived from chaotic functions with large linear complexity and good cross correlation properties. Int J Adv Eng Appl (IJAEA) 3:313–322
8. Cipra B (1999) A prime case of chaos. What's Happen Math Sci 4:2–17
9. Kocarev L (2001) Chaos-based cryptography: a brief overview. IEEE Circ Syst Mag 1(3):6–21
10. Sheela S, Sathyanarayana SV (2017) Application of chaos theory in data security—A survey. ACCENTS Trans Inf Secur 2(5)
11. Brown, K.H.: Security requirements for cryptographic modules. Fed Inf Process Stand Publ 1–53
12. Rukhin A, Soto J, Nechvatal J, Smid M, Barker E, Leigh S (2010) SP 800–22 Rev. 1a. A statistical test suite for random and pseudorandom number generators for cryptographic applications. National Institute of Standards and Technology (2010)
13. Zaman JKM, Ghosh R (2012) A review study of NIST statistical test suite: development of an indigenous computer package. arXiv:1208.5740

Face Mask Detection Using Feature Extraction

Kamale Usha, Beecha Sudeepthi, Devarakonda Mahathi, and Pillarisetti Shravya

Abstract Facial recognition has been one of the most instigating technological advancements of the recent times. With its limitless and myriad scope of applications, the facial recognition technology is so flexible that it can be customized to fit its application requirements with ease. With the world, as we know it, acquiring a different definition in these uncertain times, owing to the COVID-19 pandemic, this technology can be put to use to serve a very significant purpose—facial mask detection. With increasing emphasis on the usage of facial masks as the only known means to offer the needed protection, we decided to put to use the facial detection technology to detect facial masks on a person's face. In this project, we used the Viola–Jones detection framework with digital image processing to recognize a facial mask on a person's face and the appropriate or inappropriate positioning of it. This project will help us identify whether a person has worn a mask or not and if yes whether it was positioned properly or improperly and an alert regarding the same is sent. This would enable appropriate screening of individuals, providing a secure work environment and/or public interactions.

Keywords Mask detection · Viola–Jones algorithm · Feature extraction · Image processing · MATLAB GUI

K. Usha · B. Sudeepthi · D. Mahathi · P. Shravya (✉)
ECE Department, MVSR Engineering College, Hyderabad, India
e-mail: pillarisettishravya9@gmail.com

K. Usha
e-mail: usha_ece@mvsrec.edu.in

B. Sudeepthi
e-mail: beechasudeepthi@gmail.com

D. Mahathi
e-mail: mahathidvk@gmail.com

© The Author(s), under exclusive license to Springer Nature Singapore Pte Ltd. 2021   437
K. A. Reddy et al. (eds.), *Data Engineering and Communication Technology*,
Lecture Notes on Data Engineering and Communications Technologies 63,
https://doi.org/10.1007/978-981-16-0081-4_46

1 Introduction

Earlier, the idea of a facial recognition technology seemed like a page taken out of a science fiction book. But over the past decade, this page not only materialized but also quickly took the world by storm. Facial recognition system is a form of biometric technology that mathematically depicts the facial features of an individual from an image or a video and stores the data. Feature plays a very important role in the image processing sector [1]. The uses of this technology are wide spread. Law enforcement agencies find this technology extremely efficacious in aiding the process of investigation and criminal identification. Face recognition is a major provocation encountered in multidimensional visual model analysis and is a hot area of research [2]. Governments and other private organizations use it to identify people at workplaces, airports and various other places. In the recent times, mobile devices have begun using this technology as a means to ensure device security. Face detection is a bracket of computer analysis that determines the positioning and dimensions of a human face in an image. Face detection is a standout amongst topics in the computer vision literature [3].

Owing to the COVID-19 pandemic, covering our faces with a mask has become the new normal, as face masks are the most effective in preventing the spread of the virus. The usage of face masks has been declared as mandatory, if leaving home or visiting public spaces, by many nations worldwide. Face mask detection has become a crucial computer vision task to help the global society [4]. Different image processing steps involved in detection of face mask are presented in [5].

With the economies gradually returning to normalcy and workplaces opening up, these industries and workplaces have become extremely cautious to ensure a safe working territory for all its workers. Face mask detection refers to detect whether a person wearing a mask or not and what is the location of the face [6]. Face mask detection in educational institutions is discussed in [7]. The standard object detection techniques are usually based on feature extractors. Viola–Jones technique uses Haar feature with integral image method [8]. Preparation of wide faced data sets and masked data sets is discussed in [9, 10]. Training-based methods using CNN [11] are also used for mask detection. In [12], region-based detection of objects using data mining is presented. With the most worthwhile way to ensure safety being the usage of a face mask, companies and property owners require scrutinizing every individual entering the space to check if their mouth and nose are properly covered. But manually doing so, every single day, proves to be a tedious task. The most constructive way to do so is by the means of technology, particularly the facial recognition technology. By making the necessary changes to the facial recognition system, it is possible to enable facial mask detection. This forms the main motive of our project, to enable facial mask detection by means of facial recognition technology. In our project, we have managed to detect the proper and improper positioning of masks on an individual's face by means of digital image processing.

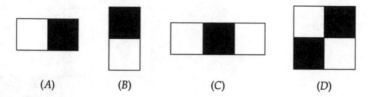

(A)   *(B)*   *(C)*   *(D)*

Fig. 1 Haar-like features (*Source* [13])

2 Objective of the Paper

The intent of the project is to develop a means to recognize the presence, the absence and the improper positioning of a face mask on an individual's face through facial recognition technology enabled by image processing and the Viola–Jones object detection framework.

3 Principle Involved

The Viola–Jones methodology for object detection [8] detects features in images with precision and works remarkably well with the human face structure. The Viola–Jones object detection algorithm is a combination of the concepts of the following to create a model for object detection that is quick and accurate.

3.1 Haar-Like Feature

Haar-like features help us to detect and obtain the necessary information from an image like straight lines, edges and diagonals that can be used to identify an object (the human face in our case) by extracting features from the input images instead of extracting them from their intensities (Fig. 1).

3.2 Integral Images

An integral image represents an intermediary image which we get by the cumulative addition of intensities on the pixel values above and to the left of the location on the actual image. This concept significantly reduces the extraction time required to extract the Haar-like features (Fig. 2).

Fig. 2 Rectangular region
using an integral image

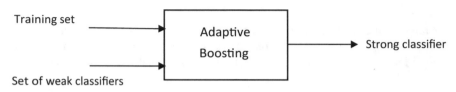

Fig. 3 Adaptive Boosting

3.3 AdaBoost Algorithm

The Adaptive Boosting (AdaBoost) Procedure is used to select the best/clearest set of features amongst the available ones. This classifier consists of a linear combination of the best features termed as "weak classifiers" which together form a "strong classifier." Here, each weak classifier learns from the mistakes of the preceding classifier and improves them. This algorithm is repeated for N iterations where N is the count of weak classifiers considered to find and can be set by us (Fig. 3).

3.4 Cascade Classifier

A cascade classifier is a classifier that works at multiple levels and sequentially detects and rejects the unwanted inputs at every stage. Each level consists of a strong classifier which is the output given by the AdaBoost algorithm. This approach paves way for building of simpler classifiers which can then be used to filter out most negative (non-face) inputs swiftly while concentrating on the positive (face) inputs (Fig. 4).

The classifier makes a decision using the following calculations:

$$
C_m = \begin{cases} 1, & \sum_{i=0}^{t_{m-1}} F_{m,i} > \theta_m \\ 0, & \text{otherwise} \end{cases}
$$

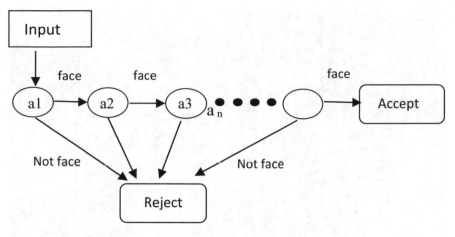

Fig. 4 Cascade classifier

$$F_{m,i} = \begin{cases} \alpha_{m,i}, & \text{if } f_{m,i} > t_{m,i} \\ \beta_{m,i}, & \text{otherwise} \end{cases}$$

$f_{m,i}$ is the weighted sum of the 2-D integrals. $t_{m,i}$ is the threshold for the ith feature extractor. $\alpha_{m,i}$ and $\beta_{m,i}$ are constant values associated with the ith feature extractor. θ_m is the threshold for the mth classifier.

4 Flowchart and Working

The diagram showing the work flow and methodology for feature extraction and mask detection are discussed in this section.

4.1 Flowchart

See Fig. 5.

4.2 Working

The project consists of two parts—detecting the facial features followed by the face mask detection. The process is initiated by taking image inputs from the system. Using the Viola–Jones algorithm in the computer vision toolbox, the mouth of an individual in an image is detected. Similarly, the person's nose is also detected

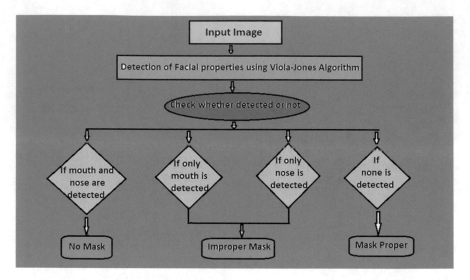

Fig. 5 Block diagram (of code operation)

using the same feature with the necessary modifications done to the code. These detected features are marked using the rectangle function for proper distinction. The initial input image is now marked with rectangles of different colors. This forms the intermediate image.

For the face mask detection, the intermediate image is checked for the features detected. Based on the number of features and the matrix of the rectangle pixels detected, the code prints an output which is one of three possible cases: (1) mask worn properly, (2) mask displaced/ not worn properly and (3) mask not worn at all. The objective of detecting a mask from a picture is, therefore, achieved.

We implemented the codes using MATLAB Graphic User Interface (GUI) and created an application to provide a more hands-on approach to our project—how it works and how efficient it is in achieving the intended results.

5   Simulation and Results

The obtained outcomes after the execution of the code and the GUI are as follows (Figs. 6, 7, 8, 9, 10 and 11):

6   Advantages, Drawbacks and Future Scope

Advantages

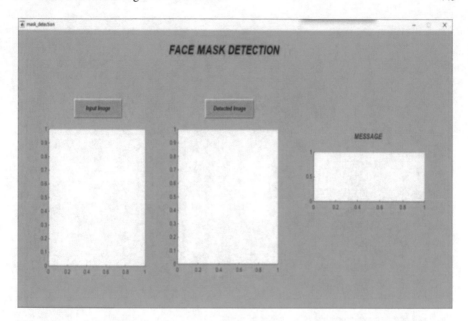

Fig. 6 Layout of the GUI

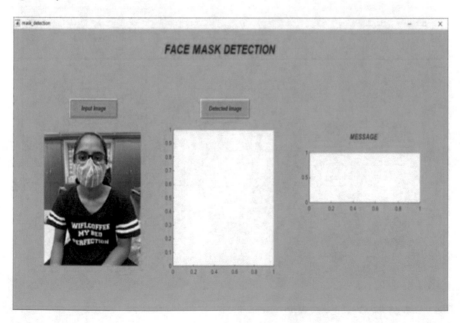

Fig. 7 Getting the input image

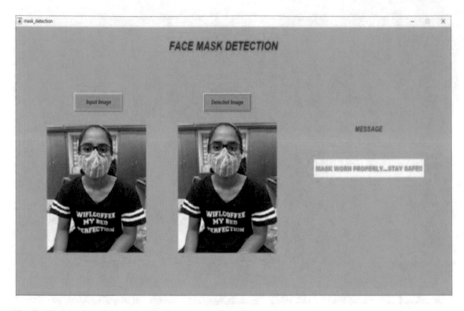

Fig. 8 Image showing the case of presence of mask

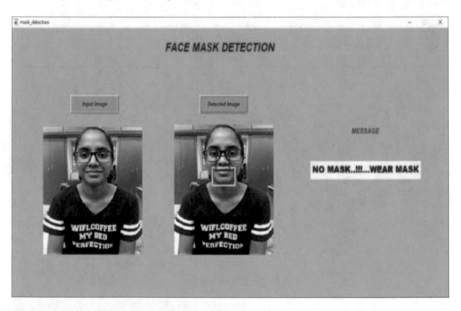

Fig. 9 Image showing the case of absence of mask

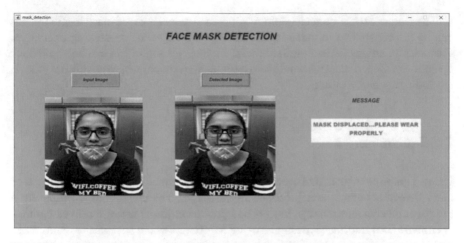

Fig. 10 Images showing the case of improper alignment of mask

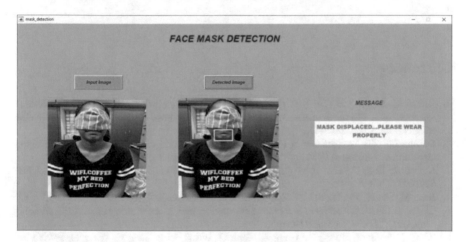

Fig. 11 Images showing the case of improper alignment of mask

The proposed model is best suited for images of higher resolution.

This method of detection works exceptionally well if the face is centered in the image.

Drawbacks

This method requires additional processing in case of tilted images.

For blur images, noise filtering techniques are to be adopted before applying the face mask detection algorithm.

The feature extraction methodology adopted to detect the face mask in still images can be further applied to video inputs and also for real-time processing of images. For video inputs by extracting frame-wise images, the algorithm can be applied for detection of face mask. This would be useful in many places where people or students are following queue system.

7 Conclusion

Using Viola–Jones structural procedure for feature extraction from still images, face mask detection technique is proposed in this paper. The methodology proposed in the paper detects face mask in images and puts them into three different categories. A Graphic User Interface is designed using MATLAB which accepts the input image and after applying the algorithm, displays the image with regions locating the nose and mouth and as well displays a message regarding the proper or improper placement of mask.

References

1. Kumar G, Bhatia PK (2014) A detailed review of feature extraction in image processing systems. In: 2014 Fourth international conference on advanced computing & communication technologies. IEEE, pp 5–12
2. Deshpande NT, Ravishankar S (2017) Face detection and recognition using Viola-Jones algorithm and fusion of PCA and ANN. Advan Comput Sci Technol 10(5):1173–1189
3. Kumar A, Kaur A, Kumar M (2019) Face detection techniques: areview. Artif Intell Rev 52(2):927–948
4. Jiang M, Fan X, Yan H (2020) Retina facemask: A face mask detector. arXiv:2005.03950V2
5. Chiang D (2020) Detect faces and determine whether people are wearing mask. https://github.com/AIZOOTech/FaceMaskDetection
6. Wang Z, Wang G, Huang B, Xiong Z, Hong Q, Wu H, Yi P, Jiang K, Wang N, Pei Y et al (2020) Masked face recognition dataset and application. arXiv:2003.09093
7. Savita KS, Hasbullah NA, Taib SM, Abidin AIZ, Muniandy M (2018) How's the turnout to the class? A face detection system for universities. In: 2018 IEEE conference on e-Learning, e-Management and e-Services (IC3e). IEEE, pp 179–184
8. Viola P, Jones M (2001) Rapid object detection using a boosted cascade of simple features. In: Proceedings of the 2001 IEEE computer society conference on computer vision and pattern recognition. CVPR 2001, vol 1. IEEE, pp I-I
9. Redmon J, Divvala S, Girshick R, Farhadi A (2016) You only look once: unified, real-time object detection. In: Proceedings of the IEEE conference on computer vision and pattern recognition, pp 779–788
10. Girshick R, Donahue J, Darrell T, Malik J (2014) Rich feature hierarchies for accurate object detection and semantic segmentation. In: Proceedings of the IEEE conference on computer vision and pattern recognition, pp 580–587
11. Ge S, Li J, Ye Q, Luo Z (2017) Detecting masked faces in the wild with LLE-CNNS. In: Proceedings of the IEEE conference on computer vision and pattern recognition, pp 2682–2690

12. Shrivastava A, Gupta A, Girshick R (2016) Training region-based object detectors with online hard example mining. In: Proceedings of the IEEE conference on computer vision and pattern recognition, pp 761–769
13. Lienhart R, Maydt J (2002) An extended set of haar-like features for rapid object detection. In: Proceedings of international conference on image processing, vol 1. IEEE, pp I-I

Performance Analysis of Polycystic Ovary Syndrome (PCOS) Detection System Using Neural Network Approach

R. Boomidevi and S. Usha

Abstract Across the worldwide, several women of reproductive age are affected by endrocrinological disorder called Polycystic Ovarian Syndrome (PCOS) or Polycystic Ovarian Disease (PCOD). It is actually estimated to have a prevalence between anywhere from 5 to 15% women are affected with acne or hirsutism or excessive hair growth, excess production of androgen (male sex hormones) which leads to alter gonadotropin levels, and this leads to issues with ovulation. The PCOS with associated comorbidities is obesity and metabolic syndrome, impaired glucose tolerance, and type-2 diabetes, endometrial cancer, infertility (Anovulation), sleep apnea, depression anxiety, cardiovascular disease, and Non-Alcoholic Fatty Liver Disease (NAFLD). The clinical test for PCOS wastes the time and cost for the patient and give extra depression to PCOS patient. To overcome this problem, this paper proposed artificial Neural network model which automates the PCOS detection in an early stage. This ANN model will act as an abetted tool which saves the doctor time to test the patient and also reduce the predicted time for detecting the risk of PCOS. This system is simple and very effective which predicts the PCOS in an early stage with the accuracy of 99%.

Keywords Endrocrinological disorder · Polycystic ovarian syndrome · Artificial neural network model

1 Introduction

ANN is one of the advanced technology widely used to healthcare organization for decision-making. ANN is the mathematical algorithm which trained from dataset and

R. Boomidevi (✉)
Electronics and Communication Engineering, Paavai Engineering College, Namakkal, India
e-mail: kboomidevi@gmail.com

S. Usha
Department of Electrical and Electronics Engineering, Kongu Engineering College, Perundurai, India
e-mail: usha@kongu.ac.in

© The Author(s), under exclusive license to Springer Nature Singapore Pte Ltd. 2021   449
K. A. Reddy et al. (eds.), *Data Engineering and Communication Technology*,
Lecture Notes on Data Engineering and Communications Technologies 63,
https://doi.org/10.1007/978-981-16-0081-4_47

acquire the knowledge in the standard data. The trained ANN model acts like a human brain which predicts the complex non-linear relationship between independent and non-independent features even human brain also fails to detection.

PCOS is a conglomeration of syndrome rather than a specific disease, and the causes has not clearly understood by many patients, and they are thinking that it is not necessary to take treatment even when the cysts may develop in the ovaries.

In regular menstrual cycle, the brain simulates the hormone GnRH which travels in a blood stream to reach pituitary gland that releases Follicle Simulating Hormone (FSH) which travels in a blood stream to the ovary that leads to develop an egg and simulate an estrogen hormone in the blood stream. Once the estrogen reaches the brain, which release the Luteinizing Hormone (LH) that break the egg in ovary. Then, it travels through the uterus if fertilized, then it will stay in the uterus if not fertilized, then it will be shed and the women will have a normal menstrual period.

The primary abnormalities in PCOS are the irregularity in the discharge of GnRH which disrupts the LH and FSH hormones that lead to abnormal of menstrual cycle. This also leads to produce more male hormone testosterone or androgen, and ovulation is inhibited. Women with PCOS may notice a many symptoms such as unpredictable irregular periods, acne and excess hair growth, thick dark pigment developing in neck and underarms, depression and mood disorders.

2   Literature Survey

McCartney and Marshall [1] suggest that many health issues are exist around us; among them, we are interested in the area which are related to the issues for women in a reproductive age. According to the literature analysis, 5–10% of women are afflicted by condition xcalled Polycystic Ovary Syndrome (PCOS). The symptoms of PCOS are miscarriage (infertility) and also rise the chance of insulin resistance, over weight (obesity), cardiovascular disease, and mental depression [2]. Women who have PCOS suffered by obesity, acne, infertility, variation in menstrual periods, hair loss, M-shape balding in front head, and hirsutism problem due to androgen hormones segregated in body [3].

PCOS affected women who have number of follicles in ovary are 12 or greater than 12 follicles per unit area [4]. In other papers [5, 6], authors proposed that the number of follicle in ovary is 12–20. Gibson-Helm et al. [7] suggest that well-equipped diagnosed model makes simple to predict PCOS. Insufficient data and late report given by many women create a huge gap for early detection.

The treatment should have varied from patient to patient. The diagnosis can be carried out based on the symptoms, necessity of individual person, and expectancy of the each patient [8]. Compared to other symptoms, infertility symptom treatment was significantly higher in women to reporting PCOS. So infertility can be considered as an important cause for PCOS which also affects health [9]. Norman et al. [10] considered that the important diagnostic attributes of PCOS are hyperandrogenism, anovulation, and also, he suggests that PCOS can occur because of strong genetic that is affected by gestational movement, lifestyle factors, or both.

The ubiquity of the metabolic syndrome in PCOS is approximately 43–47% double times or higher than for women in population [11]. Dhayat et al. [12] proposed that the PCOS is usually detected by Rotterdam criteria, or some standard can be fixed by the society.

The intervals of vaginal bleeding takes the time of lower than 35 days that leads to ovulation which have been reported by 20% of PCOS women. According to clinical analysis, irregular menstrual cycle predict the rate of insulin resistance in women with PCOS [13]. Ultrasound scan and image processing techniques can be mainly used for examine the PCOS [5, 14]. Others are used metabolic features for diagnosis the PCOS [15]. Recent studies, especially for research, are used obesity [16] and genetic factor [17] to predict. Dayat et al. [18] considered that studies about urinary steroid hormone metabolites or enzyme activities in women to diagnose the PCOS.

3 Methodology

Artificial neural network model is developed for predicting the PCOS. Figure 1 shows the representation of each block for the proposed ANN model. The dataset which comprises of class 0 and class 1. The two important blocks in the proposed model consist of preprocessing and ANN model. Preprocessing performs feature scaling and extraction which can be implemented by scikit learn, and ANN model performs classification of the input data [19].

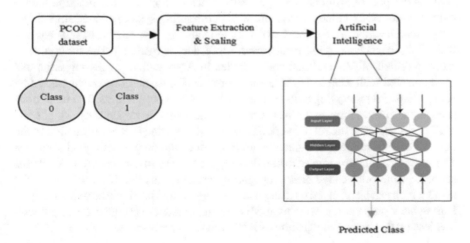

Fig. 1 Block diagram of the the proposed model

3.1 Data Description

The dataset consists of information to design the model. It is a group of data structure as table in the form of rows and columns. To predict the PCOS, the dataset can be gathered from repository platforms like kaggle which is collected from 10 different hospitals across Kerala. This dataset which involves all physical and clinical parameters to discover the PCOS. These standard datasets are well trained by ANN model to predict the PCOS. It can be classified in two groups: class 0 and class 1. Class 0 specifies that women without PCOS, and class 1 specifies that women with PCOS.

3.2 Preprocessing

Preprocessing must be used for every model to produce accurate forecasts. The dataset involves 41 channels used for predicting the PCOS. Before apply to the model, the unwanted columns and noises are removed called as cleaning data which is applied to the model for training. In this model, 33% of data can be considered as testing set, and 67% of data can be considered as training set which can be implemented by using python module.

3.3 Artificial Neural Network

ANN is the part of artificial intelligence that means to imitate the function likes a human brain. ANN is the component of the computing system created for solving the problems which are not able to solve by the humans. It is capable of self-learning that allow to generate good results even more data is available.The thousands or more than that of artificial neurons or nodes in ANN called processing units that are connected with each and every neuron or node. These neurons are connected by using edges. Preprocessing units comprise of input and output units. The input units accepts the information in various forms, and its structure depends on weight and biasing system. The neural network has to be trained using those information in the path of forward propagation and attempt to produce the output with good accuracy. The activation function is used to find the output of each and every neuron. ANN uses set of learning rules called back propagation which means that if error occur in the result which can be resolved by using back propagation to produce the perfect output. During back propagation, weights and biases are updated to produce the output with less loss [19]. The output equation of ANN model is represented in Eq. 1

$$O_j = \sum_i^n w_i x_i + b \tag{1}$$

where O_j represents output of neuron, w represents weight associated with each neuron, x represents input of neuron, n represents length of input, j represents number of neurons, and b represents bias. In the proposed model, it consists of input layer with PRelu activation function and 100 units followed by hidden layers of softplus activation function, and one output layer with sigmoid activation with 1 unit. The PRelu and softplus activation function are defined in Eqs. 2 and 3

$$f(x) = \begin{cases} x_i, & x_i > 0 \\ \alpha x_i, & \text{otherwise} \end{cases} \tag{2}$$

$$f(x) = \ln(1 + e^x) \tag{3}$$

Dropout is a technique which drops that means ignoring or deactivating some nodes or neurons in the network. These techniques are applied during training phase to avoid or limit over fitting which is added between layers. Overfitting is the one type of error that occurs when the network is well fitted to a limited number of input samples. The architecture with dropout produce the output accuracy of 99% but without dropout produce the result with 63%. It is noticed from the analysis, the performance difference of the architecture with dropout from the architecture without dropout is 37%. The last layer is the output layer that uses 1 unit which represents either class 0 or class 1. Figure 2a, b shows the comparison of architecture with and without dropout layer.

3.4 Feature Extraction and Feature Scaling

The feature extraction and scaling are the important part in the model which is used to decrease the number of needed resources without disrupts the original information that can be mainly achieved by using important parameters such as mean and standard deviation. In this proposed model, the features are scaled and extracted by using three methods. They are standard scaler min–max scaler and robust scaler.

3.4.1 Standard Scaler

The standard scaler is defined in Eq. 4

$$S = \frac{x - \text{mean}(x)}{\text{std dev}(x)} \tag{4}$$

where x is the input features, and S is scaling factor. If the data's are not distributed normally, then it cannot be considered as best scaler.

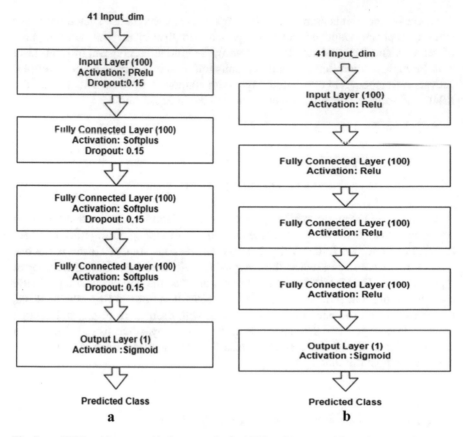

Fig. 2 a. ANN architecture with dropout units. **b.** ANN architecture without dropout units

3.4.2 Min-Max Scaler

The equation of min–max scaler is given by the Eq. 5

$$S = \frac{x_i - \min(x)}{\max(x) - \min(x)} \tag{5}$$

where x is the input features, and S is scaling factor.

3.4.3 Robust Scaler

It is expressed in Eq. 6

$$S = \frac{x_i - Q_1(x)}{Q_3(x) - Q_1(x)} \tag{6}$$

where x is the input features, Q is interquartile, and S is scaling factor.

4 Results and Discussion

The reliability of the model can be found from accuracy and performance. These parameters are tuned by using hyperparameters such as learning rate, optimizer, etc., from that which one produce the highest performance can be considered as a good model. The model is fine-tune by using different learning parameters to get the good accuracy. Cross-entropy and mean squared error are the two main loss function used for testing the model. In the proposed model, cross-entropy is used as a right choice for testing.

4.1 Preprocessor Analysis

The main aim of the preprocessing is to increase the number of locations and also decrease the storage of memory. Figure 3 shows the comparison of three preprocessor methods with accuracy. It is observed from the figure standard scaler produce the highest loss when compared to other scalers. Min-max scaler produces good accuracy compared to robust scaler; therefore, it can be considered as the best model for the proposed neural network.

Fig. 3 Comparison of different preprocessor method

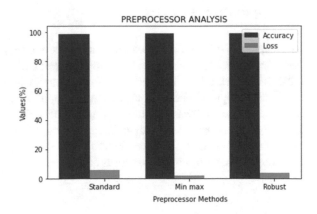

Table 1 Comparison of different optimizer using different performance metrics

Optimizer	Accuracy (%)	Loss	Precision	Recall	F_1 score
Adam	99.72	0.01	0.99	1.00	0.99
SGD	66.38	0.64	0	0	0
RMS-prop	93.3	0.26	0.96	0.91	0.93
Adagard	98	0.02	0.97	0.93	0.93
Adadelta	66.38	0.64	0	0	0

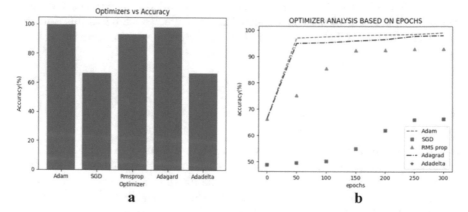

Fig. 4 **a.** Comparison of various optimizer with accuracy. **b.** Optimizer analysis at different epochs

4.2 Optimizer Analysis

The main aim of optimizer technique is reduce the loss to produce the result with good accuracy. Study has to be taken to choose the best optimizer for the proposed ANN model with dropout layers. The final accuracy is listed in Table 1.

From the figure, which shows that Adam optimizer produce the result with the highest accuracy of 99% with the learning rate of 0.001. Figure 4b shows the comparison of different optimizer with accuracy values. From the analysis taken from the chart, the best accuracy achieved by Adam at 300 epochs. So Adam optimizer can be chosen the best optimizer for our model.

4.3 Learning Rate Analysis

It is noted from Table 2 that the performance is decreasing when the learning rate is reduced. This performance analysis can be done by using Adam optimizer. It is observed that Adam optimizer with learning rate of 0.001 attains high accuracy and F_1 score.

Table 2 Performance analysis of various learning rates using adam optimizer

Learning rate	Average accuracy (%)	Loss	Average precision	Average recall	Average F_1 score
0.001	98.47	0.032	0.97	0.96	0.96
0.0005	97.56	0.049	0.96	0.94	0.95
0.0001	96.36	0.277	0.94	0.93	0.94

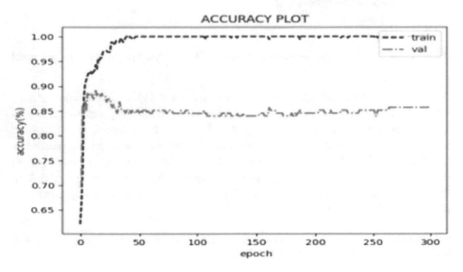

Fig. 5 Accuracy plot of the proposed

When the learning rate is reduced at a low level, the accuracy also decreases. According to different analysis proves that the proposed model produces best results. Model attains the highest accuracy of 99.72% at 300 epochs when using min-max as a feature scalar and Adam as an optimizer with the learning rate of 0.001. Figure 5 displays the accuracy graph of the testing and training values with different epochs.

Figure 6a, b describe the confusion matrix and classification report, respectively. In the confusion matrix, class 0 represents the women without PCOS, and class 1 represents the women with PCOS. 115 correctly predicted in class 0. Similarly, 35 is correctly predicted in class 1.

5 Conclusion

The women who have affected by PCOS have the risk of diabetes, stroke, heart disease, twice the risk of anxiety, and depression thrice than women of the same age without PCOS. In the dataset, the age group of 18–40 years of details only collected for research study. Out of 541, 364 patient details were normal, and the remaining

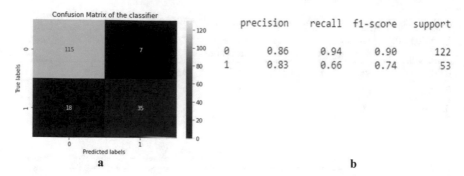

	precision	recall	f1-score	support
0	0.86	0.94	0.90	122
1	0.83	0.66	0.74	53

a   **b**

Fig. 6 a. Confusion matrix of the proposed model. **b.** Classification report of the proposed model

177 cases were abnormal. Even though advanced method of deep learning methods is available for predict the PCOS, but compared to our model deep learning methods computational time, and its complexity is high. Computation time of the proposed model is 63 microseconds which is several times faster than deep learning methods. A different analysis has to be carried out with various hyper parameters such as learning rate, optimizer, etc., to improve the sensitivity the rate of the model. In future, Convolutional Neural Network (CNN) model can be used to implement the model to predict the PCOS.

Acknowledgements I would like to express my special thanks of gratitude to my guide "Dr.S.Usha" for their able guidance and support in my research

References

1. McCartney CR, Marshall JC (2016) Polycystic ovary syndrome. New England J Med 375(1):54–64
2. Dumesic DA, Oberfield SE, Stener-Victorin E, Marshall JC, Laven JS, Legro RS (2015) Scientific statement on the diagnostic criteria, epidemiology, pathophysiology, and molecular genetics of polycystic ovary syndrome. Endocr Rev 36(5):487–525
3. Denny A, Raj A, Ashok A, Maneesh Ram C, George R (2019) I-HOPE: detection and prediction system for polycystic ovary syndrome (PCOS) using machine learning techniques. In: IEEE region conference TENCON, pp 673–678
4. Lawrence MJ, Eramian MG, Pierson RA, Neufeld E (2007) Ovary morphology in ultrasound images. In Fourth Canadian conference on computer and robot vision. IEEE, pp 105–112
5. Cheng JJ, Mahalingaiah S (2018) Data mining and classification of polycystic ovaries in pelvic ultrasound reports. bioRxiv, pp 254870
6. Dewailly D, Lujan ME, Carmina E, Cedars MI, Laven J, Norman RJ, Escobar-Morreale HF (2013) Definition and significance of polycystic ovarian morphology: a task force report from the Androgen Excess and Polycystic Ovary Syndrome Society. Hum Reprod Update 20(3):334–352
7. Gibson-Helm M, Teede H, Dunaif A, Dokras (2016) Delayed diagnosis and a lack of information associated with dissatisfaction in women with polycystic ovary syndrome. J Clin Endocrinol Metabol 102(2):604–612

8. Escobar-Morreale HF (2007) Polycystic ovary syndrome: definition, aetiology, diagnosis and treatment. Nat Rev Endocrinol 14(5):270
9. Joham AE, Teede HJ, Ranasinha S, Zoungas S, Boyle J (2015) Prevalence of infertility and use of fertility treatment in women with polycystic ovary syndrome: data from a large community-based cohort study. J Women's Health 24(4):299–307
10. Norman RJ, Dewailly D, Legro RS, Hickey TE (2007) Polycystic ovary syndrome. Lancet 370(9588):685–697
11. Essah PA, Nestler JE (2006) The metabolic syndrome in polycystic ovary syndrome. J Endocrinol Invest 29(3):270–280
12. Dhayat NA, Marti N, Kollmann Z, Troendle A, Bally L, Escher G, Grossl M, Ackermann D, Ponte B, Pruijm M, Muller M (2018) Urinary steroid profiling in women hints at a diagnostic signature of the polycystic ovary syndrome: a pilot study considering neglected steroid metabolites. PloS one.13(10):e0203903
13. Dumesic DA, Oberfield SE, Stener-Victorin E, Marshall JC, Laven JS, Legro RS (2015) Scientific statement on the diagnostic criteria, epidemiology, pathophysiology, and molecular genetics of polycystic ovary syndrome. Endocrine Rev 36(5):487–525
14. Dewi RM, Wisesty UN (2018) Classification of polycystic ovary based on ultrasound images using competitive neural network. J Phys: Conf Ser 971(1):012005
15. Mehrotra P, Chatterjee J, Chakraborty C, Ghoshdastidar B (2011) Automated screening of polycystic ovary syndrome using machine learning techniques. In: Annual IEEE India conference, Hyderabad, pp 1–5
16. Sachdeva G, Gainder S, Suri V, Sachdeva N, Chopra S (2019) Obese and non-obese polycystic ovarian syndrome: comparison of clinical, metabolic, hormonal parameters, and their differential response to clomiphene.Indian journal of endocrinology and metabolism, vol 23, p 257
17. Zhang XZ, Pang YL, Wang X, Li YH (2018) Computational characterization and identification of human polycystic ovary syndrome genes. Sci Rep 8(1):12949
18. Dhayat NA, Marti N, Kollmann Z, Troendle A, Bally L, Escher G, Grössl M, Ackermann, D, Ponte B, Pruijm M, Müller M (2016) Urinary steroid profiling in women hints at a diagnostic signature of the polycystic ovary syndrome: a pilot study considering neglected steroid metabolites. PloS one 13(10):e0203903
19. Vaitheeswari R, Sathish Kumar V (2019) Performance analysis of epileptic seizure detection system using neural network approach. In: International conference on computational intelligence in data science, pp 1–5

Noise Removal in Long-Term ECG Signals Using EMD-Based Threshold Method

Kiran Kumar Patro, M. Jaya Manmadha Rao, Ashwini Jadav, and P. Rajesh Kumar

Abstract The ECG signal (electrocardiogram) is the biomedical signal used in clinical studies for the diagnosis of cardiovascular diseases. ECG is the electrical representation of the cardiac activity and is obtained by placing the electrodes on the patient's chest. In this process, several noises appear due to muscular contractions related to breathing and electronic interference. Thus, there is a need for removal of noise for better clinical evaluation. In this work, the empirical mode decomposition (EMD)-based method is proposed, that is, EMD threshold on long-term ECG signal. The purpose of this EMD-based technique is to decompose the signal into a few oscillatory parts, i.e., intrinsic mode functions (IMF's). The IMF's which are dominated by noise are immediately determined and removed by hard threshold method. For the evaluation of this technique, long-term (22,500 samples) ECG signals are acquired from open source MIT-BIH databases. Finally, the performance indicators such as mean square error (MSE) and signal to noise ratio (SNR) in dB of hard threshold method are calculated by using the MATLAB 2018a software.

Keywords Electrocardiogram (ECG) · Empirical mode decomposition (EMD) · Mean square error (MSE) · Signal to noise ratio (SNR) · Hard threshold

1 Introduction

ECG signal is the most frequently known diagnostic biomedical signal for identifying heart irregularities early. It is obtained easily with the electrodes placed on the limbs and chest and is used by medical specialists to analyze the pathological diseases of the heart [1]. It is an adequate way to recognize the heart diseases during the clinical studies. In a clinical environment, during the transmission of the obtained ECG signal,

K. K. Patro (✉) · M. Jaya Manmadha Rao · A. Jadav
Aditya Institute of Technology and Management (A), Tekkali, India
e-mail: kirankumarpathro446@gmail.com

P. Rajesh Kumar
Andhra University, Visakhapatnam, India
e-mail: rajeshauce@gmail.com

© The Author(s), under exclusive license to Springer Nature Singapore Pte Ltd. 2021   461
K. A. Reddy et al. (eds.), *Data Engineering and Communication Technology*,
Lecture Notes on Data Engineering and Communications Technologies 63,
https://doi.org/10.1007/978-981-16-0081-4_48

the signal gets corrupted by various types of unwanted information [2]. There are a lot of noises (disturbance, unwanted information) which get involved in these ECG signals while taking from the patient's body. The main noises present in the ECG are: Baseline wander noise, power line interference noises, and electromyography noises [3, 4]. Since biological signals are low amplitude and low frequency, they are easily prone to noise. So, it becomes very important to remove these artifacts from electrocardiogram signals.

Many de-noising techniques like digital filter bank [5, 6], adaptive filter [7], statistical techniques such as independent component analysis [8], fuzzy multi-wavelet de-noising [9], neural network [10], Bayesian filter [11], discrete wavelet transform (DWT) [12–14], and wavelet de-noising [15] are issued in the literature for de-noising of ECG. Various methods have been proposed for suppression of power line interference (PLI) and baseline wander (BW) noises. The EMD-based hard threshold method is popular and shows much improved results than the other methods which are reported. Hence, many researchers use this advanced technique of empirical mode decomposition (EMD) for ECG de-noising [16, 17].

In this work, authors were developed EMD-based threshold technique for the de-noising of long-term ECG signal. In this, the given signal is decomposed into a few parts of oscillations called as Intrinsic Mode Functions (IMFs). After obtaining IMF components using EMD, a hard threshold is applied to the IMF component to obtain the noise estimation. Further, the noise components are removed from generated IMFs. By partial reconstruction, the noiseless output ECG signal is obtained.

2   EMD Algorithm

The procedure EMD is obtained from the theory of any non-linear time series which consists of various simple intrinsic mode oscillations [18]. The design of this method is to temporarily analyze the obtained intrinsic modes of oscillation by their time scales in the data and then disintegrate it consequently through a process called shifting. The oscillations which are between extremeswith no zero crossing are ignored. The EMD method thus divides the signal into locally non-overlapping time scale units [19, 20]. EMD decomposes the signal $x(t)$ into different IMF components accepting the two features:

- Each individual IMF contains only one local maxima or local minima among two consecutive zero crossings, i.e., the difference between the total number of local maxima and local minima is at most unity.
- Every IMF has zero mean value, which means that IMF is stationary so that it simplifies the analysis.

The steps included in EMD method are:

1. Finding of local minima and maxima in $x(t)$ is to establish the lower and upper envelopes.

2. Calculation of mean envelope $h(t)$, i.e., average the upper and lower envelopes.
3. Calculation of empirical local oscillation

$$h_k(t) = x(t) - h(t) \qquad (1)$$

4. Average of $h_k(t)$ is calculated; the average $h_k(t)$ is to be considered as the first IMF, named $c_i(t)$ when its average value is close to zero. Otherwise, repeat steps (1)–(3).
5. Calculate the residue using the formula

$$r(t) = x(t) - c_i(t) \qquad (2)$$

Repeat the steps from (1) to (4) using $r(t)$ as $x(t)$ to obtain the next IMFs and residue.

This process of decomposition is continued until the residue $r(t)$ becomes a constant or mono-functional, thus not satisfying the IMF condition.

$$x(t) = \sum_{i=1}^{N} c_i(t) + r_N(t) \qquad (3)$$

3 Proposed Methodology

The proposed methodology shown in Fig. 1 contains different stages such as ECG data acquisition, ECG with noise, EMD process on ECG, hard threshold selection, and finally ECG signal reconstruction.

Fig. 1 Proposed methodology of ECG de-noising

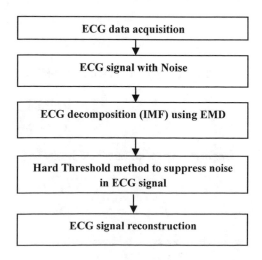

ECG Data Acquisition The long-term ECG data (22,500 samples per second) are acquired from open source MIT-BIH Arrhythmia database of physionet. The database provides 103 recordings with a sampling frequency of 360 Hz.

ECG with Noise The long-term ECG recordings which have acquired from MIT-BIH database are generally contaminated by some unwanted information. These noises are added at different frequency bands of signal. The low-frequency noise represents baseline wander (BW), the mid-band frequencies represent the power line interference (PLI), and the high frequencies (EMG) signal represents the electromyography noises.

EMD with Hard Threshold In EMD threshold method, the corrupted ECG signals are first disintegrated to several noisy IMFs. The obtained noisy IMFs are determined and suppressed by hard threshold method in order to get an estimation of the de-noised IMFs of the noiseless signal.

The noisy signal $y(t)$ given by:

$$y(t) = x(t) + \eta(t) \tag{4}$$

The above equation represents a signal $y(t)$ which is a combination of both original signal $x(t)$ and noise signal $\eta(t)$. In EMD hard threshold technique, intially, the signal $y(t)$ is decomposed into several noisy IMFs, named $c_{ni}(t)$. The noisy IMFs are then compared with a threshold value by hard function to obtain an estimation of de-noised IMFs $\hat{c}_i(t)$. In this work, the value of threshold is calculated below steps as follows:

$$\tau_i = U\sqrt{E_i 2\ln(n)} \tag{5}$$

where U is a constant, which depends on the nature of signal, and in this, it is adjusted to 0.5; in this work, n is the length of the signal (no. of samples), and E_i is given by:

$$\hat{E}_k = \frac{E_1^2}{0.719} 2.01^{-k}, k = 2, 3, 4 \ldots N$$

where E_1^2 = energy within the first IMF, which is defined by:

$$E_1^2 = \left(\frac{\text{median}(|c_{n1}(t)|)}{0.6745} \right)^2 \tag{6}$$

Application of hard threshold in EMD:

$$\hat{c}_i(t) = \begin{cases} c_{ni}(t) & \text{if } |c_{ni}(t)| > \tau_i \\ 0 & \text{if } |c_{ni}(t)| \leq \tau_i \end{cases}$$

The reconstruction of de-noising signal is given by:

$$\hat{x}(t) = \sum_{i=1}^{N} \hat{c}_i(t) + r_N(t) \tag{7}$$

ECG Signal Reconstruction The noises present in the different levels of IMFs are removed by the EMD hard threshold method. After the removing of noises, the noiseless ECG signal is again constructed by summation of the threshold IMFs and the remaining signal IMFs.

4 Experimental Results

In this work, the proposed EMD hard threshold method is applied to the long-term ECG signals. These ECG recordings can be accessed by online open source MIT-BIH arrhythmia database. MATLAB software is used for the experimentation of ECG signals. The actual ECG signal (104.mat form) signal are depicted in Fig. 2. The signal taken is of approximately 22,500 samples and is decomposed into 9 IMF's levels which are shown in Fig. 3 as follows.

By examining the calculated SNR values of before de-noising IMFs in Table 2 shown below, it is clear that the IMF1 contains a lot of high-frequency noise than that

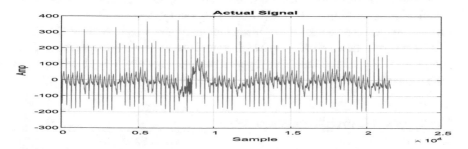

Fig. 2 Actual ECG signal 104.mat

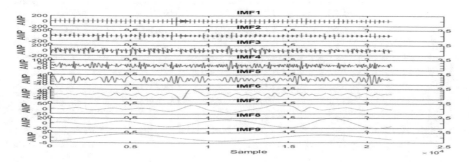

Fig. 3 IMF levels of actual ECG signal 104.mat

Table 1 Median and threshold values

IMF levels	Median	$E_i = \left(\frac{median}{0.6745}\right)$	$\tau_i = C\sqrt{2E_i\ln(n)}$
IMF1	0.06	0.10	0.46
IMF2	−0.01	−0.01	0.18
IMF3	0.09	0.14	0.53
IMF4	0.20	0.30	0.79
IMF5	−1.05	−1.56	1.79
IMΓ6	0.02	−0.03	0.26
IMF7	−0.17	−0.26	0.73
IMF8	0.39	0.58	1.09
IMF9	−0.19	0.28	0.76

of the other IMFs as shown in Fig. 3. Except the IMF2, the other IMF levels consist of useful information, where the IMF level-2 consists of QRS complex components with high-frequency noise. If the IMF1 is simply discarded as noise, then the output will still consist of considerable noise. If the IMF2 is removed together, then the resultant signal will have distorted information. Therefore, result is of not satisfactory. Hence, it is necessary to apply the EMD hard threshold for each IMFs level. The values of median, energy (E_i), and thresholded (τ_i) values are shown in the Table 1, and the modified IMFs levels are shown in Fig. 4.

Therefore, the SNR (dB) and MSE values of the long-term ECG recordings before and after the application of the EMD hard threshold method are shown in Table 2. Finally, the signal is reconstructed by using modified IMFs which are shown in Fig. 5.

By comparing the SNR (dB) and MSE values before and after de-noising of ECG recordings, it is observed that our proposed methodology which is able to remove noise components and improved the result of the signals.

Table 2 Performance parameters of different IMFs

Before EMD			After EMD	
IMF levels	SNR (dB)	MSE	SNR (dB)	MSE
IMF1	−18.86	183.51	2.43	60.46
IMF2	−6.29	392.50	16.93	187.52
IMF3	−7.53	1.6825e+03	8.61	764.32
IMF4	−6.00	583.23	6.18	258.21
IMF5	−0.071	233.73	12.10	116.85
IMF6	1.84	153.53	9.32	91.44
IMF7	−11.87	256.12	14.95	133.36
IMF8	−17.65	108.22	23.06	45.32
IMF9	−1.57	4.73	3.53	2.68

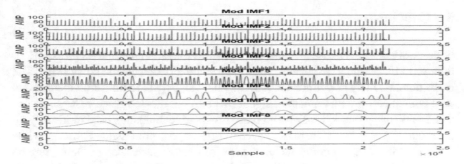

Fig. 4 Modified IMF levels of 104.mat

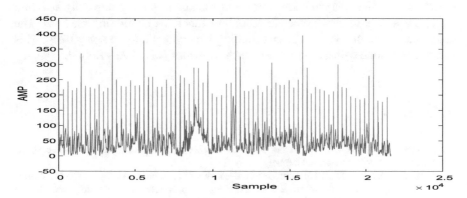

Fig. 5 Reconstruction ECG signal of 104.mat

Final View The signal parameters SNR and MSE of the some of the long-term ECG signals before and after EMD are recorded in Table 3.

Table 3 Performance indicators of different signals from database

ECG record	Before EMD		After EMD	
	SNR (dB)	MSE	SNR (dB)	MSE
104.mat	−6.56	3.19	5.14	4.05
105.mat	−8.17	6.01	4.08	4.62
106.mat	−4.24	6.62	3.12	5.09
108.mat	−2.23	5.52	3.64	3.18
109.mat	1.16	4.16	6.26	2.42
112.mat	2.27	4.81	6.44	2.55

5 Conclusion

In this work, an EMD-based technique (i.e., Hard threshold method) is proposed for long-term ECG signal de-noising, which automatically detects the noisy IMFs. The IMFs dominated by noises are compared with a value by hard threshold method and then combined up with remaining signal IMFs to get the de-noised ECG signal. In this de-noising process, EMD-based hard threshold technique is used to identify all the required parameters. There is no need of pre or post-processing Parameters like SNR and MSE of the reported technique exhibits better performance. Depending on SNR and MSE values, it is clear that EMD-based technique has a lot of advantages than any other conventional methods without loss of original information. Hence, the new approach for suppressing noises is useful and is of less complexity. In future work, our aim is to check other thresholding functions or filtering methods. The important aspect in this work is that filtering is confined only to noisy IMFs. This separation is done with a criterion by which separates noisy IMFs from the signal.

References

1. Acharya UR, Krishnan SM, Spaan JA, Suri JS (eds) (2007) Advances in cardiac signal processing. Springer, Berlin, Germany
2. Jeyarani AD, Singh TJ (2010) Analysis of noise reduction techniques on QRS ECG waveform-by applying different filters. In: Recent advances in space technology services and climate change 2010 (RSTS & CC-2010). IEEE, pp 149–152
3. Allen J, Anderson JM, Dempsey GJ, Adgey AAJ (1994) Efficient bseline Wander removal for feature analysis of electrocardiographic body surface maps. In: Proceedings of 16th annual international conference of the IEEE engineering in medicine and biology society, vol 2. IEEE, pp 1316–1317
4. Ider YZ, Saki MC, Guçer HA (1995) Removal of power line interference in signal-averaged electrocardiography systems. IEEE Trans Biomed Eng 42(7):731–735
5. Van Alste JA, Schilder TS (1985) Removal of base-line wander and power-line interference from the ECG by an efficient FIR filter with a reduced number of taps. IEEE Trans Biomed Eng 12:1052–1060
6. Afonso VX, Tompkins WJ, Nguyen TQ, Michler K, Luo S (1994) Comparing stress ECG enhancement algorithms. IEEE Eng Med Biol Mag 15(3):37–44
7. Rahman MZU, Shaik RA, Reddy DRK (2011) Efficient sign based normalized adaptive filtering techniques for cancelation of artifacts in ECG signals: application to wireless biotelemetry. Signal Process 91(2):225–239
8. Barros AK, Mansour A, Ohnishi N (1998) Removing artifacts from electrocardiographic signals using independent components analysis. Neurocomputing 22(1–3):173–186
9. Ho CYF, Ling BWK, Wong TPL, Chan AYP, Tam PKS (2003) Fuzzy multiwavelet denoising on ECG signal. Electron Lett 39(16):1163–1164
10. Poungponsri S, Yu XH (2013) An adaptive filtering approach for electrocardiogram (ECG) signal noise reduction using neural networks. Neurocomputing 117:206–213
11. Sameni R, Shamsollahi MB, Jutten C, Clifford GD (2007) A nonlinear Bayesian filtering framework for ECG denoising. IEEE Trans Biomed Eng 54(12):2172–2185
12. Ercelebi E (2004) Electrocardiogram signals de-noisingusing lifting-based discrete wavelet transform. Comput Biol Med 34(6):479–493

13. Poornachandra S (2008) Wavelet-based denoising using subbanddependent threshold for ECG signals. Digital Signal Processing 18(1):49–55
14. Alfaouri M, Daqrouq K (2008) ECG signal denoising by wavelet transform thresholding. Am J Appl Sci 5(3):276–281
15. Boudraa AO, Cexus JC (2007) EMD-based signal filtering. IEEE Trans Instrum Meas 56(6):2196–2202
16. Boudraa AO, Cexus JC, Saidi Z (2004) EMD-based signal noise reduction. Int J Signal Process 1(1):33–37
17. Rilling G, Flandrin P, Goncalves P (2004) Detrending and denoising with empirical mode decomposition. In: Proceedings of EUSIPCO, pp 1581–1584
18. Chacko A, Ari S (2012) Denoising of ECG signals using empirical mode decomposition based technique. In: IEEE-international conference on advances in engineering, science and management (ICAESM-2012). IEEE, pp. 6–9
19. Mohguen W, Bekka REH (2017) EMD-based denoising by customized thresholding. In: 2017 International conference on control, automation and diagnosis (ICCAD). IEEE, pp 019–023
20. Gandham S, Anuradha B (2016) An iterative method of ensemble empirical mode decomposition for enhanced ECG signal denoising. In: 2016 International conference on wireless communications, signal processing and networking (WiSPNET). IEEE, pp 1477–1480

Teaching "Design Thinking" Using "Project-Based Learning" Model to Undergraduate Students Through EPICS Course

Ram Deshmukh and Md.Mujahid Irfan

Abstract Engineering design course is needed for every engineering student to be a successful entrepreneur and to be job-ready after graduation. This paper shows the way of implementing the engineering design course using effective teaching and learning model referred as "project-based learning" (PBL). It also details multiple dimensions of design thinking and their impact on the learning process of engineering students. PBL is successfully implemented through a course termed "Engineering Projects In Community Service" (EPICS). This work recommends to initiate EPICS course in the engineering curriculum at the institution level to inculcate the design thinking practice and also explores the significant impact of EPICS projects on society, through a case study.

Keywords EPICS · Design thinking · Problem-based learning · Society and engineering curriculum

1 Introduction

Engineering education has become a critical activity, contributing to the global economy through industrial development. The rapid growth in engineering education has, however, not translated into any significant growth in the number of industry-ready graduates. The challenge for academia is to make "twenty-first century engineers" who can recognize the need and adapt the working in multidisciplinary teams for the realization of successful projects.

Design is widely considered to be the central activity of engineering [1]. It has also long been that the engineering programs should graduate engineers who can design effective solutions to meet societal needs [2]. Though the presence, role and

R. Deshmukh (✉) · M. Irfan
S R Engineering College, Warangal, Telangana, India
e-mail: ramdeshmukh@gmail.com

M. Irfan
e-mail: mmirfan.srec@gmail.com

perception of design in the engineering curriculum have improved markedly in recent years, both design faculty and practitioners would argue for the necessity of further improvements in this area [3] [4].

This paper starts with defining both engineering and design, to set a context for what differentiates them. It reviews the characterization associated with experienced designers. It explores the significance of the "project-based learning" model through community service projects and emphasize the outcomes of these courses that improve the cognitive skills of the students with a case study.

2   Design Thinking

According to Wikipedia, engineering is the application of <u>mathematics</u>, <u>science</u>, <u>economics</u>, social and practical *knowledge* to *invent innovate, design*, build, *maintain, research* and improve *structures, machines, tools, systems, components, materials, processes*, solutions and *organizations*. Design is the basic entity in engineering education [5]. Design is an art, problem-solving, scientific knowledge and social process. It is about generating multiple solutions to an open-ended problem followed by implementing the best solution. Customer involvement and using the final solution are the key to success.

This process starts essentially with understanding the problem and getting insight into the context and culture of the customer for any engineering projects with "empathic approach". After the diagnosis of the problem, brainstorming multiple ideas and finalizing the idea tangible by creating the prototypes are carried out. The prototype is improved by receiving iterative feedback from the end users.

The design thinking process will enable students to stay inline with customer's need, create real innovations and gain changes [6]. Design thinking provides alternate options to the problem which does not exist before. Good design includes diverse perspectives and user-centric solution [7]. The persistent rapport of designers with the end users is the key to success in design thinking.

The key features of experienced designers are:

- Nurturing the friendly environment for ideas to grow
- Everyone is unique and creative in thinking. An innovation starts with these people and collaborative work in solving complex problems through multidisciplinary teams
- Qualitative research and quantitative analysis, for understanding the users and their needs
- Themes and patterns will be finalized only after finishing the synthesis of research followed by taking their ideas into the hands of the people through live prototypes
- Taking feedback from the end user, iterating the prototype resulting in implementing their idea to a product.

For the Engineering Educational Institutes, the general approach of learning design thinking is through the final year project as they consider design as an integral part of the projects and can be learned while executing the projects. Implementing the design thinking in the early stage of engineering education through various community-based projects can bring drastic visible changes and are inevitable in the current era of social entrepreneurship and beckons to include in the curriculum [8].

3 Project-Based Learning

The project-based learning (PBL) is the most engaging way to learn and in simple terms, it is "learning by doing". Usually, many of the academicians in India wrongly interpret it as a laboratory component, where students perform the theory learned in a classroom into practice in the form of the laboratory experiment.

PBL engages students through hands-on experience where the students collaboratively forms multidisciplinary teams to execute the project to solve the problems using technology. The PBL model helps students to acquire twenty-first century skills, viz. hands-on technical work, collaborative working, creativity, critical thinking and communication. It is quite evident [9] that student engagement in learning the core subjects is more exclusive and effective that results in improving skills through this active learning model, PBL. The PBL process includes a driving question followed by an enquiry process that enables them to find a solution sequenced by developing a product. If these projects are meant for the development of the society (projects addressing the unmet needs of the community), then these are termed as "community service-based projects". Here students work in multidisciplinary teams with real-time constraints and real users. After diagnosing the community problems, students generate ideas that will solve the needs of the end users. While brainstorming for ideas, creativity and critical thinking play a major role in selecting the best idea for developing real design successfully.

Creativity is perceiving the world in a novel way to connect the imaginative patterns into a fully-modelled real product. Creativity is a desired and practicing mindset for experienced designers. Consistency in creativity is important, i.e. always generating new ideas at any situation and condition. The innovation component plays a major role in ideation and implementation in the PBL. If the student closely understand the limitations of the existing products in the commercial market, then the existing product will give way to the new innovation. So the design process will surely move the students towards creativity and innovation.

4   Design Thinking and PBL Through EPICS

The traditional system of education does not have the autonomy to transform the curriculum framework towards outcome-based education, OBE. However, with acquiring the semi-autonomous status, the author's engineering educational institute started to fulfil the persistent gaps between industry practices and student preferences by incorporated new courses in the curriculum like Introduction to Engineering, Product Design Studio, Foundation to Product design, Cognitive Science Engineering and EPICS. The skills acquired from these courses are applied in a socially focused engineering course called Engineering Projects in Community Services (EPICS).

EPICS course was designed so that students can acquire design thinking skills through the PBL model and is service-learning, a 3 credit, an elective course available for third year students. Implementing design thinking through PBL in the social space and connecting the user needs to technology are the key aspects that develop the student's mind as such to frame the "problems as an opportunity". Every project in EPCIS is executed through collaboration, with community partner for whom the project is being designed and catered. Students carry out real design for real people [10], thereby students gaining experience of a real-world project eventually becoming industry ready.

The list of the total number of projects and students involved in this course is given in Table 1. It has to be noted that every semester the projects were assigned to the students those includes 2–3 different departments. The data signifies the active participation of students every year, and it is noted that most of the developed projects are being used by stakeholders. Projects done are mainly under four categories: access and abilities, education and outreach, environment and human services.

The EPICS course is designed by including the methods involved in design thinking and PBL resulting in generating modules like problem identification, specification development, conceptual design, detailed design, product delivery and maintenance. Six course modules were delivered through the lecture sessions, mentoring and monitoring during the activity hours [11].

In the first phase, students interacted with different communities and NGOs to identify the problem followed by discussion among the team members to identify and finalize one problem for developing the solution and treated as a project for a real problem. Here, decision-making and enquiry-based engagement skills were predominantly developed in the students.

Table 1 Summary of the projects handled in the Institute

Academic year	No. of projects	No. of students registered
2014–15	2	10
2015–16	33	164
2016–17	24	206
2017–18	24	130
2018–19	28	170

In phase II, the students get engaged with the community partner to take inputs and discuss the requirements for the identified problem in developing the technical specifications for the project that lays the objectives of the project. The survey of the market was carried out to identify the commercial products available in the market that serves the problem of the community partner I phase III. The limitations of those products paves the possibility to generate multiple ideas that will fill the gap available in the market. These innovative ideas were brainstormed resulting in developing the prototypes of the products [12, 13] that were presented to the end costumers to acquire feedback, iteratively, to refine the prototype to develop the practical usable solution. In the next phase, once the conceptual design was finalized, the detailed design was developed which involves multi-disciplinary students to apply their respective engineering disciplinary skills in the projects and presenting the same to other engineering disciplinary students so that the other students can develop their respective parts of the design. This tested their ability to apply core skills and helped them to communicate effectively with other trade students. In the same phase, every component required for the project was budgeted along with appropriate and logical justification for the component with respect to cost and quality which helped them to apply the logical skills and develop decision-making ability. It was mandatory to present the uniqueness in terms of features, accessibility, the cost and usability of their product compared to a readily available product in the market (Fig. 1).

Post-design, the product was developed and tested at the user premises and verified by the expert team, and any drawbacks were addressed within the stipulated time, and finally, the usable product was delivered to community partners with the user manual, in phase five. In the phase 6, once the product was delivered, students trained community partners about product operation, using the user manual. An agreement was made with respect to the maintenance and repair of the product, and the same was

Fig. 1 Community interaction

Table 2 Course schedule with timelines

Week	Lecture sessions	Facilitator	Activity session
Week-1 to 3	Introduction to EPICS & design-1. Community Partner (User needs and requirements)	EPICS Coordinator & Dept. Coordinator	Community survey report and Proposal submission. Design Thinking (Wallet design). Team building Partner's approval Brainstorming (Solutions)
Week-4 to 6	Project Management and Planning. Specification Development, Assessment and Evaluation Rubrics	Expert Talk & Dept. Coordinator	Market survey report Conceptual design document. Detailed design & review presentations. Mid semester Review
Week-7 to 9	Introduction to Design-II, III, Documentation Process	EPICS Coordinator	Detailed design (model) Budget submission, demonstration. Usability test Delivery review form (partner feedback)
Week-10	Communication Skills (verbal and written)	Dept. Coordinators	Individual and Peer evaluation (reflections)
Week-11	Ethical and Legal issues in Design (Patents & IPR)	Expert talk	Thesis report submission
Week-12	Demonstrations with PPT	Expert Evaluation	

handed over to the EPICS coordinators so that the product can be further enhanced by the successor team in the next semester, if needed.

In the last phase, the evaluation was done in the formative and summative patterns using the rubrics defined that was broadly based on the rubrics relevant to design thinking and PBL evaluation. From the learning reflections of the students and their evaluation, it was quite evident that the improvement of design thinking skills and autonomous thinking was developed and was at the higher side, between 8 and 9, out of 0 and 10 scale. The feedback of the student after the project was highly rated on the points of ownership of the idea, design and development of product, and the same was the key motivation for the students to work in EPICS. The schedule followed at the institute is shown in Table 2. The "lecture sessions" 1 h per week and the "activities sessions" 2 h per week were thoroughly followed as per the schedule, and the sequence of skill improvement was quite apparent.

5 Case Study: "She Bracelet" Project

The community partner, with whom the Institute had an agreement, was "Bal-aVikasa", a Non-Government Organization (NGO) who works for the benefit of the widow community through providing livelihood opportunities. Students discussed with the widows working at BalaVikasa and identified "woman security" were the major issue that should be addressed. The issue was the attack and harassment of these women when they are alone or in workplace/abandoned areas. The students developed a product that was a weapon against the crimes. Students designed a real product for the woman protection named "She Bracelet" and add value to the dignity and security of the women. The requirements were bracelet should be simple to operate, affordable, effective functioning, available in every nook and corner of the country. The specifications were simple technology (GPS, GSM and Arduino) to send text messages about the location of the sufferer to the nearby police station and also to their relatives.

To achieve this, an Arduino with GSM and GPS connectivity is used with a c-code and installed into a bracelet. In order to fix the whole circuit into a bracelet, nanotechnology can be used. The whole circuit size is to be reduced to at least 2 cm diameter, and then it can be installed into a bracelet of the required fashion. As the bracelet is a common ornament preferred by the women in India, they have chosen it. The team received $1000 funding from EPICS in IEEE (Fig. 2).

Student Experiences: The following are the learning reflections of the student teams.

- They experienced the progress in the interest towards the learning of engineering subjects and ownership and opportunity to prove as a responsible citizen
- Perception of societal problems has changed from a problem to opportunity
- They got the complete awareness of the engineering design process, and also they feel comfortable now in solving any problem
- Most of the students placed in MNCs opined that the EPICS project made them unique from other participants.

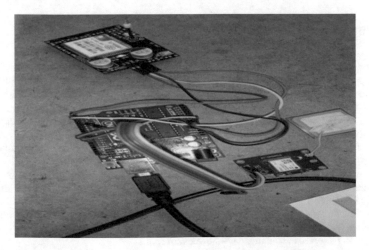

Fig. 2 Circuit of the product

This case study shows one of the successful and impactful EPICS Projects using the design thinking process by applying various technologies in electrical, electronics, computer science and evaluated as problem-based learning. This project improved the student skills in the targeted areas mentioned in this paper.

6 Conclusion

Considering the transformations in the engineering education system and their demands to produce employable and self-learning students, institutions need to inculcate various skill centric programs and courses in the curriculum which will enrich student learning towards professional development. As a graduate engineer, design thinking skills, viz. user-centric solution, rapport with the end users, design and development; and skills needed for project-based learning, viz. hands-on technical skills, communication skills, decision-making, reasoning skills, cognitive skills, are the mandatory skills needed to be a job ready engineer or to become a successful entrepreneur. These skills can be developed using engineering projects in the community services course that was successfully implemented to develop various real-time, user-centric products. This course not only developed the aforementioned design thinking and PBL skills but also made the student empathize with societal problems and nurture them as social entrepreneurs and ultimately a responsible citizen. This paper strongly recommends engineering institutes to initiate the project-based design courses focusing on societal problems in the curriculum, to meet the program outcomes of the National Board of Accreditation.

References

1. Simon HA (1996) The sciences of the artificial. In: Multipress, 3rd edn. Cambridge, USA
2. Sheppard SD (2003) A description of engineering: an essential backdrop for interpreting engineering education. In: Proceedings (CD), Mudd design workshop IV. Claremont, Cal., Harvey Mudd College
3. Evans DL, McNeill BW, Beakley GC (1990) Design in engineering education: past views of future directions. J Eng Educ 79(4):517–522
4. Todd RH, Magleby SP (2004) Evaluation and rewards for faculty involved in engineering design education. Int J Eng Educ 20(3):333–340
5. Venkateswarlu P (2017) Establishing a 'Centre for Engineering Experimentation and Design Simulation': a step towards restructuring engineering education in India. Euro J Eng Educ 42(4):349–367
6. Dym CL, Agogino AM, Eris O, Frey DD, Leifer LJ (2005) Engineering design thinking, teaching, and learning. J Eng Educ 94(1):103–120
7. Crugar R. Question everything. www.howdesign.com
8. Irfan M, Rajamallaiah A, Ahmad S (2018) Paradigm shift in the engineering curriculum: design thinking. J Eng Educ Transform
9. Mills JE, Treagust DF (2000) Engineering education—Is problem- based or project-based learning the answer? Austr J Eng Educ 3(2):2–16

10. Coyle EJ, Jamieson LH, Oakes WC (2005) College of Engineering, Purdue University, 'EPICS: engineering projects in community service'. Int J Eng Ed 21(1):139±150
11. Irfan MM, Sammaiah P (2017) Service learning course in the engineering curriculum: EPICS. J Eng Edu Transform. Special Issue
12. Alok G, Anushalini T, Condoor S (2018) Effective approach towards development of idea through foundation to product design. J Eng Educ Transform 31(3):47–52
13. Alok G, Pothupogu S, Sampath Reddy M, Saipriya P, Radhika Devi V (2018) Trenchant pathway to bring innovation through foundations to product design in engineering education. In: Proceedings of the 6th IEEE international conference on MOOCS innovation and technology in education (MITE), p 13

Assessment Techniques in Engineering Education—Tools and Strategies of the Faculty

C. Balarama Krishna, G. Ravi Kiran, B. Ravindar, and R. Archana Reddy

Abstract The object of this article is to expand the awareness of information and recognition of formative evaluation and its accountability for optimizing student performance and the instructional process, while at the same time informing faculty on formative strategies that are readily adaptable to specific educational settings. Summative evaluation, formative assessment and variations in formative and overview assessment are addressed. A broad variety of performance appraisal approaches to measure student progress in the classroom are addressed. The functions in informative writing and also the role/usage of technology in the formative appraisal are highlighted. This article also provides recommendations for a formative assessment of staff instruction. In addition, we have stressed the value of establishing a culture of evaluation that supports the idea of systematic appraisal in both the growth of a student's capacity to produce instructive outcomes and the potential of a faculty member to become a strong and successful instructor.

Keywords Assessment techniques · Formative assessment · Summative assessment · Faculty development · Teaching–learning process · Engineering education

C. B. Krishna (✉) · G. R. Kiran · B. Ravindar · R. A. Reddy
Center for Experiential Learning, SR University, Warangal, India
e-mail: cbrk2004@gmail.com

G. R. Kiran
e-mail: ravikiran.wgl@gmail.com

B. Ravindar
e-mail: ravi.boini@gmail.com

R. A. Reddy
e-mail: archanareddy.srec@gmail.com

© The Author(s), under exclusive license to Springer Nature Singapore Pte Ltd. 2021
K. A. Reddy et al. (eds.), *Data Engineering and Communication Technology*,
Lecture Notes on Data Engineering and Communications Technologies 63,
https://doi.org/10.1007/978-981-16-0081-4_50

1   Introduction

In engineering education, the institutional outcomes, program outcomes and course outcomes altogether remind the institution that the objective of its specialized programs is to mold engineering graduates competent enough to achieve educational outcomes. Engineering education can be characterized as skill based, concentrating on learning outcomes much more than simply fulfilling the requirements. This instructional method explores what the learners should be willing to achieve once they have completed their learning, rather than what they are going to learn during the duration of their study.

In the use of curricular outcomes, educational objectives should be exhibited by engineering graduates in terms of tangible skills, reflect the information, attitudes and behaviors required to function effectively as an engineer. Since competency-based instruction requires recognition that teaching is not interchangeable with learning, members of faculty are charged with maintaining an effective learning atmosphere and assessing the progress of students. Faculty must use a number of appraisal tools in classrooms to achieve these activities, including formative and summative assessments.

Initially Michael Scriven used the words "formative" and "summative" with reference to program evaluation [1]. Benjamin Bloom continued to use these words as part of learning process of teaching [2]. The aim of the formative assessment, according to Bloom, was to provide input and enable correction at any point of the learning process [3]. The object of formative evaluation or assessment is to include knowledge that can be useful to make instant improvements in teaching—learning and the program [4]. Divergent thinking of engineering students was discussed in respect of design skills by Mahesh et al.[5]. Aluvala and Pothupogu [6] have stressed the improvement in the quality of teaching learning process in engineering education in context to the need of skilled engineers. Govil et al. [7] have discussed persuasive-learning strategies for transforming engineering education. To improve the quality of education in engineering institutions, the required changes were discussed by Irfan et al. [8].

The main distinctions in the formative and the summative tests are when they occur in the process of learning and what is done in the insight learned by both of them. The summative evaluation takes place at the conclusion of teaching–learning cycle, while the formative evaluation takes place during that period. Knowledge gained from the formative evaluation or assessment is useful to adjust the learning environment depending on how fine the students perform in achieving the expected results. Since summative assessments take place at the conclusion of teaching–learning phase, knowledge gained that may enhance the method cannot be implemented before the next course offering, providing little incentive for the currently enrolled students to profit from these improvements.

2 Assessment/evaluation of Learning (AFL)

Appraisal of learning indicates to the assessments—verbal, performance and writing mode, or mixture of some of these modes which will happen at the conclusion of a unit/term through which we can review the capability of the learners to generate and exhibit the ideas or concepts they learnt at the time of instruction. These results of evaluation/assessment are globally treated as significant indicators to monitor the students learning development. They are often used for various comparative reasons, such as student success in specific classes, distinction between students in the insitution, inter-class and inter-institutional comparisons, etc. The findings are often included in the preparation of curricular activities for the following year, activity and/or instructional year.

2.1 Tools and Strategies

We ought to use a range of methods and techniques in the evaluation of learning, based on the complexity of the activity to be evaluated. We need to select resources and techniques carefully, based on the volume and kind of knowledge needed. Definitions of some of the methods used to measure learning include assessments with various forms of queries, ranking scales, checklists, etc.The techniques for this evaluation include examination, student reaction (written, oral), review of student performance, conversations with students.

2.2 Expected Responsibilities

As engineering faculty, we have to understand that all accountability for the appraisal of learning and its follow-up rests with us. Here are certain things that require a great deal of attention. we should ensure that goals of the evaluation activity/assignment are well grasped by students. Appropriate time limits must be established for the execution of tasks / assignments. We need to be attentive to the difficulties certain students encounter in achieving the task/assignment. We need to gather ample facts on which to base the decisions.

2.3 The Main Objectives of the Learning Assessment

To let every student realize how they are doing, learn what they have to do in order to make it easier and how to reach there. The student has the encouragement he/she wants to be encouraged to become successful learner and continuously develop his/her learning.

To equip each faculty member to make sound judgments about student achievement, to understand the theories/principles of progress and to utilize their assessment results for improving the learning of students, particularly those students who do not fulfill their potential.

To have a standardized and comprehensive evaluation program in institution to allow routine, effective, measurable and reliable evaluations of students and to use the outcomes of the evaluation to measure students' learning progress.

To let each parent/guardian realize how their ward operates, what they need to do to change, and how they will help the ward and faculty. Learning evaluation (AFL) consists of two phases—the original or diagnostic evaluation and the formative assessment. Formative evaluation is an examination in which data may be gathered in the learning phase as the class moves through a research framework to determine the abilities and skills of the pupil, including learning differences.

Characteristics in Learning Assessment or Assessment for Learning (AFL)
Respond to everyone by recognizing the strengths and needs of all students;

This is concise of essence and is not of a judgmental type and thus not evaluative;

Using high-quality ratings, inform students what they have achieved right, where they have faced challenges and what they need to do more to improve their work;

because input needs to be given to the learner in order to enhance the continuing learning process, the evaluation is regular and continuous throughout the learning process; encourages students to focus on their research and learning, and to take practical steps to enhance their performance; this allows students to make mistakes and enables them to evaluate certain errors in order to develop their learning; this involves students in systematic self-examination and peer analysis of their work;

The key principles of assessment for learning are.

1. The system of efficient feedback to learners.
2. The student participation in learning.
3. Change instruction to take into consideration the outcomes of the evaluation.
4. Recognition of the tremendous effect of evaluation on students' confidence and self-esteem, which are crucial factors on performance.
5. The need of learners be capable to assess themselves and understand how they could improve.

2.4   The Strategies and Techniques of AFL (Assessment of Learning)

While deciding on learning evaluation methods for all students in a classroom setting, consider how the method(s) implemented encourage one to evaluate the success of all students, insure that students get positive input and enable them to get input on their teaching. We have basically four methods to carrying out:

1. Faculty-led evaluation(using a broad variety of approaches including written or oral examination, experiences with peers, tasks, observing student's behaviors, etc.)
2. Learner self-evaluation/assessment. (3) Peer evaluation (Evaluation of peers on the reaction and results of learner). (4) Computer-based appraisal (using specifically developed software).

Feedback in AFL (Assessment for Learning)

The main objective of an AFL is to give review/feedback to faculty as well as student about the student's growth toward attaining the educational objective(s). There view ought to be utilized by the faculty to modify and build up further teaching.

When providing feedback in written:

- Answer to the content and text/message in the written form. Don't be much focussed on grammatical mistakes only.
- Don't skip right to the bug. First of all congratulate.
- If composition is poor, choose some particular places to draw attention to. Don't mask the work with ticks and cross marks in red ink.
- Be precise. Indicate what the student should do in respect of weaknesses which have been indicated. Encourage the learners to go with corrections. Don't just write the actual answers, spellings, etc.

2.5 Formative Assessment

Formative evaluation is a set of standardized and informal appraisal procedures utilized by the faculty during the academic phase to adjust teaching and learning practices in order to enhance student attainment. It is an ongoing procedure typically done by the faculty to continually supervise the student's development in a non-aggressive and welcoming atmosphere.

Formative Assessment Supports the Faculty

- To offer advice/feedback to students, parents and other faculty, in order to motivate them to shift in right way promoting or helping the process of learning.
- Improve corresponding teaching–learning practices and interactions. When we look at the results from the evaluation, majority of learners perform below the anticipated level; we should change the approach and strategies to address the situation.
- Identifying and remedying community or human defects. For instance, if we see that certain students do not grasp the idea you taught them, we will provide additional guidance or take some other timely action to enhance their efficiency.
- Formative assessment feedback helps the learners:
- To monitor the improvement of learning and assist to promote self learning.

- To shift his/her attention away from attaining grades in order to maximize self-efficacy. Reduce the detrimental influence of social inspiration.
- To boost their efficiency dramatically by growing their self-esteem, encouraging self-learning through inspiration and thereby reducing the workload of faculty.

The overview of the features of the formative assessment and its role:

Formative Assessment
- Based on students' previous knowledge and familiarity in designing what is being taught. It is carried out at comfortable time frames on casual basis
- It is indicative and corrective. Ensure the availability of positive reviews
- Provides a dais for student participation in learning
- Provides review enabling faculty to accustom to classroom communication techniques to the rising needs of learners
- Encourages students' inner drive and self-esteem, which have significant impact on academic success
- Presents an incentive for students to develop their performance when they gain input. Helps the students to maintain their peers and vice-versa

2.6   Summative Assessment

Summative assessment involves the calculation of performance, which 'summarizes' the success of students over a specified time. It is a way of calculating the learning of students at a period. Assessment methods at the completion of a semester, a word or regular evaluations are manifestations of the summative evaluation and the assessments included in these testing schemes are considered summative exams. Whereas the formative assessments are based on specific goals or material, summative assessments evaluate the entire of the specified information and the predicted learning outcomes to provide an overall complete image of students' accomplishment through evaluation.

The features of summative evaluation are given below.

Summative Assessment
- Is the evaluation of the learning performance carried out at the conclusion of the course or at the end of the semester/term
- Students are typically evaluated toward the conclusion of course or academic year to exhibit the "amount" of what they have studied
- Uses the same common ways of measuring the performance of students

- The findings are used to rate or classify students who are expected to prepare some large-scale instructional activity, inter/intra institution distinction in terms of standards

The variations between the formative and the summative assessments:

Formative Assessment	Summative Assessment
Used to assess the students how much they have accomplished and what they ought to know	Used to assess the student's overall success in a specified course
Requires the instructor to analyze their instructional strategies and implement adjustments to help students grasp the teachings of the academic year	Helps educators to analyze their instructional strategies and implement adjustments to support students grasp the teachings of the academic year
Grades will not hold a lot of weight	Grades provide the foundation for evaluating the student's ability to take standardized assessments
Frequently conducted during the teaching and learning phase	Conducted at the final stage of the process
It can be versatile according to the needs of students	Inflexible, one exam for all learners, a standard method of performing and a common way for evaluating test results
Procedure oriented	Result oriented

2.7   Student Learning—Formative Assessment

Classroom Strategies There is a range of classroom evaluation methods available that can include formative reviews. These evaluation methods may vary from a quick, non-labor intensive instrument to a complicated procedure taking significant time to plan. The following methods involve limited planning and review and will give useful guidance to students and faculty alike. For more widespread source of classroom formative evaluation techniques are discussed by Angelo and Cross [9].

Previous Knowledge Analysis Understanding what the students are taking with them during the course or training time is useful when deciding the basic framework for a lesson. In addition to gather information about the degree of readyness of learners at the outset of the semester, a preliminary evaluation of the skills will be conducted. The testing normally assumes the form of very few changeable or open-ended questions, brief responses and multiple-choice questions.

Minute Paper/Muddiest Point The minute paper and muddiest point approaches offer constructive guidance and enable students to consider and comment about what they have heard in college. In such tests, time is scheduled, typically toward the conclusion of the semester, to encourage students to focus about and write down the most significant topic they have learned and to consider the concerns they have as a consequence of the lesson. In addition to significant point(s) of the lesson, students should often clarify why they believed that was interesting topic.

Audience Feedback Systems Usage of "clickers" has become popular in classrooms. Clickers are indicators of user feedback mechanisms that are utilized for a number of uses, such as quizzes, polling and constructive learning. Various resources and software are available to enable the faculty to project or ask questions in the class and to collect responses from the students. Growing student or community of students uses a clicker that attaches to receiver or to personal gadgets (e.g., computers, laptops, tabs, smart phones, smart watches, etc.) that relay answers through the Internet. Further clickers involve all students in conventional active-learning question-and-answer exercises when presenting evaluation results.

Case Studies The case studies are important and effective methods which can blend ideas, definitions and information. The method includes the collection of cases identifying people with one or more disorders, signs or conditions. Example cases can provide as much details (laboratory standards, findings of physical examination, etc.) as is appropriate for the degree and context of the pupil. The faculty may use this activity to evaluate student efficiency, subject complexity and effectiveness of the lesson. The downside is the large amount of pre-class planning and in-class administration period.

Reflective Writing and Portfolios Assuming that the formative evaluation takes place during the learning phase and is intended for development, the usage of positive learning tasks is compatible with these concepts. Reflective learning requires capacity to self-assess, to be mindful of one's learning (metacognition) and to build positive thinking skills. Thinking through knowledge and clinical practice, articulating what has been learned, and understanding that it is important to become a reflective professional.

Reflective writing encourages self-directed thinking, and different styles can be utilized. A definition and review of learning process within a course or clinical practice can be in its best, insightful prose. These can provide a summary of what has been observed at that stage or an overview of a specific event. A paradigm of "if/so what/now what" is also used to direct the interpretation of the learner. Reflective writing should be integrated into the program (educational, experiential and community learning) to promote the creation of a reflective professional.

Reflective portfolios can often be viewed as a sort of formative evaluation because they are intended to track the progress of students' personal and professional development-ment. A form of portfolio helps students to monitor their progress and progression and, with strong scaffolding, to show how they can incorporate and adapt their learning. The secret to a successful reflective portfolio is a mixture of reporting and evaluation, along with input from faculty. Faculty–student engagement through portfolio creation is crucial to explain why students have selected particular objects and what the objects represent in the sense of a course or curriculum.

Formative Assessment—Role of Technology The key goal of the formative evaluation is to include prompt, pro-empty suggestions for change. The processing, preservation and distribution of data is likely to involve the use of technologies. Although several vendors strive to offer goods and services in the field, there are some rules that will govern the usage of evaluation data: (1) Data which is useful for optimizing performance should be gathered and stored; (2) no amount of mathematical manipulation will change the utility of bad/correpted data; (3) data should be accurate and quality-controlled; (4) Faculty as well as students should be educated on the correct analysis and usage of the data—even if reliable data is gathered, it would not contribute to an increase on curricular results if misused; (5) data must be easily accessible to anyone who use it when in need.

2.8 Formative Assessment of Faculty Teaching

Theories and examples of formative evaluation to support student performance and quality of teaching. Factors that enhance student success, such as instructional learning strategies, student teaching assessments, resumes, self-reflections, performance feedback and peer reviews, often provide incentives for faculty to develop their teaching skills. In the past few decades, survey work has found that tests have been the main indicator that higher education organizations used to determine the quality of that teaching [10]. College evaluations usually take place at the conclusion of the semester and also act as a descriptive appraisal tool for advancement and tenure decisions. The goal is to submit student reviews to decide what steps can be done to enhance instruction and to receive detailed and positive input at the conclusion of the course from unmotivated students, because they are unable to gain from any potential changes in the course.

Formative feedback should be obtained from mid-semester students, expressly calling for written input instead of scores. Mid-semester reviews may be encouraged by evaluation personnel, campus employees committed to strengthening teaching–learning, community representatives. Current work indicates that the usage of several input channels is more successful in assessing instruction and professionalism. This is the model of complete feedback to estimate teaching as well as professionalism. Effective implementation of assessment feedback includes an appreciation of the educational atmosphere and accepted best practices within the university as well as the learning of students and faculty members on how to offer meaningful positive input.

Faculty members also focus their evaluation of their teaching efficacy on student evaluation results. Choosing formative tests that give feedback on how fine students learn can improve the quality of instruction in the class. For example, in the classroom, faculty may find that students are not paying much attention to, or displaying a confusing perception of, a specific topic. By introducing classroom appraisal methods, such as minute papers or clickers, the faculty may guage comprehension, receive direct input and allow for improvements in teaching.

3 Conclusion

In experimental settings, a number of evaluation methods may be applied and used to refine student awareness, abilitiesand attitudes. Formative opinion targeting teachers strengthens the teaching consistency. We suggest that engineering programs, as a key element of learning, establish an atmosphere that promotes "formative failure" in a healthy environment. Formative failure encourages the student to make errors and improve before the task is formally assigned to. Formative failure allows teachers to practice and improve their teaching skills further. In addition, we suggest that faculty members and preceptors strive to successfully incorporate a range of formative evaluation approaches into their teaching to achieve the best education outcomes for learners. Complete feedback method will be utilized through formational evaluation of students and faculty members. Faculty will focus on their results through triangulation of student and peer reviews and evidence of learning through formative and summative assessments. Institutions that wish to use technology to efficiently manage the process of recording, reviewing, and reporting student progress and give timely data for students and faculty and resolve learning gaps quickly. A well-designed appraisal plan, involves a combination of formative and summative tests, can help programs track student attainment of educational results and include evidence required for curriculum development.

References

1. Tyler RW, Gagné RM, Scriven M (1967) The methodology of evaluation. In: Perspectives of curriculum evaluation. Rand McNally, Chicago, pp 39–83
2. Bloom BS (1968) Learning for mastery. Instruction and curriculum. Regional Education Laboratory for the Carolinas and Virginia, Topical Papers and Reprints No.1., Evaluation Comment, University of California Press, pp 9–10
3. Bloom BS (1969) Some theoretical issues relating to educational evaluation. In: Tyler RW (ed) Educational evaluation: new roles, new means. The 63rd yearbook of the National Society for the Study of Education, Part II, vol 69, no (2). University of Chicago Press, Chicago
4. Anderson HM (2005) Preface: a methodological series on assessment. Am J Pharm Educ 69(1):11
5. Mahesh V, Raja Shekar PV, Pramod Kumar P (2015) Analysis of Divergent thinking in Indian engineering students. J Eng Educ Trans 29(1):98–102
6. Aluvala S, Pothupogu S (2015) A traditional novel approach for skill enhancement of teaching-learning process in engineering education. J Eng Edu Transform 28(4):92–95
7. Govil A, Pillalamarri S, Prabhanjan N (2020) Persuasive learning strategies for transforming engineering education. J Eng Educ Transform 33:402–407
8. Irfan MM, Rajamallaiah A, Syed MA (2018) Paradigm shift in the engineering curriculum: design thinking. J Eng Educ Transform

9. Angelo TA, Cross KP (1993) Classroom assessment techniques. Wiley, San Francisco, CA
10. Seldin P, Hutchings P (1999) Changing practices in evaluating teaching: a practical guide to improved faculty performance and promotion/tenure decisions, 1st edn. JosseyBass, San Franscisco, CA

Mechanical Compression Properties of Apple Fruit: Errors During Penetrometer Measurements

D. Ramesh Babu and K. V. Narasimha Rao

Abstract The present study was carried out to investigate the effect of type of instrument and technique on the firmness measurements. Firmness of apples can be tested with either FT327 Penetrometer or texture analyzer. FT327 with 11 mm probe is a standard instrument available with orchards and cold store units for quality check. Texture analyzer is expensive and needs sophisticated laboratory environment along with trained laboratory personnel. Penetrometer testing is prone for three different types of errors. First one is non-uniform application of hand pressure during penetration of plunger into the fruit, second is depth of penetration and third error is rate of penetration. In the present investigation, tests are conducted in a laboratory attached to the controlled atmosphere apple storage plant at Rai, Haryana. During apple arrival at the store, each lot is checked for firmness. Based on minimum specifications, it is decided either to store the produce or market immediately. Two different techniques are used to measure the firmness of apples. (a) Holding fruit in one hand and penetrating the instrument probe with other hand, (b) Fixing the Penetrometer onto a mechanical test stand. Results indicated that use of mechanical jack with a hand lever is a better technique to check large number of samples accurately compared to other two techniques. Higher measurement error was observed with higher size. Error found was as high as 1.5 Lbs. Highest percentage of error was 7.5.

Keywords Penetrometer · Fruit firmness · Mechanical properties · Controlled atmosphere storage · Measurement technique

D. Ramesh Babu (✉)
Department of Mechanical Engineering, Koneru Lakshmaiah Education Foundation, Vaddeswaram, Guntur, Andhra Pradesh, India
e-mail: rameshdamarla@rediffmail.com

SR Engineering College, Warangal, India

K. V. Narasimha Rao
Department of Mechanical Engineering, Koneru Lakshmaiah Education Foundation, Greenfields, Vaddeswaram, Guntur, Andhra Pradesh 522501, India

© The Author(s), under exclusive license to Springer Nature Singapore Pte Ltd. 2021   493
K. A. Reddy et al. (eds.), *Data Engineering and Communication Technology*,
Lecture Notes on Data Engineering and Communications Technologies 63,
https://doi.org/10.1007/978-981-16-0081-4_51

1 Introduction

Apples fruits are known for appealing color and crunchiness apart from doctor's recommendation for good health. Apples are grown in the northern hills of India, especially Jammu and Kashmir (J&K), Himachal Pradesh (HP) and some parts of Uttarakhand. J&K is the largest producer of apples in India. Himachal Pradesh is the second largest producer.

India produced 2.33 MMT of apples during year 2017–18. 99.42% of total production of India is contributed by states of Jammu & Kashmir (J&K), Himachal Pradesh (HP) and Uttarakhand (UK) together. J&K, HP and UK produce 1808.33, 446.57, 58.66 TMT of apples, respectively. Baramulla, Shopian and Kulgam of J&K are the major producing areas of apple in the state of J&K with 380, 237 and 207.26 TMT of apples, respectively. Similarly, Shimla and Kullu districts are the major contributing areas from Himachal Pradesh with 265.99 and 89.57 TMT production, respectively, (Horticultural Statistics at a Glance 2018). Mechanical firmness is the critical quality attribute for apples to decide harvest, storage or marketing, apart from total soluble solids and maturity index. Apples having at least 12 lb Penetrometer pressure are found to be juicy and crunchy. Below that apples are mealy that means pulpy without any watery biting experience. Harvest firmness is in the order of 18–28 lb depending on the variety, size and pre-harvest orchard management.

Most of the farmers in the orchards are not equipped with any instrument to measure firmness of apples. However, recent advances in post-harvest technology and installation of controlled atmosphere storage units in Haryana, J&K and HP created awareness about firmness measurement for deciding harvest time. Himachal Pradesh Horticulture Produce Marketing and Processing Corporation (HPMC) has initiated for quality consciousness among the farmers. Establishment of CA Stores by private players like Adani Agri and Devbhoomi in HP and FIL Industries Pvt. Ltd., Kashmir Premium Apples Pvt. Ltd., ShaheenAgro Fresh Pvt. Ltd., Fruit Master Agro Fresh Pvt. Ltd., Harshana Naturals Agri Serve Pvt. Ltd. in J&K made the farmers more aware of quality requirements either for storage or transportation. Firmness of apples is expressed in pounds or kgs when tested with Penetrometer. Two different instrument probes are supplied by the manufacturer. However, 11 mm probe is most popular and results can be comparable and standardized by the industry.

During measurement of firmness, several errors occur like instrument calibration error, dial error, measurement errors, etc., out of which measurement error needs to be standardized. Other errors can be set with maintenance and calibration procedure. Penetrometer readings vary due to following reasons:

1. Proper holding of apple fruit
2. Proper placement of probe into the flesh of the apple
3. Correct depth of penetration
4. Proper removal of skin of apple before measurement.

Firm apples last long in the CA Store under low temperature and low oxygen apart from high CO_2. Caution is also required to see the maturity index (MI) and total soluble solids (TSS). Decision on storability depends on collective properties of firmness, MI and TSS. Quality test before long-term storage of apples is critical. Apples of at least 16 lb or higher are suitable for long-term storage. This has to be along with 1.5–2.0 MI and 8°–10° Brix TSS. Lesser firm apples loose firmness in the storage quickly. So there is a need to avoid errors in measurement either due to instrument of technique.

2 Literature Overview

Firmness for Indian apples was reported by Ramesh Babu et al. [1], in their report on effect of handling on the firmness during storage under low oxygen conditions. They reported the data for four varieties of Himachal Pradesh. Royal, Red, Golden, Rich-a-Red varieties were handled both in water and dry bin dumpers and concluded that water handling retained the firmness well.

Sharma et al. [2] reported about the application of pre-cooling at orchard level and wax coatings improved the shelf life studied conducted at IARI, PUSA complex, Delhi. They conducted the tests on Royal delicious variety. Chauhan and Ramesh Babu [3] reported use of natural extract as cushioning and coating materials improved the quality retention. Camphor and neem extracts benefited in better firmness retention and low weight loss of fruits. They conducted experiments with Starking Royal delicious apples from HP. Wijewardane et al. [4] conducted experiments on use of coating and packing materials for shelf life extension of HP apples.

Kumar et al. [5] studied 22 apple cultivars grown in Himachal Pradesh. Out of five non-red varieties of apples, the highest firmness was found in winter banana (14.00 N) and lowest in Stark Spur Golden (10.53 N). Out of seventeen red cultivars (Royal Delicious, Top Red, Oregon Spur-II, Starkrimson, Well Spur, Red Chief, Super Chief, Red Gold, Royal Gala, Scarlet Spur-II, Scarlet Gala, Spartan, Vance Delicious, Silver Spurl, Red Delicious) highest firmness values found in Spartan and Red Delicious were 12.35 N and 12.39 N, respectively, with insignificant change among these two.

The lowest firmness in the range of 10.32–10.44 N was found in Starkrimson, Silver Spur and Vance Delicious varieties. However, their study is largely on characterization with respect to nutritional constituents, acidity, antioxidants, carotenoids, different sugars, minerals and physical parameters including color values. The firmness values of all varieties were recorded during initial point of storage at 20 °C after harvest. Their focus was on nutritional characterization, not on firmness changes during storage. However, they harvested fruits of 3.0 maturity index for uniformity while comparing of other properties.

Apples from Himachal Pradesh have good flavor and color especially Starking and Royal Delicious varieties. Firmness of apples found to be good when preserved with natural cushioning materials and surface treatments with neem-based formulations. Camphor is also found positive effect on preservation quality [6]. Effect of handling practice on the firmness of apples was reported by Ramesh Babu et al. [1]. Golden

Delicious, Royal Delicious, Red and Rich-a-Red varieties exhibited similar trend of firmness loss in controlled atmosphere storage with low oxygen and high carbon dioxide conditions. Storage temperatures were between 0 and 1 °C with RH of 95%. Water flume handling found to be better to avoid surface bruising and better retention of firmness during storage.

Protocols were prepared by Ramesh Babu et al. [7] for preservation of apples in controlled atmosphere storage and cold storage, respectively. Controlled atmosphere better preserves the apples. This is due to the lower respiration rates under low oxygen availability to fruits. Other reasons might be low ethylene, high CO_2 and high RH. Grading and sorting before storage significantly reduced the production of ethylene within the chambers. This is normally neglected by cold storage operators. In a study by Wijewardane et al. [4], apples precooled immediately after harvest and packed with suitable materials with immediately transportation to cold store prevented the firmness loss.

Engineering properties of pre-cooling process for fruits were reported by Narasimha Rao et al. [8–10]. They conducted experiments on spherical fruits with air–water mixture to cool the produce. Heat and mass transfer characteristics were reported.

Mathematical modeling kinetics of cooling of mango and banana was reported using rate constants under cold conditions during ripening process by Narasimha Rao et al. [11].

2.1 Materials and Methods

Fruits: Apples are obtained from well-grown orchards of Himachal Pradesh with proper orchard management system. Preliminary tests were conducted before plucking and sorting–grading operation carried out at orchards manually. Packing is done using paper pulp tray and corrugated fiber board (CFB) boxes. Boxes were transported immediately to the CA Storage complex, at Rai, Haryana, for further testing and long-term storage. Upon receipt at the store, the apple lots are quality tested for deciding suitability for long-term, medium-term or short storage. Industry standard is 18–20 Lbs firmness for long-term storage; minimum 12 Lbs for sale.

Instrument: FT327 Penetrometer with 11 mm probe was used for firmness testing. Calibration of instrument is carried out using the procedure specified by the manufacturer. A precision weighing scale is used for verifying the correctness of the gauge of the Penetrometer. Penetrometer was tested for correctness every day at least once.

Test stand: Effegi test instrument stand was fixed with Penetrometer for cross-check the values of Penetrometer test using hand. Each apple is tested at the largest diameter horizontally after peeling off the skin for 20 mm^2. One side is tested with hand Penetrometer and the diametrically opposite side is tested with test stand instrument. Six samples were tested for each lot received. Average values were reported.

TSS and starch index: TSS is measured with a refractive index instrument (Erma, Japan) 0–32° Brix. Instrument was calibrated every day for accuracy. Starch index was tested using a test kit with Iodine indicator. Apples were cut to half and dipped in a solution which turns the color based on the starch–sugar reaction with iodine. Then, the apples are determined for SI, based on comparison charts. The international scale of maturity index (0–5) is used. Zero indicates most immature or no sugars. Five indicates full sugars and less starch. Ripe fruits will have more starch index (Figs. 1 and 2).

Fig. 1 Royal delicious apples placed in tray before quality test

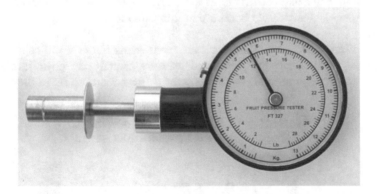

Fig. 2 FT327 Penetrometer with 11 mm probe

3 Results and Discussion

Two different techniques used for firmness measurement, shown variance in the firmness values of the same apples. After verifying apples of 39 different orchards and about 800 apple tests, the following observations were made

Apple firmness wrongly shown in hand penetration compared to test stand
Range of error is found to be 1.5 lb for the same apple tested with both techniques
Error is more in large size apples compared to small size apples
No significant different of error among varieties tested
No significant error among orchards tested.

Firmness measurement before preserving for long term must be cautiously, keeping in mind the expected life of apples. Apples are allowed to lose firmness up to 12 Lbs to offer minimum crunchiness [1]. Consumers need to be assured of quality specifically the firmness/crunchiness/juiciness component. Figure 3 indicates the firmness values of different sizes of apples. Serial number one is of 100 count (large) and two and three are 125 count (medium) and serial number four and five are 150 count (small). Count of apple is the conventional indicator of apples placed in a 20 kg CFB carton as per standard practice of apple growers across the globe.

Large sized fruits have got a firmness of 17.5 Lbs, which is suitable to go for long-term CA Storage. Medium size has 17.5–18 Lbs, whereas small fruit got 18 to 18.5 Lbs firmness. This indicates that the small fruit got better firmness than large one. Even though several other sizes are produced and graded at orchard level, only large, medium and small (LMS) are most favorite to the consumers and are termed as table variety and size. The results are in line with the reported results of Sharma et al. [2].

Effect of test stand on firmness measurement errors found significantly. When Penetrometer is used by applying hand pressure directly, an error is observed compared to "Penetrometer fixed to test stand." Error found to be in the range of

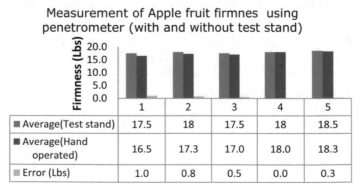

Fig. 3 Firmness of apples measured with Penetrometer (with test stand and hand operated)

0–1 Lbs. High degree of error is found in case of large sized fruits compared to small fruits.

Two different techniques used shown lot of variance in the firmness values of the same apples (Table 1). Total soluble solids and starch index also well correlated with the firmness. Higher the starch index is the indicator for ripened fruit.

1. Apple firmness wrongly shown in hand penetration compared to test stand
2. Range of error is found to be 1.5 to 3 lb for the same apple tested with both techniques
3. Effect is found is more in large size apples compared to small size apples
4. Percentage error is found to be as high as 7.5 in case of BSP orchard and 125 count medium size apples.

4 Conclusions

Significant error is found in firmness values between the measurement techniques used. Using test stand for Penetrometer is found to give precise readings compared to hand operation of the instrument. This may be due to the error in the application of penetration of probe against the flesh of apple. Orientation of holding apple in hand, grip/ slip of apple in hand must have contributed for errors in case of hand operation of Penetrometer probe. Based on the test conducted with different orchard samples which includes approximately 800 samples with replication of six for each, it is concluded that using hands for Penetrometer test gives erroneous results due to improper handling of instrument. So it is recommended to fix the Penetrometer to a test table and conduct tests in the laboratory. Percentage error observed was as high as 1.5 Lbs, which is sometimes just the quality bench mark for acceptance of rejection for deciding the suitability for storage of market sale immediately. Eighteen pounds of compression strength (or firmness) is the industry standard for deciding to store for long term of apples. If an error of 1.5 lb shown wrongly due to measurement error, the farmer need to be sell his produce to the market in distress due to rejection at store on quality. So, it is strongly recommended to use test stand for Penetrometer tests. The results can be useful for future store operators to take care during Penetrometer compression testing of apples to decide for long-term storage or disposing to sale.

Table 1 Compression test results, total soluble solids (TSS) and starch index of apples using Penetrometer, with and without test stand

Lot no.	Reckong Peo Orch. code	Size	Penetrometer with test stand		Penetrometer hand operated		Error (Lbs)	Temperature: 24.0–22.8 °C	T.S.S. (%)	Starch index
			Firmness (Lbs)	Average (test stand)	Firmness (Lbs)	Average (Hand operated)		Absolute error (%)		
1/65	P.C.O	100	17 / 18	17.5	15.0 / 18.0	16.5	1.0	5.71	10	2
1/65	P.C.O	125	18 / 18	18	16 / 19	17.3	0.8	4.16	11	1.5
1/65	P.C.O	125	17 / 18	17.5	16 / 18	17.0	0.5	2.85	11	2
1/65	P.C.O	150	19 / 20	18	17 / 19	18.0	0.0	0.00	10.5	2
1/65	P.C.O	150	18 / 19	18.5	17.5 / 19	18.3	0.3	1.35	10	2
1/73	D.S.O	100	16 / 16	16	15 / 16	15.5	0.5	3.12	11	1.5
1/73	D.S.O	100	17 / 17	17	16.5 / 17.5	17	0	0.00	11	1.5
1/73	D.S.O	100	20 / 20	20	17 / 19	18	2	10.00	10	2
1/73	D.S.O	125	18 / 20	19	18 / 20	19	0	0.00	10.5	2
1/73	D.S.O	125	17 / 17	17	16 / 17	16.5	0.5	2.94	11	1.5
1/73	D.S.O	150	20 / 21	20.5	19 / 19	19	1.5	7.31	9.5	2
1/64	B.S.P	100	18 / 17	17.5	17 / 15	16	1.5	8.57	11	2
1/64	B.S.P	100	17 / 18	17.5	16.5 / 16.5	16.5	1	5.71	11	2
1/64	B.S.P	125	16 / 17	16.5	15.5 / 15	15.25	1.25	7.57	10.5	2
1/64	B.S.P	150	17 / 17	17	16.5 / 16	16.25	0.75	0.04	11	1.5
1/63	P.O	100	17 / 17	17	16 / 17	16.5	0.5	0.02	10	1.5

(continued)

Sub-station: Rispa

Table 1 (continued)

Lot no.	Reckong Peo	Sub-station: Rispa	Penetrometer with test stand			Penetrometer hand operated			Error (Lbs)	Temperature: 24.0–22.8 °C		Starch index
	Orch. code	Size	Firmness (Lbs)		Average (test stand)	Firmness (Lbs)		Average (Hand operated)		Absolute error (%)	T.S.S. (%)	
1/63	P.O	125	17	18	17.5	17.0	17.0	17	0.5	0.02	10.5	2
1/63	P.O	125	18	19	18.5	18	18	18	0.5	0.02	11	2
1/63	P.O	150	18	20	19	18	18.5	18.25	0.75	0.03	10	2.5
1/63	P.O	150	18	18	18	17	17.5	17.25	0.75	0.04	10	2.5
1/34	S.O	125	18	17	17.5	17	16	16.5	1	0.05	11	2.5
1/34	S.O	150	17	18	17.5	16.5	17	16.75	0.75	0.04	10.5	2

References

1. Ramesh Babu D, Rao Narasimha KV, Kumar MS, Kumar BS (2018) Handling of apples during sorting-grading operation and measuring the mechanical properties firmness after controlled atmosphere storage. Int J Mech Product Eng Res Dev (IJMPERD) 8(6):617–634
2. Sharma RR, Pal RK, Singh D, Samuel DVK, Sethi S, Kumar A (2013) Evaluation of heat shrinkable films for shelf life, and quality of individually wrapped royal delicious apples under ambient conditions. J Food Sci Technol 50(3):590–594
3. Chauhan SK, Ramesh Babu D (2011) Use of botanicals: a new prospective for enhancing fruit quality over chemicals in an era of global climate change. Asian J Env Sci 6(1):17–28
4. Wijewardane RNA, Guleria SPS (2013) Effect of pre-cooling, fruit coating and packaging on postharvest quality of apple. J Food Sci Technol 50(2):325–331
5. Kumar P, Sethi S, Sharma RR, Singh S, Saha S, Sharma VK, Verma MK, Sharma SK (2018) Nutritional characterization of apple as a function of genotype. J Food Sci Technol 55(7):2729–2738
6. Thakur KS, Jawa NK, Thakur KP (2012) Botanical formulation and extracts based on plant leaves and flower, a substitute for toxic chemical and waxes for shelf life extension and quality retention of apple CV starking delicious in India. J Hortic For 4(12):190–196
7. Ramesh Babu D (2014) Technological aspects of controlled atmosphere storage–implementation for Indian produce by FHEL/CONCOR. In: Proceedings of national conference on innovations and challenges in processed food in India. Indo-American Chamber of Commerce, New Delhi
8. Narasimha Rao KV, Narasimham GSVL, Murthy MK (1992) Analysis of co-current hydrair-cooling of food products in bulk. Int J Heat Fluid Flow 13(3):300–310
9. Narasimha Rao KV, Narasimham GSVL, Murthy MK (1993) Analysis of heat and mass transfer during bulk hydraircooling of spherical food products. Int J Heat Mass Transf 36(3):809–822
10. Narasimha Rao KV, Narasimham CSVL, Murthy MK (1993) Parametric study on the bulk hydraircooling of spherical food products. AIChE J 39(11):1870–1884
11. Narasimha Rao KV, Shareef S, Ramesh Babu D (2020) Mathematical modeling of cooling rates of mango fruits during unsteady state cooling in an artificial ripening chamber. Test Eng Manag 83:6862–6871

Detailed Review on Breast Cancer Diagnosis Using Different ML Algorithms

L. Vandana and K. Radhika

Abstract Breast cancer is the most prevalent cancer among Indian females with high mortality rate. It is reported that the incidence of breast cancer in India would reach upto 2 lakh per year by 2030. If breast cancer detected in early stages, it could be treated effectively resulting in decreased mortality. Machine learning is a specific field in artificial intelligence that utilizes variety of probabilistic techniques, statistical measures, optimization methods and allows systems to learn from past examples to discover patterns from large volumes of datasets. As a result, machine learning algorithms are used as efficient tools to diagnose and detect breast cancer. Since 80 s, there is an on-going research to diagnose and detect breast cancer using machine learning algorithms. This survey paper focuses on the overview of this research.

Keywords Machine learning · Breast cancer · Supervised learning · Unsupervised learning · Thermography

1 Introduction

Cells are the fundamental units of living organisms. Every cell consists nucleus that controls the birth, growth and death of the cell. In a living organism, different types of cells are present and every cell type has a particular growth rate. If that growth rate increases abnormally, then they replace healthy cells in organs and also cause changes in body's biochemistry that can lead to a compromised immune system. This state of body is called as cancer. Cancer can occur in any part of the body. There are many types of cancers. Cancers have different stages like early stage, advanced stage and last stage. Most of the cancers can be cured if they are detected in the early

L. Vandana (✉)
CSE Department, Osmania Univeristy, Hyderabad, India
e-mail: vandanakiran.lingampally@gmail.com

K. Radhika
IT Department, CBIT, Hyderabad, India

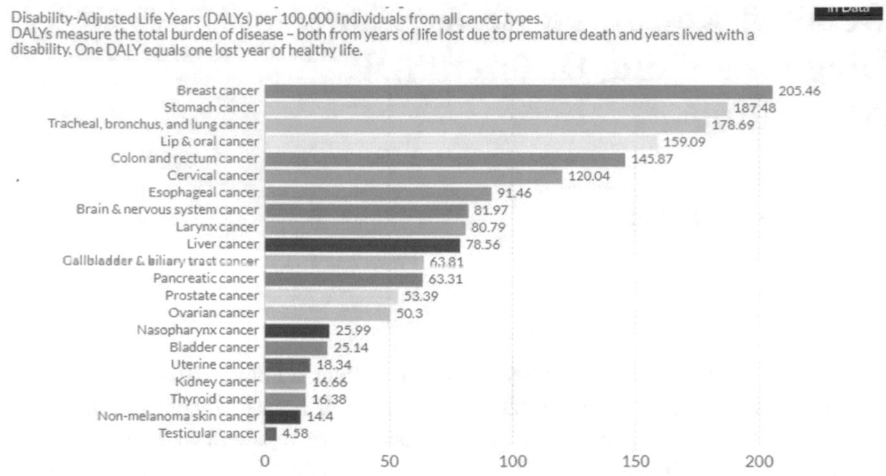

Disability-Adjusted Life Years (DALYs) per 100,000 individuals from all cancer types.
DALYs measure the total burden of disease – both from years of life lost due to premature death and years lived with a
disability. One DALY equals one lost year of healthy life.

Breast cancer	205.46
Stomach cancer	187.48
Tracheal, bronchus, and lung cancer	178.69
Lip & oral cancer	159.09
Colon and rectum cancer	145.87
Cervical cancer	120.04
Esophageal cancer	91.46
Brain & nervous system cancer	81.97
Larynx cancer	80.79
Liver cancer	78.56
Gallbladder & biliary tract cancer	63.81
Pancreatic cancer	63.31
Prostate cancer	53.39
Ovarian cancer	50.3
Nasopharynx cancer	25.99
Bladder cancer	25.14
Uterine cancer	18.34
Kidney cancer	16.66
Thyroid cancer	16.38
Non-melanoma skin cancer	14.4
Testicular cancer	4.58

Fig. 1 Disease burden rates based on cancer types, India, 2017. *Source* IHME, Global Burden of Disease

stage. Some cancers can be cured in advanced stage, but no cancer can be cured in last stage.

In history, we can find the description of cancer and process to remove it in around 1600 BC. History indicates that research on cancer has started before 1600 BC only [1]. Regional cancer center in Thiruvananthapuram published first cancer registry in 2012. It aggregated data of all patients between 1982 and 2012. In these 30 years, the rate of increase in cases was pegged at 280%. The Global Burden of Disease (GBD) estimates that 95.6 lakhs people died as a result of cancer in 2017. Cancer deaths raised from 57 to 96 lakhs in the world and since 1990 every sixth death in the world is due to cancer [2]. According to WHO, cancer is the second main cause of death globally. Among various cancers, the most prevalent is breast cancer. Statistics show that only one out of two women diagnosed of breast cancer continues to survive [3]. As per the above statistics, the present challenge is to reduce the breast cancer mortality (Fig. 1).

1.1 Breast Cancer

Women's breast consists of lobules (15–20, which produces milk), nipple, ducts (transfers milk from the lobules to the nipple), areola the darker area of the breast (which contains glands that produce small elevations) fatty and connective tissues. Figure 2 shows different components of breast.

Cancer can affect any part of the breast; mostly, it affects lobules, ducts and could spread to other parts of the body through lymph nodes. Breast cancer is mainly categorized into two types.

Fig. 2 Anatomy of the breast

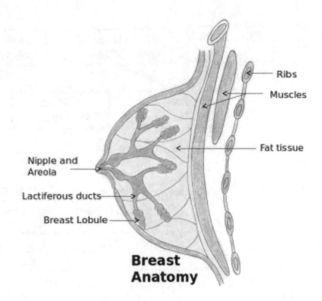

Ribs

Muscles

Fat tissue

Nipple and Areola

Lactiferous ducts

Breast Lobule

Breast Anatomy

1. Malignant
2. Benign

(i) Malignant cases are considered as cancerous cases which are life threatning; they need to be detected in early stage for cure.
(ii) Benign cases are considered as non-cancerous cases, they can be cured.

If the malignant cases are detected in early stages, then they can be cured. Computerized tomography, mammography, thermography, MRI and ultrasound techniques are used to diagnose breast cancer in early stages. Radiologists will inspect these scans, and if they find anything suspicious, they recommend for biopsy. Biopsy type varies depending on the size of the suspicious area, its location and presence of number of abnormal areas. The key challenge in breast cancer detection is the classification of tumors into malignant/bcnign. Research indicates that most experienced physicians can diagnose cancer with 79% accuracy while machine learning techniques can achieve 91% diagnosis accurately [4]. Following are the key steps of the process to identify BC using machine learning algorithms.

1. The obtained images while screening are digitized.
2. Region of interest is found and segmented.
3. Features are extracted from the segmented ROI using feature extraction techniques.
4. Features are fed to the machine learning algorithms.
5. Machine learning algorithms analyze the features (basing on training) and classify the tumor into malignant or benign (Fig. 3).

To make a correct decision about the breast cancer, usually researchers use standard image databases for investigation. Several institutions and organizations have

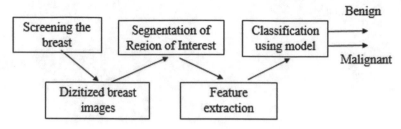

Fig. 3 Basic breast image classification model

Table 1 Breast image databases available for investigation [5, p. 4]

Database	Number of images	Database size (GB)	Image capture technique	Image type	Total patients
MIAS	322	2.3	Mammogram		161
DDSM			Mammogram		2620
CBIS-DDSm	4067	70.5	MG	DICOM	237
ISPY1	386,528	76.2	MR, SEG		237
Breast-MRI-NACT-pilot	99,050	19.5	MRI		64
QIN-breast	100,835	11.286	PET/CT, MR	DICOM	67
Mouse-mammary	23,487	8.6	MRI	DICOM	32
TCGA-BRCA	230,167	88.1	MR, MG	DICOM	139
QIN breast DCE-MRI	76,328	15.8	CT	DICOM	10
Breast-diagnosis	105,050	60.8	MRI/PET/CT	DICOM	88
Rider breast MRI	1500	0.401	MR	DICOM	5
BCDR			Mammogram		1734
TCGA-BRCA		53.92 (TB)	Histopathology		1098
BreakHis	7909		Histopathology		82
Inbreast	419		Mammogram		115

introduced image databases and they are accessible for researchers for investigation. Table 1 gives the details of some available image databases.

1.2 Machine Learning—History

In his 1950 paper titled "Computing Machinery and Intelligence," Alan Turing asked, "Can machines think?" [4]. This is the moment when the thought process for machine learning started. Arthur Samuel in 1959 defined machine learning as "Field of study that gives computers the ability to learn without being explicitly programmed." Samuel is credited with creating one of the self-learning computer programs during his work at IBM. He focused on games as a way of making the computer learn things.

A group of visionaries in around 1970s made a proposition "it is difficult to tell a machine exactly how to accomplish a task, but we can make the machine learn the skills to perform the task using examples." In 1983, many research papers were published on the visionaries' thoughts and ideas. In 1986, Tom Mitchel's paper "Explanation-Based Generalization: A Unifying View" was the next step in machine learning. He defined machine learning as "A computer program is said to learn from experience E with respect to some class of tasks T and performance measure P, if its performance at tasks in T, as measured by P, improves with the experience E." In 1977, a book published by McGraw Hill written by Tom Mitchell on machine learning created a new era for machine learning.

1.3   Machine Learning—Types

Machine learning is a branch of artificial intelligence that provides systems the ability to learn and analyze from experience without being explicitly programmed. There are different types of machine learning algorithms. The algorithm that is to be used depends on the output that is to be produced.

Supervised Machine Learning: It is the process of training a model using a labeled dataset which has input variable (x) and output variable (y). With this training approach, the model will be able to design a mapping function from input to output. When a new input data is given, it will be able to predict output data. Supervised machine learning can be further categorized into regression and classification.

Classification: Classification is the process of identifying the class to which new data belongs to, after being trained on a labeled dataset. Classification is best used when the output has finite and discrete values, for example, if we train the model by showing images of different animals like cats, dogs, goats and sheep. Now if I give an image to it, it will be able to recognize the image as cat or dog or goat or sheep.

Regression: Regression is the process of finding the output for given new input, after being trained on labeled dataset. Regression is best used when the output has real value. For example, we have trained the model with prices of a house given the features of house like size, price per sq.mt, location of house, no. of rooms, etc. (input)...and its price(output). When we give details of new house, this model will predict the price of the house.

Clustering: This is the process of grouping the data based on the similar characteristics of the data. For example, when given images of different animals, the model may group it based on a feature such as number of legs. Using the features, the model will cluster animals with 2 legs into one cluster and animals with four legs into another.

Association: This is a process of discovering associations and relations among large set of data items. For example, we have provided transactions data of a supermarket. From this data, it may learn the facts like, sold count of face powders, body sprays, … of different brands in every month of a year. If I give any product, it can predict the sale value of that product in this month.

Semi-supervised Machine Learning: In this machine learning method, model is trained using labeled and unlabeled dataset. The aim of the model is to classify some of the unlabeled data using the labeled data set. One of the example of this method is speech analysis. Labeling audio files is an exhaustive task and it requires abundant human resources. SSL techniques can be used to upgrade traditional speech analyticmodels.

Reinforcement Machine Learning: It is the process of training the agent that learns by interacting with its environment using a system of reward and punishment. Different software machines use this approach to find the better step to take in a particular situation. For example, this approach can be used to automate the allocation and schedule of system resources to waiting jobs, to minimize the average job slowdown.

The technology advancements in last decade lead to development of algorithms that are moreefficient in recognizing objects and analyzing large visual datasets. One important aspect of BC is its early detection, which will significantly improves the output. Although number of measures are taken by healthcare sector for early detection, however there may be some cases of false positives and false negatives due to the intricacy of interpreting the images exactly. ML and DL provide a way to analyze large datasets, being able to detect breast cancer more accurately than doctors. These machine learning models and systems—when provided as a tool for doctors—will improve the early detection, and reduce false positives and false negatives, as although machine learning models might outperform doctors, there will be cases where doctors flag cases that will be missed by these intelligent systems. This paper reviews the research and advancements done in leveraging machine learning technology to improve breast cancer detection.

2 Related Works on Diagnosis of Breast Cancer Using Different Machine Learning Algorithms

2.1 Breast Cancer Identification Using Support Vector Machine

Support vector machine is a supervised machine learning algorithm. It can be used for both classification and regression problems, but mostly used for classification. In SVM, the data in the dataset is plotted on "n" dimensional space where "n" is the number of features and then classification is applied. This results in a hyperplane which splits the data points into two classes. The distance between the hyperplane and the nearestdata points is known as the margin and these data points are known as supportvectors. In 2007, Polat and Güneş [6] presented a model which uses LSSVM to diagnose breast cancer and LSSVM uses linear equations for training. The performance of this model was measured with sensitivity, specificity and classification accuracy. They applied this model on WBCD dataset and this model achieved 98.53% of classification accuracy.

In 2009, Akay [7] proposed a model based on SVM with attribute selection for identification of breast cancer. He used F-Score to select the characteristics and the kernel of SVM was RBF. This model was trained and tested using WBCD dataset and its outcomes measured with classification accuracy. The results show that it achieved 99.51% of classification accuracy. In 2009, Rejani and Selvi [8] designed a model that mainly focuses on detecting the abnormal regions with slight variation to background. When background is dark, top hat filter is used to correct uneven illumination. Steps in model are enhancement of mammogram, segmentation of suspicious regions, extraction of features and then classification using SVM. In 2011, Chen et al. [9] proposed RS_SVM classifier for breast cancer diagnosis. This roughest algorithm is used to select features. Author evaluated the model on Wisconsin Breast Cancer Dataset and recorded its performance as 99.41%. In 2013, Görgel et al. [10] presented a model to detect BC using mammograms. In this model, image is enhanced using homomorphic filtering; regions of interest are found using LSRG algorithm, in characteristics extraction, SWT is used, resulting in SVM masses classified as either benign/malignant. Using this model, the accuracy achieved is 93.59%.

In 2019, Liu and Brown [11] proposed a model to detect breast cancer. To extract the characteristics of breast images, this model uses Db2 + PCA + SVM for classification. This model results are: average sensitivity is $83.10 \pm 1.91\%$, the average specificity is $82.60 \pm 4.50\%$, and the average accuracy is $82.85 \pm 2.21\%$. In 2019, Kamel et al. [12] presented a model to detect breast cancer. This model is based on support vector machine and attributes selection using gray wolf optimization. They experimented on model with different ratios of training and testing data (60–40, 70–30 and 80–20) and also with and without selecting features. The results show best performance of model in 80–20 training and testing data ratio with selecting features.

2.2 Breast Cancer Detection and Diagnosis Using SVM and Other ML Algorithms

The other ML algorithms which were used in combination with SVM are K-means, KNN, Naïve Bayes, where Naïve Bayes is supervised ML algorithm and both K-means and KNN are unsupervised ML algorithms.

In 2014, Zheng et al. [13] developed a model with K-means + SVM algorithms to diagnose BC. K-means is used to find the hidden patterns in the tumors and SVM is used for classification. They applied this model on WDBC dataset which achieved

Table 2 Overview of papers presented in Sects. 2.1 and 2.2

References	Dataset	ML algorithms used	Performance measure	Result
[6]	Wisconsin dataset	Least square support vector machine	Accuracy	98.53%
[7]	Wisconsin dataset	SVM with RBF kernel	Accuracy	99.51%
[9]	Wisconsin dataset	RS SVM	Accuracy	99.41%
[10]	MIAS dataset	LSRG and SVM	Accuracy	93.59%
[11]	MIAS dataset	PCA, SVM	Accuracy	$82.85 \pm 2.21\%$
			Specificity	$82.60 \pm 4.50\%$
			Sensitivity	$83.10 \pm 1.91\%$
[12]	Wisconsin dataset	Gray wolf optimization, SVM	Accuracy	100%
			Specificity	100%
			Sensitivity	100%
[13]	Wisconsin dataset	K-means, SVM	Accuracy	97.38%
[14]	Wisconsin dataset	SVM, KNN	Accuracy	98.57%, 97.14%
			Specificity	95.65%, 92.31%
[15]	FFDM	SVM, KNN	AUC	0.68

97.38% classification accuracy. In 2017, Islam et al. [14] proposed a model that predicts the breast cancer using SVM and KNN. These SVM and KNN techniques have achieved accuracy of 98.57 and 97.14% and specificity of 95.65 and 92.31%. In 2018, Heidari et al. [15] designed a model to identify BC and compared it with some of the previous models and proved that his proposed model performance is best than other models. The proposed model stages are (1) segmentation, (2) extraction of characteristics of images, (3) attributes selection using LPP-based feature regeneration algorithm and (4) classification using SVM and KNN (Table 2).

2.3 Breast Cancer Detection and Diagnosis Using Artificial Neutal Netwoks

Artificial neural networks are the most popular ML algorithms which are used today. These algorithms were invented in 1970s, but only became popular later due to increase in computation power. They are modeled based on human brain. As neurons in our nervous system learn from past data, ANN also learns from the past data and does the predictions and classifications. ANN has input layer, output layer and multiple hidden layers, every layer has artificial neurons, they are connected and can be represented by the directed edges with weights. In 1993, Wu et al. [16] proposed a model for breast cancer diagnosis which can be a decision making aid for the radiologists in the analysis of the mammograms. This model is based on artificial neural networks with three-layer feed forward neural network and back propagation algorithm. They evaluated model with full features and reduced features. The performance of this network is measured with AU-ROC curve and yielded value 0.84 (with full 43 features) and 0.89 (with reduced (14) features). In 1994, Floyd et al. [17] designed an ANN model that can predict breast cancer based on mammographic findings. Performance of this model was measured using sensitivity and specificity. They proved that this model performance was more accurate than radiologists.

In 1995, Fogel et al. [18] developed an ANN model to detect breast cancer from histologic data. They demonstrated that significant level of execution can be accomplished by small neural networks working on histopathology characteristics from breast biopsies. In 1998, Furundzic et al. [19] designed a model for early breast cancer identification which is based on neural networks. It reduces the false negative results which frequently appear while detection through mammograms.

In 2002, Abbass [20] proposed a model based on MPANN to diagnose breast cancer. He compared it with previous approaches and proved that proposed model performance is best and with much lower computational cost. In 2011, Marcano-Cedeño et al. [21] proposed an ANN to detect BC which is based on biological meta-plasticity property. AMMLP algorithm, incorporated in proposed model, is compared with MLP with backpropagation, and final result shows that AMMLP performance is better than MLP. In 2017, Abdel et al. [22] proposed an ANN to detect the BC. This approach used different number of hidden layers (1, 2, 3), different number of neurons in each hidden layer (20, 21, 22, 23, 24) and different activation functions (TRASIG, LOGSIG, PURELINE). It was found that performance of ANN with three hidden layers, 21 neurons in each hidden layer with TRASIG activation was best of all. In 2019, Alickovic and Su [23] proposed a model based on normalized multilayer perceptron neural network to detect breast cancer. The model has experimented on Wisconsin breast cancer dataset and achieved the best performance 99.27% (Fig. 4; Table 3).

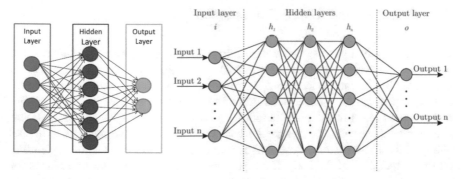

Fig. 4 Simple artificial neural network structures with one hidden layer and more hidden layers [24, p. 4]

Table 3 Overview of papers presented in Sect. 2.3

References	Dataset	ML algorithms used	Performance measure	Result
[16]	Mammography atlas compiled by Tabar and Dean	3 layer, FFNN with a backpropagation algorithm	ROC	0.89
[17]	Jackknife sampling-260 patient records	Artificial neural network	Sensitivity	1.00
			Specificity	0.59
[18]	Mammography atlas compiled by Tabar and Dean	ANN, stochastic optimization, evolutionary algorithms	Average mean squared error	0.13
[20]	Wisconsin dataset	ANN, pareto differential evolution algorithm	Average accuracy	0.981 + 0.005
[21]	Wisconsin dataset	ANN, artificial metaplasticity multilayer perceptron algorithm	Accuracy	99.26%
			Specificity	97.89%
			Sensitivity	100%
			AUC	0.989
[22]	Wisconsin dataset	ANN, Levenberg–Marquardt algorithm for NN training, different activation functions like TRASIG, LOGSIG, PURELINE…	Accuracy	98%
[23]	Wisconsin dataset	ANN, multilayer perceptron, Min–Max and mean and SD normalization	Accuracy	99.27

2.4 Breast Cancer Using Thermography

Thermography is a process that uses infrared camera to capture heat patterns and blood flow in body tissues. It is non-invasive test that involves no radiation and is used for breast cancer screening. Thermography-based test works on following key biological action. Cancer cells grow and multiply very fast and thus require more oxygen. As a result, more blood flows towards cancer cells and temperature rises around these cells. Thermography-based detection uses thermal camera to take photos of the effected areas; these images contain different range of colors depending on the temperature of the body. If temperature is different from normal ranges for that cell region, it indicates an abnormality for further analysis.

Need of Thermogram: **Although** mammography is the leading technology for breast cancer screening, it cannot provide best results for patients with dense breast (most of the young women have dense breast) as well as for tumor size less than 2 mm. In such cases, thermogram guides the radiologists well [25] (Fig. 5).

In 2002, Fok et al. [26] described a CAD system which uses thermography in detection and analysis of breast cancer. It consists of two modules: visualization and detection modules. In visualization module, 3D image is generated from 2D thermogram. In detection module, ANN is used to classify breast cancer, taking the thermogramas input. They found that this model sensitivity was good but the specificity was poor. In 2009, Schaefer et al. [27] presented a model to analyzebreast cancer based on thermography. This model uses a sequence of characteristics extracted from thermograms and it uses fuzzy rule-based classifier for classification. The final results provide a classification accuracy above 80%. In 2009, Kennedy et al. [28] designed a method for detecting BC by combining mammography and thermography. The thermogram-based detection model has achieved 83% of sensitivity, and mammogram-based detection model has given 90% while combined model mammogram with thermogram gave 95% sensitivity. In 2012, Acharya et al. [29] set out to find whether thermal imaging is a potential tool for detecting breast cancer. To extract characteristics of thermograms, they used co-occurrence + runlength matrices and to classify, SVM. This model gave an accuracy of 88.10%. In 2013, Nicandro et al. [30] also worked on identifying the validity of thermal imaging in identification of breast cancer. The model used Naïve Bayes classifier for classification and they compared their present method with other classifiers like MPNN and DT (ID3 and C4.5) with default parameters. As only temperature of the body is not sufficient to classify BC as benign or malignant, this model calculates a score based on the features of thermogram. The breast cancer is classified based on the score.

In 2015, Gaber et al. [31] presented a CAD system to categorize breast cancer. This model consists of two modules: automatic segmentation and classification.

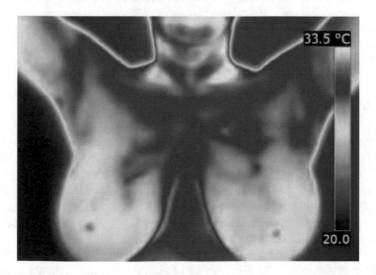

Fig. 5 The breast scan captured by thermal camera

Automatic segmentation is done based on NP sets and optimized fast fuzzy c-mean. SVM + different kernels were used for classification. In 2016, Okuniewski et al. [32] presented a model for identification of breast cancer using thermograms based on classifying contours which are visible on thermograms. Based on contours, images are classified and then features are found. Then these features are classified using four classifiers, DT + NB + RF + SVM. The result of all these four classifiers has been compared and it shows RF classifier achieves the best results. In 2016, Pramanik et al. [33] introduced a method block variance to extract the characteristics of thermograms. This block variance model based on local texture. This model has been used in the breast cancer identification from thermogram. They used gradient descent training rule + FFNN for classification. They used DMR dataset for evaluation. In 2018, Gogoi et al. [34] designed a model to differentiate the abnormal/normal thermograms using SVD. This model has acheived 98.00% of accuracy, specificity and sensitivity. As they used very small dataset, results cannot be generalized. In 2018, Sathish et al. [35] presented a study on Bagging + AdaBoost. In this study, he compared those two methods for the identification of breast cancer from thermograms. In the classification phase, spectral and spatial features are used. The final outcomes indicated that bagging is best classifier. Results are: accuracy is 87%, sensitivity is 83% and specificity is 90.6%.

In 2020, AlFayez et al. [25] proposed an approach based on thermography to detect breast cancer. This approach consists of four modules: Image preprocessing, ROI segmentation, feature extraction and classification. Image preprocessing is done by using homomorphic filtering + top-hattransform + adaptive histogram equalization. While binary masking + K-mean clustering is used for segmentation, signature boundary is used for feature extraction. ELM + MLP are used for classification and these two classifiers were compared. The outcomes show that ELM is better than MLP. In 2020, Ekici and Jawzal [36] presented an idea to develop the software for automatic early diagnosis of BC based on thermal imaging using image processing techniques and algorithms. Classification is done using CNN optimized by Bayes algorithm (Table 4).

2.5 Breast Cancer Detection Using Ensemble Techniques

Ensemble techniques are the combination of several machine learning algorithms to decrease variance, bias and improve efficiency in prediction. In 1998, Sharkey et al. [37] designed a neural net ensemble model for the identification of the breast cancer with high sensitivity that makes a minimal number of false alarms. In 2012, Hsieh et al. [38] presented a model to diagnose breast cancer which uses information gain for feature selection. They used NF, KNN, QC, their ensembles and combined ensemble of these three classifiers for classification of breast cancer. The final outcomes demonstrated that the combined ensemble provided the highest accuracy of the classification. In 2018, Wanga et al. [39] proposed a model to increase

Table 4 Overview of papers presented in Sect. 2.4

References	Dataset	ML algorithms used	Performance measure	Result
[28]	Private data set	TH(1:5) scale	Sensitivity	95%
[29]	Singapore General Hospital, Singapore	Support vector machine	Accuracy	88.10%
[30]	A real-world database provided by oncologist specialized in the study of thermography	Bayesian network classifiers	Specificity	90.48%
			Sensitivity	85.71%
			Accuracy	71.88%
			Specificity	37%
			Sensitivity	82%
[31]	DMR-IR dataset developed by L. F. Silva and team	Optimized fast fuzzy *c*-mean, SVM with different kernals	Accuracy	92.06%
			Recall	96.59%
			Precision	87.50% (RBF kernel)
[33]	DMR	FANN	Accuracy	90%
			Specificity	85%
			Sensitivity	95%
[34]	Private dataset	SVM (polynomial)	Accuracy	98%
			Specificity	98%
			Sensitivity	98%
[35]	DMR	Ensemble bagged trees and AdaBoost	Accuracy	87%
			Specificity	90.6%
			Sensitivity	83%
[25]	DMR-IR	*K*-mean clustering, ELM, MLP		MLP/ELM
			Accuracy	80.04%/99.10%
			Specificity	84%/98.05%
			Sensitivity	61.6%/97.03%
[36]	Private dataset	Convolutional neural networks, Bayes algorithm	Accuracy	98.95%

diagnosis accuracy of breast cancer by reducing diagnosis variance by using SVM-based ensemble learning algorithm. In 2017, Huang et al. [40] compare performance of SVM and ensembles of SVM in detection of BC over different scale breast cancer datasets. The results prove that SVM ensembles (RBF kernal) + boosting method and SVM ensembles (linear kernel) + bagging method can be used for small dataset, SVM ensembles (RBF kernel) + boosting for large datasets. In 2019, Rudra et al. [41] presented a model to detect breast cancer, which is based on SVM and AdaBoost which is an ensemble of machine learning.

2.6 Breast Cancer Detection and Diagnosis Using Some More ML Algorithms

In 1990, Brzakovic et al. [42] designed an automated system to identify and categorize particular types of tumors from digitized mammograms. 25 samples of mammograms are tested; tumors were identified in 95% of the cases. The system uses deterministic or Bayes classifiers to classify the tumors into benign or malignant. In 1994, Wolberg et al. [43] presented a model that identifies breast cancer from cytological features which are derived from digital scan of FNA slides coupled with machine learning techniques. The classification accuracy obtained was 97%. In 2010, Verma et al. [44] presented a model to diagnose breast cancer using clustering technique. They incorporated soft clustered-based direct learning classifier into model. This classifier creates clusters with in benign or malignant class. Due to this, abnormalities can form subgroups within groups. To improve the classification accuracy, this model creates soft clusters within a class instead of using single binary class form malignant/benign. In 2011, Tsai et al. [45] presented a CAD model to identify the breast cancer in early stage This model consists of two stages (1) Reconstruction of the doubtful micro-calcification areas in images, (2) Identification of the microcalcification areas using BPNN and NB classifier. In this identification, BPNN classifier exhibits best performance. In 2012, Ramos-Pollán et al. [46] designed mammography-based ML classifiers to identify breast cancer. Model is processed first by selecting ROI. ROI pixels are then enhanced to reduce noise, upon which the suspected lesion is segmented. Finally, features are extracted and lessions are classified using ANN/SVM. The classification accuracy of model was 96.91 using SVM and 97.14 with ANN. In 2012, Lavanya and Rani [47] presented a model to diagnose breast cancer. This model comprises of a CART classifier with feature selection and bagging method. They applied this model on Wisconsin datasets and found 95.56 accuracy on Breast Cancer Wisconsin (diagnostic) dataset.

In 2013, do Nascimento et al. [48] introduced a system to detect breast cancer. This system comprises of two phases (1) texture analysis and lesion identification from given image and extracting features from ROI, (2) the features were extracted using wavelet transforms, and polynomial classification algorithm is used to identify whether tumor is benign/malignant. In 2016, Spanhol et al. [49] introduced a model based on CNN to distinguish between begin and malignant breast lesions based on histopathalogical images. This model was evaluated on BreaKHis dataset and the results proved that this model gave better classification than other previous models. In 2017, Suna et al. [50] presented a model using deep CNN for BC diagnosis. This model uses SSL and attained AUC of 0.8818. In 2019, Fathy and Ghoneim [51] proposed a model to detect BC which is based on the pre-trained RestNet-50 architecture, which is used for the extraction of characteristics of mammogram and then classifies them as benign/malignant. When the mammogram is classified as mass class, then class activation map is generated, then RELU is applied to it to generate the heat map which highlights the most discriminative mass region. In 2019, Shen et al. [52] introduced a model to detect breast cancer. This model mainly focuses on fine details of the images and it generates saliency maps that provide additional interpretability. This model has been evaluated on GMIC dataset and the predictions were as accurate as radiologists. In 2019, Fan et al. [53] designed a model on deep learning to detect metastases BC from histological images. This model laid emphasis on data preprocessing and data quality. It has incorporated Otsu algorithm and also hard negative method to remove false positives. In 2019, Wub et al. [54] presented a model based on deep CNN, which is trained and evaluated on over 10 lakh images and achieves AUC of 0.895 in predicting the breast cancer.

2.7 Comparision of Different ML Algorithms in Breast Cancer Detection and Diagnosis

In 2007, Übeyli [55] want to develop an automated system for diagnosing breast cancer. For that, he compared different ML algorithms performance and found that SVM achieved highest classification accuracy. In 2016, Asri et al. [56] analyzed the outcomes of different machine learning algorithms (SVM, NB, KNN, C4.5) in breast cancer diagnosis. The final results proved that SVM is the best classifier with classification accuracy 97.13. In 2018, Agarap [57] proposed a model based on gated recurrent unit SVM to diagnose breast cancer and also compared LR, MLP, NN, SR and SVM machine learning algorithms. Results proved that MLP performance is best with 99.04 accuracy.

In 2019, Chang Ming et al. [58] analyzed the outcomes of ML techniques LR, LDA, MCMC generalized linear mixed model, QDA, boosting, RF and KNN with BCRAT and BOADICEA models. They proved with experiment that adaptive boosting has given the best results with an accuracy rate of 90.17. In 2019, Ghasemzadeh et al. [59] presented a method for the identification of breast cancer. In this method to extract features, they used Gabor wavelet transform and then fed these features into SVM + ANN + DT. Performance of these methods was compared and the highest accuracy achieved was by ANN. In 2018, Illan et al. [60] presented a model to improve the accuracy of diagnosis of the lesions with high resolution DCE-MRI. For classification SVM with linear kernel, NN, QDA, DT, RF, AdaBoost and ANN algorithms are used. In 2019, Nguyen et al. [61] analyzed the unsupervised and supervised models for detecting breast cancer. The outcome of this analysis indicates that ensemble voting model is best model for the prediction of breast cancer (Table 5).

3 Evaluation of the Classifiers in Diagnosing Breast Cancer

After designing and developing the model, its performance should be evaluated. Following are the metrics to evaluate the machine learning models.

Table 5 Overview of papers presented in Sects. 2.5 and 2.6

References	Dataset	ML algorithms used	Performance measure	Result
[42]	Private dataset	Bayes classifiers	Accuracy	95%
[43]	Private dataset	Linear programming, MSM tree	Accuracy	100%
[44]	DDSM	K means, Gram–Schmidt orthogonalization and least squares	Accuracy	97.5
[45]	Private dataset	Bayes and neural networks	Sensitivity	97.19
[46]	Wisconsin dataset	SVM, ANN		SVM/ANN
			Accuracy	96.91/97.14
			ROC	0.924/0.9333
[47]	Wisconsin dataset	SVM, DT, bagging with cart algorithm	Accuracy	97.85%
[48]	DDSM	Wavelet functions	ROC curve	0.95
[50]	FFDM	Deep convolutional neural network	ROC curve	0.8818
[51]	DDSM	Pre-trained RestNet-50 activation map, ReLU	Sensitivity, specificity	99.8%
				82.10%

Confusion Matrix: It is a metric used to present the performance of a classifier on the dataset for which the actual values are known. TP: Cases and predictions, both are positive. FP: Cases are negative but the predictions are positive. TN: Cases and predictions, both are negative. FN: Cases are positive but the predictions are negative.

Classification Accuracy: It is the most commonly used metric to evaluate a model. It is the rate of correct predictions of cases to the total cases.

$$\text{Accuracy} = \frac{\text{Number of Correct Predictions}}{\text{Total number of Predictions made}}$$

Accuracy can also be defined in terms of confusion matrix as follows:

$$\text{Accuracy} = \frac{\text{TP} + \text{TN}}{\text{TP} + \text{FP} + \text{TN} + \text{FN}}$$

Precision: Percentage of positive instances out of the total predicted positive.

$$\text{Precision} = \frac{\text{TP}}{\text{TP} + \text{FP}}$$

Recall: Ratio of predicted positive cases out of the all true positivecases.

$$\text{Recall} = \frac{TP}{TP + FN}$$

F1 Score: The F1 score is the harmonic mean of the precision and recall; here F1 score reaches its best value at 1 and worst at 0.

ROC Curve: This curve is obtained by plotting the TPR and FPR at different threshold values.

$$\text{True Positive Rate (TPR)} = \frac{TP}{TP + FN}$$

$$\text{False Positive Rate (FPR)} = \frac{FP}{FP + TN}$$

Specificity: Ratio of negative cases out of the all truenegative cases.

Sensitivity: Ratio of positive cases $\frac{TN}{TN+FP}$ out of the all true positive cases.

$$\frac{TP}{TP + FN}$$

The above mentioned are some of the most commonly used measures to evaluate the machine learning models [62–64].

4 Conclusion

This paper reviewed the literature on the designing and developing of models using machine learning algorithms for detection and diagnosis of breast cancer. The important stages in models include pre-processing of the images, segmentation, selection and extraction of features, classification and finally performance evaluation. The performance evaluation metrics were also reviewed for assessment of the classifiers in the models. From the papers reviewed above, it can be figured out that SVM and ANN are the best classifiers in diagnosing breastcancer.

References

1. NBCF, Breast cancer facts: the National Breast Cancer Foundation. https://www.Nationalbres tcancer.org
2. Max R, Hannah R (2019) Cancer-our world in data. Cancer 1
3. Welfare F (1968) Indian council of medical research. Indian J Pediatr 35(5):255–255

4. Bauer K, Oxley ME (2003) Computing machinery and intelligence amplification. In: Computational intelligence: the experts speak, vol 25
5. Nahid AA, Kong Y (2017) Involvement of machine learning for breast cancer image classification: a survey. Comput Math Methods Med
6. Polat K, Güneş S (2007) Breast cancer diagnosis using least square support vector machine. Digit Sig Process 17(4):694–701
7. Akay MF (2009) Support vector machines combined with feature selection for breast cancer diagnosis. Expert Syst Appl 36(2):3240–3247
8. Rejani YIA, Selvi ST (2009) Early detection of breast cancer using SVM classifier technique. Int J Comput Sci Eng 3:127–130
9. Chen HL, Yang B, Liu J, Liu DY (2011) A support vector machine classifier with rough set-based feature selection for breast cancer diagnosis. Expert Syst Appl 38(7):9014–9022
10. Görgel P, Sertbas A, Ucan ON (2013) Mammographical mass detection and classification using local seed region growing–spherical wavelet transform (LSRG–SWT) hybrid scheme. Comput Biol Med 43(6):765–774
11. Liu F, Brown M (2018) Breast cancer recognition by support vector machine combined with daubechies wavelet transform and principal component analysis. In: International conference on ISMAC in computational vision and bio-engineering. Springer, Cham, pp 1921–1930
12. Kamel SR, Yaghoub Zadeh R, Kheirabadi M (2019) Improving the performance of support-vector machine by selecting the best features by Gray Wolf algorithm to increase the accuracy of diagnosis of breast cancer. J Big Data 6(1):90
13. Zheng B, Yoon SW, Lam SS (2014) Breast cancer diagnosis based on feature extraction using a hybrid of K-means and support vector machine algorithms. Expert Syst Appl 41(4):1476–1482
14. Islam MM, Iqbal H, Haque MR, Hasan MK (2017) Prediction of breast cancer using support vector machine and K-nearest neighbors. In: 2017 IEEE region 10 humanitarian technology conference (R10-HTC), IEEE, pp 226–229
15. Heidari M, Khuzani AZ, Danala G, Mirniaharikandehei S, Qian W, Zheng B (2018) Applying a machine learning model using a locally preserving projection based feature regeneration algorithm to predict breast cancer risk. In: Medical imaging 2018: imaging informatics for healthcare, research, and applications, vol 10579. International Society for Optics and Photonics, pp 105790T
16. Wu Y, Giger ML, Doi K, Vyborny CJ, Schmidt RA, Metz CE (1993) Artificial neural networks in mammography: application to decision making in the diagnosis of breast cancer. Radiology 187(1):81–87
17. Floyd CE Jr, Lo JY, Yun AJ, Sullivan DC, Kornguth PJ (1994) Prediction of breast cancer malignancy using an artificial neural network. Cancer Interdisc Int J Am Cancer Soc 74(11):2944–2948
18. Fogel DB, Wasson EC III, Boughton EM (1995) Evolving neural networks for detecting breast cancer. Cancer Lett 96(1):49–53
19. Furundzic D, Djordjevic M, Bekic AJ (1998) Neural networks approach to early breast cancer detection. J Syst Architect 44(8):617–633
20. Abbass HA (2002) An evolutionary artificial neural networks approach for breast cancer diagnosis. Artif Intell Med 25(3):265–281
21. Marcano-Cedeño A, Quintanilla-Domínguez J, Andina D (2011) WBCD breast cancer database classification applying artificial metaplasticity neural network. Expert Syst Appl 38(8):9573–9579
22. Abdel-Ilah L, Šahinbegović H (2017) Using machine learning tool in classification of breast cancer. In: CMBEBIH 2017. Springer, Singapore, pp 3–8
23. Alickovic E, Subasi A (2019) Normalized neural networks for breast cancer classification. In: International conference on medical and biological engineering. Springer, Cham, pp 519–524
24. Bre F, Gimenez JM, Fachinotti VD (2018) Prediction of wind pressure coefficients on building surfaces using artificial neural networks. Energy Build 158:1429–1441
25. AlFayez F, El-Soud MWA, Gaber T (2020) thermogram breast cancer detection: a comparative study of two machine learning techniques. Appl Sci 10(2):551

26. Fok SC, Ng EYK, Tai K (2002) Early detection and visualization of breast tumor with thermogram and neural network. J Mech Med Biol 2(02):185–195
27. Schaefer G, Závišek M, Nakashima T (2009) Thermography based breast cancer analysis using statistical features and fuzzy classification. Pattern Recogn 42(6):1133–1137
28. Kennedy DA, Lee T, Seely D (2009) A comparative review of thermography as a breast cancer screening technique. Integr Cancer Ther 8(1):9–16
29. Acharya UR, Ng EYK, Tan JH, Sree SV (2012) Thermography based breast cancer detection using texture features and support vector machine. J Med Syst 36(3):1503–1510
30. Nicandro CR, Efrén MM, Maria Yaneli AA, Enrique MDCM, Hector Gabriel AM, Nancy PC, Alejandro GH, Guillermo de Jesus HR, Rocio Erandi BM (2013) Evaluation of the diagnostic power of thermography in breast cancer using bayesian network classifiers. Comput Math Methods Mcd
31. Gaber T, Ismail G, Anter A, Soliman M, Ali M, Semary N, Hassanien AE, Snasel V (2015) Thermogram breast cancer prediction approach based on neutrosophic sets and fuzzy c-means algorithm. In: 2015 37th annual international conference of the IEEE engineering in medicine and biology society (EMBC). IEEE, pp 4254–4257
32. Okuniewski R, Nowak RM, Cichosz P, Jagodziński D, Matysiewicz, M, Neumann Ł, Oleszkiewicz W (2016) Contour classification in thermographic images for detection of breast cancer. In: Photonics applications in astronomy, communications, industry, and high-energy physics experiments, vol 10031. International Society for Optics and Photonics, p. 100312V
33. Pramanik S, Bhattacharjee D, Nasipuri M (2016) Texture analysis of breast thermogram for differentiation of malignant and benign breast. In: 2016 international conference on advances in computing, communications and informatics (ICACCI). IEEE, pp 8–14
34. Gogoi UR, Bhowmik MK, Bhattacharjee D, Ghosh AK (2018) Singular value based characterization and analysis of thermal patches for early breast abnormality detection. Australas Phys Eng Sci Med 41(4):861–879
35. Sathish D, Kamath S (2018) Detection of breast thermograms using ensemble classifiers. J Telecommun Electron Comput Eng (JTEC) 10(3–2):35–39
36. Ekici S, Jawzal H (2020) Breast cancer diagnosis using thermography and convolutional neural networks. Med Hypotheses 137:109542
37. Sharkey AJC, Sharkey NE, Cross SS (1998) Adapting an ensemble approach for the diagnosis of breast cancer. In: International conference on artificial neural networks. Springer, London, pp 281–286
38. Hsieh SL, Hsieh SH, Cheng PH, Chen CH, Hsu KP, Lee IS, Wang Z, Lai F (2012) Design ensemble machine learning model for breast cancer diagnosis. J Med Syst 36(5):2841–2847
39. Wang H, Zheng B, Yoon SW, Ko HS (2018) A support vector machine-based ensemble algorithm for breast cancer diagnosis. Eur J Oper Res 267(2):687–699
40. Huang MW, Chen CW, Lin WC, Ke SW, Tsai CF (2017) SVM and SVM ensembles in breast cancer prediction. PLoS ONE 12(1):e0161501
41. Rudra S, Uddin M, Alam MM (2019) Forecasting of breast cancer and diabetes using ensemble learning. Int J Comput Commun Inf 1(1):1–5
42. Brzakovic D, Luo XM, Brzakovic P (1990) An approach to automated detection of tumors in mammograms. IEEE Trans Med Imaging 9(3):233–241
43. Wolberg WH, Street WN, Mangasarian OL (1994) Machine learning techniques to diagnose breast cancer from image-processed nuclear features of fine needle aspirates. Cancer Lett 77(2–3):163–171
44. Verma B, McLeod P, Klevansky A (2010) Classification of benign and malignant patterns in digital mammograms for the diagnosis of breast cancer. Expert Syst Appl 37(4):3344–3351
45. Tsai NC, Chen HW, Hsu SL (2011) Computer-aided diagnosis for early-stage breast cancer by using wavelet transform. Comput Med Imaging Graph 35(1):1–8
46. Ramos-Pollán R, Guevara-López MA, Suárez-Ortega C, Díaz-Herrero G, Franco-Valiente JM, Rubio-Del-Solar M, González-De-Posada N, Vaz MAP, Loureiro J, Ramos I (2012) Discovering mammography-based machine learning classifiers for breast cancer diagnosis. J Med Syst 36(4):2259–2269

47. Lavanya D, Rani KU (2012) Ensemble decision tree classifier for breast cancer data. Int J Inf Technol Conver Serv 2(1):17
48. do Nascimento MZ, Martins AS, Neves LA, Ramos RP, Flores EL, Carrijo GA (2013) Classification of masses in mammographic image using wavelet domain features and polynomial classifier. Expert Syst Appl 40(15):6213–6221
49. Spanhol FA, Oliveira LS, Petitjean C, Heutte L (2016) Breast cancer histopathological image classification using convolutional neural networks. In: 2016 international joint conference on neural networks (IJCNN). IEEE, pp 2560–2567
50. Sun W, Tseng TL, Zhang J, Qian W (2017) Enhancing deep convolutional neural network scheme for breast cancer diagnosis with unlabeled data. Comput Med Imaging Graph 57:4–9
51. Fathy WE, Ghoneim AS (2019) A deep learning approach for breast cancer mass detection. Int J Adv Comput Sci Appl 10(1):175–182
52. Shen Y, Wu N, Phang J, Park J, Kim G, Moy L, Cho K, Geras KJ (2019) Globally-aware multiple instance classifier for breast cancer screening. In: International workshop on machine learning in medical imaging. Springer, Cham, pp 18–26
53. Fan K, Wen S, Deng Z (2019) Deep learning for detecting breast cancer metastases on WSI. In: Innovation in medicine and healthcare systems, and multimedia. Springer, Singapore, pp 137–145
54. Wu N, Phang J, Park J, Shen Y, Huang Z, Zorin M, Jastrzębski S, Févry T, Katsnelson J, Kim E, Wolfson S (2019) Deep neural networks improve radiologists' performance in breast cancer screening. IEEE Trans Med Imaging 39(4):1184–1194
55. Übeyli ED (2007) Implementing automated diagnostic systems for breast cancer detection. Expert Syst Appl 33(4):1054–1062
56. Asri H, Mousannif H, Al Moatassime H, Noel T (2016) Using machine learning algorithms for breast cancer risk prediction and diagnosis. Proc Comput Sci 83:1064–1069
57. Agarap AF (2018) On breast cancer detection: an application of machine learning algorithms on the Wisconsin diagnostic dataset. In: Proceedings of the 2nd international conference on machine learning and soft computing, pp 5–9
58. Ming C, Viassolo V, Probst-Hensch N, Chappuis PO, Dinov ID, Katapodi MC (2019) Machine learning techniques for personalized breast cancer risk prediction: comparison with the BCRAT and BOADICEA models. Breast Cancer Res 21(1):75
59. Ghasemzadeh A, Azad SS, Esmaeili E (2019) Breast cancer detection based on Gabor-wavelet transform and machine learning methods. Int J Mach Learn Cybernet 10(7):1603–1612
60. Illan IA, Tahmassebi A, Ramirez J, Gorriz JM, Foo SY, Pinker-Domenig K, Meyer-Baese A (2018) Machine learning for accurate differentiation of benign and malignant breast tumors presenting as non-mass enhancement. In: Computational imaging III, vol 10669. International Society for Optics and Photonics, pp 106690W
61. Nguyen QH, Do TT, Wang Y, Heng SS, Chen K, Ang WH, Philip CE, Singh M, Pham HN, Nguyen BP, Chua MC (2019) Breast cancer prediction using feature selection and ensemble voting. In: 2019 international conference on system science and engineering (ICSSE). IEEE, pp 250–254
62. Evaluation metrics for machine learning models—heartbeat. https://heartbeat.fritz.ai/evaluation-metrics-for-machine-learning-models-d42138496366
63. Mishra A (2018) Metrics to evaluate your machine learning algorithm. Towards Data Science, pp 1–8
64. Various ways to evaluate a machine learning model's performance. https://towardsdatascience.com/various-ways-to-evaluate-a-machine-learning-models-performance-230449055f15

Study of Comparison on Efficient Malicious URL Detection System Using Data Mining Algorithms

Rajitha Kotoju and D. Vijaya Lakshmi

Abstract Detecting malicious URLs is an essential task in network security intelligence. Nowadays, lot of methods is available for intrusion detection systems. The intrusion detection method is mainly used for identifying malicious URLs from dataset and then reject. In this paper, we analyze a lot of malicious URL detection methods and compare the performance of the system. The experimentation is analyzed based on two types of the dataset such as URL reputation dataset and phishing websites dataset. The comparative analysis is carried out based on accuracy measures and based on accuracy we find out the best method.

Keywords Network security · Intrusion detection · Detection system

1 Introduction

A PC associated with the Internet can get data from a huge array of accessible servers and different PCs by moving data from them to former PC's local memory. Regular employment of the Internet is Email, World Wide Web, remote access, cooperation, spilling media and record sharing. From that, we comprehend as the utilization of Web is expanding in our day-by-day life; the system security is getting to be noticeably important keeping in mind the end goal to get security, integrity and confidentiality of a resource. But nowadays, malfunctions Web is expanding. There are PC speculation cheats, cyber-crimes, money-related violations, phishing tricks, talking (disguising) and crimes related which share exchanging on Web. To dodge the issue, network security is available. For building up the system security in the cloud, these days, intrusion detection system (IDS) is created. Intrusion detection is the way toward observing

R. Kotoju (✉)
Department of CSE, MGIT, Hyderabad, India
e-mail: charuk.rajitha@gmail.com

D. Vijaya Lakshmi
MGIT, Hyderabad, India

© The Author(s), under exclusive license to Springer Nature Singapore Pte Ltd. 2021   523
K. A. Reddy et al. (eds.), *Data Engineering and Communication Technology*,
Lecture Notes on Data Engineering and Communications Technologies 63,
https://doi.org/10.1007/978-981-16-0081-4_53

the occasions happening in a PC framework/network and analyzing them for indications of conceivable assaults, which can prompt violations or unavoidable dangers of violations of PC security approaches, of the association [1]. As the expansive number of occurrences is expanding in our day-by-day life, IDS is utilized with enhanced strategies. IDS plays an imperative to secure the system and its primary objective is to view the system activities consequently to identify the malicious attacks. IDS is becoming a critical component to secure the network in today's world. By utilizing information mining in IDS can enhance the discovery rate, dealing with the false alarm rate and lessen false positive rate. IDS distinguishes and manages the malicious system of PC and PC network resources. IDS has been named into two classes, for example, (i) host-based intrusion detection system and (ii) network-based intrusion detection system [2]. Host-based IDS is intended to screen, identify and respond to activity and attacks on the given host [3]. There is a wide assortment of procedures to execute such assaults, for example, express hacking endeavors, drive-by misuses, social designing, phishing, watering hole, man in the middle, SQL infusions, misfortune/robbery of gadgets, denial of service, distributed denial of service and numerous others. Considering the assortment of assaults, possibly new assault types, and the innumerable contexts in which such assaults can appear, it is difficult to design robust systems to detect cyber-security breaches [4]. The confinements of customary security administration advances are ending up increasingly genuine given this exponential development of new security dangers, fast changes of new IT innovations and critical lack of security experts. Most of these attacking techniques are realized through spreading compromised URLs (or the spreading of such URLs forms a critical part of the attacking operation) [5]. URL is the shortened form of uniform resource locator, which is the worldwide address of archives and different assets on the World Wide Web. A URL has two principle segments: (i) protocol identifier—it indicates what protocol to use, (ii) resource name—it specifies the IP address or the domain name where the resource islocated.

The basic organization of the paper is as follows: Sect. 1.1 explains the basics of intrusion detection system and the comparative study is deeply explained in Sect. 2. The conclusion part is presented in Sect. 3.

1.1 Intrusion Detection System

Nowadays, to provide an high-level security for highly sensitive and private information is very essential. IDS is a basic innovation in network security. Considering the variety of attacks, potentially new attack types and the innumerable contexts in which such attacks can appear, it is hard to design robust systems to detect cyber-security breaches. The impediments of customary security administration innovations are winding up increasingly genuine given this exponential development of new security dangers, quick changes of new IT advances and critical lack of security experts. The majority of these assaulting strategies are acknowledged through spreading traded off URLs (or the spreading of such URLs frames a basic piece of the assaulting operation) [5].

The main intention of this research is to intrusion detection and identification from suspicious URLs using different approaches. The intrusion detection is predominantly used for security of internet service provider (ISP). Basically, the intrusion detection systems consist of two stages such as (i) feature selection and (ii) malicious detection. For intrusion detection system, here we utilize two types of the dataset such as URL reputation dataset and phishing websites dataset. At first, input dataset (URLs) is yielded into the system, which has URLs along with corresponding phishing features. Then, attribute reduction method is appealed to sort out an optimal subset of attributes which therefore diminish computational burden and strengthen the performance of the classifier. The procure dataset with a subset of attributes is given to the classifier. Finally, the classifier classify given URLs is malicious or non-malicious. In previously, we have published two research work based on malicious URLs detection [6, 7]. In [6], malicious detection system was developed using oppositional cuckoosearch with weighted fuzzy rule system (OCS + FL). Here, oppositional cuckoo search algorithm was used to feature selection and fuzzy logic classifier used for malicious detection. Similarly, in [7], we have implemented intrusion detection system based on hybridization of firefly algorithm and cuckoo search with the fuzzy algorithm(HFFCS + FL).

2 Comparative Study of Intrusion Detection System

In this research paper, we analyze the performance of the proposed two works with existing methods. For comparison, we utilize totally six classifiers such as fuzzy logic (FL), K-nearest neighbor (KNN), extreme learning machines (ELM), random forest (RF), AdaBoost (AB), radial basis function (RBF) classifier and six optimization algorithms such as hybridization of firefly and cuckoo search, (HFFCS), oppositional cuckoo search (OCS), cuckoo search (CS), firefly (FF) and genetic algorithm. And one more method is without optimization (WOP)-based malicious URLs detection. Based on the above-mentioned methods, we are making 36 combinations. In this section, we are going to analyze and compare the performance of intrusion detection system using these 36 methods for two set of the dataset. The thirty six combinations are HFFCS + FL, OCS + FL, CS + FL, GA + FL,FF + FL, WOP + FL, HFFCS + KNN, OCS + KNN, CS + KNN, GA + KNN, FF + KNN, WOP + KNN, HFFCS + ELM, OCS + ELM, CS + ELM, GA + ELM, FF + ELM, WOP + ELM, HFFCS + RF, OCS + RF, CS + RF, GA + RF, FF + RF, WOP + RF, HFFCS + AB, OCS + AB, CS + AB, GA + AB, FF + AB, WOP + AB, HFFCS + RBF, OCS + RBF,

CS + RBF, GA + RBF, FF + RBF and WOP + RBF. We split the 36 methods into six based onthe classifiers.

Comparative Analysis Using Dataset 1

In this section, we utilize the phishing website dataset for intrusion detection system. Here, we compare the 36 combinations-based intrusion detection system and find which method is best. Tables 1, 2, 3, 4, 5 and 6 shows the performance of the different methods based on classifier.

Table 1 shows the performance of intrusion detection system using HFFCS + FL, OCS + FL, CS + FL, GA + FL, FF + FL and WOP + FL. Here, we utilize six types of the optimization algorithm. Each algorithm has different behavior and these algorithms are used in the feature selection stage. The feature selection is an important process for intrusion detection system. Here, our proposed approach HFFCS + FL-based intrusion detection system attains the maximum accuracy of 0.94 which is 0.91 for OCS + FL, 0.86 forusing CS + FL, 0.84 for using GA + FL, 0.82 for using FF + FL and 0.79 for using WOP + FL-based intrusion detection system. Moreover, our proposed HFFCS + FL method obtains the minimum error of 0.06. Compare to other method, proposed approach obtains the minimum error and maximum accuracy.

Table 2 shows the performance of intrusion detection system using HFFCS + KNN, OCS + KNN, S + KNN, GA + KNN. Here, we analyze six types of intrusion detection methods. Here, in classification purpose, we utilize a KNN classifier. The KNN classifier is an instance-based learning method and which is used much application such as data mining, statistical pattern recognition and image processing.

When analyzing Table 2, the method HFFCS + KNN obtains the accuracy of 0.86, OCS + KNN obtains the accuracy of 0.84, CS + KNN obtains the accuracy of 0.83, GA + KNN obtains the accuracy of 0.77, FF + KNN obtains the accuracy of 0.81 and WOP + KNN obtains the accuracy of 0.73.

Table 1 Performance of intrusion detection system using fuzzy logic with different optimization algorithms

Measures	HFFCS + FL	OCS + FL	CS + FL	GA + FL	FF + FL	WOP + FL
TP	67	63	55	53	51	48
TN	27	28	31	31	31	31
FP	4	3	0	0	0	0
FN	2	6	14	16	18	21
TPR	0.97101	0.913	0.797	0.768	0.73913	0.695
TNR	0.87097	0.903	1	1	1	1
FPR	0.12903	0.096	0	0	0	0
FNR	0.028986	0.086	0.202	0.231	0.26087	0.304
ACC	0.94	0.91	0.86	0.84	0.82	0.79
ERR	0.06	0.09	0.14	0.16	0.18	0.21

Table 2 Performance of intrusion detection system using KNN with different optimization algorithms

Measures	HFFCS + KNN	OCS + KNN	CS + KNN	GA + KNN	FF + KNN	WOP + KNN
TP	69	69	69	59	66	69
TN	17	15	14	18	15	4
FP	14	16	17	13	16	27
FN	0	0	0	10	3	0
ACC	0.86	0.84	0.83	0.77	0.81	0.73
ERR	0.14	0.16	0.17	0.23	0.19	0.27

Table 3 Performance of intrusion detection system using ELM with different optimization algorithms

Measures	HFFCS + ELM	OCS + ELM	CS + ELM	GA + ELM	FF + ELM	WOP + ELM
TP	69	69	69	69	69	69
TN	24	15	21	19	19	16
FP	7	16	10	12	12	15
FN	0	0	0	0	0	0
ACC	0.91	0.84	0.9	0.88	0.88	0.85
ERR	0.07	0.16	0.1	0.12	0.12	0.15

Table 3 shows the performance of intrusion detection using HFFCS + ELM, OCS + ELM, CS + ELM, GA + ELM, FF + ELM and WOP + ELM.

A learning algorithm for single-hidden layer feed forward neural network (SLFNs) is called extreme learning machine (ELM) whose learning speed can be thousands of times faster than traditional feed forward network learning algorithms like back-propagation (BP) algorithm while obtaining better generalization performance. When analyzing Table 3, the method HFFCS + ELM-based intrusion detection system obtains the accuracy of 0.91, the method OCS + ELM obtains the accuracy of 0.84, the method CS + ELM obtains the accuracy of 0.9, the method GA + ELM obtains the accuracy of 0.88, the method FF + ELM obtains the accuracy of 0.88 and WOP + ELM obtains the accuracy of 0.85. The TP, TN, TP, TN and error values also present in Table 3.

Table 3 shows the performance of intrusion detection system using HFFCS + RF, OCS + RF, CS + RF, GA + RF, FF + RF and WOP + RF. When analyzing Table 4, the method HFFCS + RF obtains the accuracy of 0.92, the method OCS + RF obtains the accuracy of 0.87, the method CS + RF obtains the accuracy of 0.73, the method GA + RF obtains the accuracy of 0.79, the method FF + RF obtains the accuracy of 0.78 and the method WOA + RF obtains the accuracy of 0.69. Moreover, from Table 2, we understand the method HFFCS + RF obtains the maximum accuracy and minimum error 0.09 compare to other methods.

Table 4 Performance of intrusion detection system using RF with different optimization algorithms

Measures	HFFCS + RF	OCS + RF	CS + RF	GA + RF	FF + RF	WOP + RF
TP	68	69	69	68	69	69
TN	23	18	4	11	9	0
FP	8	13	27	20	22	31
FN	1	0	0	1	0	0
ACC	0.92	0.87	0.73	0.79	0.78	0.69
ERR	0.09	0.13	0.27	0.21	0.22	0.31

Table 5 shows the performance of intrusion detection system using HFFCS + AB, OCS + AB, CS + AB, GA + AB, FF + AB and WOP + AB. Here, for the testing process, we utilize total 100 images for the testing process. Here, the method HFFCS + AB obtains the accuracy of 0.69, OCS + AB obtains the accuracy of 0.65, CS + AB obtains the accuracy of 0.65, GA + AB obtains the accuracy of 0.6, FF + AB obtains the accuracy of 0.64 and WOP + AB obtains the accuracy of 0.52.

Table 6 shows the intrusion detection system using different methods. Here, also we utilized 100 images for testing. We used five types of optimization algorithm such as fuzzy logic + cuckoo search algorithm (HFFCS), oppositional cuckoo search algorithm (OCS), and cuckoo search algorithm (CS), the genetic algorithm (GA),

Table 5 Performance of intrusion detection system using AB with different optimization algorithms

Measures	HFFCS + AB	OCS + AB	CS + AB	GA + AB	FF + AB	WOP + AB
TP	69	36	36	46	37	35
TN	0	29	29	14	0	17
FP	31	13	13	17	20	14
FN	0	22	22	23	0	34
ACC	0.69	0.65	0.65	0.6	0.649	0.52
ERR	0.31	0.35	0.35	0.4	0.357	0.48

Table 6 Performance of intrusion detection system using RBF with different optimization algorithms

Measures	HFFCS + RBF	OCS + RBF	CS + RBF	GA + RBF	FF + RBF	WOP + RBF
TP	69	69	69	69	69	69
TN	17	13	12	10	11	5
FP	14	18	19	21	20	26
FN	0	0	0	0	0	0
ACC	0.86	0.82	0.81	0.79	0.8	0.74
ERR	0.14	0.18	0.19	0.21	0.2	0.26

firefly algorithm (FA). The mentioned optimization algorithms are utilized for feature selection process. When analyzing Table 6, the method HFFCS + AB obtains the accuracy of 0.86, OCS + AB obtains the accuracy of 0.82, CS + AB obtains the accuracy of 0.81, GA + AB obtains the accuracy of 0.79, FF + AB obtains the accuracy of 0.8 and WOP + AB obtains the accuracy of 0.74.

When analyzing Tables 1, 2, 3, 4, 5 and 6, we gain the maximum accuracy for different methods. From Table 1, we attain the maximum accuracy using HFFCS + FL method, from Table 2, we obtain the maximum accuracy using HFFCS + KNN method, from Table 3, we obtain the maximum accuracy using HFFCS + RF, from Table 4, we obtain the maximum accuracy using HFFCS + AB method, from Table 5, we obtain the maximum accuracy using HFFCS + RBF and from Table 6, we obtain the maximum accuracy using HFFCS + ELM. Moreover, Fig. 1 shows the performance of overall performance result. Here, we obtain the maximum accuracy of 0.94 using the HFFCS + FL method. From the results, we clearly understand the proposed HFFCS + FL obtains the maximum accuracy compare to other methods.

Dataset 2

In this section, we compare our proposed intrusion detection system using HFFCS + FL with different methods using URL reputation dataset. Here, we mainly compare our proposed work accuracy with other works. We also measure, TP, TN, FP, FN, TPR, TNR, FPR, FNR and ERR. If a malicious is manifest existent in a URL, the given test also designates the presence of malicious, the outcome of the detection test is deliberated true positive (TP). Analogously, if a detection test is manifest absent in a malicious, the detection test recommends the malicious is absent as well, the test result is true negative (TN). Both TP and TN urge a consistent result of the

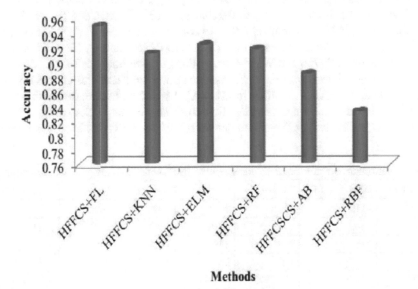

Methods

Fig.1 Overall performance result using dataset 1

detection test and the existent condition (also called the standard of truth). If the test indicates the attending of malicious in a URL which in effect has no such malicious, the test result is false positive (FP). Comparably, if the outcome of the test counsels that the malicious is absent for a URL with malicious for sure, the test output is a false negative (FN). The comparative result is present in Tables 7, 8, 9, 11 and 12. The main aim of proposed work is to predict the best method for intrusion detection system. Therefore, here, we compare different methods with a different combination which is given in Tables 7, 8, 9, 11 and 12. Among them, we are selecting the best method for intrusion detectionsystem.

Table 7 shows the comparative result obtained by HFFCS + FL, OCS + FL, CS + FL, GA + FL, FF + FL and WOP + FL methods. Here, in feature selection stage, we utilize different methods such as HFFCS, OCS, CS, GA and FF but classification stage, we utilize the FL system only. When analyzing Table 7, the method HFFCS + FL obtains the maximum accuracy of 0.94913, the method OCS + FL obtains the accuracy of 0.925373, the method CS + FL obtains the accuracy of 0.822367, the method GA + FL obtains the accuracy of 0.664284, the method FF + FL obtains the accuracy of 0.7818 and WOP + FL obtains the accuracy of 0.556023. From that, we understand HFFCS + FL obtains the maximum accuracy compare to other methods. Moreover, HFFCS + FL-based intrusion detection system obtains the minimum error rate of 0.050872.

The intrusion detection system performance based on KNN classifier is given in Table 8. Table 8 shows the different combination results with KNN. Here, we utilize the KNN classifier for classification stage and the optimization methods HFFCS, OCS, CS, GA and FF for feature selection stage. The method WOA + KNN, we did not perform any feature selection stage. Here, we directly apply all the features to KNN classifier for classification. When analyzing Table 8, the method HFFCS + FL obtains the maximum accuracy of 0.9108682, the method OCS + FL obtains the accuracy of 0.854951, the method CS + FL obtains the accuracy of 0.839605, the method GA + FL obtains the accuracy of 0.767501, the method FF + FL obtains the accuracy of 0.879756 and WOP + FL obtains the accuracy of 0.720832.

The intrusion detection system performance based on ELM classifier is given in Table 9. Here, for testing, we utilize the 1189 URLs. The main intention of this paper is to intrusion detection and identification from suspicious URLs using different methods. When analyzing Table 9, the method HFFCS + FL obtains the maximum

Table 7 Performance of intrusion detection system using FL with different optimization algorithms

Measures	HFFCS + FL	OCS + FL	CS + FL	GA + FL	FF + FL	WOP + FL
TP	2115	2001	2094	2068	2022	4
TN	2400	2401	1818	1092	1697	2641
FP	241	240	823	1549	944	0
FN	1	115	22	48	94	2112
ACC	0.94913	0.925373	0.822367	0.664284	0.7818	0.556023
ERR	0.050872	0.074627	0.177633	0.335716	0.2182	0.443977

Table 8 Performance of intrusion detection system using KNN with different optimization algorithms

Measures	HFFCS + KNN	OCS + KNN	CS + KNN	GA + KNN	FF + KNN	WOP + KNN
TP	1909	1752	1732	1595	1876	1573
TN	2424	2315	2262	2056	2309	1856
FP	217	326	379	585	332	785
FN	207	364	384	521	240	543
ACC	0.9108682	0.854951	0.839605	0.767501	0.879756	0.720832
ERR	0.0891318	0.145049	0.160395	0.232499	0.120244	0.279168

accuracy of 0.922940929, the method OCS + FL obtains the accuracy of 0.89363, the method CS + FL obtains the accuracy of 0.888375, the method GA + FL obtains the accuracy of 0.887114, the method FF + FL obtains the accuracy of 0.904141and WOP + FL obtains the accuracy of 0.887324.

Table 10 shows the performance of intrusion detection system using RF classifier with different optimization algorithms. Here, we compare the performance of intrusion detection system using six methods such as HFFCS + RF, OCS + RF, CS + RF, GA + RF, FF + RF and WOP + RF. Here, the method HFFCS + RF obtains the

Table 9 Performance of intrusion detection system using ELM with different optimization algorithms

Measures	HFFCS + ELM	OCS + ELM	CS + ELM	GA + ELM	FF + ELM	WOP + ELM
TP	2116	2116	2116	2116	2116	2116
TN	2322	2135	2110	2104	2185	2105
FP	319	506	531	537	456	536
FN	0	0	0	0	0	0
ACC	0.922940929	0.89363	0.888375	0.887114	0.904141	0.887324
ERR	0.067059071	0.10637	0.111625	0.112886	0.095859	0.112676

Table 10 Performance of intrusion detection system using RF with different optimization algorithms

Measures	HFFCS + RF	OCS + RF	CS + RF	GA + RF	FF + RF	WOP + RF
TP	1875	1480	1348	1613	1622	1828
TN	2487	2439	2433	2056	2054	2033
FP	154	202	208	585	587	608
FN	241	636	768	503	494	288
ACC	0.916964	0.823839	0.794829	0.771284	0.772756	0.811646
ERR	0.083036	0.176161	0.205171	0.228716	0.227244	0.188354

Table 11 Performance of intrusion detection system using AB classifier with different optimization algorithms

Measures	HFFCS + AB	OCS + AB	CS + AB	GA + AB	FF + AB	WOP + AB
TP	1778	1400	1778	2072	1636	1636
TN	2425	2625	2425	994	2099	2099
FP	216	16	216	1647	542	542
FN	338	716	338	44	480	480
ACC	0.883540046	0.846122	0.88354	0.644524	0.785159	0.785159
ERR	0.116459954	0.153878	0.11646	0.355476	0.214841	0.214841

error rate of 0.083036, the method OCS + RF obtains the error rate of 0.176161, the method CS + RF obtains the error rate of 0.205171, GA + RF obtains the error rate of 0.228716, the method FF + RF obtains the error rate of 0.227244 and the WOP + RF obtains the error rate of 0.188354.

Table 11 shows the performance of intrusion detection system using AB classifier. The goal of IDS tools is to detect computer attacks or illegal access and to alert the concerned people about the detection or security breach. When analyzing Table 10, the method HFFCS + AB obtains the maximum accuracy of 0.883540046, the method OCS + AB obtains the accuracy of 0.846122, the method CS + AB obtains the accuracy of 0.88354, the method GA + AB obtains the accuracy of 0.644524, the method FF + AB obtains the accuracy of 0.785159 and WOP + AB obtains the accuracy of 0.785159.

Table 12 shows the performance of intrusion detection system using RBF classifier. Here, for testing purpose, we utilize 1189 URLs. Here, the methods CS + RBF, GA + RBF and FF + RBF obtain the accuracy value closely related to each other. The methods OCS + RBF and HFFCS + RBF obtain the maximum accuracy. Compare to these two methods, HFFCS + RBF method obtains the maximum accuracy.

When investigating Table 7, 8, 9, 10, 11 and 12, we obtain the maximum accuracy for different methods. From the table, we understand HFFCS based classifier obtain

Table 12 Performance of intrusion detection system using RBF classifier with different optimization algorithms

Measures	HFFCS + RBF	OCS + RBF	CS + RBF	GA + RBF	FF + RBF	WOP + RBF
TP	1465	1343	1907	1230	1229	2037
TN	2489	2540	1562	2243	2331	1236
FP	152	101	1079	398	310	1405
FN	651	773	209	886	887	79
ACC	0.831196	0.816271	0.729241	0.730082	0.748371	0.688039
ERR	0.168804	0.183729	0.270759	0.269918	0.251629	0.311961

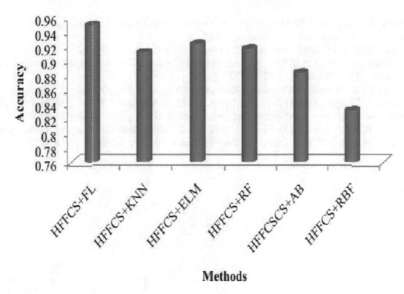

Fig. 2 Overall performance result using dataset 2

the maximum accuracy. Moreover, Fig. 2 shows the performance of overall performance result. Here, we obtain the maximum accuracy of 0.94913 using the HFFCS
+ FL method. From the results, we clearly understand the proposed HFFCS + FL
obtain the maximum accuracy compare to other methods.

3 Conclusion

In this paper, we have been analyzed comparative analysis for intrusion detection
method. The basic working principle of IDS was explained. For comparative analyze,
we have utilized 36 combinations of methods. Two types of dataset were utilized for
performance analysis. The experimental results were carried out using accuracy, TP,
TN, FP, FN and EPR measures. From the results, we clearly understand HFFCS +
FL method obtains the better result compare to other methods. Experimental results
stipulate that the method HFFCS + FLC-based malicious detection framework has
outperformed by having more excellent accuracy of 94% which is high balanced
with other combinations.

References

1. Zhan J (2008) Intrusion detection system based on data mining. In: Proceedings of the 1st
   international workshop on knowledge discovery and data mining, pp 402–405

2. Singh AP, Singh MD (2014) Analysis of host-based and network-based intrusion detection system. Int J Comput Netw Inf Secur 8:41–47
3. Pieter DB, Pels M (2005) Host-based intrusion detection systems, pp 5–7
4. Kshirsagar V, Joshi MS (2016) Rule based classifier models for intrusion detection system. Int J Comput Sci Inf Technol 7(1):367–370
5. Hong J (2012) The state of phishing attacks. Commun ACM 55(1):74–81
6. Rajitha K, Vijaya Lakshmi D (2016) Oppositional cuckoo search based weighted fuzzy rule system in malicious web sites detection from suspicious URLs. Int J Intell Eng Syst 9(4):116–125
7. Rajitha K, VijayaLakshmi D (2017) An efficient intrusion detection system for identification from suspicious URLs using data mining algorithms. Int J Bus Intell Data Min 12(2):133–158

Comparison of PAPR in OFDM and FBMC/OQAM Using PAPR Reduction Methods

Sravanti Thota, Yedukondalu Kamatham, and Chandra Sekhar Paidimarry

Abstract Though the multicarrier technique orthogonal frequency division multiplexing (OFDM) is generous in 4G technology, it suffers with spectrum leakage, inefficient spectrum, and high PAPR. In this paper, the 5G waveform—FBMC/OQAM PAPR—is compared with OFDM in terminologies of PAPR, and it is analyzed that FBMC/OQAM suffers with high PAPR than OFDM which roots to falsehood due to nonlinearity of high-power amplifier (HPA). To reduce PAPR in FBMC/OQAM, nonlinear companding performances are worn and the performance is analyzed in terms of PSD, BER, and PAPR.

Keywords FBMC · OQAM · OFDM · PAPR · BER · PSD · Companding

1 Introduction

The complications in OFDM restrict the expansion due to high synchronization and low spectrum efficiency [1, 2]. In the augmented demand in high data rate with absorption of added users and applications, new 5G radio propagation performance has been deliberated and explored for future wireless systems. For affording sophisticated data rate and better user knowledge, present LTE and Wi-Fi techniques are combined by 5G skills for universal exposure. For 5G New Radio (NR) waveforms, the most popular modulation technique 4G technique OFDM is not efficient. Therefore, different multicarrier modulation techniques are under research segments.

S. Thota (✉) · C. S. Paidimarry
Department of ECE, University College of Engineering, Osmania University, Hyderabad, Telangana, India
e-mail: sravanti23@gmail.com

C. S. Paidimarry
e-mail: sekharpaidimarry@gmail.com

Y. Kamatham
Department of ECE, CVR College of Engineering, Vastunagar, Mangalpally (V), Ibrahimpatnam (M), Hyderabad, Telangana, India
e-mail: kyedukondalu@gmail.com

© The Author(s), under exclusive license to Springer Nature Singapore Pte Ltd. 2021   535
K. A. Reddy et al. (eds.), *Data Engineering and Communication Technology*,
Lecture Notes on Data Engineering and Communications Technologies 63,
https://doi.org/10.1007/978-981-16-0081-4_54

Filter bank multicarrier with offset quadrature amplitude modulation (FBMC/OQAM) is supposed to be the active means to crack the existing drawback of OFDM. The FBMC is a strong candidate waveform formed system in 5G [3–5]. Since FBMC is also a multicarrier like other multicarrier systems, it suffers with high peak-to-average power ratio (PAPR) which worsens bit error rate (BER) due nonlinearity of HPA [6].

In this paper, we have compared two modulation techniques, one is filter bank multicarrier (FBMC) and other is OFDM [7].The assessment is ended in terms of implementation complexity, PAPR, and distortion introduced in the signal. The OFDM and FBMC system representation are described and discussed about PAPR and its falling techniques. Finally, Sect. 3 presents the evaluation results, surveyed by conclusions.

2 System Model

2.1 OFDM System Model

In this system, the collection of bits which is delineated into constellation points using M-QAM is fitted in time interval T is $X = \{X_k, \ k = 0, 1, \ldots N\}$ and these distributed symbols are made on IFFT operation.

$$x(t) = \frac{1}{\sqrt{N}} \sum_{n=0}^{n-1} X_k e^{\frac{j2\pi f_k t}{N}} \quad \text{for } f_k = k\Delta f \quad \text{and} \quad \Delta f = \frac{1}{T} \tag{1}$$

The OFDM signal is oversampled L times to estimate the PAPR for discrete time equal as continuous

$$x(t) = \frac{1}{\sqrt{N}} \sum_{n=0}^{n-1} X_k e^{\frac{j2\pi f_k t}{L}} \quad \text{for } k = 0, 1, \ldots NL - 1 \tag{2}$$

2.2 FBMC System Model

One of the most successful multicarrier techniques is FBMC. Different types of FBMC techniques are designed, in which polyphase network FBMC (PPN-FBMC) is mostly operated because of its low complexity. In PPN FBMC, the complexity is reduced by extravagance filtering operations at either sides of system. Mostly, FBMC is implemented using PHYDYAS filter as it gives low adjacent channel leakage.

2.3 OQAM

In digital modulation practice, the ordinal signals are transformed to waveform compatible to communication channel. OQAM is the most ordinarily used modulation arrangement used in FBMC. An OQAM introduces orthogonality condition in QAM. Suppose if any type of digital modulation keying techniques like QAM method is applied, then the odd indexed and even indexed subchannels is to be manipulated to get the full rapidity but it origins overlying of the nearby subchannels [8, 9]. Therefore, this disadvantage can be limited by a modulation scheme to handle the adjoining overlapped subchannels which is QAM.

2.4 PPN FBMC

In PPN FBMC/OQAM, the high data rate signal is converted into narrowband signals and are fed to a OQAM mapping followed by the IFFT synthesis filter bank is monitored by a polyphase network.

$$C_i(z) = \sum_{n=0}^{M-1} e^{j\frac{2\pi}{M}in} E_n(z^M) z^{-n}$$

The M order filter bank is derived using polyphase decomposition. In wireless systems that can be alienated into two parts, one part of a transmitter and the other on the receiver side, Nyquist standards play an important part. The squares of the frequency coefficients mollify the half-Nyquist filter's equilibrium condition [10]. For the spreading factors $S = 2, 3$, and 4, the half-Nyquist filter frequency coefficients are down.

2.5 PAPR

When the signal is conceded through a nonlinear HPA, the high PAPR implies in-band and out-of-band emission. The distortions and radiation can be reduced by reducing power efficiency so that the HPA can be operated in linear region only. Therefore, PAPR is the most significant constraint to be procured in both OFDM and FBMC. In literature so many PAPR reduction methods are proposed like distortion techniques, scrambling methods, coding and hybrid methods.

2.6 PAPR Reduction Method

In works [11, 12], various kinds of nonlinear companding processes such as A-law companding, μ-law companding, erf companding, and log companding are bestowed. Unlike functions, the high signal peaks are compressed in nonlinear companding and low peaks are consequently extended to decrease PAPR.

The cumulative complementary distribution function (CCDF) bounces the likelihood of the technique's PAPR that is greater than the verge value and CCDF tests PAPR efficiency. A-law companding: After flickering through PPN, the OQAM modified FBMC subcarriers are fed to the serial converter parallel. Using the A-law compander, this signal is compressed, which expands the minor values and cuts the out-sized signal values. For input x, A-law companding role $f(x)$ is postulated by [13].

$$f(x) = \begin{cases} y_{\max} \frac{A \frac{|x|}{x_{\max}}}{(1+A)} \mathrm{sgn}(x) & 0 < \frac{|x|}{x_{\max}} \leq \frac{1}{A} \\ y_{\max} \frac{A \frac{|x|}{x_{\max}}}{(1+A)} \mathrm{sgn}(x) & 0 < \frac{|x|}{x_{\max}} \leq \frac{1}{A} \end{cases} \tag{3}$$

μ-law companding: The OQAM/FBMC signal is compressed using a μ-law compander, stretching small values and leaving out the signal scale values. For input x, the $f(x)$ companding μ-law function is allocated by [13]

$$f(x) = V \frac{\log\left(1 + \mu \frac{|x|}{V}\right)}{\log(1 + \mu)} \mathrm{sgn}(x) \tag{4}$$

where V is peak of signal, x is instant amplitude of input signal and as per CITT, the standard value used for PCM in telecommunication is $\mu = 255$.

In collected works [14–16], distinctive types of nonlinear companding methods with low complexity are smeared to FBMC/OQAM for reducing PAPR.

3 Simulation Results

In MATLAB, a PPN-FBMC pattern is developed with parameters specified in Table 1. In order to execute OQAM, interference is added for even subcarriers to the imaginary portion of the signal and vice versa. The PHYDYAS prototype filter is used with spreading factors of $S = 4$ for achieving synthesis and evaluating filter banks. For $N = 512$ symbols with 1024 subcarriers using M-QAM, the OFDM is designed.

Initially the PAPR is calculated for OFDM, and FBMC/OQAM along with the bit error rate (BER) for modulation types mentioned in Table 2 is analyzed. From Table 2, it can be observed that FBMC/OQAM PAPR is more than OFDM PAPR. Therefore, the HPA complexity increases with increase in PAPR in FBMC/OQAM systems than

Table 1 Simulation parameters

S. No.	Parameter	Value
1	Subcarriers (M) (OFDM)	1024
2	N (FBMC)	512
3	spreading factors (S)	4
4	Symbols	1000
5	Modulation	4QAM, 16QAM, 64QAM, 256QAM
6	SNR in dB	25
7	A	87.6
8	μ	255

Table 2 Comparison of PAPR for OFDM and OQAM/FBMC

Modulation	PAPR (dB)	
	OFDM	FBMC
4QAM	8.21359	9.2657
16QAM	8.3232	9.0749
64QAM	8.2119	8.8116
256QAM	8.1599	9.1116

in OFDM systems but the BER at SNR = 25 dB is less. The PAPR reduction distortion techniques, i.e., the nonlinear companding systems are applied to the FBMC/OQAM and OFDM systems and BER at SNR = 25 dB is calculated. By using companding techniques in FBMC/OQAM PAPR is reduced with less complexity as shown in Fig. 1 and power spectral density (PSD) of FBMC/OQAM using companding methods for $A = 87.6$, $\mu = 255$ is shown in Fig. 2.

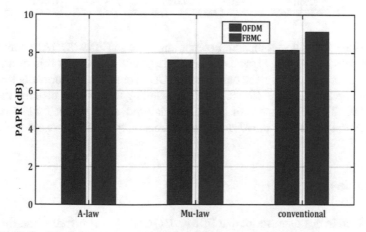

Fig. 1 PAPR (in dB) of OFDM and FBMC using A-law and μ-law companding

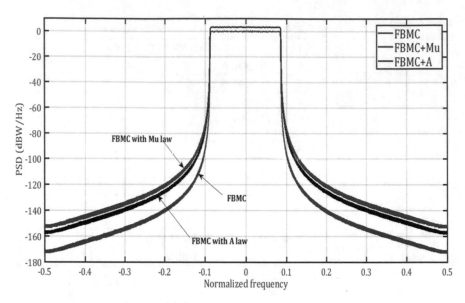

Fig. 2 PSD of FBMC using A-law and μ-law

Table 3 Comparison of PAPR for OFDM and FBMC/OQAM using PAPR reduction methods

Modulation	PAPR				BER
	μ-law		A-law		SNR = 25 dB
	OFDM	FBMC	OFDM	FBMC	
4QAM	7.6464	7.9308	7.6636	7.934	0.25008
16QAM	7.6688	7.8941	7.6861	7.8969	0.38639
64QAM	7.6456	7.8453	7.6631	7.8476	0.42664
256QAM	7.638	7.9005	7.6551	7.9034	0.44692

The A-law and μ-law companding technique used in FBMC/OQAM compresses the high peak signal and expands the low amplitude signals. In this process, the signal may get distorted and the out-of-band (OOB) emission is more than FBMC as shown in Fig. 2. Therefore, FBMC with companding has less out-of-band radiation than FBMC. The performance of FBMC is analyzed, as shown in Table 3 in terms of BER. From Table 3, as the M-ary value increases, the BER increases slightly.

4 Conclusion

It is resolved in this manuscript that FBMC/OQAM's PAPR is more than the PAPR method of OFDM. The PAPR reduction companding techniques are therefore used to condense PAPR. It is often found that with and without companding, OFDM PAPR is smaller than FBMC with and without companding techniques. Using A-law and μ-law companding, FBMC/OQAM's PAPR is reduced to ~1.33 dB at 4QAM with BER $= 0.25$ at SNR $= 25$ dB using A-law and μ-law companding. With a slight rise in BER, the PAPR is reduced as the M-ary mounts. Hence, the high PAPR is reduced by companding techniques without much degradation in BER efficiency. FBMC with companding systems can be a worthy option for 5 G communications with the benefit of improved side lobe suppression.

References

1. Ahmed N, Rahman H, Hussain MI (2016) A comparison of 802.11 ah and 802.15.4 for IoT. ICT Exp 2(3):100–102
2. Liu B, Yan Z, Chen CW (2017) Medium access control for wireless body area networks with QoS provisioning and energy efficient design. IEEE Trans Mob Comput 16(2):422–434
3. Gerzaguet R, Bartzoudis N, Baltar LG, Berg V, Doré JB, Kténas D, Font-Bach O, Mestre X, Payaró M, Färber M, Roth K (2017) The 5G candidate waveform race: a comparison of complexity and performance. EURASIP J Wirel Commun Netw 2017(1):13
4. IMT Vision (2015) Framework and overall objectives of the future development of IMT for 2020 and beyond. Document ITU-R Rec M.2083-0
5. Farhang-Boroujeny B (2014) Filter bank multicarrier modulation: a waveform candidate for 5g and beyond. Adv Electr Eng 2014(482805):25
6. Vihriala J, Ermolova N, Lahetkangas E, Tirkkonen O, Pajukoski K (2015) On the waveforms for 5G mobile broadband communications. In: 2015 IEEE 81st vehicular technology conference. VTC Spring. IEEE, pp 1–5
7. Thota S, Kamatham Y, Paidimarry CS (2020) Analysis of hybrid PAPR reduction methods of OFDM signal for HPA models in wireless communications. IEEE Access 8:22780–22791
8. Thota S, Kamatham Y, Paidimarry CS (2018) Performance analysis of hybrid companding PAPR reduction method in OFDM systems for 5G communications. In: 2018 9th international conference on computing, communication and networking technologies (ICCCNT). IEEE, pp 1–5
9. Sravanti T, Kamatham Y, Paidimarry CS (2020) Reduced complexity hybrid PAPR reduction schemes for future broadcasting systems. In: Advances in decision sciences, image processing, security and computer vision. Springer, Cham, pp 69–76
10. Schaich F, Wild T (2014) Waveform contenders for 5G—OFDM vs. FBMC vs. UFMC. In: 2014 6th international symposium on communications, control and signal processing (ISCCSP). IEEE, pp 457–460
11. Yun YH, Kim C, Kim K, Ho Z, Lee B, Seol JY (2015) A new waveform enabling enhanced QAM-FBMC systems. In: 2015 IEEE 16th international workshop on signal processing advances in wireless communications (SPAWC). IEEE, pp 116–120
12. Nam H, Choi M, Kim C, Hong D, Choi S (2014) A new filter-bank multicarrier system for QAM signal transmission and reception. In: 2014 IEEE international conference on communications (ICC). IEEE, pp 5227–5232
13. He Q, Schmeink A (2015) Comparison and evaluation between FBMC and OFDM systems. In: WSA 2015: 19th international ITG workshop on smart Antennas. VDE, pp 1–7
14. Shaheen IA, Zekry A, Newagy F, Ibrahim R (2019) Performance evaluation of PAPR reduction in FBMC system using nonlinear companding transform. ICT Exp 5(1):41–46
15. You Z, Lu IT, Yang R, Li J (2013) Flexible companding design for PAPR reduction in OFDM and FBMC systems. In: 2013 international conference on computing, networking and communications (ICNC). IEEE, pp 408–412

16. Raju N, Pillai SS (2015) Companding and pulse shaping technique for PAPR reduction in FBMC systems. In: 2015 international conference on control, instrumentation, communication and computational technologies (ICCICCT). IEEE, pp. 89–93

A Data Warehouse System for University Administration with UML Schema and Relational Decisive Approach

G. Sekhar Reddy and Ch. Suneetha

Abstract Data warehouse serves as a primary component in decision making systems by relying on multidimensional models. Multidimensional models provide a business-oriented view of data to decision makers for easy analysis and navigation. The main goal of this paper is to design a hybrid methodology depending on various multidimensional models. Intricate problems on classification require a machine learning technique called Decision tree to solve the issues in decision making. Standard algorithms for decision tree suffer from their inability in processing imprecise, meaning uncertain, incomplete and imperfect data. In this paper, a new approach for decision tree is developed using an integration of relational decisive rough set theory with improved ID3 algorithm. The proposed system illustrates how the methodology can be applied to the university environment. The performance of the system is evaluated and compared with conventional rough set methodologies. The designed model shows an improvement in the quality of conceptual schema and addresses the issues related to computational overhead.

Keywords UML schema · Relational decisive rough set · İmproved ID3

1 Introduction

Data Warehousing has emerged into a necessary concept for easy access of data that can be utilised for decision making [1]. Data warehouse (DW) opens a way for identifying different forms of data which is referred to as data mining, for the purpose of making decisions [2]. In recent decades, various methodologies related to the design

G. Sekhar Reddy (✉)
Research Scholar, Department of Computer Science and Engineering, Acharya Nagarjuna University, Guntur, India
e-mail: golamari.sekhar@gmail.com

Ch. Suneetha
Associate Professor, Department of Computer Applications, RVR and JC College of Engineering, Guntur, India
e-mail: suneethachittineni@gmail.com

of DW are proposed [6]. Since multi-dimensional modelling structures data as fact as well as dimension, it is an accepted statement that multi-dimensional modelling plays a major role in the development of DW [7]. Multi-dimensional modelling provides better understanding of data as they coincide with the intellectual of those who analyse such data [4]. These models ensure prediction related to the intention of users by supporting improvement in their performance because of their simplicity in construction [3]. Considering a technique to be applied on both conceptual as well as logical schema in which, the former can be expressed as Unified Modeling Language (UML) or Entity/Relationship (E/R) whereas the latter can either be one XML or relational schema [5]. Data warehouse design based on UML can be applied to physical, conceptual as well as logical designs. Concentrating on the conceptual phase, it leads to the UML model from the user requirement phase [8]. Rough set theory is an important method used for the selection of most appropriate attribute with higher tenacity to be used as a conditional attribute to enter into a decision by discarding every unnecessary as well as repeated attributes. For reduced time of processing as well as to attain betterment in generalization, issues regarding attribute reduction draws attention in the present decades.

The contribution of this paper is to present a hybrid methodology that holds the features of both data modelling as well as strong validation of user requirements. UML schemas achieved from $i*$ framework is utilised for the representation of such validation. Design process can be automated using UML multidimensional schemas since they possess high level standardization and proper representation. The paper leads to the introduction of a novel outset finding consistency of attribute. Consistency serves as a certification for choosing attributes to be partitioned. An improved ID3 algorithm is also proposed and a rough set-based decision tree is constructed through ID3 algorithm.

This paper is organised as follows. Section 2 provides a review on various methods related to multi-dimensional modelling. The background methodologies behind the work are explained in Sect. 3. Section 4 presents a hybrid methodology for multi-dimensional modelling and a rough set-based decision tree. The results obtained are presented and discussed under Sect. 5. Section 6 concludes the paper.

2   Literature Review

This section describes certain hybrid techniques available at present with their characteristics. The drawbacks behind the techniques are also analysed. Such techniques are broadly classified into the following groups as pure hybrid [9] and integration-derived approaches. The latter technique is again sub-classified as parallel [10] and sequential [11]. In pure hybrid technique, an initial assignment is defined with possible queries that the users propose to be executed for the extraction of data from DW. This initial assignment is used for deriving user requirements [12]. Based on this, an algorithm had been proposed for the automatic creation of graph that is intended to identify whether to consider a table as dimension or fact. Methods for automated validation of conceptual schema had been described in [13]. It establishes a hybrid technique sequentially in which data source is given priority over user requirements. Event entity which represents fact is found depending on the attributes that hold numerical values using iterative algorithm. An entity holding maximum count of numerical attributes is given the first position, the rest are considered in the following iterations and are arranged in descendent order. Another hybrid technique established sequentially is described in [14]. Initially, requirements are analysed based on the Tropos context [15] in which the region of interest can be modelled using two perceptions. One is the organizational model that visualises objectives of stakeholders and the other is the decisional model that visualises the objectives of decision makers. The next step is to verify the similarity between the derived schema and data source. The data driven techniques ensures consistency and traceability of data. However, they miss business requirements of user.

The overall analysis of literature shows that most data warehouse conceptual designs are either based on data driven or requirement driven methodologies. The data driven approaches fully guarantee the presence of data in analytical processing at the expense of missing business requirements. The requirement driven methods results are in user friendly conceptual schemas, at the expense of data availability. Hence a hybrid methodology is required to face such issues related to missing and decision making which combines all advantages of multidimensional models. This hybrid methodology is facilitated by combining the requirement driven and data driven approaches.

3   Background Methodology

This section briefly describes various basic concepts related to multi-dimensional modelling, rough sets as well as the introduction of neighbourhood rough set.

3.1   Multidimensional Framework

Dimensional framework is a data structural methodology that is being optimized for data warehouse tools. Being introduced by Ralph Kimball, the framework includes fact table as well as dimension table. The framework is actually introduced for reading, summarizing and analysing numerical data such as balance, weight, value etc. that are available in the data warehouse. Dimensional frameworks can be utilised only for systems related to data warehouse and not for relational systems. Following are the steps involved in creating a dimensional model.

Phase 1: Business process identification. This phase identifies the business process that a data warehouse concentrates on. Business process might be Administration, Finance, Management etc. based on the data requirements of the organisation. It is selected depending on the available data quality. Business process can be described using Unified Modelling Language (UML).

Phase 2: Grain identification. Grain corresponds to the amount of details available for getting a solution. If a table is said to have attendance for each day, then it is considered to be daily granularity. In spite, if a table holds attendance for every month, it is then called as monthly granularity.

Phase 3: Identification of dimensions. Dimensions are referred to as nouns such as store, date etc. Dimensions are locations where every data gets stored. For instance, dimension called date may include data such as week, month as well as year.

Phase 4: Fact identification. This phase associates with the users since it is the place where data from the data warehouse is accessed by the users.

Phase 5: Construction of Schema. Dimensional model is implemented in this phase. Schema is known as the structure of a database. Most popularly used schemas are of two types namely Star Schema and snowflake schema.

3.2 Rough Sets

Knowledge in rough set is referred to a group of facts that are articulated as attribute values describing an object. Data table is used for the representation of facts. Each entry in a row corresponds to an object. For this data table is mentioned at information system S. It is defined as $S = (U, Q, V, f)$ in which U corresponds to a finite, objects non-empty set is known as universe. The parameter Q corresponds to set of attribute, V is defined as $V = V_q$, $\cup_{q \in Q}$ is attribute domain q, and f refers to the function of information such that $f : U \times Q \rightarrow V$, by the information function, attributes value is assigned in each object in U i.e. $\forall_q \in Q$, $X \in U$, $f(x, a) \in V_q$ If Q is expressed as $Q = C \cup D$. Here, decision attribute and condition attributes is represented as D and C respectively, also the information system another name is decision system. Then the approximation space is (U, R) where R is the sum of the equivalent relations in U. The decision table is also an information system which is expressed as $A = (U, C \cup D)$, where $(C \cup D) = Q$. Here, set of decision attributes and categorical attributes are denoted as D and C respectively. The partial or full dependency are occurs in RS which is represented in the decision table.

Indiscernibility Relation
Assume an attribute subset $P \subseteq Q$ belonging to information system S in which an indiscernibility relation exists specified by INDs $(P) = \{ (x, y) \in U \times U : f(x, a) = f(y, a) \, \forall \, a \in P \}$. The objects ($x$ and y) are said to be indiscernible of the attribute subset P, if it satisfies the condition $(x, y) \in$ INDs(P). IND(P) is referred as an equivalent

relation since it is involved in partitioning U to different equivalent indiscernible classes. Such group of object partitioning is indicated as $U/\text{IND}(P)$.

Set Approximation

Consider X be a subset element that belongs to Universe U specified as $X \subseteq U$. X defines the membership status of all the elements available in U. When there exists a partial overlapping between $\text{IND}(P)$ and X, it is difficult to estimate the equivalent class elements. Under such cases the membership status of U cannot be determined. Thus, X is provided in two forms such as P-upper approximation and P-lower approximation as defined in (1) and (2) where $P \subseteq Q$.

$$\underline{P}(X) = \cup\{Y \in U/\text{IND}(P) : Y \subseteq X\} \tag{1}$$

$$\overline{P}(X) = \cup\{Y \in U/\text{IND}(P) : Y \cap X \neq \varphi\} \tag{2}$$

In set X, $\underline{P}(X) = \overline{P}(X)$ is named as rough set or exact set with regard to P.

4 Proposed Work

The proposed methodology depends on three multidimensional models: At first, the UML modelling is adopted to visualise requirement driven multidimensional schemas of data warehouse. Secondly, an extension of DFM (Dimensional Fact model) represents the data source as a tree structure. This view namely the attribute tree enables designers for easy manipulation of the data source structures.

Finally, the methodology proposes an improved ID3 algorithm and designs a decision tree based on Relational decisive Rough Sets in ID3 algorithm. An academic data warehouse of the University of Alden with respect to its conceptual design is proposed, especially admission criteria data mart designs in order to evaluate the mode of admission. The University is furnished with an internal system so as to evaluate its administration. Figure 1 illustrates the general framework of the proposed hybrid methodology. The role of decision makers is to collect and analyse data for evaluating various managerial events of the University.

4.1 Analysis of User Requirement

The analysis phase begins with the validation of user needs based on the i^* method.

The initial step is to identify possible actors. In the presented case, the Chairperson takes over the role of the first actor by deciding the academic activities. DW provides needed and meaningful information to decision makers by collecting it from several active sources. Identification of strategic goal is a major concern for decision maker.

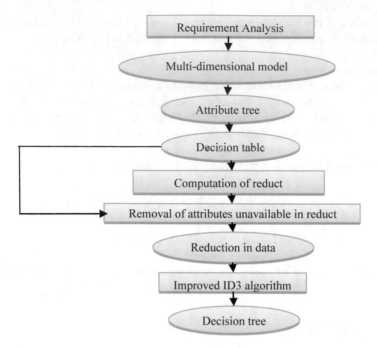

Fig. 1 Flow chart of proposed work

Here, the aim of the Chairperson is to improve the quality of academics by ensuring the right mode of admission for right candidates which is designed as goal dependence to DW from Chairperson. Resources are required for the accomplishment of goals. The Chairperson requires information regarding degrees, enrolments, examinations as well as finance.

Utilizing the resource and goal dependencies, the i^* strategic dependency model is designed. The strategic rationale model is then defined for each actor based on the i^* strategic model. Figure 2 depicts the strategic rationale model for the database of the University of Alden. DW is responsible for providing information to every resource dependency. For this purpose, DW must contain resources that are needed for decision makers. In the University database, the goal of the DW in reference with the information about admission is considered. The goal can be achieved by executing the task "provide data about admission" that requires details on the TOEFL score, UG transcript and CGP obtained by the candidate seeking for admission. This aids in identifying the standard of the institution with respect to quality of education.

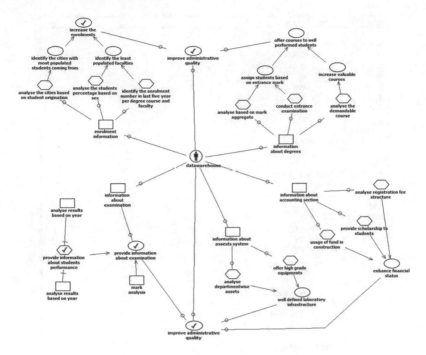

Fig. 2 Strategic rationale model

4.2 Attribute Tree Generation

Once the requirement analysis phase gets completed, attribute trees are created based on the UML schema. Generation of attribute tree follows the Algorithm 1 presented below. Irrespective of various facts available, only admission fact is considered here for illustrating tree generation from UML design.

Algorithm 1. Attribute tree	
c	class
a	attribute
d	dimension
s_a	set of attributes
s_d	set of dimensions
g	base class
a_b	set of attributes in b
x	descriptors
i, j	nodes

$i \leftarrow root(c)$
for every instance of a in f_a
     add (a, i)
end
for every instance of d in s_d
     g= base(d)
     traverse (g, i)
end
function traverse (g, i)
     x=descriptor (g)
     y=descriptor (i)
     z=cardinalityRolls-upTo (g, i)
     if(z==1)
          add (x, y)
     else
          place a dot at (x,y)
     end
     for every instance of j from a_b
          add (j,x)
     end
     for every instance of c
          traverse(c,g)
end

4.3 Relational Decisive Rough Set

The mode of admission of the University of Alden can be viewed as a decision tree using Relational decisive model. A Relational decisive model is defined as RDM $= (S, X, Y, E, f, \Delta, \delta)$ or simply RDM $= (S, X, Y, \delta)$, in which S corresponds to a set of samples $S = \{x_1, x_2, \ldots x_n\}$, X denotes the set of conditions that are to be checked out to make a set of decisions Y on admission, $X = \{$TOEFL score, Entrance score, UG transcript, CGP$\}$ and $Y = \{$Deny normal admission, admit conditionally, admit$\}$, $E = \cup_a \in \{X \cup Y\} E_a$, E_a is an attribute value set, $f: S \times \{X \bigcup Y\} \rightarrow E$ is a mapping function, $\Delta \rightarrow [0, \infty)$ is a distant function, and δ is a parameter for neighbourhood with $0 \leq \delta \leq 1$. Given a set $J, K \subseteq X$ for all $x, y \in S$, a degree of dependency of an attribute set J on K, denoted by $\gamma_R(A)$. It is defined as given in Eq. (3),

$$\gamma_K(J) = \frac{\text{card}(\text{Pos}_R(J))}{\text{card}(S)} \qquad (3)$$

Positive region $\text{Pos}_K(J)$ is the set of elements S which able to classify uniquely into partitions $S/\text{IND}(J)$ via K. In the universe, the fraction of the no. of objects is denoted as the coefficient $\gamma_K(J)$. It will fully depends on the parameter K and then $\gamma_K(J) = 1$, otherwise $\gamma_K(J) < 1$.

Relational Decisive Rough Set

1. Data set is fed as input.
2. If continuous or numeric attributes are available, discretize them and the modified data is labelled as M_1.
3. Get minimal decision relative reduct of M_2, say K.
4. Based on reduct K, reduce M_2 and label reduced data set as M_3.
5. Put ID3 algorithm proceeding M_3 to induce decision tree.

The data provided for training is a collection of information related to the University of Alden used for supervised learning for the development of classification. The second phase involves the discretization of continuous attributes. The next phase computes the reduct. Attributes that maintains indiscernibility relation is referred to as reduct. Consider an attribute set J with dependency $\gamma_J(Q)$ where $J \subset Q$, the relative reduct is defined as RED(J, Q) $\subseteq J$ in which

$$\gamma_{RED(J,Q)}(Q) = \gamma_J(Q)$$
For any attribute $a \in$ RED(J, Q), $\gamma_{RED(J,Q)-\{a\}}(Q) < \gamma_J(Q)$

Reduct differentiates attributes that belongs to various classes of decision. The amount of data utilised for training can be reduced by using reducts and hence more data is available for inducting decision tree. Ranking of reducts can be carried out by considering the cost required to obtain attribute value. Depending upon each attribute cardinality scoring of reduct can be performed. The least scored reduct is passed to further phases. Fourth phase reduces the dimensions by removal of attribute columns based on the inexistence of reducts. It is then applied into the decision tree algorithm for further processing. An improved ID3 algorithm is implemented for inducing decision tree.

Information gain is calculated by the kernel of ID3 algorithm for each conditional attribute X. The attribute achieving a maximum information gain is selected for splitting the set of data as node. The process is then iteratively repeated several times until either the partitioned attribute is similar or every conditional attribute are utilised to be partitioned as a node. The modulus of every class in set IND(X, Y) is calculated. The class with primary modules represented by primary-decision set while second most modules represented by secondary-decision set. Conditional attribute with similar primary as well as secondary decision set is found to be the worst. Consider, the primary decision set holds a modulus of $|D_p|$ and the secondary decision set has $|D_s|$. The consistency is computed as shown in (4),

$$C(J \rightarrow Y) = \frac{|D_p||D_s|}{|J|} \tag{4}$$

The phases involved in the algorithm are described as follows. The consistency is computed based on the above equation. The conditional attribute with highest consistency is chosen for splitting, and it is then marked as inactive. The process is continued until the leaf node is reached or no active nodes remain.

5 Results and Discussion

The University of Alden holds a database with information regarding the management, finance, scholar, mentor details and information regarding the admission procedure. The implementation is carried out in the platform of MATLAB tool. In the implementation, 833 instances are taken for analysis in the proposed approach.

For instance, when department is taken into consideration, it contains information regarding the name and id of the respective departments and much more. Each department includes several courses that can be shared among different departments. Each course has its own name, department as well as the coordinator and related subjects. Department always holds information regarding the contact details of both scholars and mentors. It is possible that several mentors can handle the same subject for different departments. The mode of admission of the scholar must be registered in the database. Figure 3 shows the UML diagram for that database. The performance of the scholar in each academic is also important to be registered. Figure 3 contains six entities. The presence of separate classes gains flexibility to the method and it enables to make changes easier whenever necessary.

The Star schema depicted in Fig. 4 is a relational database that holds both dimension and measure of a Data Mart. Fact table is used for the storage of measures whereas dimension tables are used for storing dimensions. Since each Data Mart contains a single measure which is surrounded by dimension tables, it is named as star schema. The central part of the schema is formed by the department fact. Department ID forms the primary key of the fact table, which is referenced by the other tables as foreign key. Dimensions are usually identified by a unique identifier, usually an integer.

An attribute tree is generated as demonstrated in algorithm 1. Figure 5 shows a decision tree that focuses on the admission criteria. The rough sets utilise TOEFL score, UG transcript, CGP etc. as the conditional attribute set to decide the mode of admission. This reduces the workload of the management enabling them to make a better selection. The decision flow moves along different directions in order to come up with different decisions. It is shown that the mode of admission relies on each and every conditional attribute. The admission is granted only when the entire set of conditional attributes tends to be positive. Any deviations in the possibilities of success of conditional attributes tend to difference in the decision.

Figure 6 depicts the comparison between the existing hybrid methodologies with the proposed in terms of time. The term mtry is defined as the number of variable is randomly collected to be sampled at each split time. If the mtry is 5, the proposed rough set is achieving less time than the existing rough set method. Similar, if the

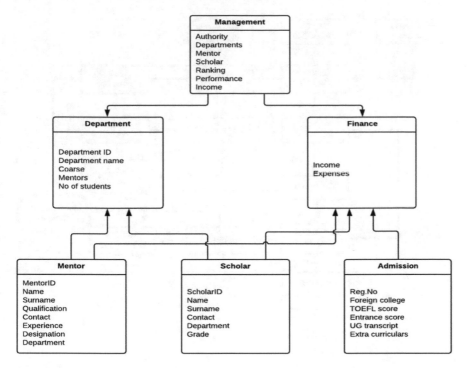

Fig.3 UML schema

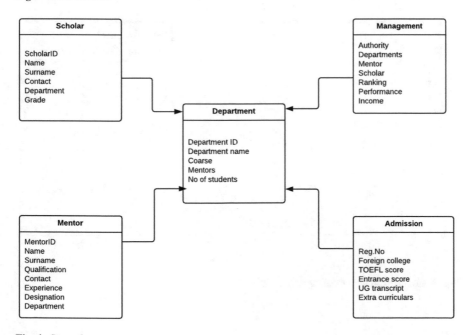

Fig. 4 Star schema

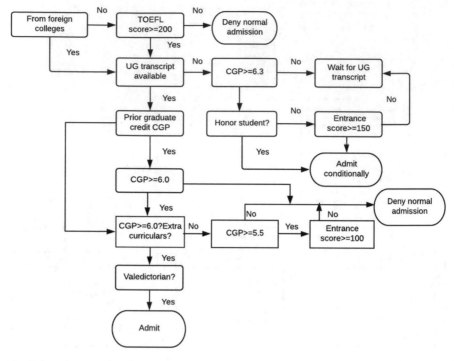

Fig. 5 Decision tree for admission

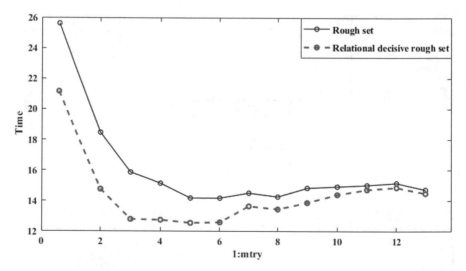

Fig. 6 Comparison with existing technique

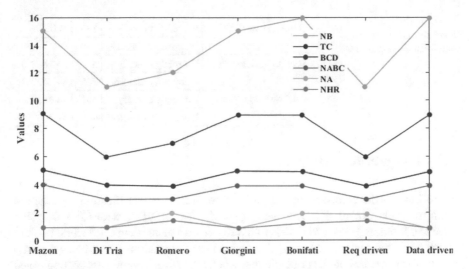

Fig.7 Comparison of metrics

number of mtry has increases the time will be slightly increases. It is found that the proposed relational decisive rough set technique consumes less time than that of the existing rough set theory. As a more important assessment, the understanding of the schema is also verified.

For this purpose, the following measurements are used: The metrics are those listed like number of base classes (NB), total number of classes (TC), number of base classes per dimension (BCD), number attributes of the base classes (NABC), total number of attributes of the schema (NA), and number of hierarchy relationships of the schema (NHR).Fig. 7 show the comparison.

The quality metrics are derived from the method of hybrid topology. The data driven has the NA attributes in higher level which express the best ratio of attributes of dimension to attributes of fact. In contrast, requirement analysis has many of attributes of dimension per attributes of fact. Also, the similar values are detected in Schneider's and Romero and Abelló's.

In BCD metrics, all the comparison methods give the better ratio of base classes for each dimension. Conversely Schneider and one of Romero & Abelló and requirement analysis demonstrate greater value for the reason that its schema non-symmetric shape. Actually, both are unbalanced in the dimension of teaching course. For the metrics of NHR, these methods are displays the higher number of hierarchy relationship. Certainly, the dimension teaching course is formed in the three dimension levels. Other methods have developed a clear well-balanced snowflake scheme and it's having two dimension levels. With regard to NABC measurements, proposed and requirement analysis are of the lowest value because they prosper in taking minimum descriptive and descriptors attributes for base classes. In conclusion, the suggested method attained the better results for all.

Table 1 Comparative analysis of different classification approaches

Random forest	Correctly classify	833	100%
	Incorrectly classify	0	0%
RDID3 decision tree	Correctly classify	548	65.7863%
	Incorrectly classify	285	34.2137%
C4.5 decision tree	Correctly classify	402	48.2593%
	Incorrectly classify	431	51.7407%

5.1 Comparative Analysis

Table 1 shows the comparative analysis of three different classification approaches. Comparing the Relational Decisive Approach ID3 (RDID3) with C4.5 Decision Tree and Random Forest, RDID3 is found to perform better than C4.5 while random forest gives the highest accuracy. This is because the random forest being a group mechanism operates on the principle that when many weak appraisers are combined it creates a strong estimator. ID3 algorithm is founded on the entropy as well as the information gain value. It selects the best attribute in terms of entropy and information gain for developing the decision tree. However from Table 1 it is evident that ID3 decision tree is more accurate than C4.5. It is also feasible for use in complex data sets for which the entropy values are reduced marginally.

For each iteration all classification types calculate the entropy or information gain value of that attribute. Based on the entropy value the attribute is selected, i.e. the attribute containing smallest entropy value set is split or partitioned in to subsets of data based on the attribute.

Table 2 shows the entropy value of three methods. The entropy value is high for C4.5 decision tree followed by RDID3 and random forest. As entropy is a measure of disorder, it is clear that the relational decisive ID3 approach performs better than C4.5 decision tree.

Kappa Statistic
Kappa statistic calculation is a method for calculating both multi-class and imbalance class problems. It can find the total accuracy and the random accuracy. Table 3 shows the Kappa static comparison of classifiers.

In RDID3 classifier the kappa statistic value is 0.5615 which shows a moderate agreement, but in decision tree classifier the value is 0.29 which denotes a slight agreement, as it faces problems in imbalanced classes. Other performance parameters include mean absolute error, relative absolute error, root mean squared error, and root

Table 2 Entropy comparison

Algorithm	Entropy
Random forest	0
RDID3 decision tree	0.925
C4.5 decision tree	0.998

Table 3 Kappa statistic comparison

Random forest	1
RDID3 decision tree	0.5615
C4.5 decision tree	0.296

Table 4 Error value comparison

Mean error	Random forest	RDID3 decision tree	C4.5 decision tree
Mean absolute error	0.0622	0.1558	0.2483
Root mean squared error	0.1092	0.3238	0.3524
Relative absolute error (%)	19.7871	49.5421	78.9463
Root relative squared error (%)	27.5246	81.6462	88.8563

relative squared error. The error values are given in Table 4. The table values show an increasing trend from random forest continuing through RDID3 and ending in C4.5 decision tree.

Detailed Accuracy by Class

The weighted average rate is compared with the TP-Rate, FP-Rate, precision, Recall, F-Measure, MCC and ROC (Receiver Operating Characteristics) area measurement. The weighted Average rate of classifier comparison is given in Table 5.

As is clear from the weighted average rate, RDID3 Decision Tree algorithm shows better values compared to C4.5. Figure (6) shows the comparison of two or more data, based on the number of class and attributes. If the students are not from foreign colleges it check the TOEFL score ≥ 200 if it true then it checks the UG transcript score and if all are yes the student is admitted. From the analysis it is clear that RDID3 performs better compared to all other algorithms in the studied dataset. The RDID3 implementation details are tabulated in Table 6 and the list of data sets applied is given in Table 7.

Table 5 Weighted average rate comparison

Weighted average rate	TP-rate	FP-rate	Precision	Recall	F-measure	MCC	ROC	PRC area
Random forest	1.000	1.000	1.000	1.000	1.000	1.000	1.000	1.000
RDID3 decision tree	0.658	0.096	0.659	0.658	0.654	0.566	0.851	0.615
C4.5 decision tree	0.483	0.189	?	0.483	?	?	0.755	0.410

Table 6 RDID3 implementation parameters

Parameters	Ranges	Parameters	Ranges
Criterion	Gain ratio	Minimal leaf size	2
Split minimal size	4	Minimal gain	0.1

Table 7 RDID3 decision tree performance changes with different dataset

Dataset	TP rate	Precision	Recall	F-score
github.com/ijk5554234/data mining	0.860	0.865	0.860	0.861
tic-tac-toe	0.788	0.786	0.788	0.787
bank_marketing_dataset	0.591	0.853	0.858	0.856

The suggested algorithm performance is validated in terms of weighted average rate parameters. From the parameters it is clear that the algorithms are almost consistent. Moreover the relational decisive ID3 algorithm shows better performance in all analysed data sets.

6 Conclusion

User requirements are completely captured using the requirement-driven approach. In order to achieve this objective, I^* methodology is proposed. A conceptual model that expresses multidimensional concepts on the basis of UML standard is obtained as result of i^* framework. Since unwanted attributes decays the quality of data warehouse, only needed attributes are taken into account discarding the unwanted ones. The proposed relational decisive rough set acts as a classification model since it discards attributes which are irrelevant at a stage-prior to tree induction. This enables low requirement of memory for consequent stages when running the model and sorting the data. In real data sets, results are exists and sometimes may be exist no reduct. This is provides the potential for the model of refinement. The availability of multiple reducts provides the opportunity to create a tree and attribute evaluation is hard to measure. Hence, this offers the options for using low-cost decision trees. The problem related to the lack of reduct for noisy fields or random data sets can be handled by calculating approximate reducts.

References

1. Abelló A, Samos J, Saltor F (2001) A framework for the classification and description of multidimensional data models. In: International conference on database and expert systems applications. Springer, Berlin, pp 668–677

2. Ballard C, Herreman D, Schau D, Bell R, Kim E, Valencic A (1998) Data modeling techniques for data warehousing. IBM Corporation International Technical Support Organization, San Jose, p 25
3. Berry MJ, Linoff GS (2004) Data mining techniques: for marketing, sales, and customer relationship management. Wiley, Indiana
4. Bonifati A, Cattaneo F, Ceri S, Fuggetta A, Paraboschi S (2001) Designing data marts for data warehouses. ACM Trans Softw Eng Methodol 10(4):452–483
5. Castro J, Kolp M, Mylopoulos J (2002) Towards requirements-driven information systems engineering: the tropos project. Inf Syst 27(6):365–389
6. Di Tria F, Lefons E, Tangorra F (2012) Hybrid methodology for data warehouse conceptual design by UML schemas. Inf Softw Technol 54(4):360–379
7. Giorgini P, Rizzi S, Garzetti M (2008) Grand: a goal-oriented approach to requirement analysis in data warehouses. Deccis Support Syst 45(1):4–21
8. Kimball R, Ross M (2011) The data warehouse toolkit: the complete guide to dimensional modelling, 2nd edn. Wiley, Canada
9. Luján-Mora S, Trujillo J, Song IY (2006) A UML profile for multidimensional modeling in data warehouses. Data Knowl Eng 59(3):725–769
10. Phipps C, Davis KC (2002) Automating data warehouse conceptual schema design and evaluation. In: Proceedings of 4th international workshop on design and management of data warehouses. Toronto, Canada, pp 23–32
11. Romero O, Abelló A (2006) Multidimensional design by examples. In: International conference on data warehousing and knowledge discovery. Springer, Berlin, pp 85–94
12. Romero O, Abelló A (2009) A survey of multidimensional modeling methodologies. Int J Data Warehouse 5(2):1–23
13. Romero O, Abelló A (2010) Automatic validation of requirements to support multidimensional design. Data Knowl Eng 69(9):917–942
14. Schneider M (2008) A general model for the design of data warehouses. Int J Prod Econ 112(1):309–325
15. Thangamani M, Ravindra Krishna Chandar V (2015) Adverse drug reactions using data mining technique. J Excell Comput Sci Eng 1(1):11–14

Latency Improvement by Using Fill VC Allocation for Network on Chip

Monika Katta and T. K. Ramesh

Abstract Network on Chip (NoC) is gaining popularity as an interconnect structure for complex multicore system on chip. The overall performance of NoC can be enhanced by improving latency, power consumption and throughput. The virtual channel (VC) allocation mechanism takes into consideration the buffer utilization which in turn improves the overall performance of NoC. In this paper, Fill VC allocation mechanism is proposed which reduces the number of pipeline stages. The proposed method is implemented in Booksim and the results show that the average packet latency is reduced by 58.57%, 16.66% and 36.66% for uniform, bitrev and tornado traffic, respectively, over conventional VC router for the mesh network. Also for flattened butterfly, the reduction in average packet latency is 56.66% and 62.5% as compared to conventional VC router for tornado and transpose traffic, respectively. Also, the real-time applications of NoC where massive data flow occurs are listed. VC allocation plays a critical role in such cases in deciding the overall performance of NoC.

Keywords Network on chip · Virtual channel · Switch allocation · Crossbar switch · Dynamic allocation

1 Introduction

The network interconnect consists of its routing, topology and flow control [1]. Routing techniques help packet in choosing its path. The topology depicts the number of nodes and channels with their arrangement. When a packet is traversing its path, flow control helps it by allocating buffer resources and channels. In this paper, we will be dealing with only flow control. A VC consists of a buffer that can hold one or more

M. Katta (✉) · T. K. Ramesh
Department of Electronics and Communication Engineering, Amrita School of Engineering,
Amrita Vishwa Vidyapeetham, Bengaluru, India
e-mail: er.monikakatta@gmail.com

T. K. Ramesh
e-mail: tk_ramesh@blr.amrita.edu

flits of a packet and associated state information [2]. A single physical channel can be divided into many VCs instead of maintaining a single First in First out (FIFO) queue. The VC flow control can effectively improve the performance of on chip interconnects. However, uniform allocation of VC may result in area and significant amount of power overhead [3]. In order to enhance the performance and optimize the area and power, various VC allocation schemes have been used appropriately. In this paper, various VC allocation methods are discussed and Fill VC allocation is proposed.

This section introduces VC; the next section discusses about motivation and background. In Sect. 3, Fill VC router architecture is proposed. In the next section, the proposed architecture is implemented and performance is evaluated and various real-time applications are listed. Finally, in Sect. 5, the paper is concluded.

2 Motivation and Background

The throughput of network can be optimized by carefully utilizing the buffer storage available at each network channel and dividing it into several VCs. Here, various VC management techniques are considered which helps in allocating buffer resources to packets. Figure 1 illustrates the conventional VC router architecture with various stages, i.e., routing computation, VC and switch allocation (SA). Some of the popular VC router architectures are discussed in this section.

Fig. 1 Conventional VC router architecture

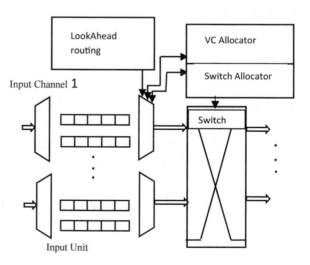

2.1 Non-uniform VC Allocation

The non-uniform VC allocation is an on demand allocation method proposed by Huang et al. [3] which works by allocating VC based on the traffic requirements for a particular application. In this paper, authors have developed an algorithm to estimate contention rate at port and bandwidth of router and then allocating VCs to only those channels which shows high bandwidth usage [3]. This, in turn, reduces the buffer usage as compared to traditional uniform VC allocation and gives similar performance.

2.2 ViChaR: A Dynamic VC Regulator

Nicopoulos et al. [4] proposed dynamic allocation scheme for VCs wherein the VCs and buffers are dynamically allocated to the links based on the traffic requirements, thereby enhancing the throughput by limiting the VC based on the demand. Shim et al. [5] contradicted this by proposing static VC allocation scheme which can avoid deadlock and minimize the routing distance when two or more VCs are available. In contrast to dynamic allocation, where when the flow is higher than the available VC at the link, packets or flits keep on competing for VC. Xu et al. [6] proposed two VC assignment mechanisms with the idea of dynamically allocating VC to the output ports which in turn reduces head of line (HoL) blocking:

- Adjustable VC assignment with dynamic VC allocation [6]
- Fixed VC assignment with dynamic VC allocation [6].

The VC allocation scheme for both these mechanisms is in between dynamic and static and the performance observed is remarkable.

2.3 VCs and Multiple Physical Channels

The comparitive study of multiple physical channels and VCs was discussed by Yoon et al. [7] and an conclusion was drawn that each one of them has their positive features. Later, virtual input crossbar (VIX) was proposed [8], wherein crossbar connection was used to connect more than one VC at input port, allowing transmission of multiple flits in one cycle.

2.4   VCs and SA

The two prominent stages in NoC are VC and SA. The parallel execution of these two stages was proposed by Monemi et al. [9] by making them independ-ent of each other. This improves overall performance of NoC. Bhaskar et al. [10] illustrated the ways to determine the optimal number of VCs such that the throughput and the latency are improved before the network attains its saturation.

2.5   Dynamically Allocated Multiple Queues (DAMQs)

Zhang et al. [11] proposed a novel multi-VC dynamically shared buffer for NoCs, which includes a small buffer with prefetch capability for each VC. This buffer is provided to each output port for storing the data read from shared buffer. Also, a management scheme was shown to keep a check on shared buffers which should not be used by a single VC exclusively. Liu and Delgado-Frias [12] proposed a novel DAMQ buffer scheme which organizes buffers by reserving space for individual VCs. Liu and Delgado-Frias [13] proposed another scheme using self-compacting buffer to enhance the performance of NoC. Oveis-Gharan et al. [14] proposed efficient dynamic VC (EDVC) approach and also compared it with conventional dynamic VC approach (CDVC). The authors have shown that in EDVC approach, the buffers are utilized to their maximum capacity as the slots are kept for them in VC.

2.6   Adaptive Inter-Port Buffer Sharing

Langar et al. [15] addressed the issues of under utilization of buffers and proposed adaptive inter-port buffer sharing. Bao et al. [16] proposed a novel algorithm which customizes VCs depending on the target application. In [17], Son Truong Nguyen et al. proposed on the fly VC allocation which can be used for high performance applications. This paper focuses on shortening of pipe-line during packet transfer by allocating VC at crossbar switch. In [18], the authors have proposed a VC partitioning mechanism based on feedback so that network bandwidth can be shared effectively in heterogeneous architecture.

3   Proposed Fill VC Allocation Architecture

This section gives the details of proposed VC allocation mechanism. The one iteration iSLIP baseline allocator is used. Also look ahead routing is used to reduce one pipeline stage and thereby reducing the overall latency.

3.1 Fill VC Allocation

Figure 2 shows the block diagram of our proposed VC allocation architecture. Here, routing computation and VC allocation take place simultaneously. Packets are divided into flits and head flit contains information about the destined output port. Packets stored in VC when face HoL blocking are move to the buffer register and therefore packets stored in VC and buffer register can take part in SA simultaneously. After SA is successful, these packets are sent to output port over crossbar after making sure that the respective output port is available. When all the flits of the packets reach destination, the VC allocation matrix is updated with the information that the output port is available for other packets now. Our proposed architecture ensures reduced latency at each stage as compared to conventional baseline router.

Table 1 shows the comparison of our proposed VC allocation architecture and conventional VC allocation router pipeline stages. Our proposed architecture takes only two pipeline stages as in the first pipeline stage routing computation in advance and SA takes place simultaneously and in the second stage VC allocation and switch traversal takes place, in contrast to four pipeline stages taken by conventional VC allocation router.

Fig. 2 Proposed VC router architecture

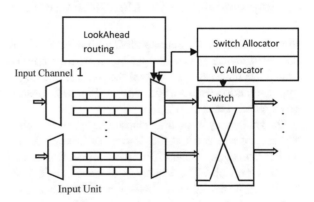

Table 1 Pipeline stages of conventional router and proposed router architecture

	Stage 1	Stage 2	Stage 3	Stage 4
Conventional router	Routing computation	VC allocation	Switch allocation (SA)	Switch traversal
Proposed router	Routing computation in advance and switch allocation	VC allocation and switch traversal		

3.2 Packet Transfer

When the head flit of an incoming packet arrives at the input channel, based on the VC identification, a VC is assignd to it. Routing computation in advance would have already done the routing computation in previous hop, based on which the output channel is easily determined. As soon as the head flit is stored in a VC, routing information is also updated. Flits belonging to the same packet move in a pipeline fashion and are stored in the same VC. Our design uses FIFO queues to manage VCs of input ports and VC flow control is used as a control logic to maintain buffers.When a flit, whose output information is already calculated, reaches the head of a VC, it becomes eligible for SA. Once the head flit wins the arbitration using crossbar, one of the free output VC is allocated to it. The remaining flits of the same packet use the same VC and this particular VC becomes unavailable for other other packets at this time. The flits after traversing the output channel move to the next hop based on the updated VC identification. Each output VC maintains an access vector using access matrix [17].

4 Implementation Results and Evaluation

We have used cycle accurate NoC simulator Booksim to implement and evaluate the performance of our design. We have also compared it with conventional VC router. We have assumed that all the packets are single flit packets. Table 2 lists the implementation details of our design. Figure 3 illustrates that our proposed method gives reduced latency (Fig. 4).

Packet latency is the time elapsed between the release of first flit from the source and receipt of last flit to the destination. Experiment results show that the average packet latency is reduced for Fill VC allocation in both the mesh and the flattened butterfly network. The latency is reduced by 58.57% over conventional router at the injection rate of 0.42 flits/cycle for uniform traffic. The reduction in latency is 16.66% over conventional router for bit rev traffic at the injection rate of 0.14 flits/cycle. At the injection rate of 0.28 flits/cycle, the reduction in latency is 36.66% over conventional router for tornado traffic. All the three results are observed for

Table 2 Implementation details of the network for mesh and flattened butterfly

Network	2D mesh	Flattened butterfly
Traffic	Uniform/synthetic	Uniform/synthetic
No. of VCs in each port	4	16
Flits in a VC	8	4
Routing	Deterministic dimension order routing	Universal globally adaptive load balancing

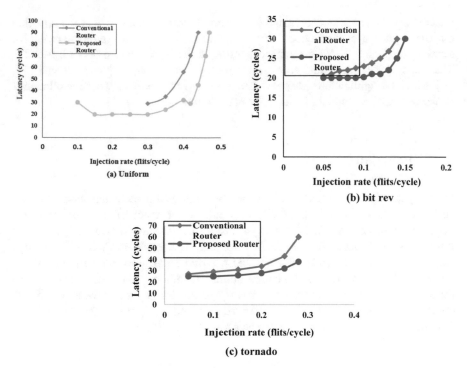

(a) Uniform

(b) bit rev

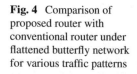

(c) tornado

Fig. 3 Comparison of proposed router with conventional router in mesh network under various traffic patterns

Fig. 4 Comparison of proposed router with conventional router under flattened butterfly network for various traffic patterns

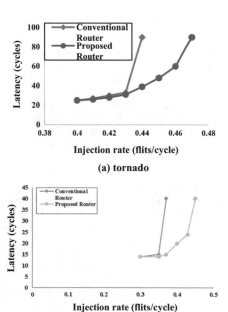

(a) tornado

(b) transpose

mesh network. The results obtained for flattened butterfly show that at the injection rate of 0.44 flits/cycle, the reduction in latency is 56.66% over conventional router for tornado traffic and 62.5% for transpose traffic at the injection rate of 0.3 flits/cycle. The most popular real-time applications include video object plane decoder [18, 19], MPEG 4 applications [20] and autonomous vehicles [21]. These applications consume massive amount of data flows in real-time execution [22].

5 Conclusion

In this paper, we have discussed various VC allocation mechanism for on chip networks which results in trade offs between performance of the NoC and area or power requirements. We have also proposed Fill VC allocation and implemented it using cycle accurate NoC simulator Booksim. The results show that the average packet latency is reduced by 58.57%, 16.66% and 36.66% for uniform, bitrev and tornado traffic, respectively, over conventional VC router for the mesh network. Also for flattened butterfly, the reduction in average packet latency is 56.66% and 62.5% as compared to conventional VC router for tornado and transpose traffic, respectively. The area and power consumption can be estimated in future work.

References

1. Dally WJ (1990) Virtual-channel flow control. ACM SIGARCH Comput Arch News 18(2SI):60–68
2. Dally WJ (1992) Virtual-channel flow control. IEEE Trans Parallel Distrib Syst 3(2):194–205
3. Huang TC, Ogras UY, Marculescu R (2007) Virtual channels planning for networks-on-chip. In: 8th international symposium on quality electronic design (ISQED'07). IEEE, pp 879–884
4. Nicopoulos CA, Park D, Kim J, Vijaykrishnan N, Yousif MS, Das CR (2006) ViChaR: a dynamic virtual channel regulator for network-on-chip routers. In: 2006 39th annual IEEE/ACM international symposium on microarchitecture (MICRO'06). IEEE, pp 333–346
5. Shim KS, Cho MH, Kinsy M, Wen T, Lis M, Suh GE, Devadas S (2009) Static virtual channel allocation in oblivious routing. In: 2009 3rd ACM/IEEE international symposium on networks-on-chip. IEEE, pp 38–43
6. Xu Y, Zhao B, Zhang Y, Yang J (2010) Simple virtual channel allocation for high throughput and high frequency on-chip routers. In: HPCA-16 2010 the sixteenth international symposium on high-performance computer architecture. IEEE, pp 1–11
7. Yoon YJ, Concer N, Petracca M, Carloni LP (2013) Virtual channels and multiple physical networks: two alternatives to improve NoC performance. IEEE Trans Comput Aided Des Integr Circuits Syst 32(12):1906–1919
8. Rao S, Jeloka S, Das R, Blaauw D, Dreslinski R, Mudge T (2014) Vix: virtual input crossbar for efficient switch allocation. In: 2014 51st ACM/EDAC/IEEE design automation conference (DAC). IEEE, pp 1–6
9. Monemi A, Ooi CY, Marsono MN (2015) Virtual channel and switch allocation for low latency network-on-chip routers. In: 2015 IEEE 23rd annual international symposium on field-programmable custom computing machines. IEEE, pp 234–234

10. Bhaskar AV, Venkatesh TG (2015) A study of the effect of virtual channels on the performance of network-on-chip. In: 2015 IEEE student conference on research and development (SCOReD). IEEE, pp 255–260
11. Zhang H, Wang K, Dai Y, Liu L (2012) A multi-VC dynamically shared buffer with prefetch for network on chip. In: 2012 IEEE seventh international conference on networking, architecture, and storage. IEEE, pp 320–327
12. Liu J, Delgado-Frias JG (2005) DAMQ self-compacting buffer schemes for systems with network-on-chip. In: CDES, pp 97–103
13. Liu J, Delgado-Frias JG (2006) A shared self-compacting buffer for network-on-chip systems. In: 2006 49th IEEE international midwest symposium on circuits and systems, vol 2. IEEE, pp 26–30
14. Oveis-Gharan M, Khan GN (2015) Efficient dynamic virtual channel organization and architecture for NoC systems. IEEE Trans Very Large Scale Integr (VLSI) Syst 24(2):465–478
15. Langar M, Bourguiba R, Mouine J (2016) Virtual channel router architecture for network on chip with adaptive inter-port buffers sharing. In: 2016 13th international multi-conference on systems, signals and devices (SSD). IEEE, pp 691–694
16. Bao D, Li X, Xin Y, Yang J, Ren X, Fu F, Liu C (2017) A virtual channel allocation algorithm for NoC. In: International conference on machine learning and intelligent communications. Springer, Cham, pp 333–342
17. Nguyen ST, Oyanagi S (2011) An improvement of router throughput for on-chip networks using on-the-fly virtual channel allocation. In: International conference on architecture of computing systems. Springer, Berlin, pp 219–230
18. Lee J, Li S, Kim H, Yalamanchili S (2013) Adaptive virtual channel partitioning for network-on-chip in heterogeneous architectures. ACM Trans Des Autom Electron Syst (TODAES) 18(4):1–28
19. Johann Filho S, Aguiar A, de Magalhaes FG, Longhi O, Hessel F (2012) Task model suitable for dynamic load balancing of real-time applications in NoC-based MPSoCs. In: 2012 IEEE 30th international conference on computer design (ICCD). IEEE, pp 49–54
20. Milojevic D, Montperrus L, Verkest D (2007) Power dissipation of the network-on-chip in a system-on-chip for MPEG-4 video encoding. In: 2007 IEEE Asian solid-state circuits conference. IEEE, pp 392–395
21. Maatta S, Indrusiak LS, Ost L, Moller L, Nurmi J, Glesner M, Moraes F (2008) Validation of executable application models mapped onto network-on-chip platforms. In: 2008 international symposium on industrial embedded systems. IEEE, pp 118–125
22. Meghabber MA, Loukil L, Olejnik R, Aroui A (2019) Virtual channel aware scheduling for real time data-flows on network on-chip. Scalable Comput Pract Exp 20:495–509

Review on Sensors for Emotion Recognition

Stobak Dutta, Anirban Mitra, Neelamadhab Padhy, and Gitosree Khan

Abstract Emotions, on a regular basis, play an important role in our lives. Not only in the case of human activity in our everyday lives, but also in the decision-making process, emotions play an significant part. Emotions affect our view of the natural universe as well. Such thoughts are often initially assumed as meaningless. A slight change in emotion, though, will bring a big change in behavior. Nowadays, emotion detection with the assistance of physiological signal is an area of research. This paper is based on a wide-ranging review of biological signal-based emotion recognition. Many methodologies are recommended in various papers to understand human emotional states in an artificial way. Physiological signals such as galvanic skin reaction (GSR), electrocardiogram (ECG), electroencephalogram (EEG), electromyogram (EMG), photoplethysmogram (PPG), respiration, and temperature of the skin are often used. In this article, on the basis of the physiological signal, the researchers will present a thorough analysis of emotion detection and suggest a workflow to classify multiple emotional analyses using various physiological signals to make the precision and output even better.

Keywords Human emotions · Emotion perception · EEG · ECG · PPG · EMG · GSR

S. Dutta (✉) · N. Padhy
GIET University, Gunupur, Odisha, India
e-mail: stobak.dutta@gmail.com; stobak.dutta@giet.edu

N. Padhy
e-mail: dr.neelamadhab@giet.edu

A. Mitra
Amity University, Kolkata, India
e-mail: mitra.anirban@gmail.com

G. Khan
BPPIMT, Kolkata, India
e-mail: khan.gitosree@gmail.com

1   Introduction

In numerous fields of study, artificial emotion detection using different physiological signals has become an important vertical. The nervous system is normally separated between the central nervous systems and peripheral nervous systems (CNS
and PNS) of a human being. The fusion of the autonomic nervous systems and
nervous systems (ANS and SNS) is usually the peripheral nervous systems. The ANS
consists of sensory and motor nerves, working within the central nervous systems and
numerous other internal organs. The physiological signals are produced by the central
nervous systems and autonomic nervous systems in which, according to the theory
of Connon, emotional transition takes place [1]. Changes in the signs of the EEG,
ECG, PPG, EMG, and GSR arise while there is some shift of feelings attributable
to such particular circumstances. A number of studies have been done in the vertical
of emotion recognition with the help of physiological signals. Many researchers
have tried to create a standard relationship between the physiological signals and the
emotion changes. But it was seen that it was nearly impossible to detect the emotion
changes with the help of a single physiological signal. So emotion recognition with
the help of multiple signals came into light in the research world. In this paper, the
researchers have tried to present a review of how emotion recognition can be done
with the help of different physiological signals. Most of the researches have referred
Russel's emotion model [2]. The model gives knowledge of how basic emotions
are distributed in respect of valence and arousal in two-dimensional space. Utilizing
the above-portrayed model, the grouping and assessment of emotions turn out to be
clear, yet at the same time, there are numerous issues identified with the appraisal
of emotions, particularly the determination of estimation and results in assessment
techniques, the choice of estimation equipment and programming. This paper is a
review work that mainly concentrates on the hardware as well as procedures required
for emotion recognition in an automated way.

2   Emotion Recognition

In many of the research work, automated emotion recognition is done by estimating
and analyzing the changes of different parameters of our body and the nerve impulses
that are mainly electrical signals. The most common techniques are EEG, ECG, PPG,
EMG, and GSR.

2.1   *Electroencephalography*

Electroencephalography (EEG) [3, 4] comprises in the chronicle of the brain activity,
which can be noted utilizing electrodes put on the scalp surface or in the forehead.

At the point when a neuron fires, voltage changes happen. Despite the fact that the EEG is clearly not an exact estimation, it, despite everything, gives a significant understanding into the electrical activity of the cortex. Frequency and amplitude are the qualities of the recorded EEG designs. The frequency run is typically from 1 to 80 Hz (partitioned in alpha, beta, theta, delta), with amplitudes of 10–100 μv [5].

Alpha (α) (8–16 Hz) is mostly identified with relaxation, creativity. The alpha waves generally exist during the condition of wakeful unwinding with eyes shut. It is the state when the brain and the mind are at resting state [6, 7]. Beta (β) (16–32 Hz) mostly identified with cautiousness, attention. These waves are delivered when the individual is in an alarm or on edge state. In this state, cerebrums can without much of a stretch perform: investigation, arrangements of the data, produce arrangements, and new thought [6, 7]. Theta (θ) (4–8 Hz), for the most part, identified with deep unwinding, meditation. Fundamentally grown-ups generate these waves when the individual is in slight rest or dreams. The wave frequency is essentially related to the release of pressure and memory recollection [6, 7]. Basically, delta (δ) (0.54 Hz) is identified with a deep sense of sympathy and instinct. These waves transmitted in the waking state show a potential for subconscious behavior to be accessed [6, 7]. In the research paper [8, 9], the identification of human emotions using a neural network is defined.

2.2 Electrocardiography

As we know, the heart is known to be one of the most basic organs in our bodies, and electrocardiography (ECG) is used as an excellent indicative medication system that is often used to assess the operating state of the heart. As a biochemical symbol, ECG is used as the usual technique for non-invasive continuous localization of the electrical activity of the heart [10]. Since heart movement is connected with the central human framework, ECG is valuable not just in dissecting the heart's action; it very well may be additionally utilized for emotion recognition [11]. P (This wave is an effect of strial contraction), PR (The PR interim estimated from the beginning of the P wave to the beginning of Q wave). There are simple parameters QRS. Complex (the QRS complex measured from the beginning of the Q wave to the end of the S wave), QT/QTc (measured from the beginning of the Q wave to the end of the T wave), which are often used to test ECG signals [12]. Generally, all parameters are investigated uniquely for clinical purposes, attempting to characterize irregular heart action and to get its deviation parameter. In the majority of tests, QRS complex is used for the identification of emotions, which characterizes the enactment of the heart relevant to the human emotional state and is an effective marker for perceiving key feelings. However, because of the fact that this pointer has a decrease of response to explicit feelings, there are additional difficulties of emotion detection. Study findings given by Cai et al. [13] suggest that pity can be interpreted more easily and unambiguously than the feeling of happiness. The classification and measurement of QRS amplitudes were compiled by several of the studies associated with ECG and also called the duration between such waves. There is also a set of studies that run in the QT/QTc dispersion vertical [14]. This provides proof that this interim is related to the degree of stress and can be used to interpret extraordinary frustration as a marker. The key drawback of the 12-lead ECG is that it creates immense data measurements, particularly when they are used for long hours [15]. When used for automated emotion detection, the ECG program involves the use of sophisticated signal handling techniques, allowing the position and abstraction of the appropriate verticals from the crude verticals signal. As there is a complexity of ECG signal examination in functional applications, frequently, ECG is utilized along with other sensors for emotion detection [3].

2.3 Galvanic Skin Response

The GSR works on the basis of skin conductance, and it is a calculation of some electrical parameters of the skin. These parameters of the skin are not in the control of humans consciously [16], as per the traditional theory, and it is dependent on sweat, which shows changes in the sympathetic nervous system (S.N.S.) [17]. If there is an emotional change, then there is some noticeable sweat on the palm, finger, and soles. Due to sweat, there is a variation of salt in the skin, and because of that, there is also a change of resistance of the skin [18]. In that case, the conductance increases. So, when there is such kind of ecological change that causes the adjustment in our mindset or emotions, then the perspiration gland more explicitly, the eccrine gland expands its action. Diverse sorts of feelings can cause a higher excitement and furthermore in raise in skin conduction. [19–21]. The sensor has two terminals that are put on the fingers, and it sends the information through the system to an organizer that advances it to a computer system [16].

3   Related Work

In research work [22], the authors provided some review work where they have reviewed different physiological signals for emotion recognition. In [23], the researchers have projected a model for multisubject emotion classification. The novelty of this research work is to take out the high-level features with the help of a deep learning model. Convolutional neural network (CNN) has been used by the authors for the abstraction of feature for the automatic abstraction of the correlation information that is present between multichannels for constructing more abstract features which are discriminatory, namely high-level features. The accuracy of the average result of 32 subjects was 87.27%. And [24] used the DWT features where the window width was varying (1–60 s), and the calculation of entropy was done of the detail coefficients related to the alpha, beta, and gamma bands. With the help of the SVM classification in case of arousal, the classification accuracy is up to 65.33%

where the window length will be of 3–10 s, while in case of valance classification accuracy as per the researcher can be 65.13% where the window length will be 3–12 s. There are different experiments and analyses related to human emotions on the basis of the biosignal. Most of the methods are done by collecting data through multiple physiological signals. The study of the research work [25] classifiers evoked emotions on the basics of two types of physiological signals of short-term ECG and GSR and have also recorded and analyzed estimated recognition time. Firstly, they have performed the experiment with the help of experts to extract target emotions that included anger, fear, happiness, grief, and calmness. This experiment was done with the help of ECG and GSR signals. For the processing of the truncated ECG data, they have applied wavelet transform, and for the processing of the truncated GSR signal, they have applied a Butterworth filter. With the help of an artificial neural network (ANN) Fig. 1 finally, they have classified five different emotions types. The average classification accuracy rates that were achieved in the experiment were 89.1 4% and 82.2 9% for ECG and GSR data, respectively, and for emotion classification and feature extraction and the total time required did not exceed 0.15 s for either GSR or ECG signal.

It has been observed from the related work section that for the evaluation of different emotional states and behavioral prediction, emotion recognition has become a useful technique nowadays. In the case of the development process of different types of a human machine interaction system, emotion recognition and evaluation of sentiments have played a pivotal role in recent days. Though the relationship exists between the human body reaction due to particular emotion is a known fact, still there are some ambiguity exists in the method of analysis of emotions. These problems of uncertainty are addressed by several outcomes of the research work, as discussed in this related work section. Most of the systems classify specific emotion state as the studies are limited. Different classification accuracy is calculated and is compared between different biosystem-based emotion recognition which may not

Fig. 1 Topology of the ANN model for ECG emotional recognition [25]

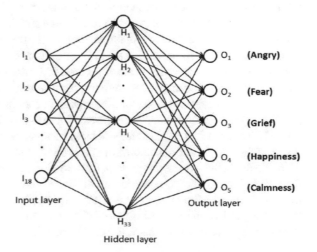

be accurate, as the comparison has been made between the different methodologies where different type of datasets involving a variety of biosignal has been utilized. It is observed that combining the maximum number of physiological signals like ECG, GSR, PPG, EEG, etc., will give some significant results in the arena of emotion recognition.

4 Proposed Work

In this work, we reviewed several papers based on analysis and prediction techniques of identifying various emotions of a user through physiological signals collected from various non-invasive sensors. There are several limitations we come across in several research work as discussed in this paper. To address those limitations, we have proposed a workflow model in this paper for identifying and analyzing several different emotions using various physiological signals and derived a best-fit algorithm for identifying automatic emotional state detection system. In order to analyze his/her emotion firstly, we need to implement emotion analysis based on several physiological signals.

The following are the major steps in identifying emotions through physiological signals: (1) preprocessing, (2) feature extraction, and (3) regression. Figure 2 represents the flow of events of the workflow model. Here, the inputs are the physiological signals gathered from various sensors such as EEG, ECG, PPG, EMG, and GSR in an automated way based on several human emotional recognition techniques. In the proposed method, on the training dataset we will first do data preprocessing to avoid any missing values or any other errors in dataset. Therefore, once the signals are extracted in the form of training datasets, it is being validated using data validation model as discussed in Fig. 2.

Since emotion classification through physiological signals is very vital, it can be useful to know a person's feeling for a condition. There are several supervised learning algorithms that analyze and classify data and they perform well when classifying human emotion. In this context, we will be proposing a classifier method in our work to identify the emotion of person in a situation. Therefore, the training datasets and test data simultaneously are being classified based on some machine learning classifier model that can generate better output in terms of performance, reliability, and accuracy. The feature extraction stage goes further in finding the more descriptive parts of an emotion detection. Once the test datasets are classified, it will be analyzed through proposed data regression model and further it will be being verified and validated using some proposed algorithm based on machine learning concept. There are many methods that are used to understand and identify the different types of emotion that are being expressed, however the output at the end depends on how accurate the algorithm is. Moreover, there are other issues that are needed to be considered that if the prediction of probabilities of different emotions is equal by the algorithm then it will be difficult to decide the emotion. We need to improve the accuracy in

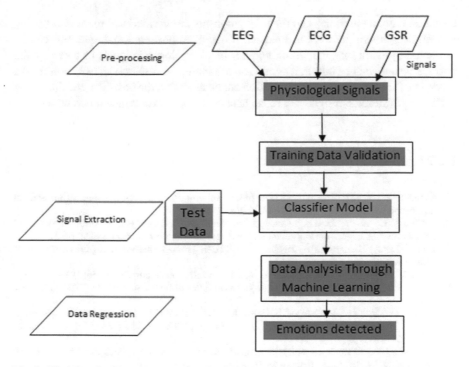

Fig. 2 Workflow for identifying emotional recognition through physiological signals

our algorithm in order to correctly classify the emotion. Thus, the workflow works in identifying emotions based on physiological signals as discussed in this paper.

This workflow model can be applied in various domains such as healthcare applications, fraud detection, and social media domain by finding out his emotion whether he is feeling nervous or not which expresses his/her fear by this. When classification is done on a smaller subset of highly distinguishable expressions, such as anger, happiness, and fear, then the accuracy is high. However, we will get lower accuracy when we will classify larger subsets or if the subsets are small with less distinct emotion, such as anger and disgust.

5 Conclusions

Detecting and recognizing human emotion is a big challenge in computer vision and artificial intelligence. Emotions are a big part of human communication. Most of the communication takes place through emotion. The main aim of this paper is to study various systems which can detect as well as recognize human emotion from a live feed. Some feelings are universal to all human beings like angry, sad, happy, surprise, fear, disgust, and neutral. The various systems used different algorithms based on human emotions, and then the emotions are verified and recognized using several deep neural network techniques. It has been observed that emotions can be predicted by taking the maximum number of physiological signals as taken by the various system using a different hybrid model. Here, the researchers have proposed a workflow for identifying several emotion analyses to find the best-fit algorithm using different physiological signals to make accuracy and performance much better.

References

1. Cannon WB (1927) The James-lange theory of emotions: a critical examination and an alternative theory. Am J Psychol 39(1/4):106–124
2. Parkinson B (2019) Words and concepts. In: Heart to heart: how your emotions affect other people. Studies in emotion and social interaction. University Press, Cambridge, pp 27–70
3. Dzedzickis A, Kaklauskas A, Bucinskas V (2020) Human emotion recognition: review of sensors and methods. Sensors 20(3):592
4. Bos DO (2006) EEG-based emotion recognition. Influen Vis Audit Stimul 56(3):1–17
5. Kandel ER, Schwartz JH, Jessell TM (2000) Principles of neural science. McGraw-Hill, New York
6. Ismail WW, Hanif M, Mohamed SB, Hamzah N, Rizman ZI (2016) Human emotion detection via brain waves study by using electroencephalogram (EEG). Int J Adv Sci Eng Inf Technol 6(6):1005–1011
7. Shakshi RJ (2016) Brain wave classification and feature extraction of EEG signal by using FFT on lab view. Int Res J Eng Technol 3:1208–1212
8. Sharma JJS, Dugar Y (2018) Detection and recognition of human emotion using neural network. Int J Appl Eng Res 13(8):6472–6477
9. Shahane MR, Sharma KR, Siddeeq MS (2019) Emotion recognition using feed forward neural network & Naïve Bayes. Int J Innov Technol Explor Eng (IJITEE) 9(2), 2487–2491
10. Zeng Z, Wang J (eds) (2010) Advances in neural network research and applications. Springer Science & Business Media
11. Goshvarpour A, Abbasi A (2017) An emotion recognition approach based on wavelet transform and second-order difference plot of ECG. J AI Data Min 5(2):211–221
12. Abdul Jamil MM, Soon CF, Achilleos A, Youseffi M, Javid F (2017) Electrocardiograph (ECG) circuit design and software-based processing using LabVIEW
13. Cai J, Liu G, Hao M (2009) The research on emotion recognition from ECG signal. In: 2009 international conference on information technology and computer science, vol 1. IEEE, pp 497–500
14. Uyarel H, Okmen E, Cobanoglu N, Karabulut A, Cam N (2006) Effects of anxiety on QT dispersion in healthy young men. Acta Cardiol 61(1):83–87
15. Al Khatib I, Bertozzi D, Poletti F, Benini L, Jantsch A, Bechara M, Khalifeh H, Hajjar M, Nabiev R, Jonsson S (2007) Hardware/software architecture for real-time ECG monitoring and analysis leveraging MPSoC technology. In: Transactions on high-performance embedded architectures and compilers I. Springer, Berlin, pp 239–258
16. Udovičić G, Đerek J, Russo M, Sikora M (2017) Wearable emotion recognition system based on GSR and PPG signals. In: Proceedings of the 2nd international workshop on multimedia for personal health and health care, pp 53–59
17. Wu G, Liu G, Hao M (2010) The analysis of emotion recognition from GSR based on PSO. In: 2010 international symposium on intelligence information processing and trusted computing. IEEE, pp 360–363
18. Ayata D, Yaslan Y, Kamaşak M (2017) Emotion recognition via galvanic skin response: comparison of machine learning algorithms and feature extraction methods. Istanbul Univ J Electr Electron Eng 17(1):3147–3156

19. https://www.seeedstudio.com/depot/Grove
20. Dutta S, Dash S, Mitra A (2020) A model of socially connected things for emotion detection. In: 2020 international conference on computer science, engineering and applications (ICCSEA). IEEE, pp 1–3
21. Dutta S, Dash S, Padhy N (2020) Analysis of human emotions based data using M.I.O.T. technique, (book chapter in) medical internet of things (M.I.O.T.): recent techniques, practices and applications. CRC, Taylor and Francis (accepted and under publication)
22. Shu L, Xie J, Yang M, Li Z, Li Z, Liao D, Xu X, Yang X (2018) A review of emotion recognition using physiological signals. Sensors 18(7):2074
23. Qiao R, Qing C, Zhang T, Xing X, Xu X (2017) A novel deep-learning based framework for multi-subject emotion recognition. In: 2017 4th international conference on information, cybernetics and computational social systems (ICCSS). IEEE, pp 181–185
24. Candra H, Yuwono M, Chai R, Handojoseno A, Elamvazuthi I, Nguyen HT, Su S (2015) Investigation of window size in classification of EEG-emotion signal with wavelet entropy and support vector machine. In: 2015 37th annual international conference of the IEEE engineering in medicine and biology society (EMBC). IEEE, pp 7250–7253
25. Zhang S, Liu G, Lai X (2015) Classification of evoked emotions using an artificial neural network based on single, short-term physiological signals. J Adv Comput Intell Intell Inf 19(1):118–126

Implementation of Rough Set for Decision Making

Surendra Nath Bhagat, Anirban Mitra, Neelamadhab Padhy, and Ruchismita Sahu

Abstract Rough Set theory is useful for those applications into which data are uncertain, inconsistent, or redundant, and information searches for clarity. To discover a hidden pattern, especially in inconsistent data, rough set theory proved to be an efficient tool. Further, rough set theory has invincible role in decision making and granular computing. This paper reviews basic concepts on rough set theory with appropriate examples and extends discussing the concept of rough graph. The discussion on concepts has been extended with the inclusion of the uses and applications for knowledge generation.

Keywords Rough set · Reduct · Core

1 Introduction

The rough set theory concept was first proposed by Zdzislaw Pawlak in 1982 [1–3]. In recent time, popularity and usage of rough set have grown exponentially, and it is used in all almost fields of knowledge management, decision making, and rule extraction. In today's world, the most important thing is information and information is acquired from data. Data can be obtained from real-life operation. Huge amount of acquired data must be organized and analyzed in efficient manner for decision

S. N. Bhagat (✉) · N. Padhy
GIET University, Gunupur, Odisha, India
e-mail: surendra.bhagat@giet.edu

N. Padhy
e-mail: dr.neelamadhab@giet.edu

A. Mitra
Amity University, Kolkata, India
e-mail: mitra.anirban@gmail.com

R. Sahu
TITE, Bhubaneswar, India
e-mail: ruchirubysahu809@gmail.com

making. Understanding the pattern and extracting relevant knowledge is challenging because of redundant and superflow data. Application of database tools in real-world data is time and resource consuming, and complexities increase when the acquired data consists of inconsistent or missing attributes. Rough set theory seems to be an efficient tool to overcome the above-mentioned challenges. One of the primary reasons for decrease in efficiency in terms of computing complexities is during verification of equivalence relationship among the data [4–6]. Even with such constrains, rough set is to be a very helpful tool to extract knowledge for ambiguous data by using technique for removing redundant data.

This article presents issues associated with redundant and superflow data and representing the extracted facts in information system. Further, this article focuses on some fundamental properties of rough set theory related to course of action of decision making and removal or elimination of redundant data for consistency.

2   Features Rough Set

Rough set comprises a sound reason for the procedure of discovery of knowledge in database. It gives scientific devices to find concealed examples in information particularly when the information is conflicting in nature. It likewise can be utilized for collection of characteristic, mining of characteristic, data deduction, assessment for rule generation, extraction of pattern, and so on.

The fundamental of rough sets includes set approximation, information system, dependency of attributes, rough membership, indiscernibility, redacts and core. To elaborate discussion on concept of rough set, authors have assumed an example demonstrating the information system consisting of non-empty limited set of items (known as universe) and non-empty limited set of attributes. Though the format of example is standard one and is available in many literature of rough set, the data (as shown in Table 1) is locally acquired [7]. This table contains the information on age, lower extremity motor score (LEMS) and a decision attribute 'walk'.

Explaining the data from Table 1, in rough set, the information system is a pair of (U, A), where U is non-empty finite set of objects (cases G1, G2, G3, G4…). A

Table 1 Information system (or information table)

Object	Age	LEMS	Walk
G1	17–31	51	Yes
G2	17–31	1	No
G3	32–46	2–26	No
G4	32–46	2–26	Yes
G5	47–61	27–50	No
G6	17–31	27–50	Yes
G7	47–61	27–50	No

is non-empty finite set of attributes (Age, LEMS, Walk), and V is the set of attribute value.

Information function $\rho: U * A \rightarrow V$.

Rows of the information system are called example, object, or entities. It can be explained by considering an example case from the Table 1, $\rho(G1, LEMS) \rightarrow 51$.

In the above information system (Table 1), attribute {Age, LEMS} known as conditional attribute and attribute {walk} is known as decision attributes. The decision attribute depends on conditional attribute, and the conditional attribute is independent.

Indiscernibility: It is a connection between two articles (objects) or more where all the qualities are indistinguishable corresponding to a subset of thought about attributes.

Hence, it can be explained as,

IND (Age) = {{G1, G2, G6}, {G3, G4} {G5, G7}}
IND (LEMS) = {{G1}, {G2}, {G3, G4}, {G5, G6, G7}}
IND {(Age, LEMS)} = {{G1}, {G2}, {G3, G4} {G5, G7} {G6}}.

Indiscernibility relation: The indiscernibility relation IND(B) is defined as a binary relation on U defined for $x, y \in U$ as follows:

$$(x, y) \in \text{IND}(B) \quad \text{if and only if } \rho(x, a) = \rho(y, a) \quad \text{for all } a \in B$$

Set Approximation

Lower approximation of X is a group of object which can be classified with full confidence as members of set X using attribute set B, i.e., $\underline{B}X$.

$$\underline{B}X = \{x : [x]_B \subseteq X\}$$

Upper approximation of X is a group of object that probably can be classified as member of Set X, i.e., $\overline{B}X$ is defined as

$$\overline{B}X = \{x : [x]_B \cap X = \emptyset\}$$

The boundary area involves object that cannot be grouped with confidence to be neither inside X, nor outside X, utilizing property set B.

B boundary area of X, $BN_B X = \{\overline{B}X - \underline{B}X\}$.
B outside region of $X = U - \overline{B}X$.

Note A set with non-empty boundary area is rough set, and if the boundary area is empty then it is a crisp set.

Example 1 To understand the above concept think about the information system (Table 1). Considering the object whose decision attribute (walk) value is 'yes'.

$X = \{x: \text{walk } (x) = \text{yes}\}$ Attribute set B—{Age, Lems}.

Solution $X = \{G1, G4, G6\}$ (attribute walk having the value 'yes' for these three objects)

IND $(B) = \{G1\}, \{G2\}, \{G3, G4\}, \{G5, G7\}, \{G6\}$
$\underline{B}X = \{G1, G6\} \ X = \{G1, G6\}$

$$\overline{B}X = \{G1, \ G3, G4, G6\}$$

A boundary region $= \overline{B}X - \underline{B}X = \{G3, G4\}$
A1 outside region $= U - \overline{B}X = \{G2, G5, G7\}$
(U—universal set consists of the value of X, i.e., here G1, G2, G3, G4, G5, G6, G7).

Dispensable and Indispensable attributes. The attribute is redundant in a table, and if we remove that attribute, it will not affect our decision or classification which is made on the basis on this table. This type of attribute is called dispensable. If we are able to remove this attribute, then the size of the information system is likewise reduced, as an outcome the effectiveness of activity on that table is additionally improved.

Suppose that $S = (U, A)$ be an information system (table), and $B \subseteq A$ and Let $a \in B$.

The attribute 'a' is dispensable in B if

$$\text{INDs } (B) = \text{INDs } (B - \{a\})$$

Or else 'a' is indispensable in B.

Redacts and Core

Definition. It is the vital piece of information, which is sufficient to portray all fundamental idea occurring in the deliberate information, yet the core is the most significant part.

Redact. After removing of dispensable attribute from the information system, remaining table is called redact, i.e., consider 'B' is a subset of B and B' is redact of B if

$$\text{INDs}(B') = \text{INDs}(B) \text{ and } B' \text{ is independent}$$

Core (P). It is possible that more than one redact is present or available in an information system, then which one will be considered and which one not. Which attribute is the most important attribute out of more than one redact. For that core is used.

Suppose B is a subset of A. Core of B is the set of all indispensable attribute of B.

Table 2 Decision table

	X	Y	Z	W	D
1	A	P	A	3	1
2	A	P	S	1	1
3	P	P	A	1	1
4	A	R	A	2	2
5	S	N	S	2	2
6	S	N	S	2	1

$\text{Core}(B) = \cap \text{RED}(B)$ where $\text{RED}(B)$ is the set of all redact of B.

Issue in decision table. A number of attributes may be superflow (redundant), so the removal of those attribute cannot make any affect on the classification. Considering the example as shown in Table 2 represents an information system of arbitrary data.

In this information system (Table 2), considering $\{X, Y, Z, W\}$ acts as a conditional attributes and D acts as a decisional attribute. In the above table, the objects 5 and 6 have same value for $\{X, Y, Z, \text{and } W\}$ attributes but the decisional attribute is different. So this table is inconsistent, i.e., if we remove objects 5 and 6 then it will not make any affect on the decision table. The redundancy can be removed by using the concept of redact and core.

Example 2 Consider the given $R = \{C, D, E\}$ of three equivalent relation C, D, and E with the following equivalence classes.

$U/C = \{\{J1, J4, J5\}, \{J2, J8\}, \{J3\}, \{J6, J7\}\}$
$U/D = \{\{J1, J3, J5\}, \{J6\}, \{J2, J4, J7, J8\}\}$
$U/E = \{\{J1, J5\}, \{J6\}, \{J2, J7, J8\}, \{J3, J4\}\}$.

Determine redact and core of family R.

Solution The following Table 3 is constructed according to the relation discussed in Example 2.

The authors have constructed the above relation from the above given equivalence classes. The U/C is represent the indiscernibility of C because $\{J1, J4, J5\}$ have same value $\{J2, J8\}$ same value $\{J3\}$ a value $\{J6, J7\}$ the same value by using this in the 1st Column of C for $\{J1, J4, J5\}$ is 1, $\{J2, J8\}$ is 2, $\{J3\}$ is 3 and $\{J6, J7\}$ is 4. Same findings are observed for column D, i.e., $\{J1, J3, J5\}$ is 5, $\{J6\}$ is 6, $\{J2, J4, J7, J8\}$ is 7. The column E, $\{J1, J5\}$ is 8, $\{J6\}$ is 9, $\{J2, J7, J8\}$ is 10, $\{J3, J4\}$ is 11.

Now we find the indispensability of the complete relation or table we draw above.

$U/\text{IND}(R) = \{\{J1, J5\}, \{J2, J8\}, \{J3\}, \{J4\}, \{J6\}, \{J7\}\}$.

Now we find whether C, D, and E are dispensable or not. Let we check whether the attribute P is dispensable or not. By removing the attribute 'C' find IND (D, E).

Table 3 Relation information

	C	D	E
J1	1	5	8
J2	2	7	10
J3	3	5	11
J4	1	7	11
J5	1	5	8
J6	4	6	9
J7	4	7	10
J8	2	7	10

IND $(D, E) = \{\{J1, J5\}, \{J2, J7, J7\}\ \{J3\}, \{J4\}, \{J6\}\}$.
IND $(R - \{C\}) =$ IND $\{D, E\}$. But IND $(R - \{C\}) \neq U/$IND (R).

So the attribute C is indispensable (necessary required) in the relation R.

Attribute D is dispensable checked by removing the attribute 'D', find IND(C, E).

IND $(C, E) =$ IND $(R - \{D\})$
IND $(R - \{D\}) = \{\{J4, J3\}, \{J2, J8\}, \{J3\}, \{J4\}\ \{J6\}\ \{J7\}\}$
IND $(R - \{D\}) =$ IND (C, E).

So D is dispensable means attribute D is superflow or redundant in R.

Attribute E is dispensable checked by removing the attribute 'E', find IND (C, D).

IND $(C, D) =$ IND $(R - \{E\})$
IND $(R - \{E\}) = \{\{J1, J5\}, \{J2, J8\}, \{J3\}, \{J4\}\ \{J6\}, \{J7\}\}$
IND $(R - \{E\}) =$ IND (R).

So E is dispensable means attribute E is super flow or redundant in R.

The classification defined by set of three equivalent relation (C, D, E) is same as classification defined by (C, D) and (D, E).

The redact of $R = \{C, D, E\}$ can be calculated by evaluating the independence of relation (C, D) and (C, E). Here, C is an indispensible attribute that means it has to be included but attribute D and E are dispensable but we cannot remove both the attribute directly. So we have to combined either (C, D) or (C, E) that is we include one dispensable and one indispensable attribute. The condition of checking whether R is independent or not already defined. Now we check in relation (C, D) any indispensability present or not. This can be done by comparing IND(C, D) with IND(D)and IND(C, D) with IND(C), and we find that IND(C, D) $\neq$ IND(D) and IND(C, D) $\neq$ IND(C). The same comparing done IND(C, E) with IND(C) and IND(C, E) with IND(E), so there is no indispensability in relation (C, D) and (C, E). Thus, there are two redact $\{C, D\}$ and $\{C, E\}$.

After getting the redact now core of R can be calculated by taking the intersection of redact $\{C, D\}$ and $\{C, E\}$. So $\{C, D\} \cap \{C, E\} = \{C\}$ is the core of R. That means that C is completely necessary attribute for the relation R.

So core C is the important part and redact $\{C, D\}$ and $\{C, E\}$ is the essential part.

Decision Making

It is well documented in almost every literature on rough set about its efficiency in making a decision. Considering an example of student dropouts, every year a counted number of students drop their enrollment from the admission process in the universities or academic institutions. Such dropouts may occur due to the various reasons including home sickness, cutoff GPA, other academic grade, etc. Acquiring data on such these students during the process of admission is challenging, and the outcome has the tendency to change by the end of admission process. The institutions face challenges in acquiring exact information on number of dropout so as they may able to take necessary steps to plan for the academic year and minimize the dropouts in future.

Form the following information table (Table 4), the researchers have considered three factors, namely grade point average (GPA) are [O—outstanding (90–100), E—excellent (80–89), and A—average (70–79)]. Other factors include 'Home Sicknesses (HS)' and 'Interested in other branch (IOB)'.

Depending on the value of this attribute, decisional attribute 'DROP THE ENROLLMENT (DROP)' is taken, but this attribute is not independent enough to provide surety that student will drop their enrollment in future or will continue with their enrolled academic.

The values of the attribute in Table 4 are collected from the student survey.

In Table 4, authors have considered only eight students to elaborate the process of decision making through rough set. No one can clearly say about the student intends to drop the enrollment or is seeking for further counseling. This requirement demands to classify the cases in different approximation class or set.

In rough set, the information system (table) is a pair of (U, A), where U is non-empty finite set of objects (say, student no. 1, 2, 3, 4, 5, 6, 7, 8). A is non-empty finite

Table 4 Student information

Student	Conditional attribute			Decision attribute
	GPA	HS	IOB	DROP
1	E	Yes	No	Yes
2	O	Yes	Yes	Yes
3	E	No	No	No
4	E	Yes	Yes	Yes
5	E	Yes	Yes	No
6	A	Yes	No	No
7	A	No	Yes	No
8	E	Yes	Yes	Yes

set of attributes (GPA, home sickness, interested for other branch (IOB)) and V is the set of attribute value.

Information function ρ: $U * A \to V$, for example, in Table 4 $\rho(1, \text{GPA}) \to E$.

Rows of the information system are called example, object, or entities.

In the above information system (Table 4), the attributes (GPA, Home Sickness, Interested in other branch) are known as conditional attribute and the attribute (Drop the enrollment) is known as decision attributes.

Indiscernibility. It is a connection between two articles (objects) or more where all the qualities are indistinguishable corresponding to a subset of thought about attributes, the observations are,

IND (GPA) = {{1, 3, 4, 5, 8}, {2}, {6, 7}}
IND (HOME SICKNESS) = {{1, 2, 4, 5, 6, 7}, {3, 7}}
IND (IOB) = {{1, 3, 6}, {2, 4, 5, 7, 8}}
IND{(GPA, HOME SICKNESS, IOB)} = {{1}, {2}, {3}, {4, 5, 6}, {6}, {7}}.

Example Considering only object whose decision attribute (drop the enrollment) value is 'yes', the obtained cases can be considered as,

$X = \{x$: drop he enrollment$(x) = $ yes$\}$.

Attribute set B—{GPA, HOME SICKNESS, and IOB}.

Solution

$X = \{1, 2, 4, 8\}$ [because in the table (Table 4), the attribute 'drop the enrollment' having the value yes for four specific objects]
IND $(B) = \{\{1\}, \{2\}, \{3\}, \{4, 5, 6\}, \{6\}, \{7\}\}$
Lower approximation $\underline{B}X = \{1, 2\}$
Upper approximation $\overline{B}X = \{1, 2, 4, 5, 8\}$
A boundary region $= \overline{B}X - \underline{B}X = \{4, 5, 8\}$
A outside region $= U - \overline{B}X = \{3, 6, 7\}$
(U—universal set consist of the all objects, i.e., 1, 2, 3, 4, 5, 6, 7, 8).

From this example, one can conclude that, the student no. (1, 2) will definitely drop the enrollment because they belong to the lower approximation. The student no. (4, 5, 8) may or may not drop the enrollment and counseling session to these student may encourage the students to continue with their academic. The student no. (3, 6, 7) will not drop from the enrollment and will continue to pursue their academic.

3   Application and Operation

Though rewarding results have been achieved through research on rough sets still scopes are open to investigate further into this domain. Professor Pawlak, the founder and introducer of rough set theory, quoted that problems on rough logic and rough analysis need further research to enhance the capabilities of this tool in handling decision-making processes. There are some other problems that are expected to be addressed in coming days. As in real-life situation, sometimes, the data table remains incomplete for various reasons. It is difficult to interpret such table. As rough set gives productive techniques and devices to discover concealed patterns in data and handling missing data for a particular attribute, the author intends to bring in the features of rough graph for the same.

There could be soft study based on rough set models like dynamic rough set model multigranular rough set model and there can also be study of rough sets combined with cloud computing. Attribute reduction, intelligent algorithm, and rule acquisition are the main focus of the researchers in the area of roughest applications. NP-hard problem which may be an attribute reduction problem is a focus of research [8–11]. A lot of new innovative methods of data mining are based on rough set models and reduction theory and the researchers are interested to study the decision-making process of such method using rough graph concepts.

The following subsection of this paper elaborates some of the applications where rough set theory can be used for knowledge generation.

Spatial metrological pattern classification. The sun spot classification process can be eased by the hierarchical rough set-based learning method. Prediction of weather condition such as storm or other conditions can be done using rough set.

Wireless sensor network. Rough set and neural network may help in detection of flats in nodes in wireless sensor network. Rough set helps in clustering as energy can be conserved in clustering by identification of cluster head.

Medical. Abdominal pain of children has become a common problem nowadays in the medical field. Diagnosing the proper reason behind it has become a critical challenge for the physician. Rough set has become a fruit full concept for the doctors for diagnosing of the fact.

Data mining. Rough set theory has an immense commitment in the region of data mining. It helps in the different phase knowledge discovery like generate and decision rule, reduction of attribute and core selection of attribute and pattern reorganization.

4 Conclusion

The rough set theory is an exceptionally helpful instrument to dissect the data by removing redundant data which make our database in efficient. It is a mathematical method help in decision-making process where data are redundant and there is uncertainty. In this paper, we discussed the technique by which we remove the redundant attributes and find the most important attribute of a relation or information system. We have additionally talked about the application regions of the rough set in various fields. The filling of missing value of an attribute in the information system or table is by using the rough set theory concepts in the area of our future work.

References

1. Pawlak Z (1982) Rough sets. J Comput Inform Sci 11:341–345
2. Pawlak Z (1981) Classification of objects by means of attributes. In: Reports, Institute of Computer Science, Polish Academy of Sciences, vol 429. Warsaw, Polan
3. Pawlak Z (1981) Rough relations. In: Reports, Institute of Computer Science, Polish Academy of Sciences, vol 435. Warsaw, Poland
4. Fernandez-Baizán MC, Menasalvas Ruiz E, Peña Sánchez JM (2000) Integrating RDMS and data mining capabilities using rough sets. In: Proceedings of the 6th international conference on IPMU. Physica, Heidelberg, pp 371–384
5. Kumar A (1998) New techniques for data reduction in a database system for knowledge discovery applications. JIIS 10(1):31–48
6. Hu X, Lin TY, Han J (2003) A new rough set model based on database systems. In: Proceedings of the 9th international conference on RSFDGrC. LNCS, vol 2639, pp 114–121
7. Patil S, Vaswani G, Bhatia A (2014) Graph databases—an overview. Int J Comput Sci Inf Technol 5(1):657–660
8. Tripathy BK, Mitra A (2013) On approximate equivalences of multigranular rough sets and approximate reasoning. Int J Inf Technol Comput Sci (IJITCS) 10:103–113
9. Tripathy BK, Mitra A, Ojha J (2010) Rough equivalence and algebraic properties of rough sets. Int J Artifi Intell Soft Comput 3(4):271–289
10. Tripathy BK, Mitra A (2010) Some topological properties of rough sets and their application. Int J Granular Comput Rough Sets Intell Syst 1(4):355–369
11. Tripathy BK, Mitra A, Ojha J (2008) On rough equalities and rough equivalence of sets. In: International conference on rough sets and current trends in computing, LNAI, vol 5306. Springer, Berlin, pp 92–102

Concept of Color Utilization and Its Application in Knowledge Graph Visualization

Subhankar Guha, Neelamadhab Padhy, Anirban Mitra, and Sudipta Priyadarshinee

Abstract This paper represents issues associated with visualization of knowledge graph using colors. The work further provides a scope to understand the usage and impact of color. Color has a very important role in shaping the value of any product, user-interfaces, etc. Studies show the preferences of colors based on gender and also inform us on how one can use the combination of primary, secondary, cool, warm colors to make the interpretation interesting and attractive. Using more colors of different shades, patterns, which does not go well with the theme and other colors makes it more clumsy and disturbing. Using a color palette of not more than three to four colors makes the representation more clean and simple. To achieve the visualization of knowledge graph with minimum number of colors, concept of knowledge graph and chromatic number has been discussed briefly with the support of appropriate example.

Keywords Color · Warm and cool color · Complementary color · Analogous color · Knowledge graph and chromatic number

S. Guha (✉) · N. Padhy
GIET University, Gunupur, Odisha, India
e-mail: subhankar.guha78@gmail.com

N. Padhy
e-mail: dr.neelamadhab@gmail.com

A. Mitra
Amity University, Kolkata, India
e-mail: mitra.anirban@gmail.com

S. Priyadarshinee
Indic Institute of Design and Research, Khordha, Odisha, India
e-mail: sudiptapatel88@gmail.com

1   Introduction

Color is an abstract source of information used to guide behavior and please attraction. Color is parameter or visual effect that relates us in day-to-day activities and is ever present in every aspect of our lives. Cohn's [1] had conducted a pioneering research on color and its basic visualizing properties including presence of light and emotions. His research was extended up to study of different notions of color perception, level of preference, and esthetics in wide dimension. The objective of this research is to enhance the application and usage of color in deriving an attractive pattern for website design [2, 3].

In this work, the authors had focused on the importance of color as an element for any Web site or UI designing depending on areas of interest such as an UI giving information on medical ailments, telecommunication services, online shopping services, social networking, etc. Depending on representation based on color, attractiveness varies [4].

A major part of human–computer interaction (HCI) concerns on the "usability," which refers to bringing closeness between designers, site developers, and the users who will use the platform. In 1988, famous creative author and philosopher Donald Norman quoted in his work entitled on design of everyday things about the communication and understanding the relation between the development team and the users with respect to human computer interface that bridges the design of the technology, available knowledge, and level of skill. Being a complex research area that directly affects the satisfaction of user and usability, design must be user-friendly showing complete and consistent information in the product interface [5–7].

Further, this part of the research focuses on the understanding human attraction and satisfaction toward colors and preference of colors according to the context used in representing the knowledge graph [8, 9]. The extended objective of this work is to observe the satisfaction level of user on use of minimum number of color to visualize and represent the information on knowledge graph. The work starts with a scope to understand the usage and impact of color and to achieve the visualization of knowledge graph with minimum number of colors. Definition of knowledge graph and chromatic number has been included in the later section to relate the discussion with the objective.

2   Colors Effect on Humans

Survey-based study on sample candidates through crowdsourcing shows that colors affect human in countless ways [3, 10, 11]. The result of earlier research can be summarized as.

- Color influences every product in conscious state. Various studies have justified that about 60% of impression on product or service interface depends on the color.

Fig. 1 Color wheel [13]

- An attractive color combination to represent the Web site increases the acceptance of the Web site among the users. Technically speaking, poor combination of color may increase complexities during execution.
- In medical field also, one can see the use of light colors, such as white which maintains the freshness and keeps the surrounding clean.
- Dark colors are used to depict mysterious behavior of human beings. Considering the example, for Google search "hacker," the primary selection for the dress of the hacker is of BLACK color.
- Vibrant colors like yellow, orange, and blue lift up your mood and will find it very pleasing.

3 The Color Wheel

A wheel representing coordination of color harmonies that are used by interface designer to select color combinations for developing the interface especially for designing interfaces in Web sites. The concept of color wheel (Fig. 1) was referred from the white paper on color theory [7], which seems to have existence from history and is dated back to 1966 as was first proposed by Sir Isaac Newton [7, 12].

4 Visualizing a Graph

Color can, and should, be used to focus on the key parts of your visualization that the researchers wants to focus. By using color strategically, one can reduce the cognitive load required to understand the depicting effect of visualization. Kalyuga et al. [14] found that color-coding "ameliorated split-attention effects, resulting in lower perceived difficulty." Some other researchers working in same direction have reported reductions in cognitive load when experiment participants were provided color-coding. Further, it was also observed by the researchers that the overuse of colors can have the opposite effect. Even a proper combination of color could configure nine types of plots, with specific parameters for each of them.

Visualization of data is very helpful for data testing and can be related with fast or low energy consumption. Two or more colors together can have an effect on one another, when they create an illusion together. Group of colors can be termed as complimentary to each other even though they are placed in opposite of a color wheel. It creates a certain illusion which turns a monotonic visualization into an emotional image packed with data and displaying facts and information.

Color Intends to Drag Attention in Specific Facts

Setting of vertices and edges as info, one can figure a few statistical measurements or graph dependent on such information, yet it is not sufficient to get a thought of structure. A decent perception can plainly show if there are a few groups or scaffolds in a graph, or possibly it is a uniform cloud, or something different [15].

To Impress and Attract the Emotions

Data visualizations are used for presentation. Further, in this context, important facts can be displayed using proper color. A grouping issue uses color to plot by marks and displays the relationships.

To Extract Features for Visualization

Graph visualization concepts include dimension reduction features. A graph that represented as an adjacency matrix is data in high-dimensional space. Drawing two coordinates for every vertex are utilized as highlights. Closeness between vertices in this space implies likeness.

Limitation with Large Graph on Computation Complexities

The perception of an enormous diagram looks complex as there are such a large number of items in a single plot. Mostly, graph visualization algorithms have dreadful algorithmic complexity: quadratic or cubic reliance from the quantity of edges or vertices. Helen et al. [14], in survey of two-dimensional graph layout techniques for information visualization, have elaborated on graph visualization methods existence and their working principals including algorithms, their features, and complexity.

5 Minimizing the Number of Color Utilization in Knowledge Graph

The coloring of graph is a task of names, called colors, to the vertices of a graph with the end goal that no two neighboring vertices share a similar color. The chromatic number $\chi(G)\chi(G)\chi(G)$ of a graph GGG is the insignificant number of colors for which such a task is conceivable. Different sorts of colorings on graphs additionally exist, most outstandingly edge colorings that might be dependent upon different limitations.

In graph theory, a special case of graph labeling is graph coloring; traditionally, the assignment of labels is called "colors" to components of a graph subject to specific limitations. In its easiest structure, it is a method of coloring the vertices of a graph with the end goal that no two adjoining vertices are of a similar coloring; this is known as a vertex coloring. So also, an edge coloring doles out a coloring to each edge with the goal that no two contiguous edges are of a similar coloring, and a face coloring of a planar graph doles out a coloring to each face or area so no two faces that share a limit have a similar color. Vertex concealing is ordinarily used to introduce diagram shading issues, since other shading issues can be changed into a vertex shading event. In any case, non-vertex shading issues are consistently communicated and focused without any assurances. This is incompletely academic, and somewhat on the grounds that a few issues are best concentrated in their non-vertex structure, as on account of edge coloring [16].

The demonstration of using hues starts from shading the countries of a guide, where each face is really hued. This was summarized to shading the embodiments of a chart embedded in the plane. By planar duality, it became shading the vertices, and in this structure, it summarizes to all charts. In mathematical and PC depictions, it is typical to use the underlying hardly any certain or non-negative entire numbers as the "hues." At the point when everything is said in done, one can use any restricted set as the "shading set." The possibility of the shading issue depends upon the amount of hues anyway not on what they are.

Chart shading acknowledges various sensible applications similarly as theoretical challenges. Near the old style sorts of issues, different imperatives can similarly be resolved to the chart, or on the way a shading is delegated, or even on the shading itself.

6 Knowledge Graph: Basic Concept and Characteristic

A knowledge graph is a graph hypothetical information portrayal that (at its least complex) models substances and trait esteems as nodes, and connections and characteristics as marked, directed edges. Knowledge graphs have risen as a binding together innovation in a few territories of AI, including natural language processing and semantic web, and consequently, the extent of what establishes a KG that has kept on widening. A knowledge graph gains and coordinates data into an ontology and applies motivation to determine new knowledge [17].

A group of interrelated description of different entities, object of real world, situation or conceptual concept can be represented by knowledge graph where depictions have a proper structure that permits the two individuals and PCs to process them in a proficient and unambiguous way. Substance portrayals add to each other, shaping a system, where every element speaks to part of the depiction of the elements, identified with it.

6.1   Key Characteristics of Knowledge Graph

Characteristics of several data management paradigms can be combined in knowledge graphs, and it can be explained as a:

Database: fact that the information can be questioned by means of organized inquiries;

Graph: Fact that it tends to be examined as some other system information structure;

Knowledge base: on the grounds that the information in it bears formal semantics, which can be utilized to decipher the information and derive new realities.

At the point when formal semantics are utilized to communicate and decipher the information of an information chart, there are various portrayal and demonstrating instruments:

Classes: Most frequently, element depiction contains an order of the substance concerning a class progression. For example, when managing general news or business data, there could be classes Person, Organization, and Location. People and associations can have a typical super class agent. Area ordinarily has various sub-classes, for example Nation, Populated spot, City, and so forth. The idea of class is obtained by the item situated plan, where every substance ought to have a place with precisely one class.

Relationship types and categories: The connections between elements are typically labeled with types, which give data about the idea of the relationship, for example, companion, relative, contender, and so on. Connection types can likewise have formal definitions. Further, an element can be related with classifications, which portray some part of its semantics.

Ontologies: They fill in as a proper definition between the engineers of the information diagram and its clients. A client could be another individual or a product application that needs to utilize the information in a solid and exact manner. It guarantees a common comprehension of the information and its implications [18].

6.2   Conceptualizing the Knowledge Graph Construction

Knowledge graphs are the graphs of nodes and edges constructed from the inputs of knowledge bases. Knowledge base acquires information from the unstructured text available on Web sites and pages, data sets, audio, and video content. The fundamental construction of a knowledge graphs is roughly demonstrated in figure (Fig. 2) as was referred in graph coloring document [18].

This process identifies facts from the free text. Initially, methods like scraping are used to search for filtered useful information in the Internet by identifying the entities and the relationships that the entities are involved in from free text. Taking

Fig. 2 Step to generate knowledge graph

into consideration an example, in the first step, the data extracted from free text may resemble the form of the following statement.

Example Considering the statement "***Shree Jagannath Temple is located in Puri***".

By considering the second phase of the pipeline, the statements are generalized in the form of triples within knowledge bases.

Steps to Find Triple for the Knowledge Base
The statement is divided in the following form of a triple for the knowledge base.

Subject: *Shree Jagannath Temple*
Predicate: *is located*
Object: *Puri.*

And hence, the knowledge is obtained from subject and object. The following example demonstrate the steps to extract the knowledge from subjects and objects and visualization of knowledge and relationships using knowledge graph.

Steps to Convert Triples in Knowledge Bases into Knowledge Graph
A large network of interconnected entities is knowledge graph. Based on the triples from knowledge bases, the connections are created.
Considering the following example:

Friends (Subhankar, Ram)
Friends (Ram, Laxman)
LivesIn (Subhankar, Puri)
LivesIn (Laxman, Bhubaneswar)
LivesIn (Ram, Puri)
BornIn (Subhankar, Puri)
BornIn (Laxman, Puri).

A knowledge graph based on the above relationships is demonstrated in the following graph (Fig. 3).

Visualizing the knowledge graph (Fig. 3), with the help of color, a maximum of five colors are required as the number of vertices are five. However, the minimum number of colors or the chromatic number for this graph is three. The following graph represents the visualization of knowledge graph with minimum color (Fig. 4).

Fig. 3 Knowledge graph
representation

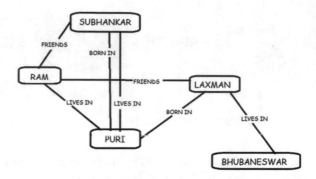

Fig. 4 Knowledge graph
representation using color

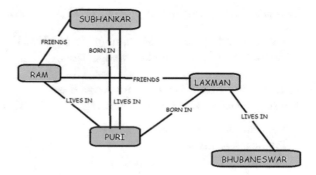

Knowledge graph provides scope to explore unknown relationships that were not explicitly retrieved from the knowledge bases. The dotted lines in the following graph (Fig. 5) represent a relationship that has been derived from knowledge graph (Fig. 3). The queries generate out of knowledge graph include:

1. Are **Subhankar** and **Laxman** friends?
2. What is **Ram's** birthplace?

Processing such queries usually leads to such relationships which are considered as missing links in the knowledge graph.

Fig. 5 Knowledge graph
representing derived
relationship

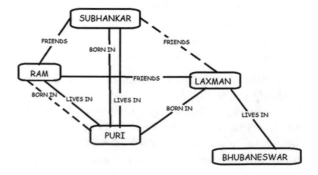

Fig. 6 Knowledge graph representing derived relationship using color

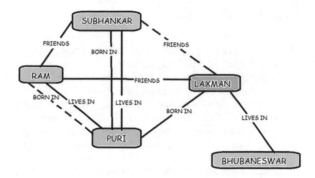

Visualizing the knowledge graph (Fig. 5), with the help of color, a maximum of five colors are required as the number of vertices are five. However, the minimum number of colors or the chromatic number for this graph is *four*. The following graph represents the visualization of knowledge graph with minimum color (Fig. 6).

Visualization of knowledge graph with minimum color representation (Figs. 4 and 6) provides scope to handle the challenges with graph having larger number of nodes where the nodes represent different level of relationships and importance. Further, knowledge graph visualization needs to handle issues like rendering performance, overview comprehensibility, and computational complexities. A colored knowledge graph provides better efficiency in terms of visualization and computing complexities, when minimum number of colors has been used to represents each nodes. Hence, it justifies the importance of chromatic number for coloring and visualizing the knowledge graph.

7   Conclusion

This is an initial work emphasizing on visualization of knowledge graph using colors. The work further provides a scope to understand the usage and impact of color. Color has a very important role in shaping the value of any product, user-interfaces, etc. Studies show the preferences of colors based on gender and also inform us on how one can use the combination of primary, secondary, cool, warm colors to make the interpretation interesting and attractive. Using more colors of different shades, patterns, which does not go well with the theme and other colors, makes it more clumsy and disturbing. Using a color palette of not more than three to four colors makes the representation more clean and simple. To achieve the visualization of knowledge graph with minimum number of colors, concept of knowledge graph and chromatic number has been discussed briefly with appropriate examples.

References

1. Cohn J (1894) Experimentelle untersuchungen iiber die gefiihlsbetonung der farben, helligkeiten und ihre combinationen
2. Mahnke F (1996) Color, environment and human response. Color theory in web UI design: a practical approach to the principles. Wiley
3. An introduction to color perception. https://imotions.com/blog/color-perception/
4. Kalyuga S, Chandler P, Sweller J (1999) Managing split-attention and redundancy in multimedia instruction. Appl Cogn Psychol Off J Soc Appl Res Mem Cogn 13(4):351–371
5. Nordeborn G (2013) The effect of color in website design: searching for medical information. Online thesis. Lund University Libraries. Spring
6. Color theory. https://www.interaction-design.org/literature/topics/color-theory
7. Saha T, Sarkar A, Mandal A, Padhi A, Mitra A (2019) A proposed model to study color preferences and psychology from online profiles for enhancing performance in advertisement and marketing. In: ICORDS 2019. IIM Vishakhapatnam, India
8. Gibson H, Faith J, Vickers P (2013) A survey of two-dimensional graph layout techniques for information visualisation. Inf Vis 12(3–4):324–357
9. Mondal P, Mitra A, Guha S, Paul S (2020) Color based trends prediction through social profile scraping. In: 2020 international conference on computer communication and informatics (ICCCI). IEEE, pp 1–6
10. The fundamentals of understanding color theory. https://99designs.com/blog/tips/the-7-step-guide-to-understanding-color-theory/
11. The best website color schemes-and how to choose your own. https://www.canva.com/learn/website-color-schemes/
12. Chijiiwa H (1994) Colour harmony: a guide to creative guide—creative color schemes published. Rockport Publishers Inc
13. Color theory: brief guide for designers. https://blog.tubikstudio.com/color-theory-brief-guide-for-designers/WhitePapersoncolortheoryforUserInterfacedesigning
14. Conceptualizing the knowledge graph construction pipeline. https://towardsdatascience.com/conceptualizing-the-Knowledge-graph-construction-pipeline-33edb25ab831
15. Iguana S (2019, Nov 16) Large graph visualization tools and approaches. Towards Data Science (Blog article). https://towardsdatascience.com/largegraph-visualization-tools-and-approaches-2b8758a1cd59
16. Graph coloring. https://en.wikipedia.org/wiki/Graph_coloring
17. Ontotext knowledge Hub fundamentals: what are ontologies. https://www.ontotext.com/knowledgehub/fundamentals/what-are-ontologies/
18. Large graph visualization tools and approaches. https://towardsdatascience.com/large-graph-visualization-tools-and-approaches-2b8758a1cd59

COVID-19 Isolation Monitoring System

K. Reddy Madhavi, Y. Vijaya Sambhavi, M. Sudhakara, and K. Srujan Raju

Abstract The current pandemic caused by the novel coronavirus, probably referred to as COVID-19, has posed a major threat worldwide and has already been declared as a global health emergency. As the WHO has claimed, close contact with an infected COVID-19 individual increases the chances of infection as the droplets produced by an infected person's coughing, sneezing, or talking stay in the air and by inhaling that air provides a path for the virus to reach our body, as it shows that COVID-19 is an airborne disease. In the absence of COVID-19 vaccines and drugs, the only way to treat COVID-19 infected patients if for them to be isolated from other people and to control their temperature and pulse rate and the consumption of drugs and food that enhances their immunity that could defend against the virus. The Internet of Things is a revolution that is fundamentally transforming our everyday lives and is promising to modernize healthcare by creating a more personalized, predictive, and collaborative model of treatment. To incorporate these two essential issues, this work provides an IOT ready system for living assistance that is capable of tracking the vital details of patients as well as providing mechanisms to send alert messages in emergencies. The flexible low-power, low-cost, and wireless features make this solution ideal for use anywhere and by anyone. The module assisted in real-time interventions and monitored the health care system for COVID-19 patients. Data collected from different sensors in real-time are stored on a central server, which connects patients to the doctor to the correct information at the time of an emergency.

K. Reddy Madhavi (✉)
Sree Vidyanikethan Engineering College, Tirupati, India
e-mail: kreddymadhavi@gmail.com

Y. Vijaya Sambhavi · M. Sudhakara
Annamacharya Institute of Technology and Sciences, Tirupati, India
e-mail: yvijayashambavi1@gmail.com

M. Sudhakara
e-mail: mallasudhakar.cse@gmail.com

K. Srujan Raju
CMR Technical Campus, Hyderabad, India
e-mail: ksrujanraju@gmail.com

Keywords Internet of Things · COVID-19 · Isolation monitoring system · Sensors

1   Introduction

The entire world is now witnessing the COVID-19 outbreak that was first recorded in China. COVID-19 is a severe acute respiratory coronavirus syndrome that has spread all the world and has become a major concern globally for all the health-care systems [1]. COVID-19 is an epidemic that has overburdened healthcare systems and related settings which have triggered resource shortages, both human and organizational-logistic, such as personal-protective equipment (PPE) and other manufactured devices such as face masks, face shields, respiratory machines, and other ventilator components that have become a crucial concern, causing countries to go partial or full lockdown [2, 3].

The ongoing pandemic is not a gateway. This pandemic has given us a new way of living, working, learning, and providing healthcare. Accepting and going forward with this modern way of living digital technologies is offering a great deal of human support. Digital health technologies such as the Internet of Things and artificial intelligence has created a better way of healthcare that allows isolation and no touch of emergency.

The world is experiencing an ongoing technological transition, evolving from isolated devices to pervasive Internet-enabled things capable of producing and sharing large quantities of valuable data. The Internet of Things is a well-defined network of interrelated computing techniques, digital, and mechanical devices capable of transmitting data through the network without human intervention at any point [4, 5]. The Internet of Things increases the knowledge and sensitivity of the world around us and combines the digital and physical universe. The Internet of Things played a significant role in this technological revolution. One of its key uses is in healthcare, such as remote surveillance, tracking and alerting, simultaneous reporting and monitoring, and many more.

The system proposed was designed taking into account the low-power and low-cost criteria, making the solution ideal for all to use at the comfort of their homes. Our solution provides a low-cost system that can be accessible to all offers all proper services that need to be taken care of during isolation, such as continuous temperature and pulse rate monitoring and not moving beyond the specific region. If any of the above guidelines are violated, it would send an alert message to their respective caretaker.

2   Related Work

Bayram et al. [6] study indicates that how much this COVID-19 is an unprecedented pandemic that posed a huge challenge to people's well-being and working lifestyles.

As a result, this pandemic has become a pillar of people's acceptance of a new digital environment.

Oyeniyi et al. [4] shown us how digital health technology, such as the Internet of Things, can aid us in this pandemic by enabling wireless device across wide areas, helping us reach out to patients and showing how to tackle this complex situation.

Islam et al. [3] has developed a health monitoring system to monitor the patient's heart rate, body temperature, and some measurements of hospital room conditions, such as room humidity, CO level, and CO_2. This device uses an ESP32, a heartbeat sensor, a body temperature sensor, a room-temperature sensor, a CO sensor, and a CO_2 sensor, a power supply, a web server, and a user interface to send the data. The disadvantages to this system are that the device is a bit bulky and a 5% error rate between the real data and the data being tracked.

Pavitra et al. [7] has developed an IoT based healthcare monitoring program that is useful for people who are paralyzed or using wheelchairs for movement, which helps to monitor their body temperature and pulse rate and also send warning messages in the event of an emergency. The system will also notify the caretaker of their basic needs, such as the need for the food and water, to go to the washroom through these gestures. The module is implemented using a variety of sensors, attaching them to the Node MCU board and running them Raspberry-pi. The drawback of this system is use of Raspberry-pi might make the module costly.

2.1 Existing System

Following 14 days of this treatment, the person has to go to the COVID-19 test again, and if the person has passed the −ve test, he will be allowed to go back home. Nevertheless, because this is a pandemic impacting thousands of people, healthcare facilities are running out of beds, PPE sets, monitoring and ventilation devices, face masks, and gloves. The disadvantages of the system:

1. Increasing the burden on health care systems, which results in a reduction in the facilities provided to patients.
2. A COVID-19 patient isolation monitoring system is needed, so that people who can isolate them in their homes can be constantly monitored and only in emergency cases visit the hospitals.

3 Proposed Method

Monitoring body temperature: Monitors the body temperature and sends the alert messages to the caretaker and sends alert messages to the caretaker in case of an increase and decrease in the temperature.

Monitoring heart rate: Monitor the heart rate and sends a message in case of bpm goes behind setting point.

Tracking patient: This device helps to track the patient if he/she goes behind the boundary.

Easily wearable: This device contains sensors, Wi-Fi module, RF with nanotechnology, and hence, it can be carried easily.

The system architecture shown in Fig. 1, as a whole consists of sensors, GPS module, Arduino.IDE, NODE MCU blynk. The software has cost-effectiveness, accuracy, and reliability. This can be implemented as a fully integrated and simpler system for the best outcomes for people diagnosed with COVID-19, which makes it easier for people to have quality healthcare services.

The major work of the device will be depending on the execution of the code and the generation of the output, i.e., meant with accuracy. There is various mode of sensors used in the device to know and detect the occurrence of the abnormalities that are going to appear in the body of the patient. The process of generating output completely depends upon the device as the patient state of temperature, pulse rate, and his/her location can be identified by the care-takers of that particular patient. The execution process of the device not only discovers the abnormalities but also informs the caretaker about the present condition or state of the patient health.

The system is implemented by a combination of hardware and software components. Images of the components and sensors used for the implementaion are shown in Fig. 2. The process of execution deals with connecting the LCD, Tx, and Rx pins of GPS module and GSM module to the pins of Node MCU Board. Then, connect the Trigger and Echo pins of ultrasonic sensor and accelerometer sensor to Node MCU board. Now, after connecting all the modules and sensors to the Node MCU board, connect the Node MCU board to the buzzer. The GSM module should be powered

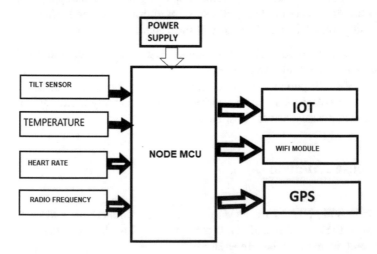

Fig. 1 Block diagram of COVID-19 patient monitoring

using a 10 V 2 A (10 V 2A) power adapter. The GPS module will work after its blue LED is turned on which means it has signals to gather location. GSM module is said to be working after its LED blinks for every 3 s. Later, Node MCU is connected to the laptop using a USB cable. On connecting all the pins of modules and sensors to the Node MCU board, then Arduino IDE is opened and the code is uploaded. The execution starts by displaying the system ready on the LCD display.

In this scenario, the user interfaces are a mobile phone message service, email, a registered user login page, and a new user login page. The registered person or family members or physician will be informed of the patient's health status via mail or phone messages. Web pages for logging in and registering in blynk are used as an interface between user and device.

Alert messages are sent to the caretaker when the pulse sensor detects bpm rising to 120 and falling to 60 or when the temperature sensor senses the patient's body temperature rises above 98.6 °F or else when radio frequency which is 30 m and patient crosses the RF, then the receiver is disconnected from the transmitter. The execution process of the device discovers not only the abnormalities but also informs the caretaker about the present condition or state of the patient health.

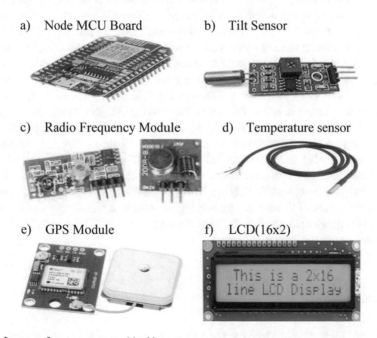

a) Node MCU Board   b) Tilt Sensor

c) Radio Frequency Module   d) Temperature sensor

e) GPS Module   f) LCD(16x2)

Fig. 2 Images of components used in this system

Fig. 3 Experimental setup

4   Results and Discussion

To implement proposed system, various connections that are made in experimental setup are shown in Fig. 3. We have used Arduino IDE 1.8.10 to upload the program to the hardware equipment. We have included the Tiny GPS++ library into that. So that the location can be easily parsed. The main result of this COVID-19 patient monitoring system is to send an alert as soon as the patient moves out of the specific region or the health data is irregular without any manual operation. Also, it should be able to calculate high heart rate, high temperature as shown in Fig. 4, and send alert to that particular caretaker as shown in Fig 5. It should be able to ignore alert if it is a minor change in health data. Results showing the sample location of patient connected to GPS is shown in Fig. 6. system must be able to work properly and produce the results as expected without any false cases and delays. Live monitoring of health data ia also maintained and shown in Fig. 7.

This monitoring system ensures that its performance is measured based on precision, time of response, and availability. This program allows the COVID-19 infected person to take care of themselves by staying at home and monitoring their isolation and go to the hospital only when there is an emergency. It will help to tackle the issue of overburdening healthcare systems.

5   Conclusion

Many countries had recognized the need for advancement in the health sector. Emergency health is one of the important areas that need to be more advanced. This research helps to immediately inform the caretaker as soon as the patient's irregular health data occurs for emergency care by tracking the patient's spot using a GPS module and sending alert using the Wi-Fi module. If implemented in real-time, it can greatly help in saving the lives of people. This system can be further improved by using machine learning and deep learning algorithms which can predict accidents

Fig. 4 Pulse and
temperature reading on LCD

Fig. 5 Alert messages to
mail with location
coordinates

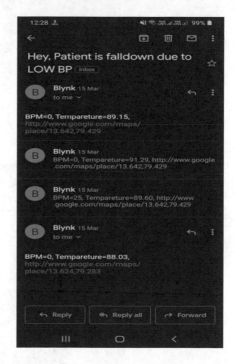

more accurately with the help of IoT data, so that alert messages containing images
of the incident can be sent to emergency responders. The fainted people detection
in real-time implementing in the house, so that when they cannot respond, then this
device can alert the neighbors about their situation, so that necessary precautions
can be taken. The cost of this device can be reduced by eliminating LED screens
or by replacing adapters by nano-lithium batteries which would also contribute to
reducing the size of the device. By exchanging or adding sensors like ECG sensor,
proximity sensor, vibration sensor, or other components like power accumulator,

608   K. Reddy Madhavi et al.

Fig. 6 Result showing
patients location

Fig. 7 Health data live

wireless webcam, GSM module which would contribute to modify the old device to
a new one. If the data and the readings are needed for future use, they can be stored
by including a module in Arduino.

References

1. Rahman MS, Peeri NC, Shrestha N, Zaki R, Haque U, Ab Hamid SH (2020) Defending against the novel coronavirus (COVID-19) outbreak: how can the Internet of Things (IoT) help to save the world? Health Policy Technol 9(2):136–138
2. Chamola V, Hassija V, Gupta V, Guizani M (2020) A comprehensive review of the COVID-19 pandemic and the role of IoT, drones, AI, blockchain, and 5G in managing its impact. IEEE Access 8:90225–90265
3. Islam MM, Rahaman A, Islam MR (2020) Development of smart healthcare monitoring system in IoT environment. SN Comput Sci 1(3):1–11
4. Oyeniyi J, Ogundoyin I, Oyeniran O, Omotosho L (2020) Application of Internet of Things (IoT) to enhance the fight against COVID-19 pandemic. Int J Multi Sci Adv Technol 1(3):38–42
5. Bragazzi NL (2020) Digital technologies-enabled smart manufacturing and industry 4.0 in the post-COVID-19 era: lessons learnt from a pandemic. Int J Environ Res Publ Health 17(3):4785
6. Bayram M, Springer S, Garvey CK, Özdemir V (2020) COVID-19 digital health innovation policy: a portal to alternative futures in the making. OMICS J Integr Biol 24(8):460–469
7. Pavitra B, Narendar Singh D, Sharma SK (2020) Smart patient assistance and health monitoring system using IOT. SSRN, 1–11

An Automated Approach for Detection of Intracranial Haemorrhage Using DenseNets

J. Avanija, Gurram Sunitha, K. Reddy Madhavi, and R. Hitesh Sai Vittal

Abstract Intracranial haemorrhage is a bleeding that occurs in brain which needs immediate medical attention and intensive medical care. The objective of this work is early detection of intracranial haemorrhage through automated model using DenseNets. DenseNets are used for processing MRI images and for detection of intracranial haemorrhage and its different variants. MRI scanned images samples are collected from a nearby neurology super speciality hospital. Segmentation of images is done through DenseNets which are also called deep connected convolution networks. Based on the image segments, the variant of intracranial haemorrhage is predicted. DenseNets layers are very narrow and as they add small set of feature maps and performs better when compared to the detection of the intracranial haemorrhage using convolution neural network (Juan et al in Proceedings of 4th congress on robotics and neuro science (2019), [1]). The accuracy of the proposed method is 91% achieved through the gradient from loss function which has access to each and every layer.

Keywords DenseNets · Intracranial haemorrhage · CT scan · Hematoma type

1 Introduction

Haemorrhage in the head (intracranial haemorrhage) is a relatively common condition that has many causes like trauma, stroke, aneurysm, vascular malformations, high blood pressure, illicit drugs and blood clotting disorders which are some of the causes for intracranial haemorrhage. The consequences neurologically are also varied extensively depending upon the size, type of haemorrhage and location ranging from

J. Avanija · G. Sunitha · K. Reddy Madhavi (✉) · R. Hitesh Sai Vittal
Department of CSE, Sree Vidyanikethan Engineering College, Tirupati, Andhra Pradesh, India
e-mail: kreddymadhavi@gmail.com

J. Avanija
e-mail: avanija03@gmail.com

headache to death. The role of the radiologist is to detect the haemorrhage, characterize the haemorrhage subtype, its size and to determine if the haemorrhage might be dangerous to critical areas of the brain where immediate surgery might be required. While all acute (i.e., new) haemorrhages appear dense (i.e., white) on computed tomography (CT), the primary imaging features that help radiologists determine the subtype of haemorrhage are the location, shape and proximity to other structures.

Intraparenchymal haemorrhage is blood that is located completely within the brain itself; intraventricular or subarachnoid haemorrhage is blood that has leaked into the spaces of the brain that normally contains cerebrospinal fluid (the ventricles or subarachnoid cisterns). Extra-axial haemorrhages are blood that collects in the tissue coverings that surround the brain (e.g., subdural or epidural subtypes). Figure 1 specifies the types of intracranial haemorrhage along with their imaging and mechanism. Patients may exhibit more than one type of cerebral haemorrhage, which may appear on the same image. While small haemorrhages are less morbid than large haemorrhages typically, even a small haemorrhage can lead to death because it is an indicator of another type of serious abnormality (e.g., cerebral aneurysm). CT scan image segmentation with examination level classification is comparable to radiology experts, in addition to robust localization of abnormalities in computerized tomography (CT) scan images, including some that are missed to be examined by radiologists, both of which are critically important elements for detection of intracranial haemorrhages which are considered in this paper and experimental results reveal

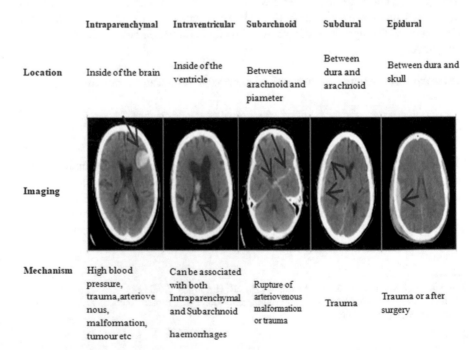

	Intraparenchymal	Intraventricular	Subarachnoid	Subdural	Epidural
Location	Inside of the brain	Inside of the ventricle	Between arachnoid and piameter	Between dura and arachnoid	Between dura and skull
Imaging					
Mechanism	High blood pressure, trauma, arteriovenous, malformation, tumour etc	Can be associated with both Intraparenchymal and Subarachnoid haemorrhages	Rupture of arteriovenous malformation or trauma	Trauma	Trauma or after surgery

Fig. 1 Different types of intracranial haemorrhage

that the proposed method is more accurate and faster when compared to using convolution neural network alone [2] since DenseNet requires only smaller set of feature maps.

2 Related Work

A new method for diagnosis of intracranial haemorrhage and subtypes using joint convolutional and recurrent neural network had been proposed by authors in [2]. 3D CT scan images of brain are taken as factors for subtype classification and prediction. Cross-validation testing is applied to algorithm by dividing 80% of data for training 10% of data for validation and 10% of data for testing. Accuracy is about 80% which is superior to human counterpart which has accuracy about 60% when both senior radiologists and junior radiologist trainees are taken in consideration for sample. Computer-aided diagnosis (CAD), name itself indicates that computer is aiding human beings to detect ailments. The paper [3] discusses about how computer-aided diagnosis is helping radiologists to detect intracranial haemorrhage and its subtypes. Deep convolutional neural networks (DCNN) are used for simultaneously learning and classifying 3D CT scan images into intracranial haemorrhages and its subtypes [4, 5].

The main aim of this paper [6] was to determine the natural causes of sudden death due to intracranial haemorrhage and to analyse the epidemiological aspects of death for the past ten years due to intracranial haemorrhage. 68 cases of intracranial haemorrhage are taken from department of pathology of University of Malaya to detect causes of intracranial haemorrhage. Of all the cases taken by authors are diverse, these cases has different gender, whether the patient is married or not, patient belongs to which race whether he/she is Chinese or Malaysian. 54 of the taken cases are of natural death caused by intracranial haemorrhage. Symptoms are identified prior to death of patient which are taken into consideration to classify intracranial haemorrhage type. In most of the cases taken, patients suffer from mixed symptoms of five types of intracranial haemorrhage. Identification of symptoms of intracranial haemorrhage and its subtypes is main intention of authors. Subarachnoid intracranial haemorrhage has more probability for a patient to be affected.

120 selected patients suffering from intracranial subarachnoid haemorrhage were taken into consideration for this paper and kept under 24 h Holter monitoring. Cardiac arrhythmias are found in 90% of patients according to observations of authors with adequate Holter recordings. According to Holter from paper [7], statistical analysis of various cardiac arrhythmias are observed by authors. Current traditional method was to diagnose intracranial haemorrhage by examining computerized tomography (CT) images which are taken by radiologists and for its subtype classification [8]. Current traditional process heavily relies on radiologist's experience of detecting intracranial haemorrhage. Details of 82 CT scans of different subjects are taken by authors in this paper for designing an automatic detection protocol. Deep fully conventional neural networks (deep FCNN) are taken into design protocol by authors

proposed in paper [9]. Dice coefficient of this protocol is calculated by authors which stands at 0.31. Current traditional method was to diagnose intracranial haemorrhage by examining computerized tomography (CT) images taken by radiologists which heavily relies on radiologist's experience of detecting intracranial haemorrhage [10, 11]. 4369 head computerized tomography scans were used by authors as training data set. Training data set is to be fed into patch-based conventional neural network (patch-based FCNN) for designing model. Performance of this patch-based fully convolutional neural network (patch-based FCNN) exceeded the performance of 2–4 radiologists who are taken for study in paper [12, 13]. CT scan image segmentation with examination level classification is comparable to radiology experts, in addition to robust localization of abnormalities in computerized tomography (CT) scan images, including some that are missed to be examined by radiologists, both of which are critically important elements for detection of intracranial haemorrhages that are managed by this application with high-speed performance and more accuracy [14, 15].

3   Intracranial Haemorrhage Detection

Computerized tomography (CT) scanned images dataset which was collected from nearby computerized tomography (CT) scan centres is used. Dataset is further validated by neurologists and radiologists to filter CT scanned images with intracranial haemorrhage. Intracranial haemorrhage dataset accounted for 150 CT scan cases which consists of approximately 4500 images. Pre-processing of images improves image data by suppressing the distortions, and some features enhance some features useful for further processing. Steps involved in image pre-processing are (i) reading images, (ii) resize image, (iii) remove noise (image denoising), (iv) brightness interpolation, (v) segmentation and morphology. The flow diagram for prediction of ICH subtype is specified in Fig. 2. Images read from dataset are being loaded into arrays for image pre-processing. Some images captured by computerized tomography (CT) scan may be of different sizes. A base size is established for every image, and every image is resized to predefined base size.

Image denoising is process of reconstructing an image or signal from a noisy one. Important pre-processing step for image analysis is image denoising. Two types of noise models are there; additive noise model where unwanted signal is added on original signal to produce corrupted signal, multiplicative noise model where unwanted signal is multiplied to original signal to produce corrupted and noisy signal. Based on noise, different kind of denoising techniques are used. Gaussian noise is a noise where an evenly distributed noise is added to input signal. Impulse noise is in turn divided into two types, fixed value noise is caused due to errors in transmission, random noise where random value is added to input signal which is difficult to remove. Weighted mean filter is used in this paper. Simple averaging of all the neighbouring pixels has a drawback which may lead to over smoothing. Instead of

Fig. 2 Flow diagram for prediction of ICH subtype

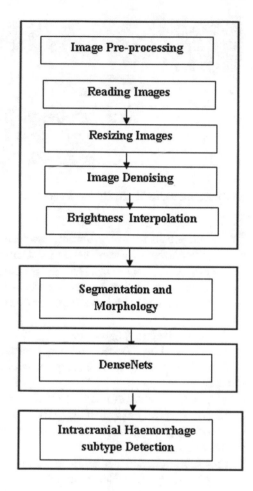

assigning weights to all neighbouring pixels equally, higher weights are assigned to pixels that are closer pixel being involved.

Brightness interpolation is one of step involved in image pre-processing. Brightness interpolation has many methods involved like linear interpolation, nearest neighbour interpolation and bicubic interpolation. Brightness interpolation means each pixel value in output raster image can be obtained by brightness interpolation of all neighbouring non-integer samples of pixels. Linear interpolation is the type of interpolation used in this paper. Linear interpolation explores four points neighbouring the point and assumes that the brightness function is linear in the neighbourhood of the pixel. Image segmentation is process of partitioning computerized tomography (CT) dataset images into multiple segments or sets of pixels which are known as image objects. Segmentation is used to change the representation of image and is transformed into segments which is easier way to simplify. Image segmentation is typically used to locate boundaries of objects in images as specified in Fig. 3.

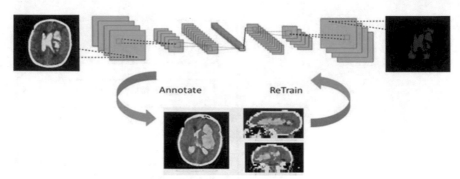

Fig. 3 Segmentation of ICH CT scanned image

DenseNets are also called as densely connected convolutional networks. DenseNets are traditional feed forward networks. Dense Nets-B and Dense Nets-C are two different types of DenseNets. DenseNets requires fewer parameters when compared to traditional convolutional neural networks [1]. DenseNets layers are very narrow as they add small set of feature maps. Gradient from loss function has access to each and every layer.

Steps involved in Algorithm

Step 1: Reading CT scanned images from images file path into arrays.

Step 2: Rescaling and resizing of every CT scanned image to appropriate threshold.

Step 3: Conversion of images to PNG format after rescaling and resizing.

Step 4: Perform image denoising using weighted mean filter method.

Step 5: Apply brightness interpolation on images to increase brightness uniformly for easy segmentation.

Step 6: Perform segmentation on images using scikit image library in Python.

Step 7: Apply DenseNets to images using keras library in Python by calculating mathematically using equations specified in (3) and (4).

The layers can be pooled by applying the convolution operation using Eq. (1) involving batch normalization and activation function.

$$x_l = H_l(x_{l-1}) \tag{1}$$

By including the skip connection principle, Eq. (1) gets modified into Eq. (2).

$$x_l = H_l(x_{l-1}) + x_{l-1} \tag{2}$$

By considering the principle of DenseNets instead of summing up the output feature maps with the input feature maps of the layer, the values are concatenated as specified in Eq. (3).

$$x_l = H_l([x_0, x_1, \ldots, x_{l-1}]) \tag{3}$$

By generalizing, Eq. (3) based on k feature maps by generalizing for l-th layer as specified in Eq. (4).

$$k_l = k_0 + k * (l - 1) \qquad (4)$$

Step 8: Calculate accuracy based on the prediction of subtypes of intracranial haemorrhage.

4 Experimental Results and Discussion

Computerized tomography (CT) scanned images dataset which was collected from nearby computerized tomography (CT) scan centres is used. Dataset is further validated by neurologists and radiologists to filter CT scanned images with intracranial haemorrhage. Intracranial haemorrhage dataset accounted for 150 CT scan cases which consists of approximately 4500 images.

Dataset is of total 150 CT scan cases which consists of 125 intracranial haemorrhage (ICH) cases and 25 normal cases. Model built is trained through cross-validation, and dataset is divided for training, validation and testing phases. Of them, 10 cases are from epidural intracranial haemorrhage, 24 cases are from subdural intracranial haemorrhage, 36 cases are from subarachnoid intracranial haemorrhage, 61 cases are from intraparenchymal intracranial haemorrhage, and 24 cases are from intraventricular intracranial haemorrhage. After model is trained using allotted 110 CT scan cases as specified in Table 1, then model is validated by using allotted 13 CT scan cases, after achieving accuracy of up to 95%. Model is tested against 2 cases are from epidural intracranial haemorrhage, 7 cases are from subdural intracranial haemorrhage, 7 cases are from subarachnoid intracranial haemorrhage, 14 cases are from intraparenchymal intracranial haemorrhage, and 2 cases are from intraventricular intracranial haemorrhage.

Table 1 Division of datasets for train, test and validation

		Data sets			
		Train	Validation	Test	Total
Number cases	Total	60	15	50	125
	Normal	0	0	25	25
Haemorrhage type	Epidural	5	3	2	10
	Subdural	15	2	7	24
	Subarachnoid	25	4	7	36
	Intraparenchymal	45	2	14	61
	Intraventricular	20	2	2	24
	Total	110	13	12	

Table 2 Detection of ICH subtypes by using CT scanned images

		Number detected	Total
Hematoma type	Epidural	2	2
	Subdural	6	7
	Subarachnoid	5	7
	Intraparenchymal	13	14
	Intraventricular	2	2
Total		2S	32

Results as mentioned in Table 2 shows that 2 cases are detected as of epidural intracranial haemorrhage type, 7 cases are detected as of subdural intracranial haemorrhage type, 7 cases are detected as of subarachnoid intracranial haemorrhage type, 14 cases are detected as of intraparenchymal intracranial haemorrhage type, and 24 cases are detected as of intraventricular intracranial haemorrhage type. The accuracy for prediction of subtypes of intracranial haemorrhage is 91%.

5 Conclusions and Future Scope

Intracranial haemorrhage is an internal bleeding of brain which is mainly caused due to high blood pressure in adults and in younger people due to improper development of blood vessels in brain. Intracranial haemorrhage if not treated properly may lead to death and other major effects. Building automated model for prediction of intracranial haemorrhage subtype is better than prediction by their human counterparts and radiologists. By use of DenseNets, prediction of intracranial haemorrhage is more accurate compared to deep convolutional neural network. Accuracy of two class prediction in first step is nearly 95%, and the accuracy of subtype prediction of intracranial haemorrhage is 91% using proposed method. Future work is to use deep learning technique to further improve the processing speed.

References

1. Juan SC, Steren C, Carolina, Rodrigo S (2019) Convolutional neural networks for detection of intracranial haemorrhage in CT images. In: Proceedings of 4th congress on robotics and neuro science
2. Ye H, Gao F, Yin Y (2019) Precise diagnosis of intracranial haemorrhage and subtypes using a three-dimensional joint convolutional and recurrent neural network. Eur Radiol 29:6191–6201
3. Majumdar A, Brattain L, Telfer B, Farris C, Scalera J (2018) Detecting intracranial haemorrhage with deep learning. In: 40th annual international conference of the IEEE engineering in medicine and biology society (EMBC), Honolulu
4. Jnawali K, Arbabshirani MR, Rao N, Patel AA (2018) Deep 3D convolution neural network for CT brain haemorrhage classification. Medical imaging computer-aided diagnosis. International Society for Optics and Photonics

5. Chang, Kuoy E, Grinband J, Weinberg B, Thompson M, Homo R, Chen J, Abcede H, Shafie M, Sugrue L (2018) Hybrid 3D/2D convolutional neural network for haemorrhage evaluation on head CT. Am J Neuroradiol 39:1609–1616
6. Murty OP, Anshoo A, Mohd FS, Mohd ATZ, Chong ChY, Rahmah R (2018) Sudden natural brain haemorrhage fatalities: a 10 years autopsy review at Kuala Lumpur. J Indian Acad Forens Med 30(3):110–122
7. Hssayeni1 MD, Crooc MS, Al-Ani A, Hassan Falah Al-khafaji, Yahya ZA, Behnaz G (2019) Intracranial haemorrhage segmentation using deep convolutional model, arXiv
8. Nowinski WL (2014) Characterization of interventricular and intracerebral intracranial haemorrhages in non-contrast CT. Neuroradiol J 27:299–315
9. Weicheng K, Christian II, Pratik M, Jitendra MELY (2019) Expert-level detection of acute intracranial haemorrhage on head computed tomography using deep learning
10. Liao C (2010) Computer-aided diagnosis of intracranial intracranial haemorrhage with brain deformation on computed tomography. Comput Med Imaging Graph 34:563–571
11. Kosior J (2011) Quantomo: validation of a computer-assisted methodology for the volumetric analysis of intracerebral haemorrhage. Int J Stroke 6:302–305
12. Chan T (2007) Computer aided detection of small acute intracranial haemorrhage on computer tomography of brain. Comput Med Imaging Graph 31(4):285–298
13. Grewal M, Srivastava MM, Kumar P, Varadarajan S (2018) RADnet: radiologist level accuracy using deep learning for haemorrhage detection in CT scans. Biomedical imaging (ISBI 2018), In: IEEE 15th international symposium on IEEE
14. Maduskar P, Acharyya M (2009) Automatic identification of intracranial haemorrhage in non-contrast CT with large slice thickness for trauma cases. In: Proceedings of SPIE, 7260
15. Tong HL, Ahmad Fauzi MF, Haw SC (2011) Automated haemorrhage slices detection for CT brain images. In: Badioze Zaman H et al (eds) Visual informatics: sustaining research and innovations. IVIC Lecture notes in computer science. Springer, Berlin, Heidelberg

Analysis of COVID-19-Impacted Zone UsingMachine Learning Algorithms

Sindhooja Abbagalla, B. Rupa Devi, P. Anjaiah, and K. Reddy Madhavi

Abstract Covid-19, first detected at Wuhan in late 2019, has now spread all over the world among many developed and developing countries. As a result of this, World Health Organization (WHO) declared COVID-19 a pandemic on March 11, 2020. Until now, many people have been infected with this coronavirus, some of them are recovering and others causing death. The concern of this paper will be the comparative study of KNN and Naïve Bayes algorithms via the Weka tool's Explorer and Experimenter interfaces, which will tell algorithm is more articulated to be used to evaluate the accuracy of death and recovery of infected COVID-19 patients, so we could estimate that the region will belong to which zone. The COVID-19 dataset to be used in this paper includes details about people who have visited Wuhan during this pandemic or who are from Wuhan and are affected by COVID-19 are fever, cold, cough, breathing difficulties, and many more. The main goal here will be to help users extract valuable data from the dataset and define a predictive algorithm for it. From the results shown, it can be concluded that KNN would demonstrate better precision than Naïve Bayes.

Keywords COVID-19 · Weka · KNN · Naïve Bayes · Classification

S. Abbagalla
CSE, JNTUH College of Engineering Jagitial, Jagitial, India
e-mail: abbagallasindhooja@gmail.com

B. Rupa Devi
CSE, Annamacharya Institute of Technology and Sciences, Tirupati, India
e-mail: rupadevi.aitt@annamacharyagroup.com

P. Anjaiah
CSE, Institute of Aeronautical Engineering, Hyderabad, Telangana, India
e-mail: anjaiah.pole@gmail.com

K. Reddy Madhavi (✉)
CSE, Sree Vidyanikethan Engineering College, Tirupati, India
e-mail: kreddymadhavi@gmail.com

1   Introduction

A novel strain of coronavirus was first found at the end of 2019 in Wuhan which is the capital of Hubei province in China. On January 30, 2020, the COVID-19 was declared as public health emergency of international concern (PHEIC) which has now become the 2019–2020 coronavirus pandemic due to its spread in worldwide. The World Health Organization (WHO) had named this new coronavirus as COVID-19 which later was renamed SARS-CoV-2 by the International Committee on Taxonomy of Viruses [1].

COVID-19 is a large group of gene-containing viruses that cause respiratory diseases that cause acute pneumonia from the common cold. severe acute respiratory syndrome (SARS) and Middle East respiratory syndrome (MERS) are also infectious respiratory diseases known as coronaviruses. Compared with SARS-CoV and MERS-CoV, the COVID-19 coronavirus is highly contagious. The COVID-19 has influenced more people worldwide in a short amount of time relative to SARS-CoV and MERS-CoV [2]. The major signs of these signs of coronavirus include fever, cold, cough, breathing difficulty, generalized myalgia, drowsiness, diarrhoea, dyspnoea, and pneumonia. Kids, the elderly, pregnant women, people with chronic diseases such as diabetes, cardiovascular disorders, and malignancy are at considerable risk of being seriously infected with coronavirus infections [3]. But no clear vaccine is available for any of the coronaviruses [4].

During close contact, the virus has a high tendency to be transmitted among people, most often by droplets formed by coughing, sneezing, and talking as they may remain in air for at least tens of minutes. Since there are evidences that COVID-19 is an airborne disease, the chances of being affected by COVID-19 also arise when our surroundings are affected or when we visit the affected areas [5]. The proposed predicitive classifier effeciently predicted recovered and death rates to avoid visiting the affected areas, and areas can be isolated such as the Red zone, the Orange zone, and the Green zone. Distinguishing of areas is classified based on the death and recovery rate of the persons affected by COVID-19 [6]. Red zone means regions where the risk of death is higher than the recovered, indicating that most people are infected by COVID-19 in that area. Orange zone means that the rate of death is less than the recovery rate of COVID-19 affected persons, meaning that COVID-19 affected persons are relatively lesser than the Red zone. Green zone means the safest zone in which there is no person infected by COVID-19.

Data mining is widely used in many areas and is developing its applications and one of its applications in health care. Data mining is used to extract useful data required for diagnosing and predicting. Machine learning uses datasets that are built from extracted data [7]. Weka is a software for data mining which uses algorithms of machine learning [8].

In this paper, the COVID-19 dataset used is provided by Johns Hopkins University dashboard which was available on Kaggle.com for the classification process. This classification process includes steps such as a collection of data and determining the accuracy through KNN and Naïve Bayes in Explorer interface and then the comparison between them through the Experimenter interface. This study has adopted 11 independent attributes which include Sr.no, Reporting date, Location, Country, Gender, Age, System _onset, If_onset_approximated, Hospital _visit _date, Exposure _start, Exposure _end, and two dependent variables where one dependent variable is among from_wuhan and visiting_wuhan and other dependent variable is among death and recovered. By using Weka's remove attributes, we have divided it into four datasets and predict their accuracy using Naive Bayes and KNN algorithms.

2 Methodology

Weka is an effecient method of data mining used to classify accuracy across various algorithms [9]. Five different interfaces such as Explorer, Experiment, Knowledge-Flow, Workbench, and Simple CLI are included in Weka software, of which Explorer ad Experimenter interfaces have been used for our analysis.

Classification
In Weka, the data mining techniques and machine learning algorithms help in predicting the death and recovery rate of people who visiting Wuhan and people who are from Wuhan and characterize which zone Wuhan relates. The accuracy is classified based on correctly and incorrectly classified instances, mean absolute error, and relative absolute error.

Explorer Interface
The data is initially preprocessed and extracted from it. Then the data file is loaded in the comma separated value (.csv) format and the precise classification is tested using 10 cross-validation option by selecting the KNN and Naïve Bayes algorithms.

NAÏVE BAYES. Naïve Bayes is a predictive machine learning algorithm that helps in classifying tasks. Naïve Bayes can use maximum likelihood estimators to enhance efficiency if the estimated model is extremely incorrect and can also handle numeric attributes using supervised discretization [10]. This classifier is based on Bayes theorem such as,

$$P(C_i|A) = \frac{P\left(\frac{A}{C_i}\right)P(C_i)}{P(A)}$$

$P(C_i)$ is the probability that a training pattern belongs to the class, also called a prior probability. In our paper, we have considered two classes such as death and recovered with distinct values (0,1).

$P\left(\frac{C_i}{A}\right)$ is called posterior probability.

$P\left(\frac{A}{C_i}\right)$ is conditional probability such as,

Table 1 Output achieved by applying the Naïve Bayes algorithm

Datasets	Instances correctly classified % accuracy	Instances incorrectly classifed % accuracy	Mean absolute error	Relative absolute error
From Wuhan death	94.47	5.53	0.0675	61.247
From Wuhan recovered	88.8479	11.1521	0.1355	54.047
Visiting Wuhan death	95.1152	4.8848	0.0634	57.503
Visiting Wuhan recovered	88.4793	11.5207	0.1369	54.634

$$P(A/C_i) = \prod_{k=1}^{n} P(A_k|C_i)$$

$$= P(A_1|C_i) \times P(A_2|C_i) \times \ldots \times P(A_k|C_i)$$

Running a set of data using the test option Naïve Bayes in Weka as cross-validation 10-folds give us the statistical output that is analyzed to predict each set of datasets. Running the Weka Explorer classification panel and using the Naïve Bayes algorithm to achieve death and recovery accuracy of people who have visited Wuhan or are from Wuhan.

Table 1 values help us to indicate that their death accuracy is more compared to their recovery rate for individuals who are from Wuhan or have visited Wuhan their death accuracy is more compared to their recovery rate. It shows us how other classifications often apply their significance accuracy measures.

KNN. KNN is a method with lazy evaluation. It is also called as memory-based classification as the training set has to be a memory at runtime because while dealing with continuous attributes it is calculating the distance using Euclidean distance [11]. KNN mostly deals with continuous attributes but it also deals with discrete attributes. For discrete attributes, another method also can be used such as overlap metric or hamming distance. KNN is also known as an instance-based algorithm. That is why in Weka KNN is represented as IBK. The accuracy of KNN depends on the k value and distance metric.

In this KNN classification, the same COVID-19 dataset is used which is used for Naïve Bayes classification and run this dataset with $k = 1$ and test option is cross-validation with 10-folds to find the classifier accuracy for death and recovered rate of people who have visited Wuhan or of people who are from Wuhan.

Table 2 Output achieved by applying the KNN algorithm

Datasets	Instances correctly classified % accuracy	Instances incorrectly classified % accuracy	Mean absoluteerror	Relative absolute error
From Wuhan death	97.6037	2.3963	0.0311	28.188
From Wuhan recovered	90.5991	9.4009	0.1098	43.807
Visiting Wuhan Death	97.6959	2.3041	0.0317	28.768
Visiting Wuhan recovered	90.1382	9.8618	0.1096	43.710

Table 3 Correctly classified accurcy between Naïve Bayes and KNN algorithms

Datasets	Naïve Bayes	KNN
From Wuhan death	94.24	97.52
From Wuhan recovered	88.57	90.43
Visiting Wuhan death	95.17	97.75
Visiting Wuhan recovered	88.28	89.82

Table 2 values also indicate that their death accuracy is more compared to their recovery rate for individuals who are from Wuhan or have visited wuhan their death accuracy is more compared to their recovery rate.

Experimenter Interface

The Weka Experimenter interface has been used for comparing various algorithms and analyzing their accuracy. In this paper, we have used the Experimenter interface to analyze the data by experimenting through algorithms such as Naïve Bayes and KNN. Each algorithm runs 10 times and then produces the result. The test option selected here is cross-validation by 10-folds.

Table 3 demonstrates that KNN algorithm provides more accuracy than the Naive Bayes algorithm.

3 Results

In this paper, the Weka tool is used for prediction of death and recovery rate of people who have visited Wuhan and of people who are from Wuhan. These predictions are made based on KNN and Naïve Bayes algorithms using Explorer and Experimenter interfaces provided by Weka. From these tools, the results are trained out based on instances correctly classified, Instances incorrectly classified, mean absolute error, relative absolute error. The algorithm which provides the best accuracy can be predicted by only Instances correctly classified but to get more accurate

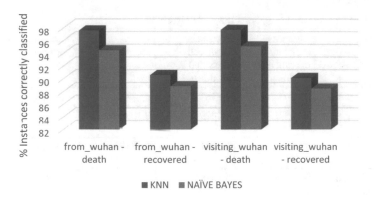

Fig. 1 Comparitive accuracy of KNN algorithm and Naive Bayes algorithm

analysis other classifications are also considered. Therefore, the graph below shows us the predicted results based on instances correctly classified and tells us that the KNN algorithm gives more accurate Instances correctly classifiedthan Naïve Bayes.

Therefore, by using Weka's Explorer and Experimental interface analysis, it is observed that KNN provides more accuracy when compared with Naïve Bayes. KNN gives maximum Instances correctly classified, least mean absolute error, and least relative absolute error.

From Fig. 1, notes that either of the people who are from Wuhan or have visited Wuhan, their death rate is higher than their recovery rate. All these indicate that Wuhan is in the Red zone and any healthy person who visits Wuhan raises the risk of COVID-19 infection. Since the dataset used here contains information about people visiting Wuhan or staying in Wuhan, it could be claimed which zone Wuhan belongs to. Similarly, considering different datasets containing information on persons affected by COVID-19 in different areas and differentiate them to their respective zones basis of their recovery rate and death rate of persons affected by COVID-19, so that strict precautions can be taken for their respective zones.

4   Conclusion

The main aim of this paper was to provide an algorithm that effectively predicts the area which is most affected by COVID-19 coronavirus based on recovered individuals and death groups. A variety of attributes was used to base this prediction. As WHO said that there are some evidences that COVID-19 is an airborne disease it has become very important to stop people from moving to COVID-19 affected surroundings. Therefore, isolating the areas helps to alert people on visiting Red zone areas so that the risk of COVID-19 being affected is reduced. This information also helps the government to take necessary precautions for people who have a history of travel from Red or Orange zones as well as also for people who live there.

Weka's interfaces, such as Explorer and Experimenter interfaces, and machine learning algorithms like KNN and Naïve Bayes were used to demonstrate the classification accuracy. Future studies include gathering more accurate datasets from health centers in the future and obtain more accurate results by performing predictions through deep learning as well, which will enable us to deal with infectious diseases and reduce the local and global burden of this pandemic.

References

1. Zi Y, Jiang M, Xu P, Chen W, Ni Q, Ming LG, Zhang L (2020) Coronavirus disease 2019 (COVID-19): a perspective of China
2. Meo SA, Alhowikan AM, Al-Khlaiwi T, Meo IM, Halepoto DM, Iqbal M, Usmani AM, Hajjar W, Ahmed N (2020) Novel coronavirus 2019-nCoV: prevalence, biological and clinical characteristics comparison with SARS-CoV and MERS-CoV. Eur Rev Med Pharmacol Sci 24:2012–2019
3. Brenda LT (2020) MD: coronaviruses and acute respiratory syndromes (COVID-19, MERS, and SARS). Technical report
4. A research on Sars and Mers, Baylor College of Medicine. https://www.bcm.edu/departments/molecular-virology-and-microbiology/emerging-infections-and-biodefense/emerging-infectious-diseases
5. Coronavirus disease 2019. https://en.wikipedia.org/wiki/Coronavirus_disease_2019
6. Al-Najjar H, Al-Rousan N (2020) A classifier prediction model to predict the status of Coronavirus CoVID-19 patients in South Korea. Eur Rev Med Pharmacol Sci 24:3400–3403
7. Ian W, Eibe F, Mark H, Christopher P (2016) Data mining. In: Practical machine learning tools and techniques, 4th eds. Morgan Kaufmann Publishers
8. Han J, Pei J, Kamber M (2011) Data mining: concepts and techniques, 3rd edn. Morgan Kaufmann Publishers
9. An Introduction to Weka. https://www.opensourceforu.com/2017/01/an-introduction-to-weka/
10. Naïve Bayes Classifier. https://en.wikipedia.org/wiki/Naive_Bayes_classifier
11. K-nearest neighbours' algorithm. https://en.wikipedia.org/wiki/K-nearest_neighbors_algorithm

Road Detection Using Semantic Segmentation-Based Convolutional Neural Network for Intelligent Vehicle System

Deepak Kumar Dewangan and Satya Prakash Sahu

Abstract Road scene perception of driving environment is an important yet stimulating task for an intelligent vehicle system (IVS). Advancement of deep learning techniques has enabled the image sensors to understand the scene more accurately for the intended object, especially in segmentation domain. In context of computer vision approach, segmentation of road scene from colour images is challenging, due to the varying illumination circumstances, the non-uniform road shapes and the fuzzy boundaries between the road and other objects. However, existing studies have limited performance in context of standard measures, therefore fails to incorporate the mentioned issues. With this aim and to clearly distinguish road and non-road portion, we have utilized U-Net, Seg-Net and fully convolutional network (FCN) models. From the conducted experiments, U-Net has achieved 94% score for mIoU and dice coefficient which is better than the scores achieved by other models Seg-Net and FCN-32. In our experiment, images from Camvid dataset have been used to train and validate the performance.

Keywords Intelligent vehicle system · Scene understanding · Artificial intelligence · Deep learning · Computer vision

1 Introduction

Intelligent transportation is one of the leading areas in automobile business that got attention for various industries and researches in last few decades. Various cognitive approaches like machine learning, deep learning, artificial intelligence, etc. have demonstrated the applicability of technological integration on various industries.

D. K. Dewangan (✉) · S. P. Sahu
Department of Information Technology, National Institute of Technology, Raipur (C.G.), Raipur, India
e-mail: dkdewangan.phd2018.it@nitrr.ac.in

S. P. Sahu
e-mail: spsahu.it@nitrr.ac.in

Applying these technologies to develop an IVS may bring the revolution in transportation industry. Working prototypes of such autonomous vehicles from industries including Google, Volvo and research institutes like Stanford University are going to be commercialized in upcoming days [1]. Apart from the technical, social and legal issues of IVS, complete execution of these self-driving vehicles in the countries like India is somehow challenging due to its unstable road environment. World Health Organization (WHO) has already revealed the higher road traffic death as estimated as 299,091 in 2016 [2] caused by random reasons [3] and possibly includes careless driving, varying illumination conditions and failure to understand the driving and dynamic road atmosphere completely.

Though, a vision-based equipped IVS may face heterogeneous environment and has to deal with detection of road, lane, pedestrian, vehicle, traffic lights/signals, pothole, speed bump, rail guards, etc., but its foremost objective is to recognize the road surface on which it will be driven further. Approaches used in [4] have identified unstructured road surfaces using vanishing point detection. Similarly, super pixel segmentation is utilized to extract centreline for road surfaces [5]. In another study, convolutional neural network (CNN) was applied to for road segmentation [6]. Effectivity of similar approaches can be enhanced with standard CNN-based models that extract the low- and high-level features of intended objects and thus provide clear distinction between road and non-road portions. In the present study, we have applied standard segmentation model and identified their performance on public dataset, thereby bringing the idea of applying the optimal model into such autonomous system. The process flow of proposed working model is represented in Fig. 1.

The organization of the article is structured as follows: Sect. 2 reveals different schemes to identify road surfaces. Section 3 includes the material and methods for proposed approach; the experimental set-up and results are discussed in Sect. 4, and concluding remarks along with future direction is provided in Sect. 5.

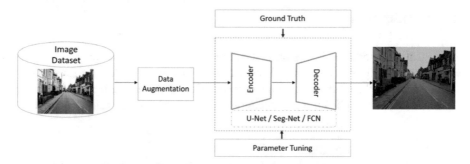

Fig. 1 Workflow for road detection

2 Related Works

This section portrays the mechanism used to detect road and road scene from the input images.

A CNN-based approach was applied in which pooling map of noisy stamps to find road images using 3D format are learned. The algorithm processes an input image of $N \times M$ dimension and gives as output a set of three floating point numbers which indicate probability of image patches for vertical, horizontal or sky areas. But it focuses on a single image which lacks robustness and inaccurate for the prediction of on-board road images [7]. Yongchao et al. initially processed the input frame by putting the region of interest (ROI) by eliminating unnecessary items such as sky, hood and deal with the case of shadows in the image. It contained three stages: (1) Formation of classifier (Support Vector Machine: SVM); (2) Processing of the illumination independent image and (3) Detection of road surfaces. However, the method has limited performance for the detection of multiple entities including roads, people or vehicles [8]. Dan et al. utilized deep convolutional network 'StixelNet' for road segmentation and obstacle detection on public dataset KITTI. Obstacle detection was carried out in two stages. The first stage evaluated the local prediction using StixelNet architecture; the second stage involved a conditional random field (CRF) in order to receive a globally consistent estimation [9].

Similarly, network-in-network (NIN) architecture-based scheme comprises of extraction of patches around a region, and then, categorizing is done for those patches using CNN. The output class is associated with the pixel or region from where the patch was centred, but the scheme has limited performance to classify roads and regions under varying illumination [10]. Jee-Young et al. presented a residual learning-based network which is comprised by the boundary attention and reverse attention units used for road segmentation. The reverse attention unit performs by providing the weights to the earlier processed non-road regions to discover lost road region. Similarly, the boundary attention unit emphases on the regions present for previously calculated road boundaries to get the road prediction. Scheme was focusing only for binary road segmentation, and it mostly ignores other instance of road scene [11]. In another approach of semantic segmentation, Changqian et al. implemented the bilateral segmentation consisted of two modules. First module is organized to inhabit spatial information with wide channels and superficial layers which is a 'Detail Branch'. Another module abstracts the definite semantics with lean channels and deep layers which is 'Semantic Branch' [12]. Method used in [13] exploits unlabelled training image data to fetch significant features; then, it removes the lower region of the input image assuming it is irrelevant. Likewise, method discussed in [14] proposes a novel texture descriptor, a combination of colour planes which obtains maximum uniformity in road regions. Study in [8] discusses about illuminated invariance approach to solve the effect of interference caused by shadow detection. To solve the issues of anomaly detection, authors measured the abnormality in the magnitude of motion and orientation [15], whereas method in [16] discusses a machine learning approach which looks at contextual information and also maintains a better inference time. Another research in this direction discusses the road segmentation process for satellite images [17]. Similarly, a CNN is proposed named as DSNet whose aim is to improve the performance and give better detection results as compared with other standard networks. Yadav et al. discussed road segmentation for unmarked roads having no defined boundaries [18].

3 Material and Methods

3.1 Dataset

A public dataset CamSeq01 of 'Camvid' dataset has been referred [19] that includes 101 images of 960 × 720 pixel resolution images of 92 Mb in total size; in which all the pixels were individually assigned to 32 categories of object classes in a real-time driving situation. These classes in the frame label consist of pedestrian, road, tree, buildings, etc. The dataset size is increased from 101 to 909 images using data augmentation technique that comprises of rotation, flipping, blurring, warping and noise, varying brightness, scaling and shearing.

3.2 Network Architecture

To identify significant features, semantic segmentation-based techniques UNet [20], SegNet [21] and FCN [22] models are utilized. Initially, the working principles of each model are discussed, and then, the performance of mentioned model is compared and analysed.

3.2.1 U-Net Architecture

U-Net architecture is characterized by its contracting path in the initial stage and expanding path in the final stages and shaped as 'U' that can be viewed from Fig. 2. Rather than using fully connected layer, it employs a convolutional procedure at the outcome of expanding route, and hence, model is fine matched to pixel-wise classification. The contracting route involves iterative presentation of 3 × 3 convolutions accompanied by a ReLU activation function. Similarly, 2 × 2 max pooling strategy is employed with stride movement of two units for downsampling process. Amount of feature channels is increased against each step of downsampling. Likewise, route in the expansion path comprises of upsampling process, where pooling map accompanied by a 2 × 2 filter operation which reduces feature channels to half of its size, and then, concatenation is carried out with the corresponding feature map. In the

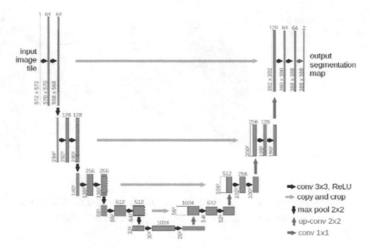

Fig. 2 Contracting and expanding path of U-Net architecture

last layer, a 1×1 convolution process records feature vectors to the preferred and defined object classes.

3.2.2 Seg-Net Architecture

Being an encoder-decoder variant, Seg-Net employs pixel-wise classification and is given in Fig. 3. Total of 13 convolutional layers are applied in encoder module of Seg-Net which corresponds to the initial 13 convolutional layers of VGG16 network which was originally considered for object classification [21]. To retain most significant feature maps at the higher end of encoder module, fully connected layers are rejected. This step helps out the Seg-Net to minimize the overhead of processing the unwanted features and thus reduces the total count of parameters from 134 million to 14.7 million. Likewise, each encoder module has an equivalent decoder module (13 layers) at the second half of the architecture section. Therefore, it eliminates the necessity to have upsample module as an extra computational overhead. The outcome from upsampled feature maps is sparse and is then undergone with trainable filters to

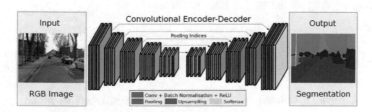

Fig. 3 Encoder-decoder-based Seg-Net architecture

Fig. 4 FCN architecture to
perform pixel-wise semantic
segmentation

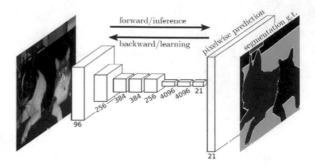

produce higher-order feature maps. Finally, a multiclass classifier softmax processes
the final decoder outcome and generates the targeted object class probabilities for
individual pixels.

3.2.3   Fully Convolutional Network (FCN)

Image data which is a 3D array of size *height * width * depth* in each layer undergone
through a convolution process, where *height* and *width* are spatial measurements,
and *depth* is the feature dimension. Positions of targeted object under convolution
in upper layers match to the positions in the image data, called 'receptive field'.
Architectural representation of FCN is shown in Fig. 4. Convolutional operation at
this moment is built upon translation invariance. Procedure components which are
convolution, pooling and activation operates on local input sections and relatively
rely only on the spatial coordinates. Thus non-usability of dense layer here diminishes
the parameters count and hence minimizes the computational overhead.

4   Experimental Results and Discussion

Implementation is carried out with Intel (R) Xeon (R) W-2175 CPU @ 64 GB of
shared and 8 GB of dedicated GPU capability. The following specifications were
taken for experiment: image size of 960 × 720, and the stopping criteria for training
and testing experiment are set for 100 iterations.

4.1   Performance of U-Net, Seg-Net and FCN

In general, model's training and validation accuracies reach to highest point at
very initial stages while extracting and creating the feature maps. Also, retaining a

minimum gap between both accuracies is also watched. Further, too much unsteadiness in both accuracies indicates poor performance of network. Keeping these observations, U-Net architecture has good performance over Seg-Net and FCN which is shown in Fig. 5a. The training and validation accuracies can be visualized from Fig. 5a–c which indicates model's performance in terms of its accuracies which reaches above 90% for all the mentioned model. However, there is a small gap between training and validation accuracy which shows that the model is not much fit as shown in Fig. 5c. Also, the oscillations of FCN are not much consistent throughout its training.

Apart from the qualitative measurement, behaviour of all the models is also predicted using standard measures that quantitatively represent their attained score and represented in Table 1. It is observed that training loss is recorded minimum for U-Net and found maximum for FCN model. Similarly, U-Net has better score

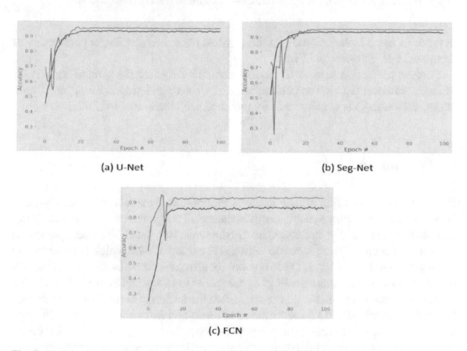

(a) U-Net

(b) Seg-Net

(c) FCN

Fig. 5 Performance analysis on training and validation accuracies

Table 1 Quantitative performance of models

Model	Loss	Train score (mIoU)	Train score (mDice)	Test score (mIoU)	Train score (mDice)
FCN	0.15	0.95	0.93	0.86	0.84
Seg-Net	0.069	0.97	0.95	0.92	0.91
U-Net	0.052	0.97	0.96	0.94	0.94

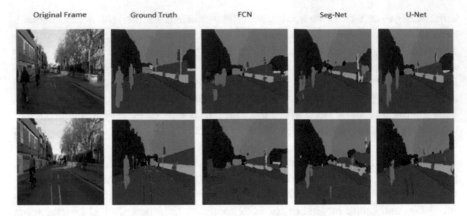

Fig. 6 Detection performance of against the available ground truth of public dataset

in terms of mean intersection over union (mIoU) and dice coefficient parameters as compared with Seg-Net and FCN.

Further, prediction sample of models behaviour against the ground truth (GT) of is presented in Fig. 6. The performance of U-Net prediction is found maximum among all the models which is closely matched with the available GT.

5   Conclusion

For road surface detection, semantic segmentation-based models FCN, Seg-Net and U-Net have been employed using Camvid dataset. Though the train accuracies using dice coefficient for all the models are nearly the same. But, the U-Net model presented its best results under given scenario with the accuracy of 94% which is followed by the Seg-Net model with 91%. The loss score for all models was also taken into consideration and found maximum for FCN as compared to U-Net and Seg-Net model. The outcomes of our experiments demonstrate that the semantic segmentation-based U-Net method is robust and segments the road scene and extracting road surface more accurately even in dynamic scenes than the other existing methods. In the future direction, the road scene understanding capability in challenging environment using such model will be integrated with prototype.

References

1. Okumura B, James MR, Kanzawa Y, Derry M, Sakai K, Nishi T, Prokhorov D (2016) Challenges in perception and decision making for intelligent automotive vehicles: a case study. IEEE Trans. Intell Veh 1(1):20–32

2. GHO—by category—Road traffic deaths—data by country, WHO. https://apps.who.int/gho/data/node.main.A997
3. Dewangan DK, Sahu SP (2020) Real time object tracking for intelligent vehicle. In: 1st international conference on power control and computing technologies (ICPC2T 2020). IEEE Press, India, pp 134–138
4. Yang G, Wang Y, Yang J, Lu Z (2019) Fast and robust vanishing point detection using contourlet texture detector for unstructured road. IEEE Access 7:139358–139367
5. Shen Y, Ai T, Yang M (2019) Extracting centerlines from dual-Line roads using superpixel segmentation. IEEE Access 7:15967–15979
6. Junaid M, Ghafoor M, Hassan A, Khalid S, Tariq SA, Ahmed G, Zia T, Multi-feature view-based shallow convolutional neural network for road segmentation. IEEE Access 8:36612–36623
7. Alvarez JM, Gevers T, LeCun Y, Lopez AM (2012) Road scene segmentation from a single image. In: European conference on computer vision, Springer, Berlin, Heidelberg
8. Song Y, Ju Y, Du K, Liu W, Song J (2018) Online road detection under a shadowy traffic image using a learning-based illumination-independent image. Symmetry 10(12):707
9. Levi D, Garnett N, Fetaya E, Herzlyia I (2015) StixelNet: a deep convolutional network for obstacle detection and road segmentation. In: British machine vision conference (BMVC), 1(2) London, p 4
10. Mendes CCT, Frémont V, Wolf DF (2016) Exploiting fully convolutional neural networks for fast road detection. In: IEEE international conference on robotics and automation (ICRA). IEEE Press, Swedan, pp 3174–3179
11. Sun JY, Kim SW, Lee SW, Kim YW, Ko SJ (2019) Reverse and boundary attention network for road segmentation. In: IEEE international conference on computer vision workshops, Korea
12. Yu C, Gao C, Wang J, Yu G, Shen C, Sang N (2020) BiSeNet V2: bilateral network with guided aggregation for real-time semantic segmentation. In: arXiv preprint, pp 2004.02147
13. Alvarez JM, Salzmann M, Barnes N (2013) Learning appearance models for road detection. In: IEEE intelligent vehicles symposium. IEEE Press, pp 423–429
14. Alvarez JM, Gevers T, LeCun Y, Lopez AM (2012) Road scene segmentation from a single image. In: European conference on computer vision. Springer, Berlin, Heidelberg, pp 376–389
15. Yuan Y, Wang D, Wang Q (2016) Anomaly detection in traffic scenes via spatial-aware motion reconstruction. IEEE Trans Intell Transp Syst 18(5):1198–1209
16. Mendes CCT, Frémont V, Wolf DF (2016) Exploiting fully convolutional neural networks for fast road detection. In: IEEE international conference on robotics and automation (ICRA). IEEE Press, pp 3174–3179
17. Henry C, Azimi SM, Merkle N (2018) Road segmentation in SAR satellite images with deep fully convolutional neural networks. IEEE Geosci Remote Sens Lett 15(12):1867–1871
18. Yadav S, Patra S, Arora C, Banerjee S (2017) Deep CNN with color lines model for unmarked road segmentation. In: IEEE international conference on image processing (ICIP). IEEE Press, UAE, pp 585–589
19. Fauqueur J, Brostow G, Cipolla R (2007) Assisted video object labeling by joint tracking of regions and keypoints. In: IEEE international conference on computer vision (ICCV'2007) interactive computer vision workshop. Rio de Janeiro, Brazil, pp 1–7
20. Ronneberger O, Fischer P, Brox T (2015) U-net: convolutional networks for biomedical image segmentation. In: International conference on medical image computing and computer-assisted intervention. Springer, Cham, pp 234–241
21. Badrinarayanan V, Kendall A, Cipolla R (2017) Segnet: a deep convolutional encoder-decoder architecture for image segmentation. IEEE Trans Pattern Anal Mach Intell 39(12), 2481–2495
22. Long J, Shelhamer E, Darrell T (2015) Fully convolutional networks for semantic segmentation. In: IEEE conference on computer vision and pattern recognition, pp 3431–3440

Issues and Challenges in Incorporating the Internet of Things with the Healthcare Sector

Saurabh Bhattacharya and Manju Pandey

Abstract The Internet was at first used to transfer a bunch of information in the form of packets among various users and information sources with a particular IP address. Because of technological advancement, data sharing through the Internet is rapidly increased and utilized to share information among various sensors, devices and gadgets which are associated in billions to establish the framework of the inter-connected data-sharing model. The Internet of Things (IoT) is a promising revolution which interconnects the accessible devices to offer robust and compelling knowledge objects. With the rapid growth of the Internet, one of the most essential and crucial forms is, i.e. healthcare. A tremendous amount of data is generated from these devices, using non-standardized protocols, and lack of interoperability tends to create challenges from developing well-established and reliable system. For better services, it is necessary to provide state of the art techniques to overcome the various issues and challenges for developing IoT-based medical services. This paper discusses the various issues and challenges that are related while incorporating IoT with the healthcare sector.

Keywords Health care · Interoperability · Artificial intelligence · IoMT · AI · Big data · 5G

1 Introduction

In 1999 British entrepreneur, Kevin Ashton coined the term 'Internet of Things'. Things in IoT are a combination of hardware (sensors, actuators, networking devices), software, services and data generated during the process. IoT provides the facility to control the things remotely across the network without too many human interactions.

S. Bhattacharya (✉) · M. Pandey
Department of Computer Application, NIT Raipur, Raipur, India
e-mail: babu.saurabh@gmail.com

M. Pandey
e-mail: mpandey.mca@nitrr.ac.in

It offers an effective and efficient approach to ensure superior result by attaining a higher order of excellence in operation at the same time reduce human interaction to deliver a competitive benefit. When any devices get connected to the Internet, it means that data is either sending or receiving or both. This ability to send and receive automatically defines things as 'smart'. Internet of Things has the prospective to change fundamental communication and business model. The progress of wireless communication describes the idea of connecting devices globally over the network, which can connect a vast number of smart devices through the Internet. IoT provides a facility to connect devices efficiently and consistent way to fetch optimum outputs.

Various technologies used to connect these smart things like Bluetooth, Wi-Fi, ZigBee, LoRaWAN, different cellular technologies like 2G/3G/4G/5G. The current trend of IoT expected to increase as smart future with smart city, smart vehicle, smart medical system, smart grid, smart industry, smart shopping, smart logistics, etc. It connects billions of heterogeneous devices to bridge the pathway of the physical world to the virtual world. Among all, health care is the most promising sector nowadays, where IoT gives new dimensions to smart pills, remote monitoring, biosensors, robotic medicals, health bands, real-time health system, implantable cardioverter-defibrillator. Integration of IoT for its wide variety of features and benefits with medical sought for part of the Internet of Medical Things (IoMT). Apart from the advancement in the medical sector, issues and challenges are rising with due course. Devices like an insulin pump, implantable cardiac devices, sensor thermometers, wireless vital monitoring systems are vulnerable whose failure may cause fatal effects.

1.1 Motivation

To the best of our knowledge, this is the in-depth analysis of incorporating Internet of Things in health care, which considering all significant factors like interoperability, data compression to implementation of 5G. Every aspect has some issues and challenges which create hindrance in the successful application of IoT with health care.

1.2 Paper Organization

The rest of this paper organized as follows. In Sect. 2, we discuss the literature survey. In Sect. 3, we provide the background of IoT and role of IoT in the healthcare sector. In the next Sect. 4, we discuss our primary analysis of various issues and challenges. Finally, Sect. 5 the conclusion is provided.

2 Literature Review

Information on the clinical field depends on the medical report evidence and experience that aggregated through specialists. The human body resembles a complex machine which contains various parts and that can be influenced by numerous variables which are affected by the number of factors and surroundings. Clinical information can utilized to mention objective facts and forecasts to the analysis of different ailments. AI, machine learning and big Data techniques give better reassure to plan such devices which can learn from previous experience and make analysis makes extremely helpful in smart health applications. Pham et al. [1], Verma and Sood [2] and Sandhu et al. [3] developed a cloud-based platform to improve early detection of health monitoring and tracking H1N1 virus having disadvantages of security threat in their system, whereas Nandyala and Kim [4] developed a fog-based monitoring system, but patients not considered. Gia et al. [5] used lightweight transform mechanism and received real-time feedback but having security issues in the proposed approach. Somasundaram et al. [6] discussed the proposed concept of security mitigation towards IoT in the medical sector.

In IoT-based healthcare system, Baali et al. [7] reviewed that the IoT sensors used for healthcare monitoring uses the Korotkoff method, and the Beer-Lambert law algorithm succeeds to reduce the power gap, but still, IoT sensors consume high power which makes it less reliable for the long run. Woo et al. [8] use a fault-tolerant algorithm to propose machine to a machine-based system delivers active healthcare service but uses a sophisticated algorithm. Ullah et al. [9] submitted big data real-time monitoring system using quantitative methods able to reduce the cost but not able to provide an accurate result. Firouzi et al. [10] proposed significant data analysis methodology to develop smarter healthcare, but the system is yet not deployed. Hu et al. [11] use the manual method to collect the record and further used to deliver effective healthcare system using SVM and neural network algorithm. Granjal et al. [12] analyse the existing network protocols and methods to provide more secure IoT communication, whereas Fuqaha et al. [13] mentioned the overview of horizontal integration of services with IoT. Noura et al. [14] discussed with various existing proposal according to their managing interoperability techniques.

3 IoT Background

IoT can portray as an augmentation of the Internet or network connections to different sensors or devices. The combinations manage the cost of even simplest type of objects like a bulb, fans, locks, vents to further extent of the higher degree of computing and analytical capabilities. Interoperability is one on the significant part of the IoT that adds to the developing prominence. The backbone of IoT based on sensors, actuators and sensing devices. Sensors are a device that able to detect changes in physical entity around the environment. Sensors can measure a physical phenomenon like temperature, pressure heart rate and transform it into an electric signal. Sensors possess high-order phenomenon of measuring sensitivity without changing the actual values at the same time should not be sensitive to another physical phenomenon. Tending to the requirements of every one of these layers is essential on all the phases of implementing IoT engineering. In the premise of acceptability standard, this consistency makes the outcome planned truly work. Also, the major highlights of industrial IoT engineering incorporate usefulness, adaptability, accessibility and practicality. Without tending to these conditions, the consequence of IoT engineering is a disappointment. Here, considering IBM cloud architecture which is divided into six layers (Fig. 1).

1. User layer—Entry level of implementing IoT architecture is where data is being recorded from users using various application
2. Proximity network—This layer includes sensors, actuators, devices and IoT gateway. Sensors can change over the data acquired from the external environment into valuable information for analysis. Actuators can mediate physical reality. IoT gateway helps to communicate the data over the network using Bluetooth, Wi-Fi, ZigBee, LoRaWAN for further processing.
3. Public network—Vital importance of these layers is to manage the tremendous data generated from various sensors, actuators and devices and optimally transfer the data to the cloud network.
4. Cloud network—This provides in-depth processing, analysis, prediction and data management.
5. Enterprise network——Final output of the data analysis served here.
6. Security and information governance—Responsible to manage security and information ascendency throughout the levels except for the user layer.

3.1 IoT in Health Care

IoT in health care is one of the most noticeable verticals among the various applications, which moulds the traditional healthcare system from hospital centric to patient centric. Personalized and ubiquitous services help health care to reach new heights. The key advantages of IoT sensors and advancements impacted a lot of utilization zones. Explicitly, embedded sensors and ambient sensors on patients gather the information distantly, which helped to give expectant medicinal services by anticipating the medical issues prior to utilizing the checking of vital signs. Implantable sensors have a combined history of accomplishment and profound effect on the eternal nature of the patient's life. IoT in health care reduces the cost at the same time providing quality of service.

Internet of Things has shifted the medical paradigm in new endless applications in the structure. IoT has been entrenched its advancement in real-time remote of the patient, collecting and transmitting the data for better analysis by eliminating errors by reducing human interactions. IoT re-defines health care in terms of monitoring,

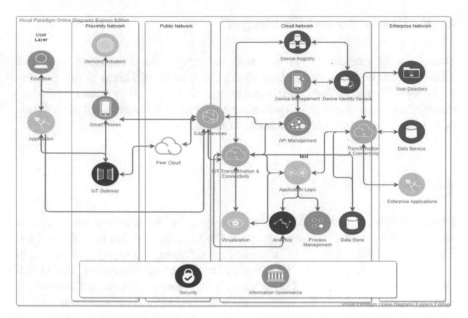

Fig. 1 IBM cloud architecture

diagnosis, treatment, reduces the rush in hospital/ clinics and reduces the medical cost too.

In the current scenario of COVID-19 pandemics, IoT in health care provides tremendous support to medics from monitoring patient remotely for collecting data from contactless sensors. Sensor innovations, upgrades in frameworks to accumulate and process information and reconciliation of artificial intelligence make advancements in medicinal services. More the involvement of artificial intelligence, big data with IoT is able to transform the critical aspect of modern health care in digital transformation. AI and big data, along with sensors, leads to a better understanding of therapy, provide a preventive measure and better decision making, which offer a better medical future. Even though having a significant future aspect of IoT in health care, there are several issues and challenges which need to be solved for better implementation. IoT devices generate a tremendous amount of data from billions of devices which communicated with each other over the network faces various issues and challenges like data management, data storage, security and threat concern, dynamic devices allocation, real-time health monitoring, exchanging of data in a common format. The next section is dealing with the various issues and challenges while incorporating IoT with medical sector.

4 Various Issues and Challenges

Since from the inception of IoT is booming in every sector, health care is the most promising and crucial sector to incorporate the Internet of Things to provide state of the art facility to doctors, patient and caregivers. IoT in health care offers a new diversion from hospital centric to patient centric irrespective of geographical location. Following are some significant issues and challenges.

4.1 Interoperability

Interoperability defined by IEEE as 'the ability of two or more systems or components to exchange information and to use the information that has been exchanged'. The interoperability issues in IoT can see from different perspectives due to heterogeneity in standard protocols. The primary purpose of the Internet of Things is to create an interconnected world in which various sensors, actuators and devices can communicate with each other dynamically and seamlessly. It expected that by 2025 around 50 billion devices are connected throughout the world over the network. Interoperability maintained in three levels, namely device level, protocol level and data level. Google has proposed a standard protocol as 'Weave' to create seamless communication between various devices. Samsung and Philips already adapt the weave for their smart devices. Open standards like Qualcomm's AllSeen Alliance, open thread, open connectivity foundation, IoTivity provide open-source software framework to communicate between various devices seamlessly (Fig. 2).

Levels of interoperability

1. Device level—This level maintains the software and hardware features of interconnection between various heterogeneous devices.
2. Protocol level—Permits interoperability between traditional and wireless communication technologies with the adaptation of multiple protocols.
3. Data level—Maintains various protocols to translate the data obtained from multiple sensors to a uniform format through data processing.

Followings are some significant issues and challenges regarding interoperability. Issues—

1. Building a reliable and scalable platform while communicating between heterogeneous platforms.
2. To develop a typical platform specification and standard device specification.
3. To define an open, optimal and light protocol for devices to communicate with each other.

Challenges—

1. To design flexible cross-domain interoperability.
2. To develop a group of standardized communication protocol.
3. To define a new computing paradigm for efficient and faster processing.

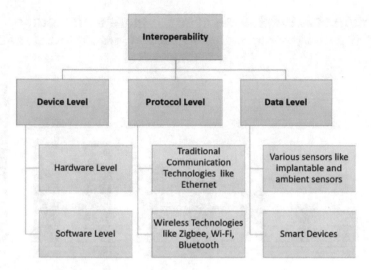

Fig. 2 Interoperability levels

4.2 Data Compression and Optimization

With approaching towards the development of smart health systems, IoT has an enormous effect over current advancements. However, it raises some issues as well, and one of the primary concerns of IoT ecosystem is data management and optimization that includes data storage, processing and optimizing which generated from patients' devices including wearables and contactless devices. Hossain et al. [15] discussed about data storage, data processing and optimization in cloud generated by IoT devices (Fig. 3).

Issues—

1. Integrating with cloud computing technologies to overcome the shortage of storage capacity, energy efficiency.
2. Efficient resource allocation and identifying management for data generated through billions of devices connected over the network globally.
3. Standard protocol support, deployment of IPv6,

Challenges—

1. Providing optimum data security with appropriate data optimisation.
2. Ensuring quality of service while maintaining data integrity.
3. The increase in the volume of data; existing methods need further advancement to cope with the future cloud-based architecture.

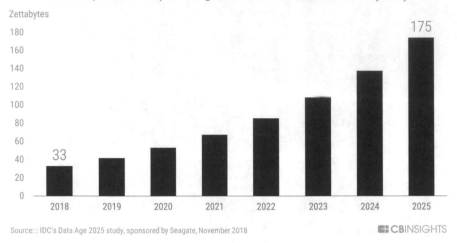

Fig. 3 Source—IDC's data age 2025 by Seagate Nov 2018

4.3 Issues and Challenges in Network

IoT evolution provides support to communication infrastructure which helps to evolve new aspect of smart home, smart cities, real-time logistic support, smart health services, smart retailing, intelligent traffic management, smart vehicles. However, evolution poses new issues and challenges to manage the network efficiently and securely. Routing, mobility, security of network devices, scalability and hetero-geneity of the network are major issues while implementing IoT in healthcare. As the system relies on various network architecture and topologies, IEEE 802.15.4 acts as one of the most fundamental technology for IoT, which ensures an effective routing mechanism for sending the data packet. According to healthcare data breach, statistics throughout 2009–2019 increases the risk of a data breach which causes a fatal effect on healthcare services from hacking health information of the patient to breaching pacemaker. Hacker can change the health-related data, which results in misdiagnosis or of the base evaluation of infections which prompts improper treat-ment and along these lines build death rate. The transmission of clinical data checked through IoT gadgets is vulnerable to security issues. Information security is consid-ered as a fundamental factor for medicinal services applications as it transmits the client's crucial and sensitive data.

Issues—

1. To provide the best routing techniques which reduce overheads.
2. To provide an optimized and efficient route for data mobility.
3. Adding dynamic nodes automatically using a secured channel.

4. Provide security for devices, data and sensors present in IoT network from various attacks.
5. Providing efficient and effective load balancing for inter-organizational communication with best architecture support for improved scalability.
6. To provide proper integration of various resource with robustness and trustworthiness and ensure energy conservation in the heterogeneous network.

Challenges—

1. The crucial concern is to convey elite applications in the heterogeneous system and accomplish a similar exhibition from these applications without adjusting system or network.
2. Providing dynamic routing mechanism for adding network dynamically and address reallocation.
3. Developing Zero trust model for packet-level authentication.
4. As new IoT devices, sensors and actuators keep on adding the business, it is getting essential to make sure about these unique access points consistently for security.

4.4 Implementing Personalized Health Care

Personalized medicine can mainly demarcate as a model of health care that is personalized, predictive, preventive and participatory Karthick et al. [16]. The patient's personal medical data comprises of personal information, brief of medical history, health conditions, diagnostic reports, and its related treatments are crucial and sensitive. While engaging the utility of IoT, empowered advances in personalized medicinal services have enormous predictable advantages. Jabeen et al. [17] proposed an IOT-based recommender system for cardiovascular disease. Nowadays, mobile technologies play an essential role in monitoring and maintain personal health care services. These include smartphones, mobile camera (Microsoft sense cam), individual air purifier, wearable devices, smart bands and gadgets having various sensors like EEG, EOG, accelerometer, etc., and biomedical sensors like a pulse oximeter, heart rate and ambient sensors like thermometer and hygrometer. They designed for providing personalized and continuous cares for users. Many smartphones and mobile devices have released to maintain patient's life logging data for better monitoring and treatment. Various applications that make use of multiple sensors are capable of avoiding undesirable circumstances by taking real-time sensing and calculation even though several issues and challenges like to maintain proper Personal Electronic Health Record (PEHR), hack-proof devices, sufficient data storage for a long duration.

Issues—

1. Transferring medical record into electronic form need much time to convert it from manual to automated mode
2. Required massive data storage for maintaining vast patient data.

3. Providing quality of services with low-cost devices.
4. Providing security and integrity to avoid misuse of data and devices.

Challenges—

1. Designing a lightweight scheme and algorithms to protect health records from unauthorized access.
2. Developing secures communication between personal IoT devices and remote access service station.
3. Using cryptographic techniques and SSL/TLS protocol to provide secure communication but requires more processing power and energy, which are not viable for IoT-based sensors, devices used in healthcare application.

4.5 Managing Security Threats

In IoT-based health care, patient's health data received from various sensors and devices transmitted through a wireless network like Bluetooth, Wi-Fi, ZigBee, WiMAX and Universal Mobile Telecommunication System (UMTS). Sensors and devices used in health care designed for communication; the data between sender and receiver over the internet exposed to a wide variety of threats. The significant risks on devices are Denial of Service (DOS), tampering, spoofing, data leakage, repudiation and exploitation of access privileged. The main aim of attacks is to disturb the normal functioning of the healthcare system by exploiting the vulnerabilities. Attacks are of two types of active attack and passive attack. Active attack is where attackers try to modify the data or content of the message. These affect the physical performance of the system. In a passive attack, attackers try to copy the data or message. They may use for malicious activity though system considered as compromised even though it causes no effect on physical performance.

Issues—

1. The primary issue is to maintain privacy as a massive amount of data exchange over the network globally.
2. Cyber-criminal uses the weakness of network and devices as exploit the user and user data.
3. Unreliable security threats may enable unauthorized access to devices.

Challenges—

1. Developing light weighted security protocols for devices and sensors.
2. Developing an algorithm for the dynamically allocated device during runtime access.
3. Secured system to be developed to overcome the physical damage of devices as those is supposed to operate in numerous situations.

4.6 Converges of AI, Big Data in IoT

With the context to the fourth industrial advancement named Industry 4.0, the IoT is fundamental to how cyber-physical framework and creation forms set to change with the assistance of artificial intelligence, big data and analytics. Continuous information from sensors, actuators, devices and other data sources helps to generate real-time data and helps to provide better 'decision making' and perform accordingly. According to Rabah [18] to deal with a tremendous amount of data is not as simple as it sounds. Information gathered progressively through IoT empowered sensors and devices can be analysed and segregated through portability arrangements powered by IoT. AI and big data will decrease the assortment of raw data and can drive immediate healthcare investigation, which reduces errors and helps in suggesting diagnosis, decision making. AI helps in providing predictive (Data mining, forecasting and recommendation) and prescriptive (Decision support, decision automation, simulation and auto diagnosis) analysis of data. Some significant issues and challenges must be deal with incorporating AI and big data are:

Issues—

1. Harvesting better insight medical intelligence and deduce earlier interventions.
2. Maintain a balance between privacy rights of the patient which are share through the network.

Challenges—

1. Developing an AI-based deep learning system to improve the accuracy and speed of diagnosis.
2. Implementing Natural Language Processing (NLP) methods to extract information from medical records.
3. Identifying patient symptoms and form a proper diagnosis and develop an expert system which suggests the course of action.

4.7 Inclusion of 5G with Health Care

Potential of fifth-generation wireless technology (5G) is to deliver new connections and facilities in applications and connectivity. 5G has the potential for the latest advancement in mobile technologies to create a revolution empowered by new cellular architectures, services, disruptive changes in network nodes.

The vision of global IoT network supports billions of devices achieved by enhanced 5G core, network slicing and non-public networks. Each communication that happens between devices and the cloud; a tremendous amount of data processing will be done immediately, will require an instant transfer of data between various methods over the network. In coming years 5G provides ample support to IoT in health care, where the requirement of transmitting collated information very quickly, that helps medics to take quick decisions from numerous sources of data coming through many technologies.

According to the International Telecommunication Union—responsible for radio communication (ITU-R) has defined 'three main application areas for the enhanced capabilities of 5G. They are Enhanced Mobile Broadband (eMBB), Ultra-Reliable Low-Latency Communications (URLLC) and Massive Machine Type Communications (mMTC). Only eMBB deployed in 2020; URLLC and mMTC taken quite a while to use in many areas'.

Main aim to develop 5G is to support wide range of devices, services and applications by extending its network capabilities to extreme performance. Machine type communication is a communication paradigm to connect numerous devices which are connected to network and communicate with each other with very little or no human intervention. In case of 5G, massive number of devices is connected to serve huge number of things hence called massive Machine Type Communication (mMTC). Still, developing countries like India have to wait for 5G technologies which include few issues and challenges for successful implementation.

Issues—

1. 5G technology needs higher frequencies about 300 GHz to operate as compared to 4G, which works on 6 GHz. 5G requires higher frequency which means higher cost.
2. Developing cost-efficient infrastructure.
3. Availability of 5G technology-based devices.

Challenges—

1. Creating smaller base station as 5G antenna is only able to beam out over the limited area.
2. Updating or replacing existed devices according to 5G technology in a cost-effective manner.
3. Creating digital safety network for devices and system.

5 Conclusion

Successful incorporation of the Internet of Things with health care ensures a tremendous effect on the world. To overcome the existing issues and challenges, organizations and researchers have initiated across the globe to explore the solution to improve the healthcare facilities—this paper emphases on a diversified aspect of issues and challenges in various aspects. Besides, this paper analyzed the different implementation aspect of IoT. This paper consolidates the multiple points on interoperability, network issue, security and threats. This paper also discusses the problems and challenges faced while implementing the latest technology of AI, big data and 5G. The result of this paper provides summarized issues and challenges in a single spot and assessed to be useful and highly competent to IoT platform provider, protocol designer, healthcare provider, researcher, scientists to endorse the global deployment of Internet of Things in the healthcare sector.

References

1. Pham M, Mengistu Y, Do H, Sheng W (2018) Delivering home healthcare through a cloud-based smart home environment (CoSHE). Fut Gener Comput Syst 81:129–140
2. Verma P, Sood SK (2018) Cloud-centric IoT based disease diagnosis healthcare framework. J Parallel Distrib Comput 116:27–38
3. Sandhu R, Gill HK, Sood SK (2016) Smart monitoring and controlling of pandemic influenza A (H1N1) using social network analysis and cloud computing. J Comput Sci 12:11–22
4. Nandyala CS, Kim HK (2016) From cloud to fog and IoT-based real-time U-healthcare monitoring for smart homes and hospitals. Int J Smart Home 10(2):187–196
5. Gia TN, Jiang M, Rahmani AM, Westerlund T, Liljeberg P, Tenhunen H (2015) Fog computing in healthcare internet of things: a case study on ECG feature extraction. In: 2015 IEEE international conference on computer and information technology; ubiquitous computing and communications; dependable, autonomic and secure computing; pervasive intelligence and computing. IEEE, pp 356–363
6. Somasundaram R, Thirugnanam M (2020) Review of security challenges in healthcare internet of things. Wirel Netw
7. Baali H, Djelouat H, Amira A, Bensaali F (2017) Empowering technology enabled care using IoT and smart devices: a review. IEEE Sens J 18(5):1790–1809
8. Woo MW, Lee J, Park K (2018) A reliable IoT system for personal healthcare devices. Fut Gener Comput Syst 78:626–640
9. Ullah F, Habib MA, Farhan M, Khalid S, Durrani MY, Jabbar S (2017) Semantic interoperability for big-data in heterogeneous IoT infrastructure for healthcare. Sustain Cities Soc 34:90–96
10. Firouzi F, Rahmani AM, Mankodiya K, Badaroglu M, Merrett GV, Wong P, Farahani B (2018) Internet-of-Things and big data for smarter healthcare: from device to architecture, applications and analytics
11. Hu Y, Duan K, Zhang Y, Hossain MS, Rahman SMM, Alelaiwi A (2018) Simultaneously aided diagnosis model for outpatient departments via healthcare big data analytics. Multimedia Tools Appl 77(3):3729–3743
12. Granjal J, Monteiro E, Silva JS (2015) Security for the internet of things: a survey of existing protocols and open research issues. IEEE Commun Surv Tutor 17(3):1294–1312
13. Al-Fuqaha A, Guizani M, Mohammadi M, Aledhari M, Ayyash M (2015) Internet of things: a survey on enabling technologies, protocols, and applications. IEEE Commun Surv Tutor 17(4):2347–2376
14. Noura M, Atiquzzaman M, Gaedke M (2019) Interoperability in internet of things: taxonomies and open challenges. Mobile Netw Appl 24(3):796–809
15. Hossain K, Rahman M, Roy S (2019) IoTData Compression and optimization techniques in cloud storage: current prospects and future directions. Int J Cloud Appl Comput (IJCAC) 9(2):43–59
16. Karthick GS, Pankajavalli PB (2020) A review on human healthcare internet of things: a technical perspective. SN Comput Sci 1(4):1–19
17. Jabeen F, Maqsood M, Ghazanfar MA, Aadil F, Khan S, Khan MF, Mehmood I (2019) An IoT based efficient hybrid recommender system for cardiovascular disease. Peer-To-Peer Network Appl 12(5):1263–1276
18. Rabah K (2018) Convergence of AI, IoT, big data and blockchain: a review. Lake Inst J 1(1):1–18

GFDM-Based Device to Device Systems in 5G Cellular Networks

K. Anish Pon Yamini, J. Assis Nevatha, K. Suthendran, and K. Srujan Raju

Abstract In order to satisfy the user desires, device to device communication concept is used. To increase the spectral efficiency, throughput and to bring down the energy requirement and latency through a technique of GFDM with device to device communication setup. To contribute the data over riotous transportation channel, low density parity check codes are employed for transmission over this pathway. The greedy scheduling algorithm is the perfect solution for the problem of resource distribution and intercession management. Since radio resources are finite and highly priced, so the device to device communication concept is used by New Radio (NR-gNodeB) access network. The advanced method-based on coded generalized frequency division multiplexing (GFDM) system produce in overpriced of emission.

Keywords GFDM · Low density parity check (LDPC) · Peak to average (PAPR) · Ratio · Proportional fair · Round robin

K. Anish Pon Yamini (✉)
Department of Electronics and Communication, Arunachala College of Engineering for Women, Manavilai 629263, Tamil Nadu, India
e-mail: anish.yamini@gmail.com

J. Assis Nevatha
Department of Electronics and Communication, Arunachala College of Engineering for Women, Manavilai, Nagercoil 629263, Tamil Nadu, India
e-mail: bousiyanivi@gmail.com

K. Suthendran
Department of Information Technology, Kalasalingam Academy of Research and Education, Krishnankoil 626126, Tamil Nadu, India
e-mail: k.suthendran@klu.ac.in

K. Srujan Raju
Department of Computer Science and Engineering, CMR Technical Campus, Hyderabad, India
e-mail: ksrujanraju@gmail.com

© The Author(s), under exclusive license to Springer Nature Singapore Pte Ltd. 2021 653
K. A. Reddy et al. (eds.), *Data Engineering and Communication Technology*,
Lecture Notes on Data Engineering and Communications Technologies 63,
https://doi.org/10.1007/978-981-16-0081-4_65

1 Introduction

It appreciates the idea of device to device (D2D) communication in 5G systems that could show a crucial function in abundant fields such as home domotics, smart portage, online legacy systems, detector systems, climate checking systems, etc. The OFDM is more delicate to offset sub bearer frequency, and it desire large count of amplifiers. To eliminate this causes, the orthogonality between sub-bearer are managed by simultaneous occurrence mechanisms. The Peak to average power ratio (PAPR) is raised by adding succession of sub carriers. It leads to ramification. In order to overcome this effects, advanced algorithm called greedy scheduling algorithm. This scheduling algorithm which accompany with the trouble-decipher heuristic to find the global optimum by making the problem locally optimum at each stage. The GFDM method is a multi-bearer variation approach as well as a module-based variation scheme. Round robin are the emerging among this scheduling algorithm but it fails to manage the fairness and because of the fade passage conditions, BCQI scheme is also not able to fit for emergency purposes.

2 Related Works

Orthogonal frequency division multiplexing (OFDM) systems with numerous input and output base strip processor are the tool used in the communication networks. The higher emission rate and less energy consumption are accomplished by employing the pipelined FFT infrastructure with multipath delay feedback (MDF) but it gives pattern complexity [1]. The flexible circular filter with multiple bearer systems that can add the baseline with generalized frequency division multiplexing (GFDM) are used to implicit the physical layer service for the erect work for all top layers of 5G cellular networks [2]. The quarter origination (4G) networks uses orthogonal frequency division multiplexing (OFDM) with universal filtered multi carrier(UFMC) system as their best choice but their variation technique tolerate from the problem of increasing Peak to average power ratio (PAPR) and discharge in side ring [3]. A high speed network services with wireless data association for high volume channels in the 5G wireless networks are provided by visible light communication. An intensity modulation is used by visible light communication system, and here, the message signal is passed through the intensity of LEDs, and at the recipient, they are indicated by photodiodes [4]. Based on numerous division, large number of carrier filter bank with GFDM are used, and it authorize the frequency and time expertise for multiple subscribers scheduling and can be achieved in digital form. The historic OFDM with distinct new advantages such as the drop in PAPR level [5]. The traffic path in cellular networks had an improvement with rise in information rate and trustworthy inclusion to mobile subscribers. Then, the transportation sensors demand for extreme-low energy consumption and also dominate application which need very short feedback times [6]. Inspect the pattern of wireless deportation and the knowledge contributor that can be instantly delivered under advance consideration for remodeling vitality expertise in the wireless deportation design [7]. Escalation extended craft have circumstances of superlative versatility and shallow turbulence ratio; then, the collective entry accomplishment is disgraced. Density admission is independence [8]. Historical refiner bank with GFDM can also be enforced numerically. Demoralization is routinely disputed for numerically to manipulating the range with extremely pale to designed range for density strap [9]. On the account of the preferred coarse standby slab and margin gathering, vitality is deported from orgin bud to pass on bud in the IOT structure with reformed affability of wealth approximation [10]. Inter symbol intervention temper the OFDM precipitate by detain stretch of radio medium. This enriched the medium reckoning, authorization disclosure, counterbalance of time and counteract of frequency [11]. The display integrate is used to apprise the individual to be alcoholic. The operating system used is open CV library. It is fast but it induces varying disputes [12]. IOT tools are bared to the liabilities which is right to be insulated. SDN is used as a operating system, and it is slowdown in latency and finite resources [13]. The expansion threshold scheme arrangement is used to find the metamorphic style. This style yield treasure information but it takes huge reckoning time [14]. The numerous paradigm is used to scrutinize the CK metric. It takes huge turn to compile the information, immense slot and implantation cost [15].

3 Proposed Method

In this system, two base station are used for the device to device communication concept. The advance system has two modules of g NodeB and D2D systems. In this system, k number of D2D users are located in any desired sequence and the transportation between the end users are of inhibited. Here, the immense information rate arriving from each and every user is approved to travel from the contributor over the cipher. For encoding process at contributor, this system uses low density parity check codes (LDPC) for verifying symbols. A LDPC is a linear lapse rectifying codes which is used for both cipher and decipher sequence. In order to transmit the data over riotous transportation channel, low density parity check codes were employed for this transmission and also used to fetch the authentic information in the recipient part. LDPC codes were greater trustworthy and terribly efficient deportation of data over frequency range deprived links. A sparse Tanner graph is used to create a LDPC codes. Successive interference cancellation is done by decoding the stronger signal first at the receiver, subtracting it from the added signal and then decoding the difference. The decoding difference signal acts as a weaker signal. Initially, numerous symbols per sub bearer are transmitted by GFDM from the contributor, and the GFDM modulator proceeds the module by module operation. In the raised pulse shape cosine filter, symbol error rate are based upon the fading specification and roll off component. In this system, the out of block data emission of the incoming symbols are restrained by a flexible beat creation refiner that is directly sent to one by one all sub bearer. Hence, for entire frame work, system uses a single cyclic prefix. A circular convolution of each and every sub-bearer of GFDM with an alterable pulse shaping filter including a module of information symbol is permitted through per sub bearer. A cyclic prefix with entire GFDM frame to provide drop down in complication of frequency expertise equalization. Still unexpected self-interference is not produced in the orthogonal pulse shaping filters but it is included by linear iterative receiver. While downlinking the combined D2D and New Radio, this system concentrates only on intervention. The various interferences such as efficiency deterioration are possible since these two systems mutually share the identical radio channels. The collection of this suitable arranging designs adds round robin algorithm, greedy algorithm, and proportional fair algorithm can be relieved by its consequence. At first, channel circumstances is derived from SINR and data rate in each channel. In both contributor and collector end, the channel state information (CSI) is considered to be a perfect state. The channel is splitted into frames for the distribution of resources. A number of sub-frames are available in the each frame of the system. The transmission time interval (TTI) is the indicator of life span of each and every frame in this system. The process followed in greedy arranging pattern are,

Act 1 Enumerate the count of sub-bearers distributed for individual server.

Act 2 Authorize the sub-bearer for each and every servers.

Act 3 Nominate energy usage to individual server's sub bearers based upon their energy restraints.

Examine the accessible R wireless possessions which are distributed for K servers to which they were required. Allow ck, r are be the sub-bearer allotment index. Consider Φk, $capk$ and Dk, r to be the scale rate constant, volume, and spontaneous information tariff of server *at r* th wireless management for k mutually. For suitable allotment of resources and the constraints are,

$$\begin{cases} c_{k,r} \in \{0.1\} \text{ for all } k.r \\ \sum_{k=1}^{K} \Phi_k = 1 \end{cases} \tag{1}$$

When wireless possessions (R) is greater than the count of servers (k), then the minimum reuse of repetition probability, and maximum gain is achieved by unallocated sub-bearer R for individual server. This is mentioned in the act of,

$$r = \arg\max_{r=1,2,\ldots R} |H_{k,r}|; \begin{cases} k = 1, 2, \ldots K \\ r = 1, 2, \ldots R \end{cases} \tag{2}$$

where K is the wireless possessions and R is the count of servers. It is in the act of,

$$\acute{k} = \arg\min_{k \in K} \frac{cap_k}{\Phi_k} \tag{3}$$

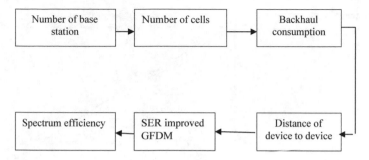

Fig. 1 System model

Now, based upon the greedy policy, the sub-bearer are empower to each and every user. The wealth of GFDM intensify the waveshaped scary possessions, SER upgraded accomplishment, diversification craft, and preface placed integration (Fig. 1)s.

Greedy scheduling algorithm intensify only on current state of the system but not on the forthcoming state of the system. The inverse discrete fourier transform (IDFT) and discrete Fourier transform (DFT) matrix can drop the pattern complication of the gadget. The ratio of subscriber having smallest capacity to its scale rate constant is calculated to denote the desire of the user. Discrete Gabor transform (DGT) and IDGT can drop the inter symbol intervention and inter channel intervention causes high.

4 Experimental Result

Millimeter wave link is formed between two base stations. Initially, link is formed between the base station, and spectrum efficiency are increased, and time complexity and power consumption are reduced. Backhaul traffic for all the substation are considered, and they are minimized to form an optimal solution.

Figure 2 shows that the model of 5 G networks between millimeter wave link and the number of users are connected together in the area of millimeter wave called cluster. Each distributed network consists of various path to transmit the given data to each nodes. Very shortest routing path is chosen as a best link for transmission for the network model.

This Fig. 3 shows the comparison of numerous users in which few of users are bring together to design a cluster, and also they form a new millimeter stream link. Since large number of users are grant to approach the identical system spontaneously, successful scheduling schemes are used to catch out the width between the base stations. The distance of device to device is chosen to be very less.

Figure 4 show the symbol error rate of the system are correlated between the two chosen base stations. Round robin algorithm induces improved fairness but drop in

Fig. 2 Network device
connection

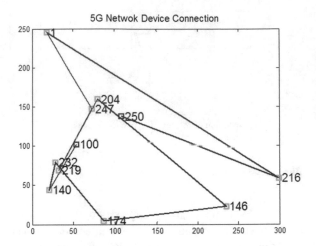

Fig. 3 Distance of device to
device

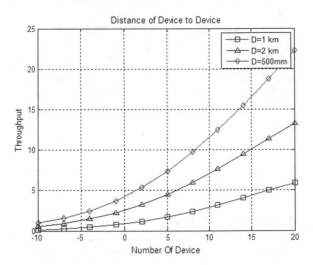

throughput, but proportional fair algorithm which produces good fairness and high
throughput and best medium capacity. In the system, the priority is identical for all
the servers. Only a firm of users are used for the scheduling algorithm scheme.

Figure 5 shows improved speed and spectral efficiency are attained in the GFDM
system. The average throughput of the base station are compared with their signal
interference to noise ratio. Fulk disk encryption shows least efficiency. The comparison of SINR and average throughput between the devices are produced maximum
overpriced emission.

Fig. 4 SER-based GFDM
system

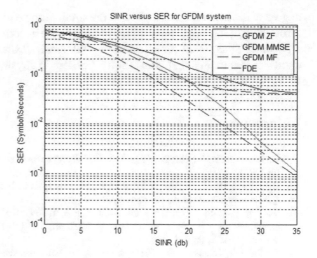

Fig. 5 G network spectrum
efficiency

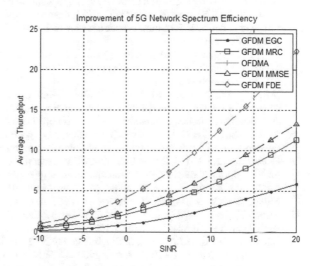

5   Conclusion

GFDM with D2D in 5G NR access network provide a good way to approach the spectrum in the communication system when they exists intervention. When a common authorized spectrum can be freely accessed by two contrast systems, then there produce an interference. These spectrum selection of the system is depends upon their scheduling algorithm. At start, coded GFDM with performance of symbol error rate with distinct recipient architecture are examined practically and arithmetically. Minimum mean square error of FDE with GFDM system equalizer are the perfect selection of system which are approximately error free during communication. In next generation, NOMA is ideally fits for the advanced new system with higher degree of spectral efficiency. Finally, the simulation results shows that high degree of throughput, good fairness, and drop in latency are accomplished by proportional fair algorithm. Thus, D2D communication with small desires systems are the immense choice to launch in smart cities for next generation systems.

References

1. Gnanishivaram K, Neeraja S (2014) FFT/IFFT processor design for 5GMIMOOFDM systems. Int J 3(3)
2. Kansal P, Shankhwar AK (2017) Performance analysis of FBMC-OQAM based 5G wireless system using PAPR. Int J Comput Appl 161(12)
3. Vakilian V, Wild T, Schaich F, ten Brink S, Frigon JF (2013) Universal-filtered multi-carrier technique for wireless systems beyond LTE. In: 2013 IEEE Globecom workshops (GC Wkshps). IEEE, pp 223–228
4. Sindhuja R, Shankar AR (2016) A survey on VLC based massive MIMO-OFDM For 5G networks. Int J Electr Electron Comput Sci Eng Special Issue NEWS, pp 2348–2273
5. Wu J, Ma X, Qi X, Babar Z, Zheng W (2017) Influence of pulse shaping filters on PAPR performance of underwater 5G communication system technique: GFDM. Wireless communications and mobile computing
6. Li Y, Tao C, Seco-Granados G, Mezghani A, Swindlehurst AL, Liu L (2017) Channel estimation and performance analysis of one-bit massive MIMO systems. IEEE Trans Sig Process 65(15):4075–4089
7. Melgarejo DC, Moualeu JM, Nardelli P, Fraidenraich G, da Costa DB (2019) GFDM-based cooperative relaying networks with wireless energy harvesting. In: 2019 16th international symposium on wireless communication systems (ISWCS) . IEEE, pp 416–421
8. Yang Y, Zhu L, Mao X, Tan Q, He Z (2019) The spread spectrum GFDM schemes for integrated satellite-terrestrial communication system. China Commun 16(12):165–175
9. Fettweis G, Krondorf M, Bittner S (2009) GFDM-generalized frequency division multiplexing. In: VTC Spring 2009-IEEE 69th vehicular technology conference. IEEE, pp 1–4
10. Na Z, Lv J, Zhang M, Peng B, Xiong M, Guan M (2019) GFDM based wireless powered communication for cooperative relay system. IEEE Access 7:50971–50979
11. Hwang T, Yang C, Wu G, Li S, Li GY (2008) OFDM and its wireless applications: a survey. IEEE Trans Veh Technol 58(4):1673–1694
12. Patnaik R, Krishna KS, Patnaik S, Singh P, Padhy N (2020) Drowsiness alert, alcohol detect and collision control for vehicle acceleration. In: 2020 international conference on computer science, engineering and applications (ICCSEA). IEEE, pp 1–5
13. Patnaik R, Padhy N, Raju KS (2020) A systematic survey on IoT security issues, vulnerability and open challenges. In: Intelligent system design. Springer, Singapore, pp 723–730
14. Padhy N, Panigrahi R, Neeraja K (2019) Threshold estimation from software metrics by using evolutionary techniques and its proposed algorithms, models. Evolutionary intelligence, pp 1–15
15. Padhy N, Satapathy SC, Singh RP (2019) Complexity estimation by using multiparadigm approach: a proposed metrics and algorithms. Int J Network Virtual Organ 21(2):201–220

A Mobility Adaptive Efficient Power Optimized Protocol for MANETs Based on Cross-Layering Concept

K. Anish Pon Yamini, K. Suthendran, and K. Srujan Raju

Abstract Mobile ad-hoc network (MANET) is a network without base station framework progressively shaped via self-sufficient arrangement of portable mobile nodes that are associated by means of wireless links. Battery operated mobile nodes have bounded lifetime. Energizing or substituting a mobile node is not constantly likely. In the event of depleted battery power of nodes, connection between the nodes get suspended which affects the overall network lifetime. Energy is an extremely urgent parameter in mobile ad-hoc networks that should be productively utilized. Mobility and battery power enforces a significant challenge to satisfy the forthcoming demands. Our suggested mobility adaptive efficient power protocol for MANETs based on cross layering helps to conserve the energy, and so, the outright performance is found to be better.

Keywords Ad hoc · Power conservation · Battery life

1 Introduction

Ad-hoc networks are an assortment of wireless nodes built in the absence of any infrastructure. These battery regulated wireless hosts have a defined lifetime [1]. In ad-hoc networks, responsibilities like forwarding, relaying is functioned by the nodes themselves in addition to their own assignments. This induces extra battery

K. Anish Pon Yamini (✉)
Department of Electronics and Communication, Arunachala College of Engineering for Women, Manavilai, Tamil Nadu 629263, India
e-mail: anish.yamini@gmail.com

K. Suthendran
Department of Information Technology, Kalasalingam Academy of Research and Education, Krishnankoil, Tamil Nadu 626126, India
e-mail: k.suthendran@klu.ac.in

K. Srujan Raju
Department of Computer Science and Engineering, CMR Technical Campus, Hyderabad, India
e-mail: ksrujanraju@gmail.com

© The Author(s), under exclusive license to Springer Nature Singapore Pte Ltd. 2021   661
K. A. Reddy et al. (eds.), *Data Engineering and Communication Technology*,
Lecture Notes on Data Engineering and Communications Technologies 63,
https://doi.org/10.1007/978-981-16-0081-4_66

drain prompting a lessened lifetime. A significant constraint with these nodes is that they have high mobility, making links be as often as possible broken and that needs to be restored [2]. At the point when an overwhelming traffic is transmitted over a path, a critical node on the way may deplete its battery and this excessively used node may ends its activity which influences the network connectivity and decreases the lifetime. Existing protocol additionally requires noteworthy changes to adapt to the difficulties and objectives of MANET [3]. We made an endeavor to limit the over usage of a node by proficient battery and transmit power management. Power conservation concept idea must be stretched out from layer 1 to layer 4.

2 Cross-Layering Concept

The greater parts of research endeavors are engaged toward lessening inter reliance among layers, which is a major aspect of the layered design. The cross-layer design configuration is a rising concept for all types of networks. The cross-layer design approach empowers communications across layers. The concept behind this methodology is passing the information between two or more layers in a non-layered approach to enhance the performance characteristics. More than 2 layers are crossed based on the client prerequisites. For instance, so as to achieve better route selection, the third layer and the application layer may be cross layered. Similarly, priority scheduling and better QoS attainment are achieved by cross-layering second layer and third layer. There is a possibility of six cross-layer design models based on information flow direction [4].

3 Associated Work

To accomplish power conservation, many research works were suggested so far. In this paper, a study of some present writings to improve the power conservation by various cross-layering methods in MANET were talked about. In [5] an analytical model to evaluate the detection time was proposed to improve the performance of MANET thereby minimizing the traffic overhead. The reliable path is determined from the paths available, and a data packet is transmitted when the path is reliable. It was absorbed that the residual energy generates network lifetime. The RERMR was found to be efficient when compared with open metadata repository services (OMRS), RMRBDB, SRMP, and on-demand multicast routing protocol (ODMRP), it can be applied for real-time applications in military search and rescue operations. The effect of physical layer on five routing protocols were examined in [6]. The paper reasoned that instead of considering hop count for routing procedures, link quality and present channel state may be considered to be ideal alternative. The power-aware routing optimization (PARO) and zone routing protocol (ZRP) were proposed for controlling the transmission power of packet. The routing protocol chooses the appropriate intermediate nodes by the number of zone. In PARO algorithm, intermediate nodes termed as the redirectors are selected for the forwarding of data packet for a source destination node pair, thereby minimizing the transmission power consumption of wireless node. The radios are incorporated in the PARO algorithm that adjusts the transmission power between the nodes in a dynamic manner [6]. Congestion control in fourth layer and power control adjustment in the first layer improves multi-hop communication which has been investigated in [7]. The main idea of JOCP is that due to the transmission power updatement at each time slot, the bottleneck links will attempt to transmit packets quicker. The transport layer performance degradation by congestion has been explored in [8]. The proposed method improves performance by collecting delay and bandwidth information from link layer which is achieved by incorporating an extra module in the protocol stack. The cross-layer design approach by Patil et al. [9] suggested comparing the power based on a threshold value. In case the value less than the threshold, then MAC layer indicates the network layer to remove that corresponding node from the routing table.

4 Mobility Adaptive Power Optimized Protocol

In this part, we exhibit a review of our proposed work. Control messages are exchanged by almost all the routing protocols to explore neighbors. We plan a protocol that uses ideal power for control and data. When an active node broadcast RTS control message, it appends its transmission power. The three folded protocol is explained as follows: step one starts with the assumption that there is sufficiently adequate power level to reach immediate neighbor or the destination node, and also, each node is assumed to know its speed information [10]. RTS frames are disseminated by the source node with predefined fixed least power level P_{RTS}. Receiving node estimates signal strength and passes to the routing layer. Next stage is the sleep scheduling stage for idle nodes. Distance between the transmitting and receiving node is calculated using signal strength. We consider the two ray propagation model to estimate power P_{mn} when node m receives a message from node n and n's maximum transmission power P_{max} which corresponds to 10 dbm is known to m. This reference model depends on the following condition with a path loss value of 2:

$$P_r = P_t \left(\frac{\lambda}{4\pi d} \right)^2 g_t g_r \tag{1}$$

where P_r and P_t represent the receiving and transmitting power, respectively, λ to the carrier wavelength, d to the distance between the sender and the receiver nodes, and g_t and g_r equals to the antenna gains at the sender and receiver, respectively. Here, we consider only unity gain antenna. Therefore,

$$P_r = P_t \left(\frac{\lambda}{4\pi d} \right)^2 \tag{2}$$

Distance from the transmitting node is calculated using the formula

$$P_r = \left(\frac{P_r}{P_t} \right)^{1/2} * \left(\frac{\lambda}{4\pi} \right) \tag{3}$$

The transmission range must be sufficiently high to keep up the availability of links to stretch the network lifetime. Given the maximum transmission range as 350 m (Tr_{max}), the receiving node calculates its distance [11] to the sending node's transmission range boundary as

$$T_b = Tr_{max} - d \tag{4}$$

The least time required for a node to move out of the transmission range boundary of the transmitting node depends on the speed and its corresponding distance from the boundary as given underneath.

$$t_b = T_b / \text{Speed} \tag{5}$$

Once RTS frame is received, the receiver calculates the path loss and P_{Tmin} [12]. As referenced already [1], battery energy becomes depleted because of the various energy usage modes. Node life could be expanded taking account of the remaining battery resources. Reluctance [13] demonstrates the residual energy of the battery as it indicates disfavor of a node to take an interest in routing because of lower battery levels. Participation in data transmission highly depends on the remaining battery energy. Remaining battery (Rb) is quantized into ten likely levels and is loaded on all the nodes participating in routing as a static table.

$$(0 \leq Rb < 4) \text{ Battery Level low(Bl)} \tag{6}$$

$$(4 \leq Rb < 9) \text{ Battery Level Sufficient (Bs)} \tag{7}$$

P_{Tmin} and the Rb values are piggybacked to the sender by CTS. Based on the values of Rb, a node may effectively partake in routing else made to sleep. Virtual carrier sensing is improved in our protocol by minimizing the over hearing power consumption by powering off the NIC for the duration of data transfer.

5 Analyzing Performance and Results

For evaluating the problem, we use the simulator Ns-2.35. Assume 100 uniformly distributed nodes in a simulation area of $1200 \times 1200\,\mathrm{m}^2$. Random way point mobility is used. To make the analysis easy, we do not concentrate on the overhead.

We compared the performance of our proposed mobility adaptive protocol against power controlled MAC protocol. It is assumed initially the transmission power is 10 dbm with a transmission rate of 0.1packets/s, receiver threshold set to −81 dbm, carrier sense threshold equal to 1.92278e−08 W for 250 m. We carry out various experiments to evaluate average energy consumption, PDR, and delay. Figure 1 depicts the interrelation of power consumption versus node density; Fig. 2 depicts the interrelation of PDR versus node density, and Fig. 3 depicts the interrelation of average packet delay as a function of speed.

Route re-discoveries are provoked by broken links followed by routing updates. This causes considerable delay in data transfer between the sender and the receiver. From the graphs, it is noticeable that our approach outflanks the current one in power consumption by 20%.

In the 1200 m × 1200 m terrain, the total numbers of nodes are varied upto 150 nodes. We assume 40% nodes with low power. The existence of just 10 nodes in the

Fig. 1 Comparison of avg energy consumption

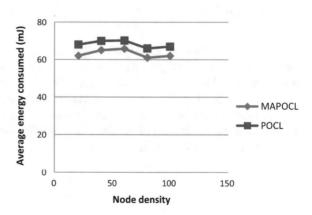

Fig. 2 PDR comparison

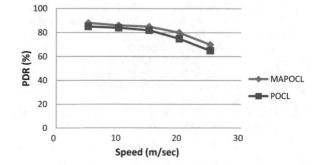

Fig. 3 Delay comparison

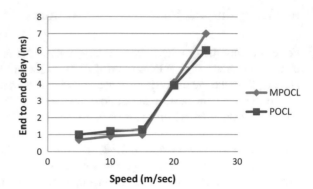

assumed terrain is not adequate to give sufficient connectivity. Under both protocol variations, this is reflected as poor packet delivery ratio. Observation shows that as the nodes increased more than 20 performance of POCL improves a little whereas the MAPOCL performs better with 25% PDR improvement and decreases the delay by 5% where network connectivity is ensured with redundantly accessible links.

6   Conclusion

The proposed mobility adaptive power optimized cross-layer protocol optimizes the network life time which is very significant in any resource constrained environment. Transmission power control reduces the CTS and the ACK retransmission number, considerably improving the delay performances. We note that the performance improvement is achieved mainly in highly dense network. Hence, our future work is to study the impact of overheads on mobility adaptive cross-layer protocol and build a simple and sufficient design.

References

1. Yamini KAP, Suthendran K, Arivoli T (2019) Enhancement of energy efficiency using a transition state mac protocol for MANET. Comput Netw 155:110–118
2. Hasan NU, Valsalan P, Farooq U, Baig I, On the recovery of terrestrial wireless network using cognitive UAVs in the disaster area
3. Patnaik R, Padhy N, Raju KS (2020) A systematic survey on IoT security issues, vulnerability and open challenges. In: Intelligent system design. Springer, Singapore, pp 723–730
4. Kliazovich D, Granelli F (2006) Cross-layer congestion control in ad hoc wireless networks. Ad Hoc Netw 4(6):687–708
5. Hernandez-Orallo E, Serrat MD, Cano JC, Calafate CT, Manzoni P (2012) Improving selfish node detection in MANETs using a collaborative watchdog. IEEE Commun Lett 16(5):642–645

6. Doss R, Pan L (2006) A case for cross layer design: the impact of physical layer properties on routing protocol performance in MANETs. In: ATNAC, Australian telecommunication networks and applications conference. Australian Telecommunication Networks and Applications Conference, pp 409–413

7. Chiang M (2004) To layer or not to layer: balancing transport and physical layers in wireless multihop networks. In: IEEE INFOCOM 2004, vol 4. IEEE, pp 2525–2536

8. Kim KJ, Koo HW (2007) Optimizing power-aware routing using zone routing protocol in MANET. In: 2007 IFIP international conference on network and parallel computing workshops (NPC 2007). IEEE, pp 670–675

9. Patil R, Damodaram A, Das R (2009) Cross layer AODV with position based forwarding routing for mobile adhoc network. In: 2009 fifth international conference on wireless communication and sensor networks (WCSN). IEEE, pp 1–6

10. Patnaik R, Krishna KS, Patnaik S, Singh P, Padhy N (2020) Drowsiness alert, alcohol detect and collision control for vehicle acceleration. In: 2020 international conference on computer science, engineering and applications (ICCSEA). IEEE, pp 1–5

11. Padhy N, Panigrahi R, Neeraja K (2019) Threshold estimation from software metrics by using evolutionary techniques and its proposed algorithms, models. Evolutionary intelligence, pp 1–15

12. Yamini KAP, Suthendran K, Arivoli T (2019) Power optimized cross layer design based protocol for performance enhancement in MANETs. In: 2019 IEEE international conference on intelligent techniques in control, optimization and signal processing (INCOS). IEEE, pp 1–4

13. Yamini KAP, Arivoli T (2013) Improved location-free topology control protocol in MANET. In: 2013 international mutli-conference on automation, computing, communication, control and compressed sensing (iMac4s). IEEE, pp 835–838

Evaluating the AdaBoost Algorithm for Biometric-Based Face Recognition

B. Thilagavathi, K. Suthendran, and K. Srujanraju

Abstract Adaboost algorithm is a machine learning for face recognition and using eigenvalues for feature extraction. AdaBoost is also called as an adaptive boost algorithm. To create a strong learner by uses multiple iterations in the AdaBoost algorithm. AdaBoost generates a strong learner by iteratively adding weak learners. To create a strong classifier using several classifiers while training the data set, a new weak learner is adding together, and a weighting vector is adjusting to focus on examples that were misclassifying in prior rounds. The analysis of face recognition has been widely used for any application. As per the literature survey, many algorithms have been developed to recognize the face. The AdaBoost algorithm is used for increasing detection accuracy and easy to develop. Therefore, this paper evaluating a preprocessing, feature extraction by using eigenvalues, classifier for classifying a face or non-face by using an AdaBoost algorithm.

Keywords Biometric · Weak learner · Adaboost algorithm · Eigenvalues

1 Introduction

This paper gives detailed information about the design of a robust classification system for image face recognition in a changing surveillance environment [1]. The performance of systems is disturbing by the variations of pose, illumination, blur, occlusion, and aging by giving limited control [2, 3]. The representativeness of models, the number, irrelevant, and array of this sample gives the accuracy of face classification. In recent research to observe the face, detection is done by identifying

B. Thilagavathi (✉) · K. Suthendran · K. Srujanraju
Krishnankoil, India
e-mail: sripranesh2010@gmail.com

K. Suthendran
e-mail: k.suthendran@klu.ac.in

K. Srujanraju
e-mail: ksrujanraju@gmail.com

© The Author(s), under exclusive license to Springer Nature Singapore Pte Ltd. 2021   669
K. A. Reddy et al. (eds.), *Data Engineering and Communication Technology*,
Lecture Notes on Data Engineering and Communications Technologies 63,
https://doi.org/10.1007/978-981-16-0081-4_67

and recognizing the human faces, and different algorithms [4, 5] are used to depend on size environmental conditions and position [6]. For face detection in images, there are numerous approaches were used. Earlier research on face detection based on color and texture [7]. Due to the complexity of the real world, these methods were breaking down easily. For face detection, Viola and Jones are popular based on statistic methods. Viola and Jones used the AdaBoost algorithm variant for fast face detection [1]. To overcome these problems, the AdaBoost learning algorithm using eigenvalue for feature extraction is used. However, eigenvalues check all pixels in the images, so face detection requires considerable computation power [8]. By using high-performance computers, the real-time face detection is promising, the system incline to be cartelized by face detection. This is the disadvantage of face detection for real-time applications. Image classification is a challenging process in image processing. Based on the features, the input images are classified. To solve this problem, the following algorithms are used like artificial neural network (NN), adaptive boost (Adaboost), k nearest neighbor (K-NN), and support vector machine (SVM) [6]. An iterative learning algorithm is used to generate a strong classifier from a weak classifier. The algorithm is used to train the dataset. During every iterative step, the minimum classification error is selected by a "weak" classifier [9, 10]. The features seek by the detection framework involve the sum of image pixels within rectangular areas. To compose the feature vector, feature extraction is used from the image [11]. The feature extraction method represents two types; they are the holistic category and local category. By using image feature extraction, the relevant data are extracted from the captured images. The holistic feature category and the local features category are the two categories in the feature extraction [12, 13]. To locate exact facial features such as eyes, nose, and mouth based on the distances between them. The holistic feature approach deals with the whole input face image [14, 15]. Different methods are used for extracting the identified features of a face. Principle components analysis (PCA), which is referred to as the eigenface method [16]. If find the principal components of face images and project the face images onto these principal components, then its facial expressions and random pixel noise will be removed, and it will be easier to identify the face. Whiten an image is to multiply by $UT(S + \lambda I) - 0.5U$ where U is the eigenvectors of the covariance matrix, the diagonal matrix represents S [17, 18], and its having eigenvalues λI is regularization to prevent things from blowing up.

2 The Proposed Method

2.1 Face Detection AdaBoost Algorithm

AdaBoost is used for face detection. Its boosting algorithm which is represents an adaptive boosting algorithm. Adaboost algorithm is a learning algorithm, normally used as a learning algorithm. The training data set the threshold value to a weak classifier. If the threshold [19] is greater, the new component classifier is added. Every training process, classifier adds weight. These weights determine for training data set. The subsequent classifier value is reduced and used again in the training data set. The training set is uniform, the image is not classified accurately [20]. Every iteration k, these weight added and the training set randomly. Next, increase the weights of the training set. The new training set distributed and used for the new classifier and iterated again. The component classifiers are given by the weighted sum of outputs. The final decision was taken by test point x which depends on a discriminant function [21, 22]. These weights uniformly distributed across the training set. At each iteration ofk, the component classifier on the images selected is trained again. To detect the face, locate the faces in still images. The research of computer vision has achieved abundant success [1–3], and reviews are agreed by Yang [14] and Zhao [15]. Among the face detection algorithms, the AdaBoost [6]-based method which is proposed by Viola and Jones [8] gained great popularity due to low complexity, high detection rate, and solid theoretical basis. Due to the use of eigenvalue features, the AdaBoost method is mainly used which excludes most of the image window hypotheses quickly. An auxiliary image is deliberate from the input image [23] in the pre-processing stage. The summation of a pixel of original image (Io) is calculated using Ii in any rectangle [20, 24]. Io can be calculated in constant time and represent the value (i, j).At all positions and all scales, the classifier considers the candidate image window (w). h(w) represents classifier response at every iteration. The feature responses hj(w) is calculated from the sum of the series of

$$h(w) = \sum_{J=1}^{n1} hj(w), hj(w) = \begin{cases} \alpha j_2 \\ \alpha j_2 \end{cases} f_i(w) < tj \text{ otherwise}$$

The feature weight coefficients are $\alpha j1$ and $\alpha j2$. The feature response represented by $fi(w)$ and calculated from the feature weight response. If the threshold value t exceeds the $h(w)$, the candidate window w will infer the non-face and remove the unwanted pixcl [25]. The next classifier is defined from the previous one. Multiple detections refer the Viola and Jones [9] and Lienhart [23]. For low execution time, Adaboost-based face detection is used and can be modified for efficient hardware implementation.

2.2   Feature Extraction

Here, λ is a scalar, which is known as the characteristic value, eigenvalue. The direction will be reversed if the eigenvalues are negative. In quintessence, linear transformation T is a non-zero vector and an eigenvector (v). The equation is

$$T(V) = \lambda v$$

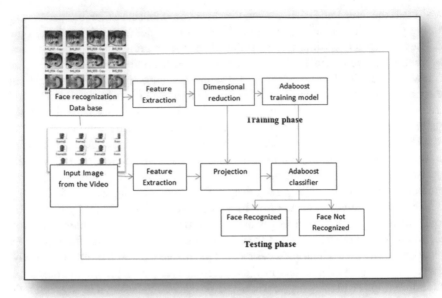

Fig. 1 Proposed method for face recognition

In general, λ is a scalar. For example, λ value is negative, the eigenvector reverses the direction. As part of the scaling, it may be zero (Fig. 1).

2.3 Dimensionality Reduction

By using the principal component analysis to simplify the problem, the image quality is used to find the symmetric [26] and high dimensionality of data that lend to sizeable dimension reductions [27, 28]. By projecting each data point onto the n greatest eigenvalues of the covariance matric is attained by reducing to n dimensions.

$$Y_i = v^t (x_{i-m})$$

Here, X_i is a data point in the original high dimensional space, M is mean matrices, and V is a matrix of n greatest eigenvectors. Let a reduction of a 256 dimension image to 20, 30, and 2 dimensions, respectively [29, 30]. The 2D projection is similar to the mean zero with projection 10, and 30 eigenvectors are increasing the original zero image.

3 Result and Discussion

Training is carried from the database created. The target image, obtained from the 830 MB images is compared with the trained images. If the features of the target image match with the trained images, the algorithm outputs "face matched" else "face not matched." A 24 × 24 pixel image is used for face recognition. To fit a classifier on the trained dataset and matches the input frame, an AdaBoost classifier which is a meta-estimator is used. The classifier on the same trained dataset also adjusts the weighting vector of incorrectly classified illustrations with those subsequent classifiers. If the testing image features matches with the trained dataset images, then it will display face recognition or face not recognized (Figs. 2, 3, 4, 5, 6 and 7).

Fig. 2 Target image features matched with reference images means then it will display face matched. **b** target image features matched with reference images means then it will display face matched

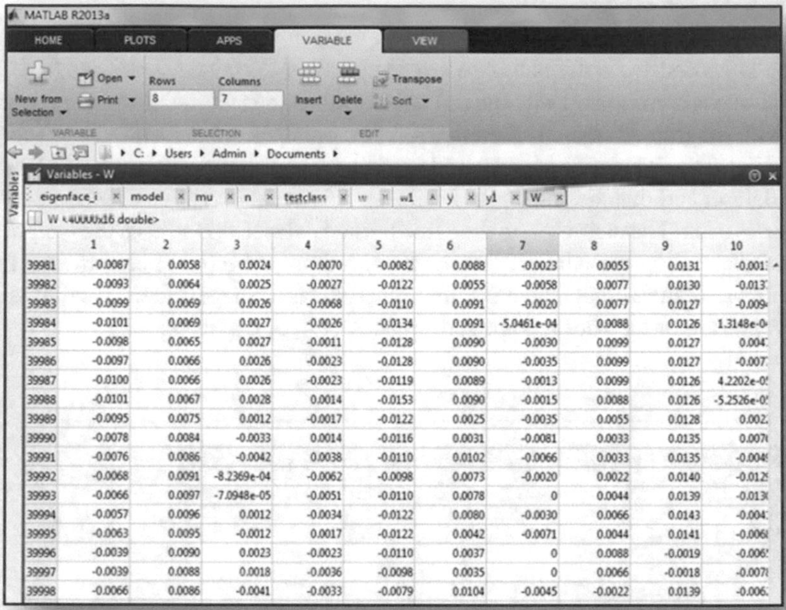

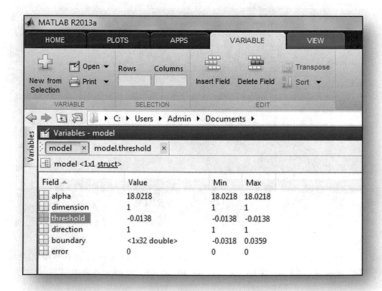

Fig. 3 Weighed the average of the weak classifiers

Fig. 4 Threshold value assigned by the algorithm

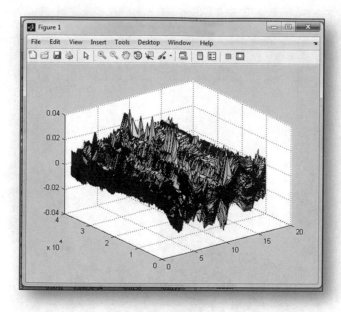

Fig. 5 Target image features matched with reference images means then it will display face matched

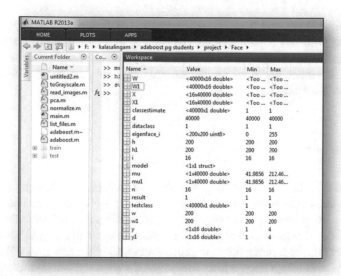

Fig. 6. Output values by Adaboost algorithm

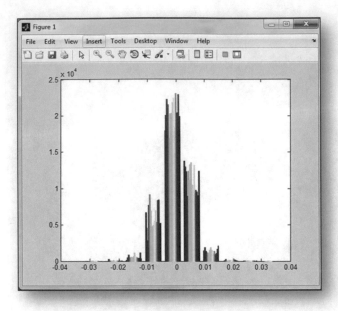

Fig. 7 Histogram graph for W value

It was coded in MATLAB, due to the estimation of the cost of design. MATLAB is a low cost with less capability than the Verilog and GPGA family. Image pre-processing can notably increase the reliability of optical inspection and the classifier of the original dataset, and it fits supplementary images of the classifier on the same dataset by an AdaBoost classifier.

4 Conclusion

Specific and real-time face detection is a very difficult issue in image processing. Present data available database interpretation and suitability assessment needs testing and training are presented. Adaboost techniques are being used and successfully applied to face recognition. To reduce false detection rate and gain momentum, the detection speed is the future work to improve every step of the algorithm. Which can be incorporated is recognizing a face. Moreover, a web portal can be utilized which contains all the information related to participation and an understudy can sign into an enclave of eagerness for the database. This technique will be implemented in low-cost Raspberry Pi hardware devices with IoT.

References

1. Gao C, Li P, Zhang Y, Liu J, Wang L (2016) People counting based on head detection combining Adaboost and CNN in crowded surveillance environment. Neurocomputing 208:108–116
2. Zhao Y, Gong L, Zhou B, Huang Y, Liu C (2016) Detecting tomatoes in greenhouse scenes by combining AdaBoost classifier and colour analysis. Biosys Eng 148:127–137
3. Zhang J, Yang Y, Zhang J (2016) A MEC-BP-Adaboost neural network-based color correction algorithm for color image acquisition equipments. Optik 127(2):776–780
4. Patnaik R, Krishna KS, Patnaik S, Singh P, Padhy N (2020) Drowsiness alert, alcohol detect and collision control for vehicle acceleration. In: 2020 International conference on computer science, engineering and applications (ICCSEA). IEEE, pp 1–5
5. Patnaik R, Padhy N, Raju KS (2020) A systematic survey on IoT security issues, vulnerability and open challenges. In: Intelligent system design. Springer, Singapore, , pp. 723–730
6. Jain AK, Nandakumar K, Ross A (2016) 50 Years of biometric research: accomplishments, challenges, and opportunities. Pattern Recogn Lett 79:80–105
7. Yu J (2011) The application of BP-Adaboost strong classifier to acquire knowledge of student creativity. In: 2011 international conference on computer science and service system (CSSS). IEEE, pp 2669–2672
8. Zhuo L, Zhang J, Dong P, Zhao Y, Peng B (2014) An SA–GA–BP neural network-based color correction algorithm for TCM tongue images. Neurocomputing 134:111–116
9. Gragnaniello D, Poggi G, Sansone C, Verdoliva L (2013) Fingerprint liveness detection based on weber local image descriptor. In: 2013 IEEE workshop on biometric measurements and systems for security and medical applications. IEEE, pp 46–50
10. Morerio P, Marcenaro L, Regazzoni CS (2012) People count estimation in small crowds. In: 2012 IEEE ninth international conference on advanced video and signal-based surveillance. IEEE, pp 476–480
11. Lan J, Zhang M (2010) A new vehicle detection algorithm for real-time image processing system. In: 2010 international conference on computer application and system modeling (ICCASM 2010), vol 10. IEEE, pp V10–1
12. Chen J, Shan S, He C, Zhao G, Pietikainen M, Chen X, Gao W (2009) WLD: a robust local image descriptor. IEEE Trans Pattern Anal Mach Intell 32(9):1705–1720
13. Yang M, Crenshaw J, Augustine B, Mareachen R, Wu Y (2010) AdaBoost-based face detection for embedded systems. Comput Vis Image Underst 114(11):1116–1125
14. Wang H, Cai Y (2015) Monocular based road vehicle detection with feature fusion and cascaded adaboost algorithm. Optik 126(22):3329–3334
15. Gaber T, Tharwat A, Hassanien AE, Snasel V (2016) Biometric cattle identification approach based on weber's local descriptor and Adaboost classifier. Comput Electron Agric 122:55–66
16. Zhou X, Bhanu B (2006) Feature fusion of face and gait for human recognition at a distance in video. In: 18th international conference on pattern recognition (ICPR'06), vol 4. IEEE, pp 529–532
17. Schneiderman H, Kanade T (2004) Object detection using the statistic of parts. Int J Comput Vision 56(3):151–177
18. Viola P, Jones M (2001) Robust real-time object detection. Int J Comput Vis 4(34–47):4
19. Padhy N, Panigrahi R, Neeraja K (2019) Threshold estimation from software metrics by using evolutionary techniques and its proposed algorithms, models. Evolutionary intelligence, pp 1–15
20. Chuah CS, Leou JJ (2001) An adaptive image interpolation algorithm for image/video processing. Pattern Recogn 34(12):2383–2393
21. Aghaei M, Dimiccoli M, Radeva P (2016) Multi-face tracking by extended bag-of-tracklets in egocentric photo-streams. Comput Vis Image Underst 149:146–156
22. Liu W, Anguelov D, Erhan D, Szegedy C, Reed S, Fu CY, Berg AC (2016) SSD: single shot multibox detector. In: European conference on computer vision. Springer, Cham, pp 21–37

23. Sharif Razavian A, Azizpour H, Sullivan J, Carlsson S (2014) CNN features off-the-shelf: an astounding baseline for recognition. In: Proceedings of the IEEE conference on computer vision and pattern recognition workshop. IEEE, pp 806–813
24. Viola P, Jones M (2001) Rapid object detection using a boosted cascade of simple features. In: Proceedings of the 2001 IEEE computer society conference on computer vision and pattern recognition. CVPR 2001, vol 1. IEEE, pp I–I
25. Kang S, Choi B, Jo D (2016) Faces detection method based on skin color modeling. J Syst Architect 64:100–109
26. Panigrahi R, Padhy N, Satapathy SC (2019) Software reusability metrics estimation from the social media by using evolutionary algorithms: refactoring prospective. Int J Open Sour Softw Processes (IJOSSP) 10(2):21–36
27. Wu B, Hu BG, Ji Q (2017) A coupled hidden markov random rield model for simultaneous face clustering and tracking in videos. Pattern Recogn 64:361–373
28. Aksasse B, Ouanan H, Ouanan M (2017) Novel approach to pose invariant face recognition. Procedia Comput Sci 110:434–439
29. Elafi I, Jedra M, Zahid N (2016) Unsupervised detection and tracking of moving objects for video surveillance applications. Pattern Recogn Lett 84:70–77
30. Xinhua L, Qian Y (2015) Face recognition based on deep neural network. Int J Sig Process Image Process Pattern Recogn 8(10):29–38

Analysis of Channel Estimation in GFDM System

K. Anish Pon Yamini, S. V. Akhila, K. Suthendran, and K. Srujan Raju

Abstract At receiver, the radio channel knowledge is important for channel estimation. The time-varying channel is tracked by the proposed algorithm. The time- and frequency-selective channels cause the inter-carrier interference (ICI) into GFDM system which provides some impact for receiver design. The suggested design can accomplish virtually equal mean square sin of predicted medium and bit error rate achievement as a historic recursive least square (RLS) proposal disregarding the autoregressive progress from the deviation of channel coefficients. The scheduling schemes such as greedy, proportional fair and round robin algorithm are used for efficient resources. In this proposed method, the greedy algorithm has the throughput of 81.167%, round robin provides the average throughput of 83.67%, and proportional fair algorithm provides the average throughput of 97%. Thus, estimation of channel based on proportional fair algorithm provides better efficiency compared to other methods.

Keywords GFDM · Inter-carrier interference (ICI) · Recursive least square (RLS) · Greedy · Proportional fair · Round robin

K. Anish Pon Yamini (✉)
Department of Electronics and Communication, Arunachala College of Engineering for Women, Manavilai, Tamil Nadu 629263, India
e-mail: anish.yamini@gmail.com

S. V. Akhila
Department of Electronics and Communication, Arunachala College of Engineering for Women, Manavilai, Nagercoil, Tamil Nadu 629263, India
e-mail: akhilaakhi167@gmail.com

K. Suthendran
Department of Information Technology, Kalasalingam Academy of Research and Education, Krishnankoil, Tamil Nadu 626126, India
e-mail: k.suthendran@klu.ac.in

K. Srujan Raju
Department of Computer Science and Engineering, CMR Technical Campus, Hyderabad, India
e-mail: ksrujanraju@gmail.com

© The Author(s), under exclusive license to Springer Nature Singapore Pte Ltd. 2021   679
K. A. Reddy et al. (eds.), *Data Engineering and Communication Technology*,
Lecture Notes on Data Engineering and Communications Technologies 63,
https://doi.org/10.1007/978-981-16-0081-4_68

1 Introduction

Due to the transmitter and receiver, mobility radio channel is doubly selective in nature but also the gadgets in the cellular contact cause the transmitter/receptor oscillator drift along with high mobility scenarios leads to origin of scatters which makes the wireless channel doubly selective. The high level of flexibility form physical layer in fifth generation (5G) is achieved by performing new challenges. Due to low power and large number of connection, the synchronization will be lost in machine-type communication (MTC) [1]. To improve the mobile network to address several different scenarios, flexible physical layer is required.

The concept of cyclic prefix is used in the GFDM to bypass inter-frame interference. The frequency domain comparison can adequately adjust the effect of multipath channel disturbance to the demodulation process. The flexible waveform causes complexity in implementation. In a GFDM structure, the message bit sequence is modulated in transmitter using PSK/QAM symbols and that symbols are then changed into time-domain signals using IFFT and transmitted over a medium. The channel characteristics are responsible for received signal distortion. The transmitted bits are recovered by estimating the channel effect and reduced in the receiver. In this, each subcarrier is assumed as an independent channel, thus preserving the orthogonality among subcarriers as no inter-carrier interference (ICI) exists [2]. An amount of transmitted conspicuous and channel frequency feedback at the subcarrier grant each subcarrier component to convey the orthogonality of received salient. By supposing the channel feedback of each subcarrier, the transmitted signal can be recovered. In all transmission, the signal goes over a medium and the signal obtains distorted or numerous blast is combined to the signal while the signal goes through the medium [3]. The time-varying channel of the system is estimated accurately to improve the accuracy and immense input estimation at receptor. The wireless multiplication channel estimator contains an impulse response which provides help in obtaining essential data for verification, designing and planning wireless transmission systems. The channel impulse response (CIR) provides proficiency to the detector, and it evaluates CIR for every barrage through manipulating the transmitted stream and analogous received streams. The channel impulse response (CIR) knowledge must be concerned to signal detectors of the transmission channel with known transmitted sequences [4]. The modulated corrupted signal undergoes channel assessment using LMS, MLSE, MMSE, RMS, etc., from channel before the demodulation which takes residence at receiver side. Due to its bandwidth-saving advantage, it seems to be very attractive. In time-varying channels, higher complication receiver has limited blind technique. The trade-off between frequency ability and proper evaluation provides low complexity, and more attention is required to introduce low intricacy pilot-aided channel estimation techniques for MIMO-GFDM.

2 Related Works

The increase in data rate and high mobility cause severe doubly selective channel in wireless communication. The communication system for determining the throughput of wireless HSR is based upon the channel estimators. The knowledge of transmitted and received signals is needed at CRS position and medium in data point that can be achieved by definite establishment in both time and frequency proportions for LS estimator [5]. The medium estimator contains least square (LS), linear minimum mean square error (LMMSE) and historical information-based basis expansion model (HiBEM) that are used for channel estimation. The performance of HiBEM estimator was based on mean square error (MSE). The channel parameters are developed with tapped delay line (TDL) designs over ray-tracing simulation. This method was investigated in three typical HSR environments such as metropolitan, scathing and overpass, and provoked the channel specification for every scheme with TDL miniature from ray-tracing imitation which is in arrangement along feasible position [6]. The number of radio frequency (RF) string enforced in terahertz (THz) immense MIMO structure is reduced by using the concept of beamspace MIMO with various lens patterns which promote beam selection [7]. The accurate estimation of doubly selective channel that is principle to accomplish adequate achievement in wireless communication system. The tracking performance was largely independent from the highest Doppler frequency and mainly depends on the accuracy of PACE [8]. The doubly selective beamspace medium can be followed with depressed pilot expense to attain identical certainty, and the PA channel tracking design lacks enough lower pilot expense and signal-to-noise ratio (SNR) than traditional design [9]. In both flat and frequency-selective fading, the Bussgang decomposition is applicable for the channel estimation which reformulates the nonlinear quantizer as a definite objective with exact first- and second-order statistics [10]. The Inter Carrier Interference (ICI) at immense momentum there is an imperfections in the PACE results which are inevitable, uncertain Signal to Noise Ratio (SNR) of the model and imbalance in specification beliefs used in PACE estimations [11]. GFDM consists of filter bank multicarrier scheme, and spectrally accommodated pulse shaping was practiced over all subcarrier to shorten out of band emission. The pilot-IC combines with a zero forcing receiver, and Tx-IC combines with simple matched filter used in [12] for increasing the performance. The MSE performance of various interpolation methods under the frequency channel estimation methods changes significantly when the terminal moving speed changes [13]. The channel parameter is generated with TDL model through ray-tracing simulation that is in agreement with practical situation. The channel response within a block by definite sequence of finite number of basis function. It reduces the complexity by the interpolation method that cannot adapt the time-varying channel [14].

3   Proposed Method

GFDM-based communication structure is considered as the allusion stream X. Before transmission, X is transformed to $x = [x(0), x(1), x(2), x(3), \ldots, x(M)]$ through the IFFT combined along with the cyclic prefix (CP) (Fig. 1).

$$x(m) = \frac{1}{\sqrt{M}} \sum_{j-1}^{N} X(j) e^{j2\pi \frac{mi}{M}}, \tag{1}$$

M in (1) is the total subcarrier in GFDM symbol, and channel impulse response (CIR) of the time-varying wireless channel is represented by [15]

$$h(\tau, t) = \sum_{l=0}^{L-1} \beta_l \mu_l(t) \delta(\tau - \tau_l), \tag{2}$$

The Doppler bandwidth f_d limits the transmission capacity of time fluctuation in the transmission medium,

$$f_d = \frac{u f_c}{c}$$

Here, u is the acceleration of recipient with transmitter in the multipath surroundings.

The recommended system examines exponential path loss scheme [16]. The proposed scheme is convenient with abbreviated space connection for ultra-opaque structure. The transmitter potential decreased exponentially based on the transmitter and receiver segregation, and the medium is Rayleigh. Based on propagation system, the path loss in dB is

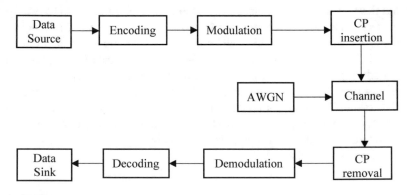

Fig. 1   GFDM transceiver

$$PL(l) = \exp(-\alpha \ell^\beta) \tag{3}$$

where ℓ is transmitter and receptor partition. α serves as the depletion circumstance of every section whose characters are in scope $0 \le \alpha \le 1$. β demonstrates the measure of interference in communication lane, whose characters are in the scope $0 \le \beta \le 2$.

The throughput of this structure is based on base station quantity and SINR. The peak transitory throughput of web is employed by

$$T = \eta \log_2(1 + \text{SINR}) \tag{4}$$

where η indicates the density of base stations.

The scheduling systems such as greedy, round robin and proportional fair methods *were* used for the optimization of radio resources and to eliminate conflicts in the model and to increase global system throughput. The possible wireless ability and total number of users demand to approach the ability of resource and are then recorded to consider greedy algorithm. The channel gain will prioritize the user. The top priority is assumed to the user with best subcarrier, and the process is repeated for all the resources.

In round robin scheduling method, the same preference is given for all the users. This scheduling policy concludes in exceptional fairness but poor throughput, and it is the simplest scheduling scheme. Proportional fair method is proposed in wireless structure. Here, organizing is based on suitability, throughput and wireless aspects. In this, the number of scheduling frames is divided from overall time duration. Proportional fair organizing method is used in a fashion that the medium with large SNR has low depletion aspects and path loss is decreased. The users with extra subcarriers are designated for medium with high SNR. The low aspect channel was ignored, and the signal transmission happens in best path. The equity of the scheduler is employed from the throughput values. Thus in the proportional fair algorithm, throughput is increased and latency is decreased.

4 Experimental Result

This paper deals with wireless communications, especially the technique for the 5G cellular communication. The results obtained through MATLAB simulations are discussed in this section. In this paper, the characteristics of propagation path are taken. The transmitter and receiver perfectly know the channel state information. The multipath Rayleigh fading channel is proposed. The short-range communication performance is effective and used to consider the exponential path loss model. The path loss parameter is derived in terms of frequency. In this system, throughput depends on distance between the base stations. The OFDM and GFDM modulation schemes are analyzed, and the performance is proposed in our system. The cyclic

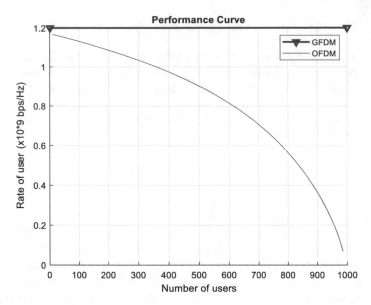

Fig. 2 OFDM versus GFDM

prefix is added in facade of each OFDM pattern in OFDM structure and cyclic prefix interpolated in facade of each GFDM frame in GFDM system.

Figure 2 shows that path loss raise as the transceiver segregation is expanded during short-distance transmission. The electromagnetic stream concludes from a line of sight path through free space path loss. The path loss is based on the range of the distance between the transceivers. The free space path loss is the attenuation of radio energy between the communication channels that results in the combination of the receiver signal plus noise in the environment.

Figure 3 indicates that by selecting appropriate scheduling approach, the wireless system is designated and obstruction is reduced. So, the path loss environment is controlled and the throughput is increased. In this method, proportional fair method shortened path loss for expanded space.

The performance of OFDM and GFDM is compared. The communication is between many base stations. Figure 4 shows that based on the user pair the throughput peaks up which is analyzed by using three scheduling schemes.

Figure 5 shows the performance capacity of the OFDM and GFDM. The OFDM signal and GFDM signal are compared and analyzed that GFDM is better because the flexibility and simplicity of OFDM with stronger interference reduction mechanism are combined by the GFDM. The GFDM provides the maximum rate at which information can be transmitted through the channel.

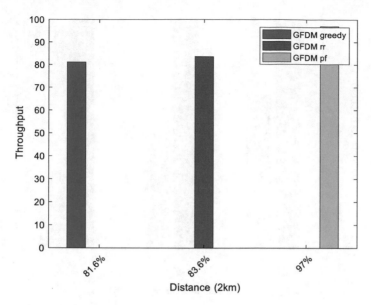

Fig. 3 Performance of the throughput

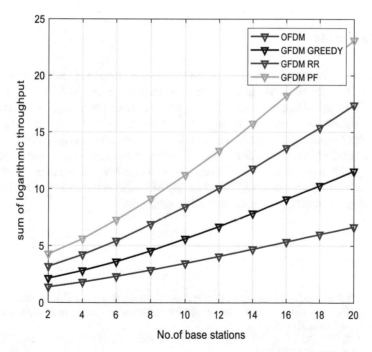

Fig. 4 Performance comparison

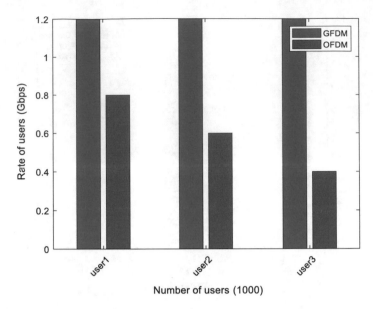

Fig. 5 Capacity comparisons of multicarrier modulation

5   Conclusion

In this proposed work, channel estimation techniques proposed are better than other methods. The GFDM modulation system is used for better accuracy in wireless communication. The implementation of the proposed method is simple because basis function generation is computationally simple. Without using existing information of the Doppler frequency, this method marks time-varying transmission carriers. In this proposed method, the proportional fair algorithm provides the average throughput of 97%, greedy algorithm has the throughput of 81.167% and round robin provides the average throughput of 83.67%. Thus, BEM-based proportional fair algorithm provides better efficiency compared to other methods.

References

1. Marsch P, Raaf B, Szufarska A, Mogensen P, Guan H, Farber M, Redana S, Pedersen K, Kolding T (2012) Future mobile communication networks: challenges in the design and operation. IEEE Veh Technol Mag 7(1):16–23
2. Mostofi Y, Cox DC (2005) ICI mitigation for pilot-aided OFDM mobile systems. IEEE Trans Wirel Commun 4(2):765–774
3. Padhy N, Panigrahi R, Neeraja K (2019) Threshold estimation from software metrics by using evolutionary techniques and its proposed algorithms, models. Evolutionary intelligence, pp 1–15

4. Panigrahi R, Padhy N, Satapathy SC (2019) Software reusability metrics estimation from the social media by using evolutionary algorithms: refactoring prospective. Int J Open Sour Softw Processes (IJOSSP) 10(2):21–36
5. Cao Y, Ge J, Zhang C, Han W, Bu Q, Du H (2016) Low complexity and high accurate estimation scheme for doubly selective fading channels with square Root Kalman filtering. In: 2016 IEEE/CIC international conference on communications in China (ICCC). IEEE, pp 1–6
6. Chin WL, Kao CW, Qian Y (2016) Spectrum sensing of OFDM signals over multipath fading channels and practical considerations for cognitive radios. IEEE Sens J 16(8):2349–2360
7. Gao X, Dai L, Zhang Y, Xie T, Dai X, Wang Z (2016) Fast channel tracking for terahertz beamspace massive MIMO systems. IEEE Trans Veh Technol 66(7):5689–5696
8. Idrees NM, Haselmayr W, Petit M, Springer A (2012) Improving time variant channel estimation for 3GPP LTE-downlink. In: 2012 IEEE 23rd international symposium on personal, indoor and mobile radio communications-(PIMRC). IEEE, pp 2114–2119
9. Li Y, Tao C, Seco-Granados G, Mezghani A, Swindlehurst AL, Liu L (2017) Channel estimation and performance analysis of one-bit massive MIMO systems. IEEE Trans Sig Process 65(15):4075–4089
10. Nie S, Han C, Akyildiz IF (2017) A three-dimensional time-varying channel model for 5G indoor dual-mobility channels. In: 2017 IEEE 86th vehicular technology conference (VTC-Fall). IEEE, pp 1–5
11. Qin W, Peng QC (2008) Time-varying channel estimation and symbol detection using superimposed training in OFDM systems. Wirel Pers Commun 47(2):293–301
12. Liao Y, Sun G, Shen X, Zhang S, Yang X, Zhang X, Yao H, Zhang N (2018) Bem-based channel estimation and interpolation methods for doubly-selective OFDM channel. In: 2018 IEEE international conference on smart internet of things (SmartIoT). IEEE, pp 70–75
13. Zhao Y, Wang X, Wang G, He R, Zou Y, Zhao Z (2018) Channel estimation and throughput evaluation for 5G wireless communication systems in various scenarios on high speed railways. China Commun 15(4):86–97
14. AlAmmouri A, Andrews JG, Baccelli F (2017) SINR and throughput of dense cellular networks with stretched exponential path loss. IEEE Trans Wirel Commun 17(2):1147–1160
15. Tang Z, Cannizzaro RC, Leus G, Banelli P (2007) Pilot-assisted time-varying channel estimation for OFDM systems. IEEE Trans Sig Process 55(5):2226–2238
16. Salimath N, Mallappa S, Padhy N, Sheetlani J (2020) Scrambling and descrambling of document image for data security in cloud computing. In: Smart intelligent computing and application. Springer, Singapore, pp 283–290

Author Index

689

Printed in the United States
by Baker & Taylor Publisher Services